Table of Symbols

+	add
−	subtract
$\cdot$, $(a)(b)$	multiply
$\frac{a}{b}$, ÷	divide
( )	parentheses, a grouping symbol
[ ]	brackets, a grouping symbol
π	pi, a number approximately equal to $\frac{22}{7}$ or 3.14
$-a$	the opposite, or additive inverse, of a
$\frac{1}{a}$	the reciprocal, or multiplicative inverse, of a
=	is equal to
≈	is approximately equal to
≠	is not equal to
<	is less than
≤	is less than or equal to
>	is greater than
≥	is greater than or equal to

(a, b)	an ordered pair whose first component is a and whose second component is b
°	degree (for angles and temperature)
$\sqrt{a}$	the principal square root of a
$\varnothing$, { }	the empty set
$|a|$	the absolute value of a
∪	union of two sets
∩	intersection of two sets
∈	is an element of (for sets)
∉	is not an element of (for sets)

reference to The Computer Tutor™

reference to The Math ACE

indicates calculator topics

4 reference to the Video Tapes

BEGINNING
ALGEBRA

with Applications

BEGINNING ALGEBRA

with Applications

THIRD EDITION

Richard N. Aufmann
Palomar College, California

Vernon C. Barker
Palomar College, California

Joanne S. Lockwood
Plymouth State College, New Hampshire

HOUGHTON MIFFLIN COMPANY   Boston   Toronto

Dallas   Geneva, Illinois   Palo Alto   Princeton, New Jersey

Sponsoring Editor: Maureen O'Connor
Development Editor: Toni Haluga
Senior Project Editor: Toni Haluga
Assistant Design Manager: Karen Rappaport
Production Coordinator: Frances Sharperson
Manufacturing Coordinator: Sharon Pearson
Marketing Manager: Michael Ginley

Most Point of Interest Art designed and illustrated by Daniel P. Derdula, with airbrushing by Linda Phinney on Chapters 6 and 11, and calligraphy by Susan Fong on Chapter 8. Linda Phinney is credited for illustrating Chapter 7, and Network Graphics is credited for illustrating Chapter 9.

All Interior Math Figures rendered by Network Graphics (135 Fell Court, Hauppauge, New York 11788).

Cover concept and design by Daniel P. Derdula and Libby Plaisted.

Interior design by George McLean.

Printed in the U.S.A.

Library of Congress Catalog Card Number: 91-72000

ISBN Numbers:
Text: 0-395-58883-9
Instructor's Annotated Edition: 0-395-58884-7
Solutions Manual: 0-395-58886-3
Student's Solutions Manual: 0-395-58887-1
Instructor's Resource Manual with Chapter and Cumulative Tests: 0-395-58885-5
Test Bank: 0-395-58888-X
Transparencies: 0-395-60127-4

CDEFGHIJ-D-95432

CONTENTS

PREFACE

The third edition of *Beginning Algebra with Applications* provides mathematically sound and comprehensive coverage of the topics considered essential in an introductory algebra course. Our strategy in preparing this revision has been to build on the successful features of the second edition, features designed to enhance the student's mastery of math skills. *Beginning Algebra with Applications* provides a complete, integrated learning system organized by objectives and linked to the ancillary package. All of the components of the package were written by the authors.

Features

The Interactive Approach

Instructors have long recognized the need for a text that requires the student to use a skill as it is being taught. *Beginning Algebra with Applications* uses an interactive technique that meets this need. Each section is divided into objectives, and every objective contains one or more sets of matched-pair examples. The first example in each set is worked out; the second example is not. By solving this second problem, the student interacts with the text. The complete worked-out solutions to these problems are provided in an appendix at the end of the book, so the student can obtain immediate feedback on and reinforcement of the skill being learned.

Emphasis on Problem-Solving Strategies

Beginning Algebra with Applications features a carefully developed approach to problem solving that emphasizes developing strategies to solve problems. For each type of word problem contained in the text, the student is prompted to use a "strategy step" before performing the actual manipulation of numbers and variables. By developing problem-solving strategies, the student will know better how to analyze and solve those word problems encountered in an introductory algebra course.

Applications

The traditional approach to teaching or reviewing algebra covers only the straightforward manipulation of numbers and variables and thereby fails to teach students the practical value of algebra. By contrast, *Beginning Algebra with Applications* emphasizes applications. Wherever appropriate, the last objective in each section presents applications that require the student to use the skills covered in that section to solve practical problems. Also, all of Chapter 4, *"Solving Equations: Applications,"* and portions of several other chapters are devoted entirely to applications. This carefully integrated applied approach generates awareness on the student's part of the value of algebra as a real-life tool.

Complete, Integrated Learning System Organized by Objectives

Each chapter begins with a list of the learning objectives included within that chapter. Each of the objectives is then restated in the chapter to remind the student of the current topic of discussion. The same objectives that organize the text organize each ancillary. The Solutions Manual, Student's Solutions Manual, Computerized Test Generator, Computer Tutor™, Videos, Transparencies, Test Bank, and the Printed Testing Program have all been prepared so that both the student and instructor can easily connect all of the different aids.

Exercises

There are more than 6000 exercises in the text, grouped in the following categories:

- **End-of-section exercise sets,** which are keyed to the corresponding learning objectives, provide ample practice and review of each skill.
- **Supplemental exercise sets,** designed to increase the student's ability to solve problems requiring a combination of skills, have been added at the end of each section.
- **Chapter review exercises,** which appear at the end of each chapter, help the student integrate all of the skills presented in the chapter.
- **Chapter tests,** which appear at the end of each chapter, are typical one-hour exams that the student can use to prepare for an in-class test.
- **Cumulative review exercises,** which appear at the end of each chapter (beginning with Chapter 2), help the student retain math skills learned in earlier chapters.
- **The final exam,** which follows the last chapter, can be used as a review item or practice final.

Calculator and Computer Enrichment Topics

Each chapter also contains optional calculator or computer enrichment topics. Calculator topics provide the student with valuable key-stroking instructions and practice in using a hand-held calculator. Computer topics correspond directly to the programs found on the Math ACE (Additional Computer Exercises) Disk. These topics range from solving first-degree equations to graphing linear equations in two variables.

New To This Edition

Topical Coverage

Chapter 5 includes a reorganization of the material on integer exponents. Negative exponents are defined prior to division of monomials; therefore, a single rule for division of monomials is stated.

In Chapter 6, factoring by grouping is presented in Section 1. This sequencing enables trinomials of the form $ax^2 + bx + c$ to be factored by grouping as well as by trial and error in Section 3.

In Chapter 7, proportions are included in the objective on solving equations containing fractions.

Chapter 8 includes scatter diagrams as an application of the rectangular coordinate system.

In Chapter 11, the section on radical equations has been expanded to include exercises that require squaring each side of an equation twice in order to find the solutions.

The application problems throughout the text have been rewritten. Many have been updated to reflect contemporary situations.

New Applications Feature

Each chapter now includes an expanded application feature entitled Something Extra. Topics include such concepts as density, acceleration, scientific notation, and break-even analysis.

Challenge Exercises

Challenge exercises are now denoted in the Instructor's Annotated Edition. An asterisk (*) is printed next to those Supplemental Exercises which require more analytical thought.

Chapter Review and Chapter Test

The Chapter Review at the end of each chapter has been expanded to include nearly twice the number of review exercises, and a Chapter Test has been added. The objective references for each of these features are provided in the Answer Section at the back of the book so that the student can determine which objectives require restudy.

New Testing Program

Both the Computerized Testing Program and the Printed Testing Program have been completely rewritten to provide instructors with the option of creating countless new tests. The computerized test generator contains high quality graphics and editing capabilities for all nongraphic questions.

New Transparencies

Approximately 150 transparencies containing the worked-out solutions to the "student problems" in the text have been added.

Supplements for the Student

Two computerized study aids, the Computer Tutor™ and the Math ACE (Additional Computer Exercises) Disk have been carefully designed for the student.

The COMPUTER TUTOR™

The Computer Tutor™ is an interactive instructional microcomputer program for student use. Each learning objective in the text is supported by a lesson on the Computer Tutor™. As a reminder of this, a small computer icon appears to the right of each objective title in the text. Lessons on the tutor provide additional instruction and practice and can be used in several ways: (1) to cover material the student missed because of absence from class; (2) to repeat instruction on a skill or concept that the student has not yet mastered; or (3) to review material in preparation for examinations. This tutorial program is available for the IBM PC and compatible computers, the Apple II family of computers, and the Macintosh. The IBM and Macintosh versions of the Computer Tutor™ have been expanded to include nine "you-try-it" examples for each lesson.

Math ACE (Additional Computer Exercises) Disk

The Math ACE Disk contains a number of computational and drill-and-practice programs that correspond to selected Calculator and Computer Enrichment Topics in the text. These programs are available for the Apple II family of computers and the IBM PC and compatible computers.

Student's Solutions Manual

The Student's Solutions Manual contains the complete worked-out solutions for all the odd-numbered exercises in the text. Also included are the complete solutions to the Chapter Reviews, Chapter Tests, and Cumulative Reviews.

Videotapes

Over 30 half-hour videotape lessons accompany *Beginning Algebra with Applications*. These lessons follow the format and style of the text and are closely tied to specific sections of the text.

Supplements for the Instructor

Beginning Algebra with Applications has an unusually complete set of teaching aids for the instructor.

Instructor's Annotated Edition

The Instructor's Annotated Edition is an exact replica of the student text except that the answers to all of the exercises are printed in color next to the problems.

Solutions Manual

The Solutions Manual contains worked-out solutions for all end-of-section exercise sets, chapter reviews, chapter tests, cumulative reviews, and the final exam.

Instructor's Resource Manual with Chapter and Cumulative Tests

The Instructor's Resource Manual/Testing Program contains the printed testing program, which is the first of three sources of testing material available to users of *Beginning Algebra with Applications*. Eight printed tests (in two formats—free response and multiple choice) are provided for each chapter, as are cumulative and final exams. In addition, the Instructor's Manual includes the documentation for all the software ancillaries (ACE, the Computer Tutor™, and the Instructor's Computerized Test Generator) as well as suggested course sequences.

Instructor's Computerized Test Generator

The Instructor's Computerized Test Generator is the second source of testing material for use with *Beginning Algebra with Applications*. The database contains over 1800 new test items. These questions are unique to the test generator and do not repeat items provided in the Instructor's Resource Manual/Testing Program. Organized according to the keyed objectives in the text, the Test Generator is designed to produce an unlimited number of tests for each chapter of the text, including cumulative tests and final exams. It is available for the Apple II family of computers and the Macintosh. It is also available for the IBM PC or compatible computers with editing capabilities for all nongraphic questions.

Test Bank

The Printed Test Bank, the third component of the testing materials, is a print-out of all items in the Instructor's Computerized Test Generator. Instructors using the Test Generator can use the test bank to select specific items from the database. Instructors who do not have access to a computer can use the test bank to select items to be included on a test being prepared by hand.

Transparencies

Approximately 150 transparencies accompany *Beginning Algebra with Applications*. These transparencies contain the complete solution to every "student problem" in the text.

Acknowledgments

The authors would like to thank the people who have reviewed this manuscript and provided many valuable suggestions:

Barbara Brook
Camden County College, NJ

Sharon Edgmon
Bakersfield College, CA

Ervin Eltze
Fort Hays State University, KS

Gerald D. Fischer
Northeast Iowa Community College, IA

Carol L. Grover
Carlow College, PA

Frank Gunnip
Oakland Community College, MI

Tim Hall
Central Texas College, TX

John A. Heublein
Kansas College of Technology, KS

Katherine J. Huppler
St. Cloud State University, MN

Buddy A. Johns
The Wichita State University, KS

Ellen Milosheff
Triton College, IL

Allan Newhart
West Virginia University at Parkersburg, WV

Doris Nice
University of Wisconsin-Parkside, WI

Donald Perry
Lee College, TX

Judith A. Pokrop
Cardinal Stritch College, WI

Diane Shores
Phillips County Community College, AR

Dean Stowers
Nicolet Area Technical College, WI

James M. Sullivan
Massachusetts Bay Community College, MA

Lana Taylor
Siena Heights College, MI

Robert A. Tolar
College of the Canyons, CA

Beverly Weatherwax
Southwest Missouri State University, MI

Warren Wise
Blue Ridge Community College, WA

Wayne Wolfe
Orange Coast College, CA

TO THE STUDENT

Many students feel that they will never understand math while others appear to do very well with little effort. Oftentimes what makes the difference is that successful students take an active role in the learning process.

Learning mathematics requires your *active* participation. Although doing homework is one way you can actively participate, it is not the only way. First, you must attend class regularly and become an active participant. Second, you must become actively involved with the textbook.

Beginning Algebra with Applications was written and designed with you in mind as a participant. Here are some suggestions on how to use the features of this textbook.

There are 12 chapters in this text. Each chapter is divided into sections and each section is subdivided into learning objectives. Each learning objective is labeled with a number from 1–5.

First, read each objective statement carefully so you will understand the learning goal that is being presented. Next, read the objective material carefully, being sure to note each bold word. These words indicate important concepts that you should familiarize yourself with. Study each in-text example carefully, noting the techniques and strategies used to solve the example.

You will then come to the key learning feature of this text, the paired Examples and Problems. These Examples and Problems have been designed to assist you in a very specific way. Notice that the Examples are completely worked-out and explanations are given for certain steps within the solutions. The solutions to the Problems are not given; *you* are expected to work these Problems, thereby testing your understanding of the material you have just studied.

Study the Examples carefully by working through each step presented. Then use the worked-out example as a model for solving the Problems. When you have completed your solution, check your work by turning to the page in the Appendix where the complete solution is given. The page number on which the solution appears is printed on the solution line below the Problem statement. By checking your solution, you will know immediately whether or not you fully understand the skill just studied.

When you have completed studying an objective, do the exercises in the exercise set that correspond with that objective. The exercises are labeled with the same number as the objective. Algebra is a subject that needs to be learned in small sections and practiced continually in order to be mastered. Doing the exercises in each exercise set will help you master the problem-solving techniques necessary for success.

Once you have completed the exercises for an objective, you should check your answers to the odd-numbered exercises with those found in the back of the book.

After completing a chapter, read the Chapter Summary. This summary highlights the important topics covered in the chapter. Following the Chapter Summary are Chapter Review Exercises, a Chapter Test, and a Cumulative Review (beginning with Chapter 2). Doing the review exercises is an important way of testing your understanding of the chapter. The answer to each review exercise is in an appendix at the back of the book. Each answer is followed by a reference that tells which objective that exercise was taken from. For example, (4.2.2) means Section 4.2, Objective 2. After checking your answers, restudy any objective that corresponds to an exercise you answered incorrectly. It may be very helpful to retry some of the exercises for that objective to reinforce your problem-solving techniques.

The Chapter Test should be used to prepare for an exam. We suggest that you try the Chapter Test a few days before your actual exam. Take the test in a quiet place and try to complete the test in the same amount of time you will be allowed for your exam. When taking the Chapter Test, practice the strategies of successful test takers: 1) scan the entire test to get a feel for the questions; 2) read the directions carefully; 3) work the problems that are easiest for you first; and perhaps most importantly, 4) try to stay calm.

When you have completed the Chapter Test, check your answers. If you missed a question, review the material in that objective and rework some of the exercises from that objective. This will strengthen your ability to perform the skills in that objective.

The Cumulative Review allows you to refresh the skills you have learned in previous chapters. This is very important in mathematics. By consistently reviewing previous materials, you will retain the previous skills as you build new ones.

Remember, to be successful, attend class regularly; read the textbook carefully; actively participate in class; work with your textbook using the Examples and Problems for immediate feedback and reinforcement of each skill; do all the homework assignments; review constantly; and work carefully.

BEGINNING
ALGEBRA
with Applications

1

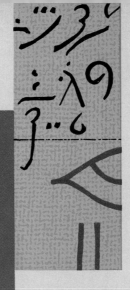

Real Numbers

Objectives

- Order relations
- Opposites and absolute value
- Add integers
- Subtract integers
- Multiply integers
- Divide integers
- Application problems
- Write rational numbers as decimals
- Add and subtract rational numbers
- Multiply and divide rational numbers
- Exponential expressions
- The Order of Operations Agreement

Early Egyptian Fractions

One of the earliest written documents of mathematics is the Rhind Papyrus. This tablet was found in Egypt in 1858, but it is estimated that the writings date back to 1650 B.C.

The Rhind Papyrus contains over 80 problems. Studying these problems has enabled mathematicians and scientists to understand some of the methods by which the early Egyptians used mathematics.

Evidence gained from the Papyrus shows that the Egyptian method of calculating with fractions was much different from the methods used today. All fractions were represented in terms of what are called unit fractions. A unit fraction is a fraction in which the numerator is 1. This fraction was symbolized with a bar over the number. Examples (using modern numbers) include

$$\bar{3} = \frac{1}{3} \qquad \overline{15} = \frac{1}{15}$$

The early Egyptians also tended to deal with powers of two (2, 4, 8, 16, . . .). As a result, representing fractions with a 2 in the numerator in terms of unit fractions was an important matter. The Rhind Papyrus has a table giving the equivalent unit fractions for all odd denominators from 5 to 101 with 2 as the numerator. Some of these are listed below.

$$\frac{2}{5} = \bar{3} \ \overline{15} \qquad \left(\frac{2}{5} = \frac{1}{3} + \frac{1}{15} \right)$$

$$\frac{2}{7} = \bar{4} \ \overline{28}$$

$$\frac{2}{11} = \bar{6} \ \overline{66}$$

$$\frac{2}{19} = \overline{12} \ \overline{76} \ \overline{114}$$

Introduction to Integers

1 Order relations

The **natural numbers** are 1, 2, 3, 4, 5, 6, 7, 8, ...

The three dots mean that the list continues on and on, and that there is no largest natural number.

The natural numbers can be shown on the **number line.**

The number line
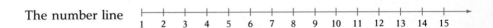

The **graph** of a natural number is shown by placing a heavy dot directly above that number on the number line.

The graph of 6 on the number line

The **integers** are . . . −4, −3, −2, −1, 0, 1, 2, 3, 4, ...

Each integer can be shown on the number line.

The integers to the left of zero on the number line are called **negative integers.** The integers to the right of zero are called **positive integers,** or natural numbers. Zero is neither a positive nor a negative number.

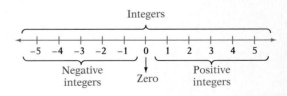

Just as the word "it" is used in language to stand for an object, a letter of the alphabet can be used in mathematics to stand for a number. Such a letter is called a **variable.**

A number line can be used to visualize the relative order of two integers. If a and b are two integers, and a is to the left of b on the number line, then a is **less than** b ($a < b$). If a is to the right of b on the number line, then a is **greater than** b ($a > b$).

Negative 4 is less than negative 1.

$-4 < -1$

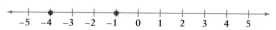

5 is greater than 0.

$5 > 0$

Example 1   Place the correct symbol, $<$ or $>$, between the two numbers.
A. $-17 \quad 6$ B. $-30 \quad -3$ C. $-12 \quad -20$ D. $-40 \quad 0$

Solution   A. $-17 < 6$ B. $-30 < -3$ C. $-12 > -20$ D. $-40 < 0$

Problem 1   Place the correct symbol, $<$ or $>$, between the two numbers.
A. $5 \quad -13$ B. $-8 \quad -22$ C. $17 \quad -6$ D. $-13 \quad -15$

Solution   See page A3.

2   Opposites and absolute value

Two numbers that are the same distance from zero on the number line but are on opposite sides of zero are **opposite numbers,** or **opposites.**

The opposite of 5 is -5.
The opposite of -5 is 5.

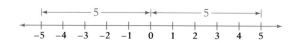

The negative sign can be read "the opposite of."

$$-(2) = -2 \qquad \text{The opposite of 2 is } -2.$$

$$-(-2) = 2 \qquad \text{The opposite of } -2 \text{ is 2.}$$

Example 2   Find the opposite number.
A. 6   B. -51

Solution   A. -6 B. 51

Problem 2   Find the opposite number.

A. -9 B. 62

Solution   See page A3.

The **absolute value** of a number is its distance from zero on the number line. Therefore, the absolute value of a number is a positive number or zero. The symbol for absolute value is $|\ |$.

The distance from 0 to 3 is 3. Therefore, the absolute value of 3 is 3.

$|3| = 3$

The distance from 0 to -3 is 3. Therefore, the absolute value of -3 is 3.

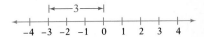

$|-3| = 3$

The absolute value of a positive number is the number itself. The absolute value of a negative number is the opposite of the negative number. The absolute value of zero is zero.

Example 3   Evaluate.

A. $|-4|$ B. $-|-10|$

Solution   A. $|-4| = 4$ B. $-|-10| = -10$ ▶ The absolute value sign does not affect the negative sign in front of the absolute value sign.

Problem 3   Evaluate.

A. $|-5|$ B. $-|-9|$

Solution   See page A3.

EXERCISES 1.1

1   Place the correct symbol, $<$ or $>$, between the two numbers.

1. 3   5

2. 7   4

3. -2 -5

4. -6 -1

5. -16 1

6. -2 13

7. 3 -7

8. 5 -6

9. -11 -8

10. −4 −10	**11.** −1 −6	**12.** −9 −4
13. 0 −3	**14.** 8 ⟩ 0	**15.** 6 −8
16. 8 −6	**17.** −14 16	**18.** −12 1
19. 35 28	**20.** 42 19	**21.** −42 27
22. −36 ⟨ 49	**23.** 21 −34	**24.** 53 −46
25. −27 −39	**26.** −51 −20	**27.** −87 63
28. −75 92	**29.** 68 −79	**30.** 95 −71
31. −62 −84	**32.** −91 ⟨ −70	**33.** 94 83
34. 76 81	**35.** 59 −67	**36.** 48 −66
37. −93 −55	**38.** −64 ⟩ −86	**39.** −88 57
40. −58 82	**41.** 0 129	**42.** −136 0
43. −131 101	**44.** 127 ⟩ −150	**45.** −194 −180

2   Find the opposite number.

46. 4	**47.** 16	**48.** −2
49. −3	**50.** 22	**51.** 45
52. −31	**53.** −88	**54.** −168
55. −97	**56.** 630	**57.** 450

Evaluate.

58. $|2|$	**59.** $|-2|$	**60.** $|-6|$
61. $|6|$	**62.** $|8|$	**63.** $|5|$
64. $|-9|$	**65.** $|-1|$	**66.** $-|-1|$
67. $-|-8|$	**68.** $-|-5|$	**69.** $-|0|$
70. $|16|$	**71.** $|19|$	**72.** $|-12|$
73. $|-22|$	**74.** $-|29|$	**75.** $-|20|$
76. $-|-14|$	**77.** $-|-18|$	**78.** $|-15|$
79. $|-23|$	**80.** $-|33|$	**81.** $-|27|$
82. $-|-36|$	**83.** $-|-41|$	**84.** $|32|$
85. $|25|$	**86.** $|-38|$	**87.** $|-30|$
88. $-|37|$	**89.** $-|34|$	**90.** $-|-42|$
91. $-|-45|$	**92.** $|44|$	**93.** $|36|$

94. $|-74|$ **95.** $|-61|$ **96.** $-|88|$

97. $-|52|$ **98.** $-|-81|$ **99.** $-|-93|$

100. $|-107|$ **101.** $|-119|$ **102.** $-|150|$

103. $|85|$ **104.** $-|-128|$ **105.** $-|48|$

SUPPLEMENTAL EXERCISES 1.1

Place the correct symbol, $<$ or $>$, between the two numbers.

106. 5 $|10|$ **107.** $-|-12|$ 9   **108.** $-|11|$ 13

109. -8 $|14|$ **110.** $|-6|$ -7 **111.** -19 $-|20|$

112. $-|-15|$ -8 **113.** $|-9|$ -4 **114.** $-|17|$ $|-10|$

115. $|8|$ $|-2|$ **116.** $-|-9|$ $|13|$ **117.** $|0|$ $-|5|$

118. $|-16|$ $-|16|$ **119.** $|-28|$ $-|-31|$ **120.** $-|-52|$ $-|43|$

121. $-(-8)$ $-|8|$ **122.** $-|-6|$ $-(-6)$ **123.** $-(5)$ $|-9|$

Write the given numbers in order from least to greatest.

124. $0, -12, 9, -17$ **125.** $-3, -21, 14, -1$

126. $|-5|, 6, -|-8|, -19$ **127.** $-4, |-15|, -|-7|, 0$

128. $-(-3), -22, |-25|, |-14|$ **129.** $-|-26|, -(-8), |-17|, -(5)$

Complete.

130. On the number line, the two points that are four units from 0 are _____ and _____.

131. On the number line, the two points that are six units from 0 are _____ and _____.

132. On the number line, the two points that are seven units from 4 are _____ and _____.

133. On the number line, the two points that are five units from -3 are _____ and _____.

134. If a is a positive number, then $-a$ is a _____ number.

135. If a is a negative number, then $-a$ is a _____ number.

136. An integer that is its own additive inverse is _____.

137. True or False. If $a \geq 0$, then $|a| = a$.

138. True or False. If $a \leq 0$, then $|a| = -a$.

SECTION **1.2**

Addition and Subtraction of Integers

1 Add integers

A number can be represented anywhere along the number line by an arrow. A positive number is represented by an arrow pointing to the right, and a negative number is represented by an arrow pointing to the left. The size of the number is represented by the length of the arrow.

Addition of integers can be shown on the number line. To add integers, find the point on the number line corresponding to the first addend. At that point, draw an arrow representing the second addend. The sum is the number directly below the tip of the arrow.

$4 + 2 = 6$

$-4 + (-2) = -6$

$-4 + 2 = -2$

$4 + (-2) = 2$

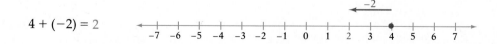

The pattern for addition shown on the number lines on page 8 is summarized in the following rules for adding integers.

Same Sign	To add two numbers with the same sign, add the absolute values of the numbers. Then attach the sign of the addends.	$2 + 8 = 10$ $-2 + (-8) = -10$
Different Signs	To add two numbers with different signs, find the difference between the absolute values of the numbers. Then attach the sign of the number with the greater absolute value.	$-2 + 8 = 6$ $2 + (-8) = -6$

Example 1   Add.

A. $162 + (-247)$ B. $-14 + (-47)$ C. $-4 + (-6) + (-8) + 9$

Solution   A. $162 + (-247)$
$\quad\quad -85$

▶ The signs are different. Subtract the absolute value of the smaller number from the absolute value of the larger number $(247 - 162)$. Attach the sign of the number with the greater absolute value.

B. $-14 + (-47)$
$\quad -61$

▶ The signs are the same. Add the absolute values of the numbers $(14 + 47)$. Attach the sign of the addends.

C. $-4 + (-6) + (-8) + 9$
$\quad -10 + (-8) + 9$
$\quad -18 + 9$
$\quad -9$

▶ To add more than two numbers, add the first two numbers. Then add the sum to the third number. Continue until all the numbers have been added.

Problem 1   Add.

A. $-162 + 98$ B. $-154 + (-37)$ C. $-36 + 17 + (-21)$

Solution   See page A3.

2   ## Subtract integers

Subtraction of an integer is defined as addition of the opposite integer.

Subtract $8 - 3$ by using addition of the opposite.

$$\text{Subtraction} \longrightarrow \text{Addition of the Opposite}$$

$$8 \;-\; (+3) \;=\; 8 \;+\; (-3) \;=\; 5$$

$$\underbrace{}_{\text{Opposites}}$$

To subtract one integer from another, add the opposite of the second integer to the first number.

$$
\boxed{\begin{array}{c}\text{First}\\\text{number}\end{array}} \quad - \quad \boxed{\begin{array}{c}\text{second}\\\text{number}\end{array}} \quad = \quad \boxed{\begin{array}{c}\text{first}\\\text{number}\end{array}} \quad + \quad \boxed{\begin{array}{c}\text{the opposite of}\\\text{the second number}\end{array}}
$$

40	−	60	=	40	+	(-60)	$= -20$
−40	−	60	=	−40	+	(-60)	$= -100$
−40	−	(-60)	=	−40	+	60	$= 20$
40	−	(-60)	=	40	+	60	$= 100$

Example 2   Subtract: $-12 - 8$

Solution $-12 - 8$
$-12 + (-8)$ ▶ Rewrite subtraction as addition of the opposite.
-20 ▶ Add.

Problem 2   Subtract: $-8 - 14$

Solution   See page A3.

When subtraction occurs several times in a problem, rewrite each subtraction as addition of the opposite. Then add.

Example 3   Subtract: $-12 - 4 - (-15)$

Solution $-12 - 4 - (-15)$
$-12 + (-4) + 15$ ▶ Rewrite each subtraction as addition of the opposite.
$-16 + 15$ ▶ Add.
-1

Problem 3   Subtract: $3 - (-4) - 15$

Solution   See page A3.

Example 4   Subtract: $-8 - 30 - (-12) - 7 - (-14)$

Solution $-8 - 30 - (-12) - 7 - (-14)$
$-8 + (-30) + 12 + (-7) + 14$ ▶ Rewrite each subtraction as addition of the opposite.

$-38 + 12 + (-7) + 14$ ▶ Add the first two numbers. Then add the sum to the third number. Continue until all the numbers have been added.
$-26 + (-7) + 14$
$-33 + 14$
-19

Problem 4   Subtract: $4 - (-3) - 12 - (-7) - 20$

Solution   See page A3.

EXERCISES 1.2

1   Add.

1. $3 + (-5)$

2. $-4 + 2$

3. $8 + 12$

4. $16 + 23$

5. $-3 + (-8)$

6. $-12 + (-1)$

7. $-4 + (-5)$

8. $-12 + (-12)$

9. $6 + (-9)$

10. $4 + (-9)$

11. $-6 + 7$

12. $-12 + 6$

13. $2 + (-3) + (-4)$

14. $7 + (-2) + (-8)$

15. $-3 + (-12) + (-15)$

16. $9 + (-6) + (-16)$

17. $-17 + (-3) + 29$

18. $13 + 62 + (-38)$

19. $-3 + (-8) + 12$

20. $-27 + (-42) + (-18)$

21. $13 + (-22) + 4 + (-5)$

22. $-14 + (-3) + 7 + (-6)$

23. $-22 + 10 + 2 + (-18)$

24. $-6 + (-8) + 13 + (-4)$

25. $-16 + (-17) + (-18) + 10$

26. $-25 + (-31) + 24 + 19$

27. $-126 + (-247) + (-358) + 339$

28. $-651 + (-239) + 524 + 487$

2 Subtract.

29. $16 - 8$ 30. $12 - 3$

31. $7 - 14$ 32. $7 - (-2)$

33. $3 - (-4)$ 34. $-6 - (-3)$

35. $-4 - (-2)$ 36. $6 - (-12)$

37. $-12 - 16$ 38. $-4 - 3 - 2$

39. $4 - 5 - 12$ 40. $12 - (-7) - 8$

41. $-12 - (-3) - (-15)$ 42. $4 - 12 - (-8)$

43. $13 - 7 - 15$ 44. $-6 + 19 - (-31)$

45. $-30 - (-65) - 29 - 4$ 46. $42 - (-82) - 65 - 7$

47. $-16 - 47 - 63 - 12$ 48. $42 - (-30) - 65 - (-11)$

49. $-47 - (-67) - 13 - 15$ 50. $-18 - 49 - (-84) - 27$

SUPPLEMENTAL EXERCISES 1.2

Simplify.

51. $|-7 + 12|$ 52. $|13 - (-4)|$

53. $|-13 - (-2)|$ 54. $|18 - 21|$

55. $1 - 2 + 3 - 4 + 5 - 6 + 7 - 8$ 56. $2 - 4 + 6 - 8 + 10 - 12$

57. $-1 + 2 - 3 + 4 - 5 + 6 - 7 + 8$ 58. $-2 + 4 - 6 + 8 - 10 + 12$

Assuming the pattern is continued, find the next three numbers in the pattern.

59. $-7, -11, -15, -19, \ldots$ 60. $16, 11, 6, 1, \ldots$

61. $12, 11, 9, 6, 2, -3, \ldots$ 62. $-14, -13, -11, -8, -4, 1, \ldots$

In each exercise, determine which statement is false.

63. **a.** $|3 + 4| = |3| + |4|$ **b.** $|3 - 4| = |3| - |4|$ **c.** $|4 + 3| = |4| + |3|$ **d.** $|4 - 3| = |4| - |3|$

64. **a.** $|5 + 2| = |5| + |2|$ **b.** $|5 - 2| = |5| - |2|$ **c.** $|2 + 5| = |2| + |5|$ **d.** $|2 - 5| = |2| - |5|$

Determine which statement is true for all real numbers.

65. **a.** $|x + y| \le |x| + |y|$ **b.** $|x + y| = |x| + |y|$ **c.** $|x + y| \ge |x| + |y|$

66. **a.** $\left||x| - |y|\right| \le |x| - |y|$ **b.** $\left||x| - |y|\right| = |x| - |y|$ **c.** $\left||x| - |y|\right| \ge |x| - |y|$

SECTION 1.3
Multiplication and Division of Integers

1 Multiply integers

Multiplication is the repeated addition of the same number.

Several different symbols are used to indicate multiplication.

$$3 \times 2 = 6$$
$$3 \cdot 2 = 6$$
$$(3)(2) = 6$$

When 5 is multiplied by a sequence of decreasing integers, each product decreases by 5.

$$(5)(3) = 15$$
$$(5)(2) = 10$$
$$(5)(1) = 5$$
$$(5)(0) = 0$$

The pattern developed can be continued so that 5 is multiplied by a sequence of negative numbers. The resulting products must be negative in order to maintain the pattern of decreasing by 5.

$$(5)(-1) = -5$$
$$(5)(-2) = -10$$
$$(5)(-3) = -15$$
$$(5)(-4) = -20$$

This illustrates that the product of a positive number and a negative number is negative.

When -5 is multiplied by a sequence of decreasing integers, each product increases by 5.

$$(-5)(3) = -15$$
$$(-5)(2) = -10$$
$$(-5)(1) = -5$$
$$(-5)(0) = 0$$

The pattern developed can be continued so that -5 is multiplied by a sequence of negative numbers. The resulting products must be positive in order to maintain the pattern of increasing by 5.

$$(-5)(-1) = 5$$
$$(-5)(-2) = 10$$
$$(-5)(-3) = 15$$
$$(-5)(-4) = 20$$

This illustrates that the product of two negative numbers is positive.

The pattern for multiplication shown above is summarized in the following rules for multiplying integers.

Same Sign   To multiply two numbers with the same sign, multiply the absolute values of the numbers. The product is positive.

$$4 \cdot 8 = 32$$
$$(-4)(-8) = 32$$

Different Signs   To multiply two numbers with different signs, multiply the absolute values of the numbers. The product is negative.

$$-4 \cdot 8 = -32$$
$$(4)(-8) = -32$$

Example 1   Multiply.
A. $-42 \cdot 62$ B. $2(-3)(-5)(-7)$

Solution   A. $-42 \cdot 62$ ▶ The signs are different. The product is negative.
-2604

B. $2(-3)(-5)(-7)$ ▶ To multiply more than two numbers, multiply the
$-6 \cdot (-5)(-7)$ first two numbers. Then multiply the product by
$30 \cdot (-7)$ the third number. Continue until all the numbers
-210 have been multiplied.

Problem 1   Multiply.
A. $-38 \cdot 51$ B. $-7(-8)(9)(-2)$

Solution   See page A3.

2   Divide integers

For every division problem there is a related multiplication problem.

$$\text{Division: } \frac{8}{2} = 4 \qquad\qquad \text{Related multiplication: } 4 \cdot 2 = 8$$

This fact can be used to illustrate the rules for dividing signed numbers.

Same Sign

The quotient of two numbers with the same sign is positive.

$\dfrac{12}{3} = 4$ because $4 \cdot 3 = 12$.

$\dfrac{-12}{-3} = 4$ because $4(-3) = -12$.

Different Signs

The quotient of two numbers with different signs is negative.

$\dfrac{12}{-3} = -4$ because $-4(-3) = 12$.

$\dfrac{-12}{3} = -4$ because $-4 \cdot 3 = -12$.

Note that $\dfrac{12}{-3}, \dfrac{-12}{3}$, and $-\dfrac{12}{3}$ are all equal to -4.

If a and b are two integers, then $\dfrac{a}{-b} = \dfrac{-a}{b} = -\dfrac{a}{b}$.

Zero and One in Division

Zero divided by any number other than zero is zero.	$\dfrac{0}{a} = 0$		because $0 \cdot a = 0$.
Division by zero is not defined.	$\dfrac{4}{0} = ?$	$? \times 0 = 4$	There is no number whose product with zero is 4.
Any number other than zero divided by itself is 1.	$\dfrac{a}{a} = 1$		because $1 \cdot a = a$.
Any number divided by 1 is the number.	$\dfrac{a}{1} = a$		because $a \cdot 1 = a$.

Example 2   Divide.   A. $(-120) \div (-8)$ B. $95 \div (-5)$

Solution   A. $(-120) \div (-8) = 15$ B. $95 \div (-5) = -19$

Problem 2   Divide.   A. $(-135) \div (-9)$ B. $84 \div (-6)$

Solution   See page A3.

3   Application problems

To solve an application problem, first read the problem carefully. The **Strategy** involves identifying the quantity to be found and planning the steps needed to find that quantity. The **Solution** involves performing each operation stated in the Strategy and writing the answer.

Example 3   The temperature at which mercury freezes is $-39°C$. Mercury boils at $360°C$. Find the difference between the temperature at which mercury freezes and the temperature at which it boils.

Strategy   To find the difference, subtract the temperature at which mercury freezes from the temperature at which it boils.

Solution $360 - (-39)$
$360 + 39$
399

The difference is $399°C$.

Problem 3   The temperature at which radon freezes is −71°C. Radon boils at −62°C. Find the difference between the temperature at which radon freezes and the temperature at which it boils.

Solution   See page A3.

Example 4   The daily low temperatures, in degrees Celsius, during one week were recorded as follows: −8°, 2°, 0°, −7°, 1°, 6°, −1°. Find the average daily low temperature for the week.

Strategy   To find the average daily low temperature:
■ Add the seven temperature readings.
■ Divide by 7.

Solution $-8 + 2 + 0 + (-7) + 1 + 6 + (-1)$
$-6 + 0 + (-7) + 1 + 6 + (-1)$
$-6 + (-7) + 1 + 6 + (-1)$
$-13 + 1 + 6 + (-1)$
$-12 + 6 + (-1)$
$-6 + (-1)$
-7

$-7 \div 7 = -1$

The average daily low temperature was −1°C.

Problem 4   The daily high temperatures, in degrees Celsius, during one week were recorded as follows: −5°, −6°, 3°, 0°, −4°, −7°, −2°. Find the average daily high temperature for the week.

Solution   See page A4.

EXERCISES 1.3

1   Multiply.

1. $14 \cdot 3$	**2.** $62 \cdot 9$	**3.** $-4 \cdot 6$
4. $-7 \cdot 3$	**5.** $-2 \cdot (-3)$	**6.** $-5 \cdot (-1)$
7. $(9)(2)$	**8.** $(3)(8)$	**9.** $5(-4)$
10. $4(-7)$	**11.** $-8(2)$	**12.** $-9(3)$
13. $(-5)(-5)$	**14.** $(-3)(-6)$	**15.** $(-7)(0)$
16. $-32 \cdot 4$	**17.** $-24 \cdot 3$	**18.** $19 \cdot (-7)$

19. $6(-17)$

20. $-8(-26)$

21. $-4(-35)$

22. $-5 \cdot (23)$

23. $-6 \cdot (38)$

24. $9(-27)$

25. $8(-40)$

26. $-7(-34)$

27. $-4(39)$

28. $4 \cdot (-8) \cdot 3$

29. $5 \cdot 7 \cdot (-2)$

30. $8 \cdot (-6) \cdot (-1)$

31. $(-9)(-9)(2)$

32. $-8(-7)(-4)$

33. $-5(8)(-3)$

34. $(-6)(5)(7)$

35. $-1(4)(-9)$

36. $6(-3)(-2)$

37. $4(-4) \cdot 6(-2)$

38. $-5 \cdot 9(-7) \cdot 3$

39. $-9(4) \cdot 3(1)$

40. $8(8)(-5)(-4)$

41. $(-6) \cdot 7 \cdot (-10)(-5)$

42. $-9(-6)(11)(-2)$

43. $-6(-5)(12)(0)$

44. $7(9) \cdot 10 \cdot (-1)$

45. $-19(28)(-43)(-11)$

46. $-65(13)(-47)(-92)$

2   Divide.

47. $12 \div (-6)$

48. $18 \div (-3)$

49. $(-72) \div (-9)$

50. $(-64) \div (-8)$

51. $0 \div (-6)$

52. $-49 \div 7$

53. $45 \div (-5)$

54. $-24 \div 4$

55. $-36 \div 4$

56. $-56 \div 7$

57. $-81 \div (-9)$

58. $-40 \div (-5)$

59. $72 \div (-3)$

60. $44 \div (-4)$

61. $-60 \div 5$

62. $-66 \div 6$

63. $-93 \div (-3)$

64. $-98 \div (-7)$

65. $(-85) \div (-5)$

66. $(-60) \div (-4)$

67. $120 \div 8$

68. $144 \div 9$

69. $78 \div (-6)$

70. $84 \div (-7)$

71. $-72 \div 4$

72. $-80 \div 5$

73. $-114 \div (-6)$

74. $-91 \div (-7)$

75. $-104 \div (-8)$

76. $-126 \div (-9)$

77. $57 \div (-3)$

78. $162 \div (-9)$

79. $-136 \div (-8)$

80. $-128 \div 4$

81. $-130 \div (-5)$

82. $(-280) \div 8$

83. $(-92) \div (-4)$

84. $-196 \div (-7)$

85. $-150 \div (-6)$

86. $(-261) \div 9$ **87.** $204 \div (-6)$ **88.** $165 \div (-5)$

89. $-132 \div (-12)$ **90.** $-156 \div (-13)$ **91.** $-182 \div 14$

92. $-144 \div 12$ **93.** $143 \div 11$ **94.** $168 \div 14$

95. $-180 \div (-15)$ **96.** $-169 \div (-13)$ **97.** $154 \div (-11)$

3   Solve.

98. Find the temperature after a rise of 9°C from −6°C.

99. Find the temperature after a rise of 7°C from −18°C.

100. The high temperature for the day was 10°C. The low temperature was −4°C. Find the difference between the high and low temperatures for the day.

101. The low temperature for the day was −2°C. The high temperature was 11°C. Find the difference between the high and low temperatures for the day.

The elevation, or height, of places on the earth is measured in relation to sea level, or the average level of the ocean's surface. The table below shows height above sea level as a positive number and depth below sea level as a negative number.

Continent	Highest Elevation (in meters)		Lowest Elevation (in meters)	
Africa	Mt. Kilimanjaro	5895	Qattara Depression	−133
Asia	Mt. Everest	8848	Dead Sea	−400
Europe	Mt. Elbrus	5634	Caspian Sea	−28
North America	Mt. McKinley	6194	Death Valley	−86
South America	Mt. Aconcagua	6960	Salinas Grandes	−40

102. Use the table to find the difference in elevation between Mt. Elbrus and the Caspian Sea.

103. Use the table to find the difference in elevation between Mt. Aconcagua and Salinas Grandes.

104. Use the table to find the difference in elevation between Mt. Kilimanjaro and the Qattara Depression.

105. Use the table to find the difference in elevation between Mt. McKinley and Death Valley.

106. Use the table to find the difference in elevation between Mt. Everest and the Dead Sea.

107. The daily low temperatures, in degrees Celsius, during one week were recorded as follows: 4°, −5°, 8°, 0°, −9°, −11°, −8°. Find the average daily low temperature for the week.

108. The daily high temperatures, in degrees Celsius, during one week were recorded as follows: −8°, −9°, 6°, 7°, −2°, −14°, −1°. Find the average daily high temperature for the week.

SUPPLEMENTAL EXERCISES 1.3

Assuming the pattern is continued, find the next number in the pattern.

109. 7, −14, 28, −56, . . .   **110.** 256, −64, 16, −4, . . .

Assuming the pattern is continued, find the sum of the first six numbers in the pattern.

111. 2, −6, 18, −54, . . .   **112.** −243, 81, −27, 9, . . .

Solve.

113. 32,844 is divisible by 3. By rearranging the digits, find the largest possible number that is still divisible by 3.

114. 4563 is not divisible by 4. By rearranging the digits, find the largest possible number that is divisible by 4.

115. How many three-digit numbers of the form 8__4 are divisible by 3?

SECTION 1.4

Rational Numbers

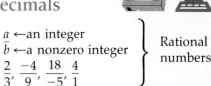

1 ▪ Write rational numbers as decimals

A **rational number** is the quotient of two integers. Therefore, a rational number is a number that can be written in the form $\frac{a}{b}$, where a and b are integers, and b is not zero. A rational number written in this way is commonly called a **fraction.**

a ←an integer
b ←a nonzero integer

$\dfrac{2}{3}, \dfrac{-4}{9}, \dfrac{18}{-5}, \dfrac{4}{1}$

$\left.\right\}$ Rational numbers

Because an integer can be written as the quotient of the integer and 1, every integer is a rational number.

$5 = \dfrac{5}{1}$ $-3 = \dfrac{-3}{1}$

A number written in **decimal nota-tion** is also a rational number.

three-tenths $0.3 = \dfrac{3}{10}$

thirty-five hundredths $0.35 = \dfrac{35}{100}$

negative four tenths $-0.4 = \dfrac{-4}{10}$

A rational number written as a fraction can be written in decimal notation.

Example 1   Write $\dfrac{5}{8}$ as a decimal.

Solution   The fraction bar can be read "÷". $\dfrac{5}{8} = 5 \div 8$

$$
\begin{array}{r}
0.625 \leftarrow \text{This is called a \textbf{terminating decimal}.} \\
8\overline{)5.000} \\
\underline{-4\,8} \\
20 \\
\underline{-16} \\
40 \\
\underline{-40} \\
0 \leftarrow \text{The remainder is zero.}
\end{array}
$$

$\dfrac{5}{8} = 0.625$

Problem 1   Write $\dfrac{4}{25}$ as a decimal.

Solution   See page A4.

Example 2   Write $\dfrac{4}{11}$ as a decimal.

Solution
$$
\begin{array}{r}
0.3636 \ldots \leftarrow \text{This is called a \textbf{repeating decimal}.} \\
11\overline{)4.0000} \\
\underline{-3\,3} \\
70 \\
\underline{-66} \\
40 \\
\underline{-33} \\
70 \\
\underline{-66} \\
4 \leftarrow \text{The remainder is never zero.}
\end{array}
$$

$\dfrac{4}{11} = 0.\overline{36}$ The bar over the digits 3 and 6 is used to show that these digits repeat.

Problem 2 Write $\frac{4}{9}$ as a decimal. Place a bar over the repeating digits of the decimal.

Solution See page A4.

Every rational number can be written as a terminating or repeating decimal. Some numbers (for example, $\sqrt{2}$ and π) have decimal representations that never terminate or repeat. These numbers are called **irrational numbers.**

$$\sqrt{2} = 1.414213562 \ldots \qquad\qquad \pi = 3.141592654 \ldots$$

The rational numbers and the irrational numbers taken together are called the **real numbers.**

▪ 2 ▪ Add and subtract rational numbers

To add or subtract fractions, first rewrite the fractions as equivalent fractions with a common denominator, using the least common multiple (LCM) of the denominators as the common denominator. Then add the numerators, and place the sum over the common denominator. Write the answer in simplest form.

Example 3 Simplify: $-\dfrac{5}{6} + \dfrac{3}{10}$

Solution Prime factorization of 6 and 10:
$$6 = 2 \cdot 3$$
$$10 = 2 \cdot 5$$
$$\text{LCM} = 2 \cdot 3 \cdot 5 = 30$$

▶ Find the LCM of the denominators 6 and 10. The LCM of denominators is sometimes called the **least common denominator** (LCD).

$$-\frac{5}{6} + \frac{3}{10} = -\frac{25}{30} + \frac{9}{30}$$

▶ Rewrite the fractions as equivalent fractions, using the LCM of the denominators as the common denominator.

$$= \frac{-25 + 9}{30}$$

$$= \frac{-16}{30}$$

▶ Add the numerators, and place the sum over the common denominator.

$$= -\frac{8}{15}$$

▶ Write the answer in simplest form.

Problem 3 Simplify: $\dfrac{5}{9} - \dfrac{11}{12}$

Solution See page A4.

Example 4   Simplify: $-\dfrac{3}{4} + \dfrac{1}{6} - \dfrac{5}{8}$

Solution $-\dfrac{3}{4} + \dfrac{1}{6} - \dfrac{5}{8} = -\dfrac{18}{24} + \dfrac{4}{24} - \dfrac{15}{24}$ ▶ The LCM of 4, 6, and 8 is 24.

$$= \dfrac{-18}{24} + \dfrac{4}{24} + \dfrac{-15}{24}$$

$$= \dfrac{-18 + 4 + (-15)}{24}$$

$$= \dfrac{-29}{24}$$

$$= -\dfrac{29}{24}$$

Problem 4   Simplify: $-\dfrac{7}{8} - \dfrac{5}{6} + \dfrac{1}{2}$

Solution   See page A4.

To add or subtract decimals, write the numbers so that the decimal points are in a vertical line. Then proceed as in the addition or subtraction of integers. Write the decimal point in the answer directly below the decimal points in the problem.

Example 5   Simplify: $14.02 + 137.6 + 9.852$

Solution
$$\begin{array}{r} 14.02 \\ 137.6 \\ +\quad 9.852 \\ \hline 161.472 \end{array}$$

▶ Write the decimals so that the decimal points are in a vertical line.

▶ Write the decimal point in the sum directly below the decimal points in the problem.

Problem 5   Simplify: $3.097 + 4.9 + 3.09$

Solution   See page A4.

Example 6   Simplify: $-114.039 + 84.76$

Solution
$$\begin{array}{r} {\scriptstyle 10\,13\ 9} \\ {\scriptstyle 0\ \cancel{0}\ \cancel{3}\,10\,13} \\ \cancel{1}\,1\,\cancel{4}.\cancel{0}\,\cancel{3}\,9 \\ -\quad 8\,4\,.7\,6 \\ \hline \end{array}$$

▶ The signs are different. Subtract the absolute value of the smaller number from the absolute value of the larger number.

$-114.039 + 84.76 = -29.279$

▶ Attach the sign of the number with the greater absolute value.

Problem 6   Simplify: $16.127 - 67.91$

Solution   See page A4.

3   Multiply and divide rational numbers

The product of two fractions is the product of the numerators divided by the product of the denominators.

Example 7   Simplify: $\dfrac{3}{8} \cdot \dfrac{12}{17}$

Solution $\dfrac{3}{8} \cdot \dfrac{12}{17} = \dfrac{3 \cdot 12}{8 \cdot 17}$ ▶ Multiply the numerators. Multiply the denominators.

$= \dfrac{3 \cdot \overset{1}{\cancel{2}} \cdot \overset{1}{\cancel{2}} \cdot 3}{2 \cdot \underset{1}{\cancel{2}} \cdot \underset{1}{\cancel{2}} \cdot 17}$ ▶ Write the prime factorization of each factor. Divide by the common factors.

$= \dfrac{9}{34}$ ▶ Multiply the numbers remaining in the numerator. Multiply the numbers remaining in the denominator.

Problem 7   Simplify: $-\dfrac{7}{12} \cdot \dfrac{9}{14}$

Solution   See page A5.

To divide fractions, invert the divisor. Then proceed as in the multiplication of fractions.

Example 8   Simplify: $\dfrac{3}{10} \div \left(-\dfrac{18}{25}\right)$ *Invert & multiply*

Solution $\dfrac{3}{10} \div \left(-\dfrac{18}{25}\right) = -\left(\dfrac{3}{10} \div \dfrac{18}{25}\right)$ ▶ The signs are different. The quotient is negative.

$= -\left(\dfrac{3}{10} \cdot \dfrac{25}{18}\right)$

$= -\left(\dfrac{3 \cdot 25}{10 \cdot 18}\right)$

$= -\left(\dfrac{\overset{1}{\cancel{3}} \cdot \overset{1}{\cancel{5}} \cdot 5}{2 \cdot \cancel{5} \cdot 2 \cdot \underset{1}{\cancel{3}} \cdot 3}\right)$

$= -\dfrac{5}{12}$

Problem 8   Simplify: $-\dfrac{3}{8} \div \left(-\dfrac{5}{12}\right)$

Solution   See page A5.

To multiply decimals, multiply as in integers. Write the decimal point in the product so that the number of decimal places in the product equals the sum of the decimal places in the factors.

Example 9   Simplify: $(-6.89)(0.00035)$

Solution

6.89	2 decimal places
$\times$ 0.00035	5 decimal places
3445	
2067	
0.0024115	7 decimal places

▶ Multiply the absolute values.

$(-6.89)(0.00035) = -0.0024115$

▶ The signs are different. The product is negative.

Problem 9   Simplify: $(-5.44)(3.8)$

Solution   See page A5.

To divide decimals, move the decimal point in the divisor to make it a whole number. Move the decimal point in the dividend the same number of places to the right. Place the decimal point in the quotient directly over the decimal point in the dividend. Then divide as in whole numbers.

Example 10   Simplify $1.32 \div 0.27$. Round to the nearest tenth.

Solution $0.27\overline{)1.32.}$

▶ Move the decimal point two places to the right in the divisor and in the dividend. Place the decimal point in the quotient.

$$
\begin{array}{r}
4.88 \approx 4.9 \\
27\overline{)132.00} \\
-108 \\
\hline
24\ 0 \\
-21\ 6 \\
\hline
240 \\
-216 \\
\hline
24
\end{array}
$$

▶ The symbol $\approx$ is used to indicate that the quotient is an approximate value that has been rounded off.

Problem 10   Simplify $-0.394 \div 1.7$. Round to the nearest hundredth.

  Solution   See page A5.

EXERCISES 1.4

1   Write as a decimal. Place a bar over the repeating digits of a repeating decimal.

1. $\dfrac{1}{3}$	**2.** $\dfrac{2}{3}$	**3.** $\dfrac{1}{4}$	**4.** $\dfrac{3}{4}$
5. $\dfrac{2}{5}$	**6.** $\dfrac{4}{5}$	**7.** $\dfrac{1}{6}$	**8.** $\dfrac{5}{6}$
9. $\dfrac{1}{8}$	**10.** $\dfrac{7}{8}$	**11.** $\dfrac{2}{9}$	**12.** $\dfrac{8}{9}$
13. $\dfrac{5}{11}$	**14.** $\dfrac{10}{11}$	**15.** $\dfrac{7}{12}$	**16.** $\dfrac{11}{12}$
17. $\dfrac{4}{15}$	**18.** $\dfrac{8}{15}$	**19.** $\dfrac{9}{16}$	**20.** $\dfrac{15}{16}$
21. $\dfrac{7}{18}$	**22.** $\dfrac{17}{18}$	**23.** $\dfrac{1}{20}$	**24.** $\dfrac{13}{20}$
25. $\dfrac{6}{25}$	**26.** $\dfrac{14}{25}$	**27.** $\dfrac{7}{30}$	**28.** $\dfrac{19}{30}$
29. $\dfrac{9}{40}$	**30.** $\dfrac{21}{40}$	**31.** $\dfrac{13}{36}$	**32.** $\dfrac{29}{36}$
33. $\dfrac{15}{22}$	**34.** $\dfrac{19}{22}$	**35.** $\dfrac{11}{24}$	**36.** $\dfrac{19}{24}$
37. $\dfrac{5}{33}$	**38.** $\dfrac{25}{33}$	**39.** $\dfrac{3}{37}$	**40.** $\dfrac{14}{37}$

2   Simplify.

41. $\dfrac{2}{3} + \dfrac{5}{12}$	**42.** $\dfrac{1}{2} + \dfrac{3}{8}$	**43.** $\dfrac{5}{8} - \dfrac{5}{6}$
44. $\dfrac{1}{9} - \dfrac{5}{27}$	**45.** $-\dfrac{5}{12} - \dfrac{3}{8}$	**46.** $-\dfrac{5}{6} - \dfrac{5}{9}$

47. $-\dfrac{6}{13} + \dfrac{17}{26}$

48. $-\dfrac{7}{12} + \dfrac{5}{8}$

49. $-\dfrac{5}{8} - \left(-\dfrac{11}{12}\right)$

50. $\dfrac{1}{3} + \dfrac{5}{6} - \dfrac{2}{9}$

51. $\dfrac{1}{2} - \dfrac{2}{3} + \dfrac{1}{6}$

52. $-\dfrac{3}{8} - \dfrac{5}{12} - \dfrac{3}{16}$

53. $-\dfrac{5}{16} + \dfrac{3}{4} - \dfrac{7}{8}$

54. $\dfrac{1}{2} - \dfrac{3}{8} - \left(-\dfrac{1}{4}\right)$

55. $\dfrac{3}{4} - \left(-\dfrac{7}{12}\right) - \dfrac{7}{8}$

56. $\dfrac{1}{3} - \dfrac{1}{4} - \dfrac{1}{5}$

57. $\dfrac{2}{3} - \dfrac{1}{2} + \dfrac{5}{6}$

58. $\dfrac{5}{16} + \dfrac{1}{8} - \dfrac{1}{2}$

59. $\dfrac{5}{8} - \left(-\dfrac{5}{12}\right) + \dfrac{1}{3}$

60. $\dfrac{1}{8} - \dfrac{11}{12} + \dfrac{1}{2}$

61. $-\dfrac{7}{9} + \dfrac{14}{15} + \dfrac{8}{21}$

62. $1.09 + 6.2$

63. $-32.1 - 6.7$

64. $5.13 - 8.179$

65. $-13.092 + 6.9$

66. $2.54 - 3.6$

67. $5.43 + 7.925$

68. $-16.92 - 6.925$

69. $-3.87 + 8.546$

70. $6.9027 - 17.692$

71. $2.09 - 6.72 - 5.4$

72. $16.4 + 3.09 - 7.93$

73. $-18.39 + 4.9 - 23.7$

74. $19 - (-3.72) - 82.75$

75. $-3.07 - (-2.97) - 17.4$

76. $-3.09 - 4.6 - 27.3$

77. $317.09 - 46.902 + 583.0714$

78. $71.0235 - 86.0974 + 254.309$

3   Simplify.

79. $\dfrac{1}{2} \cdot \left(-\dfrac{3}{4}\right)$

80. $-\dfrac{2}{9} \cdot \left(-\dfrac{3}{14}\right)$

81. $\left(-\dfrac{3}{8}\right)\left(-\dfrac{4}{15}\right)$

82. $\dfrac{5}{8} \cdot \left(-\dfrac{7}{12}\right) \cdot \dfrac{16}{25}$

83. $\left(\dfrac{1}{2}\right)\left(-\dfrac{3}{4}\right)\left(-\dfrac{5}{8}\right)$

84. $\left(\dfrac{5}{12}\right)\left(-\dfrac{8}{15}\right)\left(-\dfrac{1}{3}\right)$

85. $\dfrac{3}{8} \div \dfrac{1}{4}$

86. $\dfrac{5}{6} \div \left(-\dfrac{3}{4}\right)$

87. $-\dfrac{5}{12} \div \dfrac{15}{32}$

88. $\dfrac{1}{8} \div \left(-\dfrac{5}{12}\right)$

89. $-\dfrac{4}{9} \div \left(-\dfrac{2}{3}\right)$

90. $-\dfrac{6}{11} \div \dfrac{4}{9}$

91. $(1.2)(3.47)$ **92.** $(-0.8)(6.2)$ **93.** $(-1.89)(-2.3)$

94. $(6.9)(-4.2)$ **95.** $(1.06)(-3.8)$ **96.** $(-2.7)(-3.5)$

97. $(1.2)(-0.5)(3.7)$ **98.** $(-2.4)(6.1)(0.9)$ **99.** $(-0.8)(3.006)(-5.1)$

Simplify. Round to the nearest hundredth.

100. $-24.7 \div 0.09$ **101.** $-1.27 \div (-1.7)$ **102.** $9.07 \div (-3.5)$

103. $0.0976 \div 0.042$ **104.** $-6.904 \div 1.35$ **105.** $-7.894 \div (-2.06)$

106. $-354.2086 \div 0.1719$ **107.** $-2658.3109 \div (-0.0473)$ **108.** $(-3.92)(-27.1)(45.008)$

SUPPLEMENTAL EXERCISES 1.4

Classify each of the following numbers as a natural number, an integer, a positive integer, a negative integer, a rational number, an irrational number, and a real number.

109. -1 **110.** 28

111. $-\dfrac{9}{34}$ **112.** -7.707

113. $5.2\overline{6}$ **114.** $0.174359 \ldots$

Simplify.

115. $|2.4 - 8.7|$ **116.** $|-8.4 - 6.8|$ **117.** $|2.35 - (-4.82)|$

Computers can be programmed to perform integer arithmetic, whereby the result of each arithmetic operation is truncated (for example, $\dfrac{11}{3}$ is truncated to 3 because 3 is the integer part of $11 \div 3$). Simplify each of the following using integer arithmetic.

118. $\dfrac{16}{6} + \dfrac{21}{5}$ **119.** $4\left(\dfrac{3}{4}\right)$ **120.** $\dfrac{3(4)}{4}$

Solve.

121. Find the average of $\dfrac{5}{8}$ and $\dfrac{3}{4}$.

122. Postage for first-class mail is \$.29 for the first ounce or fraction of an ounce, and \$.23 for each additional ounce or fraction of an ounce. Find the cost of mailing a $4\dfrac{1}{2}$-ounce letter by first-class mail.

123. The price of a pen was 60 cents. During a storewide sale, the price was reduced to a different whole number of cents, and the entire stock was sold for $54.59. Find the price of the pen during the sale.

Palindromic numbers are natural numbers that remain unchanged when their digits are written in reverse order. For example, 585 is a palindromic number.

124. Find the smallest two-digit multiple of 4 that is a palindromic number.

125. Find the smallest three-digit multiple of 4 that is a palindromic number.

SECTION 1.5

Exponents and the Order of Operations Agreement

1 Exponential expressions

Repeated multiplication of the same factor can be written using an exponent.

$$2 \cdot 2 \cdot 2 \cdot 2 \cdot 2 = 2^5 \longleftarrow \textbf{exponent}$$
$$\underline{} \textbf{base}$$

$$a \cdot a \cdot a \cdot a = a^4 \longleftarrow \textbf{exponent}$$
$$\underline{} \textbf{base}$$

The **exponent** indicates how many times the factor, called the **base,** occurs in the multiplication. The multiplication $2 \cdot 2 \cdot 2 \cdot 2 \cdot 2$ is in **factored form.** The exponential expression 2^5 is in **exponential form.**

2^1 is read "the first power of two" or just "two." $\longrightarrow$ Usually the exponent 1 is not written.

2^2 is read "the second power of two" or "two squared."
2^3 is read "the third power of two" or "two cubed."
2^4 is read "the fourth power of two."
2^5 is read "the fifth power of two."
a^5 is read "the fifth power of a."

To evaluate an exponential expression, write each factor as many times as indicated by the exponent. Then multiply.

$$3^5 = 3 \cdot 3 \cdot 3 \cdot 3 \cdot 3 = 243$$

$$2^3 \cdot 3^2 = (2 \cdot 2 \cdot 2) \cdot (3 \cdot 3) = 8 \cdot 9 = 72$$

Example 1   Evaluate $(-4)^2$ and -4^2.

Solution $(-4)^2 = (-4)(-4) = 16$

$-4^2 = -(4 \cdot 4) = -16$ ▶ The -4 is squared only when the negative sign is *inside* the parentheses.

Problem 1   Evaluate $(-5)^3$ and -5^3.

Solution   See page A5.

$(-5)^3 = 5 \cdot 5 \cdot 5 = -125$

$-5^3 = -125$

Example 2   Evaluate $(-2)^4$ and $(-2)^5$.

Solution $(-2)^4 = (-2)(-2)(-2)(-2)$
$= 4(-2)(-2)$
$= -8(-2)$
$= 16$ ▶ The product of an even number of negative factors is positive.

$(-2)^5 = (-2)(-2)(-2)(-2)(-2)$
$= 4(-2)(-2)(-2)$
$= -8(-2)(-2)$
$= 16(-2)$
$= -32$ ▶ The product of an odd number of negative factors is negative.

Problem 2   Evaluate $(-3)^3$ and $(-3)^4$.

Solution   See page A5.

$(-3)^3 = -27$

$(-3)^4 = 81$

Example 3   Evaluate $(-3)^2 \cdot 2^3$ and $\left(-\frac{2}{3}\right)^3$.

Solution $(-3)^2 \cdot 2^3 = (-3)(-3) \cdot (2)(2)(2) = 9 \cdot 8 = 72$

$\left(-\frac{2}{3}\right)^3 = \left(-\frac{2}{3}\right)\left(-\frac{2}{3}\right)\left(-\frac{2}{3}\right) = -\frac{2 \cdot 2 \cdot 2}{3 \cdot 3 \cdot 3} = -\frac{8}{27}$

Problem 3   Evaluate $(3^3)(-2)^3$ and $\left(-\frac{2}{5}\right)^2$.

Solution   See page A5.

$27 \cdot (-8) = -216$

$\frac{2}{5} \cdot \frac{2}{5} = \frac{4}{25} = \frac{1}{3}$

2   The Order of Operations Agreement

Evaluate $2 + 3 \cdot 5$.

There are two arithmetic operations, addition and multiplication, in this problem. The operations could be performed in different orders.

Add first.	$\underline{2 + 3} \cdot 5$	Multiply first.	$2 + \underline{3 \cdot 5}$
Then multiply.	$\underline{5 \cdot 5}$	Then add.	$\underline{2 + 15}$
	25		17

In order to prevent there being more than one answer to the same problem, an Order of Operations Agreement has been established.

The Order of Operations Agreement

> **Step 1**   Perform operations inside grouping symbols. Grouping symbols include parentheses ( ), brackets [ ], and the fraction bar.
>
> **Step 2**   Simplify exponential expressions.
>
> **Step 3**   Do multiplication and division as they occur from left to right.
>
> **Step 4**   Do addition and subtraction as they occur from left to right.

Example 4   Simplify: $12 - 24(8 - 5) \div 2^2$

Solution $12 - 24(8 - 5) \div 2^2$

$12 - 24(3) \div 2^2$ ▶ Perform operations inside grouping symbols.

$12 - 24(3) \div 4$ ▶ Simplify exponential expressions.

$12 - 72 \div 4$ ▶ Do multiplication and division as they occur from left to right.

$12 - 18$

-6 ▶ Do addition and subtraction as they occur from left to right.

Problem 4   Simplify: $36 \div (8 - 5)^2 - (-3)^2 \cdot 2$

Solution   See page A5.

One or more of the above steps may not be needed to simplify an expression. In that case, proceed to the next step in the Order of Operations Agreement.

Example 5   Simplify: $\dfrac{4+8}{2+1} - (3-1) + 2$

Solution $\dfrac{4+8}{2+1} - (3-1) + 2$

$\dfrac{12}{3} - 2 + 2$ ► Perform operations inside grouping symbols.

$4 - 2 + 2$ ► Do multiplication and division as they occur from left to right.

$2 + 2$ ► Do addition and subtraction as they occur from left to right.

4

Problem 5   Simplify: $27 \div 3^2 + (-3)^2 \cdot 4$

Solution   See page A5.

When an expression has grouping symbols inside grouping symbols, first perform the operations inside the *inner* grouping symbols by following Steps 2, 3, and 4 of the Order of Operations Agreement. Then perform the operations inside the *outer* grouping symbols by following Steps 2, 3, and 4 in sequence.

Example 6   Simplify: $6 \div [4 - (6 - 8)] + 2^2$

Solution $6 \div [4 - (6 - 8)] + 2^2$
$6 \div [4 - (-2)] + 2^2$ ► Perform operations inside inner grouping symbols.
$6 \div 6 + 2^2$ ► Perform operations inside outer grouping symbols.
$6 \div 6 + 4$ ► Simplify exponential expressions.
$1 + 4$ ► Do multiplication and division.
5 ► Do addition and subtraction.

Problem 6   Simplify: $4 - 3[4 - 2(6 - 3)] \div 2$

Solution   See page A5.

EXERCISES 1.5

1   Evaluate.

1. 6^2

2. 7^4

3. -7^2

4. -4^3

5. $(-3)^2$

6. $(-2)^3$

7. $(-3)^4$ **8.** $(-5)^3$ **9.** $\left(\dfrac{1}{2}\right)^2$

10. $\left(-\dfrac{3}{4}\right)^3$ **11.** $(0.3)^2$ **12.** $(1.5)^3$

13. $\left(\dfrac{2}{3}\right)^2 \cdot 3^3$ **14.** $\left(-\dfrac{1}{2}\right)^3 \cdot 8$ **15.** $(0.3)^3 \cdot 2^3$

16. $(0.5)^2 \cdot 3^3$ **17.** $(-3) \cdot 2^2$ **18.** $(-5) \cdot 3^4$

19. $(-2) \cdot (-2)^3$ **20.** $(-2) \cdot (-2)^2$ **21.** $2^3 \cdot 3^3 \cdot (-4)$

22. $(-3)^3 \cdot 5^2 \cdot 10$ **23.** $(-7) \cdot 4^2 \cdot 3^2$ **24.** $(-2) \cdot 2^3 \cdot (-3)^2$

25. $\left(\dfrac{2}{3}\right)^2 \cdot \dfrac{1}{4} \cdot 3^3$ **26.** $\left(\dfrac{3}{4}\right)^2 \cdot (-4) \cdot 2^3$ **27.** $8^2 \cdot (-3)^5 \cdot 5$

2   Simplify by using the Order of Operations Agreement.

28. $4 - 8 \div 2$ **29.** $2^2 \cdot 3 - 3$

30. $2(3 - 4) - (-3)^2$ **31.** $16 - 32 \div 2^3$

32. $24 - 18 \div 3 + 2$ **33.** $8 - (-3)^2 - (-2)$

34. $16 + 15 \div (-5) - 2$ **35.** $14 - 2^2 - (4 - 7)$

36. $3 - 2[8 - (3 - 2)]$ **37.** $-2^2 + 4[16 \div (3 - 5)]$

38. $6 + \dfrac{16 - 4}{2^2 + 2} - 2$ **39.** $24 \div \dfrac{3^2}{8 - 5} - (-5)$

40. $96 \div 2[12 + (6 - 2)] - 3^3$ **41.** $4 \cdot [16 - (7 - 1)] \div 10$

42. $16 \div 2 - 4^2 - (-3)^2$ **43.** $18 \div (9 - 2^3) + (-3)$

44. $16 - 3(8 - 3)^2 \div 5$ **45.** $4(-8) \div [2(7 - 3)^2]$

46. $\dfrac{(-10) + (-2)}{6^2 - 30} \div (2 - 4)$ **47.** $16 - 4 \cdot \dfrac{3^3 - 7}{2^3 + 2} - (-2)^2$

48. $(0.2)^2 \cdot (-0.5) + 1.72$ **49.** $0.3(1.7 - 4.8) + (1.2)^2$

50. $(1.8)^2 - 2.52 \div (1.8)$ **51.** $(1.65 - 1.05)^2 \div 0.4 + 0.8$

52. $\dfrac{3}{8} \div \left(\dfrac{5}{6} + \dfrac{2}{3}\right)$ **53.** $\left(\dfrac{3}{4}\right)^2 - \left(\dfrac{1}{2}\right)^3 \div \dfrac{3}{5}$

SUPPLEMENTAL EXERCISES 1.5

Place the correct symbol, $<$ or $>$, between the two numbers.

54. $(0.9)^3 \quad 1^5$ **55.** $(-3)^3 \quad (-2)^5$ **56.** $(-1.1)^2 \quad (0.9)^2$

Simplify.

57. $1^2 + 2^2 + 3^2 + 4^2$

58. $1^3 + 2^3 + 3^3 + 4^3$

59. $(-1)^3 + (-2)^3 + (-3)^3 + (-4)^3$

60. $(-2)^2 + (-4)^2 + (-6)^2 + (-8)^2$

Complete.

61. The product of an even number of negative numbers is a _____ number.

62. The product of an odd number of negative numbers is a _____ number.

Solve.

63. A computer can do 600,000 additions in one second. To the nearest second, how many seconds will it take the computer to do 10^7 additions?

64. The sum of two natural numbers is 41. Each of the two numbers is the square of a natural number. Find the two numbers.

Determine the ones' digit when the expression is evaluated.

65. 34^{202}

66. 23^{502}

67. 27^{622}

Calculators and Computers

 Extended Precision on an Electronic Calculator

Consider the decimal equivalents of the following fractions:

$$\frac{7}{33} = 0.\overline{21} \qquad \text{and} \qquad \frac{15}{37} = 0.\overline{405}$$

These decimal equivalents were calculated on an electronic calculator that displays seven decimal places. Some fractions, however, do not have decimal equivalents that repeat until well after seven places.

Two examples of fractions with repeating cycles that are longer than seven places are

$$\frac{4}{17} = 0.\overline{2352941176470588} \qquad \text{and} \qquad \frac{9}{23} = 0.\overline{3913043478260869565217}.$$

A calculator that displays seven decimal places was used to determine each decimal equivalent on the previous page. The procedure for these calculations is illustrated below.

Find the repeating decimal expression for $\frac{8}{17}$.

1. Divide the numerator by the denominator. $\frac{8}{17} = 0.4705882$

   The decimal approximation is the first four digits.

$$\frac{8}{17} \approx 0.4705$$

2. Take the last three digits (as a decimal) and form the product with the denominator.

$$0.882 \times 17 = 14.994 \approx 15$$

   Round this product to the nearest integer and divide by the denominator.

$$\frac{15}{17} \approx 0.8823529$$

   The new decimal approximation is the approximation from Step 1 and the first four digits from Step 2.

$$\frac{8}{17} \approx 0.47058823$$

3. Repeat Step 2. Continue to repeat Step 2 until the decimal representation repeats.

   $0.529 \times 17 = 8.993 \approx 9$ $0.117 \times 17 = 1.989 \approx 2$ $0.470 \times 17 = 7.99 \approx 8$

   $\frac{9}{17} \approx 0.5294117$ $\frac{2}{17} \approx 0.1176470$ $\frac{8}{17} \approx 0.4705882$

   $\frac{8}{17} \approx 0.470588235294$ $\frac{8}{17} \approx 0.4705882352941176$

   The decimal begins to repeat with the digits 4705.

   The decimal equivalent of $\frac{8}{17}$ is $0.\overline{4705882352941176}$.

Find the decimal equivalents of each of the following.

1. $\dfrac{3}{17}$ 2. $\dfrac{10}{19}$ 3. $\dfrac{9}{23}$

Something Extra

The Kelvin Scale

The Celsius temperature scale was devised by Anders Celsius, a Swedish astronomer. On the Celsius scale, the temperature at which water freezes is indicated as 0°C, and the temperature at which water boils is indicated as 100°C. The interval between 0°C and 100°C is divided into 100 equal parts. Temperatures below 0°C are negative quantities.

Theoretically, there is a temperature that is the lowest possible temperature. Scientists refer to the lowest possible temperature as **absolute zero.** An English physicist, Lord William Kelvin, devised a temperature scale called the **Kelvin scale,** on which the zero point is absolute zero and each degree (which in the Kelvin scale is called a kelvin) is the same size as the Celsius degree. The letter K denotes a temperature on the Kelvin scale.

It is estimated that the temperature of absolute zero is −273.15°C, or 273.15 degrees below 0°C, the temperature at which water freezes. Thus, on the Kelvin scale, the temperature at which water freezes is 273.15 K, and the temperature at which water boils is 373.15 K. Generally, these values are rounded to 273 K and 373 K, respectively.

To convert temperatures from the Celsius scale to the Kelvin scale, add 273 to the Celsius temperature.

$$-50°C = (-50 + 273) \text{ K} = 223 \text{ K}$$

To convert temperatures from the Kelvin scale to the Celsius scale, subtract 273 from the Kelvin temperature.

$$250 \text{ K} = (250 - 273)°C = -23°C$$

Convert to Kelvin temperature.

1. 67°C   **2.** −15°C   **3.** −38°C   **4.** −200°C

Convert to Celsius temperature.

5. 320 K   **6.** 245 K   **7.** 189 K   **8.** 76 K

Chapter Summary

Key Words

The *natural numbers* are 1, 2, 3, 4, 5, 6, 7, ...

The *integers* are . . . −4, −3, −2, −1, 0, 1, 2, 3, 4, ...

Two numbers that are the same distance from zero on the number line but on opposite sides of zero are *opposite numbers,* or *opposites.*

The *absolute value* of a number is its distance from zero on the number line.

A *rational number* is a number of the form $\frac{a}{b}$, where a and b are integers and b is not equal to zero. A rational number written in this form is commonly called a *fraction.*

An *irrational number* is a number that has a decimal representation that never terminates or repeats.

The rational numbers and the irrational numbers taken together are called the *real numbers.*

An expression of the form a^n is in *exponential form,* where a is the base and n is the exponent.

Essential Rules

Addition of Integers with the Same Sign   To add two numbers with the same sign, add the absolute values of the numbers. Then attach the sign of the addends.

Addition of Integers with Different Signs   To add two numbers with different signs, find the difference between the absolute values of the numbers. Then attach the sign of the number with the greater absolute value.

Subtraction of Integers   To subtract one integer from another, add the opposite of the second integer to the first integer.

Multiplication of Integers with the Same Sign   To multiply two numbers with the same sign, multiply the absolute values of the numbers. The product is positive.

Multiplication of Integers with Different Signs   To multiply two numbers with different signs, multiply the absolute values of the numbers. The product is negative.

Division of Integers with the Same Sign   The quotient of two numbers with the same sign is positive.

Division of Integers with Different Signs   The quotient of two numbers with different signs is negative.

Order of Operations Agreement
Step 1   Perform operations inside grouping symbols.
Step 2   Simplify exponential expressions.
Step 3   Do multiplication and division as they occur from left to right.
Step 4   Do addition and subtraction as they occur from left to right.

Chapter Review

1.   Evaluate $-|-4|$.

2.   Subtract: $16 - (-30) - 42$

3.   Divide: $-561 \div (-33)$

4.   Write $\frac{7}{9}$ as a decimal. Place a bar over the repeating digits of the decimal.

5.   Simplify: $(6.02)(-0.89)$

6.   Simplify: $\frac{-10 + 2}{2 + (-4)} \div 2 + 6$

7.   Find the opposite of -4.

8.   Subtract: $16 - 30$

9.   Divide: $-72 \div 8$

10.   Write $\frac{17}{20}$ as a decimal.

11.   Simplify: $\frac{5}{12} \div \left(-\frac{5}{6}\right)$

12.   Simplify: $3^2 - 4 + 20 \div 5$

13.   Place the correct symbol, $<$ or $>$, between the two numbers.
$-1 \quad 0$

14.   Add: $-22 + 14 + (-8)$

15.   Multiply: $(-5)(-6)(3)$

16.   Simplify: $6.039 - 12.92$

17.   Evaluate $\frac{3}{4} \cdot (4)^2$.

18.   Place the correct symbol, $<$ or $>$, between the two numbers.
$-2 \quad -40$

19.   Add: $13 + (-16)$

20.   Multiply: $(-4)(12)$

21.   Simplify: $-\frac{2}{5} + \frac{7}{15}$

22.   Evaluate $(-3^3) \cdot 2^2$.

23. Find the opposite of -2.

24. Subtract: $7 - 21$

25. Divide: $96 \div (-12)$

26. Write $\frac{7}{20}$ as a decimal.

27. Simplify: $-\frac{7}{16} \div \frac{3}{8}$

28. Simplify: $2^3 \div 4 - 2(2 - 7)$

29. Evaluate $|-3|$.

30. Subtract: $12 - (-10) - 4$

31. Divide: $(-204) \div (-17)$

32. Write $\frac{7}{11}$ as a decimal. Place a bar over the repeating digits of the decimal.

33. Simplify: $0.2654 \div (-0.023)$
Round to the nearest tenth.

34. Simplify: $(7 - 2)^2 - 5 - 3 \cdot 4$

35. Place the correct symbol, $<$ or $>$, between the two numbers.
$8 \quad -10$

36. Add: $-12 + 8 + (-4)$

37. Multiply: $2(-3)(-12)$

38. Simplify: $-\frac{5}{8} + \frac{1}{6}$

39. Evaluate $-4^2 \cdot \left(\frac{1}{2}\right)^2$.

40. Simplify: $-1.329 + 4.89$

41. Evaluate $-|17|$.

42. Subtract: $-5 - 22 - (-13) - 19 - (-6)$

43. Simplify: $\left(\frac{1}{3}\right)\left(-\frac{4}{5}\right)\left(\frac{3}{8}\right)$

44. Place the correct symbol, $<$ or $>$, between the two numbers.
$-43 \quad -34$

45. Write $\frac{18}{25}$ as a decimal.

46. Evaluate $(-2)^3 \cdot 4^2$.

47. Add: $14 + (-18) + 6 + (-20)$

48. Multiply: $-4(-8)(12)(0)$

49. Simplify: $2^3 - 7 + 16 \div (-3 + 5)$

50. Simplify: $\frac{3}{4} + \frac{1}{2} - \frac{3}{8}$

51. Divide: $-128 \div (-8)$

52. Place the correct symbol, $<$ or $>$, between the two numbers.
$-57 \quad 28$

53. Evaluate $\left(-\frac{1}{3}\right)^3 \cdot 9^2$.

54. Add: $-7 + (-3) + (-12) + 16$

55. Multiply: $5(-2)(10)(-3)$

56. Find the temperature after a rise of 14°C from −6°C.

57. The daily low temperatures, in degrees Celsius, for a three-day period were recorded as follows: −8°, 7°, −5°. Find the average low temperature for the three-day period.

58. The high temperature for the day was 8°C. The low temperature was −5°C. Find the difference between the high and low temperatures for the day.

59. Find the temperature after a rise of 7°C from −13°C.

60. The temperature on the surface of the planet Venus is 480°C. The temperature on the surface of the planet Pluto is −234°C. Find the difference between the surface temperatures on Venus and Pluto.

Chapter Test

1. Subtract: $-9 - (-6)$

2. Write $\frac{17}{20}$ as a decimal.

3. Simplify: $\frac{3}{4}\left(-\frac{2}{21}\right)$

4. Divide: $-75 \div 5$

5. Evaluate $\left(-\frac{2}{3}\right)^3 \cdot 3^2$.

6. Add: $-7 + (-3) + 12$

7. Evaluate $|-29|$.

8. Place the correct symbol, $<$ or $>$, between the two numbers.
$-47 \quad -68$

9. Simplify: $-\frac{4}{9} - \frac{5}{6}$

10. Multiply: $-6(-43)$

11. Simplify: $8 + \frac{12-4}{3^2-1} - 6$

12. Simplify: $-\frac{5}{8} \div \left(-\frac{3}{4}\right)$

13. Subtract: $13 - (-5) - 4$

14. Write $\frac{13}{30}$ as a decimal. Place a bar over the repeating digits of the decimal.

15. Simplify: $(-0.9)(2.7)$

16. Divide: $-180 \div (-12)$

17. Evaluate $2^2 \cdot (-4)^2 \cdot 10$.

18. Add: $15 + (-8) + (-19)$

19. Evaluate $-|-34|$.

20. Place the correct symbol, $<$ or $>$, between the two numbers.
    53   -92

21. Simplify: $-18.354 + 6.97$

22. Multiply: $-4(8)(-5)$

23. Simplify: $9(-4) \div [2(8 - 5)^2]$

24. Find the temperature after a rise of 12°C from -8°C.

25. The daily high temperatures, in degrees Celsius, for a four-day period were recorded as follows: $-8°$, $-6°$, $3°$, $-5°$. Find the average high temperature for the four-day period.

2

Variable Expressions

Objectives

- Evaluate variable expressions
- The Properties of the Real Numbers
- Simplify variable expressions using the Properties of Addition
- Simplify variable expressions using the Properties of Multiplication
- Simplify variable expressions using the Distributive Property
- Simplify general variable expressions
- Translate a verbal expression into a variable expression given the variable
- Translate a verbal expression into a variable expression by assigning the variable
- Translate a verbal expression into a variable expression and then simplify the resulting expression

History of Variables

Prior to the 16th century, unknown quantities were represented by words. In Latin, the language in which most scholarly works were written, the word *res*, meaning "thing," was used. In Germany, the word *zahl*, meaning "number," was used. In Italy, the word *cosa*, also meaning "thing," was used.

Then in 1637, René Descartes, a French mathematician, began using the letters x, y, and z to represent variables. It is interesting to note, upon examining Descartes's work, that toward the end of the book the letters y and z were no longer used and x became the choice for a variable.

One explanation of why the letters y and z appeared less frequently has to do with the nature of printing presses during Descartes's time. A printer had a large tray that contained all the letters of the alphabet. There were many copies of each letter, especially those letters used frequently. For example, there were more e's than q's. Because the letters y and z do not occur frequently in French, a printer would have few of these letters on hand. Consequently, when Descartes started using these letters as variables, it quickly depleted the printer's supply and x's had to be used instead.

Today, x is used by most nations as the standard letter for a single unknown. In fact, x-rays were so named because the scientists who discovered them did not know what they were and thus labeled them the "unknown rays," or x-rays.

Evaluating Variable Expressions

1   Evaluate variable expressions

Often we discuss a quantity without knowing its exact value, for example, the price of gold next month, the cost of a new automobile next year, or the cost of tuition for next semester. In algebra, a letter of the alphabet is used to stand for a quantity that is unknown, or that can change or *vary*. The letter is called a **variable.** An expression that contains one or more variables is a **variable expression.**

A variable expression is shown at the right. The expression can be rewritten by writing subtraction as the addition of the opposite.

$$3x^2 - 5y + 2xy - x - 7$$

$$3x^2 + (-5y) + 2xy + (-x) + (-7)$$

Note that the expression has five addends. The **terms** of a variable expression are the addends of the expression. The expression has five terms.

$$\underbrace{3x^2 \quad - \quad 5y \quad + \quad 2xy \quad - \quad x}_{\text{Variable terms}} \quad \underbrace{- \quad 7}_{\substack{\text{Constant} \\ \text{term}}}$$

5 terms

The terms $3x^2$, $-5y$, $2xy$, and $-x$ are **variable terms.**

The term -7 is a **constant term,** or simply a **constant.**

Each variable term is composed of a **numerical coefficient** and a **variable part** (the variable or variables and their exponents).

When the numerical coefficient is 1 or -1, the 1 is usually not written ($x = 1x$ and $-x = -1x$).

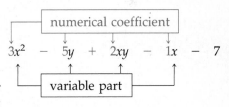

Example 1   Name the variable terms of the expression $2a^2 - 5a + 7$.

Solution $2a^2$, $-5a$

Problem 1   Name the constant term of the expression $6n^2 + 3n - 4$.

Solution   See page A6.

43

Variable expressions occur naturally in science. In a physics lab, a student may discover that a weight of 1 pound will stretch a spring $\frac{1}{2}$ inch. A weight of 2 pounds will stretch the spring 1 inch. By experimenting, the student can discover that the distance the spring will stretch is found by multiplying the weight by $\frac{1}{2}$. By letting W represent the weight attached to the spring, the distance the spring stretches can be represented by the variable expression $\frac{1}{2}W$.

With a weight of W pounds, the spring will stretch $\frac{1}{2} \cdot W = \frac{1}{2}W$ inches.

With a weight of 10 pounds, the spring will stretch $\frac{1}{2} \cdot 10 = 5$ inches.

With a weight of 3 pounds, the spring will stretch $\frac{1}{2} \cdot 3 = 1\frac{1}{2}$ inches.

Replacing the variable or variables in a variable expression and then simplifying the resulting numerical expression is called **evaluating the variable expression.**

Example 2   Evaluate $ab - b^2$ when $a = 2$ and $b = -3$.

Solution $ab - b^2$
$2(-3) - (-3)^2$ ▶ Replace each variable in the expression with the number it represents.
$2(-3) - 9$ ▶ Use the Order of Operations Agreement to simplify
$-6 - 9$ the resulting numerical expression.
-15

Problem 2   Evaluate $2xy + y^2$ when $x = -4$ and $y = 2$.

Solution   See page A6.

Example 3   Evaluate $\frac{a^2 - b^2}{a - b}$ when $a = 3$ and $b = -4$.

Solution $\dfrac{a^2 - b^2}{a - b}$
$\dfrac{3^2 - (-4)^2}{3 - (-4)}$ ▶ Replace each variable in the expression with the number it represents.
$\dfrac{9 - 16}{3 - (-4)}$ ▶ Use the Order of Operations Agreement to simplify the resulting numerical expression.
$\dfrac{-7}{7}$
-1

Problem 3   Evaluate $\frac{a^2 + b^2}{a + b}$ when $a = 5$ and $b = -3$.

Solution   See page A6.

Example 4   Evaluate $x^2 - 3(x - y) - z^2$ when $x = 2$, $y = -1$, and $z = 3$.

Solution $x^2 - 3(x - y) - z^2$
$2^2 - 3[2 - (-1)] - 3^2$ ▶ Replace each variable in the expression with
                                     the number it represents.
$2^2 - 3(3) - 3^2$ ▶ Use the Order of Operations Agreement to
$4 - 3(3) - 9$ simplify the resulting numerical expression.
$4 - 9 - 9$
$-5 - 9$
-14

Problem 4   Evaluate $x^3 - 2(x + y) + z^2$ when $x = 2$, $y = -4$, and $z = -3$.

Solution   See page A6.

2   The Properties of the Real Numbers

The Properties of the Real Numbers describe the way operations on numbers can be performed. Here are some of the properties of real numbers and an example of each.

Properties of the Real Numbers

The Commutative Property of Addition

If a and b are real numbers, then $a + b = b + a$.

$$4 + 3 = 3 + 4$$
$$7 = 7$$

The Commutative Property of Multiplication

If a and b are real numbers, then $a \cdot b = b \cdot a$.

$$(5)(-2) = (-2)(5)$$
$$-10 = -10$$

The Associative Property of Addition

If a, b, and c are real numbers, then $(a + b) + c = a + (b + c)$.

$$2 + (3 + 4) = (2 + 3) + 4$$
$$2 + 7 = 5 + 4$$
$$9 = 9$$

The Associative Property of Multiplication

If a, b, and c are real numbers, then $(a \cdot b) \cdot c = a \cdot (b \cdot c)$.

$$(2 \cdot 3) \cdot 4 = 2 \cdot (3 \cdot 4)$$
$$6 \cdot 4 = 2 \cdot 12$$
$$24 = 24$$

The Addition Property of Zero

If a is a real number, then $a + 0 = 0 + a = a$.

$$4 + 0 = 0 + 4 = 4$$

The Multiplication Property of Zero

If a is a real number, then $a \cdot 0 = 0 \cdot a = 0$.

$$(5)(0) = (0)(5) = 0$$

The Multiplication Property of One

If a is a real number, then $a \cdot 1 = 1 \cdot a = a$.

$$6 \cdot 1 = 1 \cdot 6 = 6$$

The Inverse Property of Addition

If a is a real number, then $a + (-a) = (-a) + a = 0$.

$$8 + (-8) = (-8) + 8 = 0$$

The sum of a number and its opposite is zero.
The opposite of a number is called its **additive inverse.**

The Inverse Property of Multiplication

If a is a real number and $a \neq 0$, then $a \cdot \frac{1}{a} = \frac{1}{a} \cdot a = 1$.

$$7 \cdot \frac{1}{7} = \frac{1}{7} \cdot 7 = 1$$

$\frac{1}{a}$ is the **reciprocal** of a. $\frac{1}{a}$ is also called the **multiplicative inverse** of a.
The product of a number and its reciprocal is 1.

The Distributive Property

If a, b, and c are real numbers, then $a(b + c) = ab + ac$
or $(b + c)a = ba + ca$.

$$2(3 + 4) = 2 \cdot 3 + 2 \cdot 4 \qquad (4 + 5)2 = 4 \cdot 2 + 5 \cdot 2$$
$$2 \cdot 7 = 6 + 8 \qquad\qquad 9 \cdot 2 = 8 + 10$$
$$14 = 14 \qquad\qquad 18 = 18$$

Example 5   Complete the statement by using the Commutative Property of Multiplication.

$(6)(5) = (?)(6)$

Solution $(6)(5) = (5)(6)$ ▶ The Commutative Property of Multiplication
states that $a \cdot b = b \cdot a$.

Problem 5   Complete the statement by using the Inverse Property of Addition.

$7 + ? = 0$

Solution   See page A6.

Example 6   Identify the property that justifies the statement.

$2(8 + 5) = 16 + 10$

Solution   The Distributive Property   ▶ The Distributive Property states that
$a(b + c) = ab + ac$.

Problem 6   Identify the property that justifies the statement.

$5 + (13 + 7) = (5 + 13) + 7$

Solution   See page A6.

EXERCISES 2.1

1   Name the terms of the variable expression. Then underline the constant term.

1. $2x^2 + 5x - 8$

2. $-3n^2 - 4n + 7$

3. $6 - a^4$

Name the variable terms of the expression. Then underline the variable part of each term.

4. $9b^2 - 4ab + a^2$

5. $7x^2y + 6xy^2 + 10$

6. $5 - 8n - 3n^2$

Name the coefficients of the variable terms.

7. $x^2 - 9x + 2$

8. $12a^2 - 8ab - b^2$

9. $n^3 - 4n^2 - n + 9$

Evaluate the variable expression when $a = 2$, $b = 3$, and $c = -4$.

10. $3a + 2b$

11. $a - 2c$

12. $-a^2$

13. $2c^2$

14. $-3a + 4b$

15. $3b - 3c$

16. $b^2 - 3$

17. $-3c + 4$

18. $16 \div (2c)$

19. $6b \div (-a)$

20. $bc \div (2a)$

21. $-2ab \div c$

22. $a^2 - b^2$

23. $b^2 - c^2$

24. $(a + b)^2$

25. $a^2 + b^2$

26. $2a - (c + a)^2$

27. $(b - a)^2 + 4c$

28. $b^2 - \dfrac{ac}{8}$

29. $\dfrac{5ab}{6} - 3cb$

30. $(b - 2a)^2 + bc$

Evaluate the variable expression when $a = -2$, $b = 4$, $c = -1$, and $d = 3$.

31. $\dfrac{b + c}{d}$

32. $\dfrac{d - b}{c}$

33. $\dfrac{2d + b}{-a}$

34. $\dfrac{b + 2d}{b}$

35. $\dfrac{b - d}{c - a}$

36. $\dfrac{2c - d}{-ad}$

37. $(b + d)^2 - 4a$

38. $(d - a)^2 - 3c$

39. $(d - a)^2 \div 5$

40. $(b - c)^2 \div 5$

41. $b^2 - 2b + 4$

42. $a^2 - 5a - 6$

43. $\dfrac{bd}{a} \div c$

44. $\dfrac{2ac}{b} \div (-c)$

45. $2(b + c) - 2a$

46. $3(b - a) - bc$

47. $\dfrac{b - 2a}{bc^2 - d}$

48. $\dfrac{b^2 - a}{ad + 3c}$

49. $\frac{1}{3}d^2 - \frac{3}{8}b^2$

50. $\frac{5}{8}a^4 - c^2$

51. $\frac{-4bc}{2a - b}$

52. $\frac{abc}{b - d}$

53. $a^3 - 3a^2 + a$

54. $d^3 - 3d - 9$

55. $-\frac{3}{4}b + \frac{1}{2}(ac + bd)$

56. $-\frac{2}{3}d - \frac{1}{5}(bd - ac)$

57. $(b - a)^2 - (d - c)^2$

58. $(b + c)^2 + (a + d)^2$

59. $4ac + (2a)^2$

60. $3dc - (4c)^2$

Evaluate the variable expression when $a = 2.7$, $b = -1.6$, and $c = -0.8$.

61. $c^2 - ab$

62. $(a + b)^2 - c$

63. $\frac{b^3}{c} - 4a$

2   Use the given property to complete the statement.

64. The Commutative Property of
Multiplication
$2 \cdot 5 = 5 \cdot ?$

65. The Commutative Property of Addition
$9 + 17 = ? + 9$

66. The Associative Property of
Multiplication
$(4 \cdot 5) \cdot 6 = 4 \cdot (? \cdot 6)$

67. The Associative Property of Addition
$(4 + 5) + 6 = ? + (5 + 6)$

68. The Distributive Property
$2(4 + 3) = 8 + ?$

69. The Addition Property of Zero
$? + 0 = -7$

70. The Inverse Property of Addition
$8 + ? = 0$

71. The Inverse Property of Multiplication
$\frac{1}{-5}(-5) = ?$

72. The Multiplication Property of One
$? \cdot 1 = -4$

73. The Multiplication Property of Zero
$12 \cdot ? = 0$

Identify the property that justifies the statement.

74. $-7 + 7 = 0$

75. $(-8)\left(-\frac{1}{8}\right) = 1$

76. $23 + 19 = 19 + 23$

77. $-21 + 0 = -21$

78. $2 + (6 + 14) = (2 + 6) + 14$

79. $(-3 + 9)8 = -24 + 72$

80. $3 \cdot 5 = 5 \cdot 3$

81. $-32(0) = 0$

82. $(4 \cdot 3) \cdot 5 = 4 \cdot (3 \cdot 5)$

83. $\frac{1}{4}(1) = \frac{1}{4}$

SUPPLEMENTAL EXERCISES 2.1

84. Use the expressions $(6 - 3) - 2$ and $6 - (3 - 2)$ to show that subtraction is not associative.

Evaluate the variable expression when $a = -2$ and $b = -3$.

85. $|2a + 3b|$ 86. $|-4ab|$ 87. $|5a - b|$

Evaluate the variable expression when $a = \frac{2}{3}$ and $b = -\frac{3}{2}$.

88. $\frac{1}{3}a^5b^6$ 89. $\dfrac{(2ab)^3}{2a^3b^3}$ 90. $|5ab - 8a^2b^2|$

S E C T I O N **2.2**

Simplifying Variable Expressions

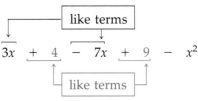

1   Simplify variable expressions using the Properties of Addition

Like terms of a variable expression are the terms with the same variable part. (Because $x^2 = x \cdot x$, x^2 and x are not like terms.)

$$\overset{\text{like terms}}{\overbrace{3x \quad + \quad 4}} \quad - \quad \underset{\text{like terms}}{\underbrace{7x \quad + \quad 9}} \quad - \quad x^2$$

Constant terms are like terms.
4 and 9 are like terms.

To **combine** like terms, use the Distributive Property $ba + ca = (b + c)a$ to add the coefficients.

$2x + 3x$
$(2 + 3)x$
$5x$

Example 1   Simplify.   A. $-2y + 3y$ B. $5x - 11x$

Solution   A. $-2y + 3y$
$(-2 + 3)y$ ▶ Use the Distributive Property $ba + ca = (b + c)a$.
$1y$ ▶ Add the coefficients.
y ▶ Use the Multiplication Property of One.

B. $5x - 11x$
$[5 + (-11)]x$ ▶ Use the Distributive Property $ba + ca = (b + c)a$.
$-6x$ ▶ Add the coefficients.

Problem 1 Simplify.   A. $9x + 6x$ B. $-4y - 7y$

Solution See page A6.

In simplifying more complicated expressions, the Properties of Addition are used.

The Commutative Property of Addition can be used when adding two like terms. The terms can be added in either order. The sum is the same.

$$2x + (-4x) = -4x + 2x$$
$$[2 + (-4)]x = (-4 + 2)x$$
$$-2x = -2x$$

The Associative Property of Addition is used when adding three or more terms. The terms can be grouped in any order. The sum is the same.

$$3x + 5x + 9x = (3x + 5x) + 9x = 3x + (5x + 9x)$$
$$8x + 9x = 3x + 14x$$
$$17x = 17x$$

By the Addition Property of Zero, the sum of a term and zero is the term.

$$5x + 0 = 0 + 5x = 5x$$

By the Inverse Property of Addition, the sum of a term and its additive inverse is zero.

$$7x + (-7x) = -7x + 7x = 0$$

Example 2 Simplify.   A. $8x + 3y - 8x$ B. $4x^2 + 5x - 6x^2 - 2x$

Solution A. $8x + 3y - 8x$
$3y + 8x - 8x$ ▶ Use the Commutative Property of Addition to rearrange the terms.
$3y + (8x - 8x)$ ▶ Use the Associative Property of Addition to group like terms.
$3y + 0$ ▶ Use the Inverse Property of Addition.
$3y$ ▶ Use the Addition Property of Zero.

B. $4x^2 + 5x - 6x^2 - 2x$
$4x^2 - 6x^2 + 5x - 2x$ ▶ Use the Commutative Property of Addition to rearrange the terms.
$(4x^2 - 6x^2) + (5x - 2x)$ ▶ Use the Associative Property of Addition to group like terms.
$-2x^2 + 3x$ ▶ Combine like terms.

Problem 2   Simplify.   A. $3a - 2b + 5a$ B. $x^2 - 7 + 9x^2 - 14$

Solution   See page A6.

2   Simplify variable expressions using the Properties of Multiplication

The Properties of Multiplication are used in simplifying variable expressions.

The Associative Property is used when multiplying three or more factors.

$$2(3x) = (2 \cdot 3)x = 6x$$

The Commutative Property can be used to change the order in which factors are multiplied.

$$(3x) \cdot 2 = 2 \cdot (3x)$$

By the Multiplication Property of One, the product of a term and 1 is the term.

$$(8x)(1) = (1)(8x) = 8x$$

By the Inverse Property of Multiplication, the product of a term and its reciprocal is 1.

$$5x \cdot \frac{1}{5x} = \frac{1}{5x} \cdot 5x = 1, \; x \neq 0$$

Example 3   Simplify.   A. $2(-x)$ B. $\dfrac{3}{2}\left(\dfrac{2x}{3}\right)$ C. $(16x)2$

Solution   A. $2(-x)$
$\quad\;\; 2(-1 \cdot x)$ ▶ $-x = -1x = -1 \cdot x$.
$\quad\;\; [2 \cdot (-1)]x$ ▶ Use the Associative Property of Multiplication to group factors.

$\quad\;\; -2x$ ▶ Multiply.

B. $\dfrac{3}{2}\left(\dfrac{2x}{3}\right)$

$\quad\;\; \dfrac{3}{2}\left(\dfrac{2}{3}x\right)$ ▶ Note that $\dfrac{2x}{3} = \dfrac{2}{3} \cdot \dfrac{x}{1} = \dfrac{2}{3}x$.

$\quad\;\; \left(\dfrac{3}{2} \cdot \dfrac{2}{3}\right)x$ ▶ Use the Associative Property of Multiplication to group factors.

$\quad\;\; 1x$ ▶ Use the Inverse Property of Multiplication.

$\quad\;\; x$ ▶ Use the Multiplication Property of One.

C. $(16x)2$
$\quad\;\; 2(16x)$ ▶ Use the Commutative Property of Multiplication to rearrange factors.
$\quad\;\; (2 \cdot 16)x$ ▶ Use the Associative Property of Multiplication to group factors.

$\quad\;\; 32x$ ▶ Multiply.

Problem 3   Simplify.   A. $-7(-2a)$ B. $-\frac{5}{6}(-30y^2)$ C. $(-5x)(-2)$

   Solution   See page A6.

3   Simplify variable expressions using the Distributive Property

The Distributive Property is used to remove parentheses from a variable expression.

$3(2x - 5)$
$3(2x) - 3(5)$
$6x - 15$

An extension of the Distributive Property is used when an expression contains more than two terms.

$4(x^2 + 6x - 1)$
$4(x^2) + 4(6x) - 4(1)$
$4x^2 + 24x - 4$

Example 4   Simplify.
   A. $-3(5 + x)$ B. $-(2x - 4)$ C. $(2y - 6)2$ D. $5(3x + 7y - z)$

   Solution   A. $-3(5 + x)$
              $-3(5) + (-3)x$ ▶ Use the Distributive Property.
              $-15 - 3x$ ▶ Multiply.

   B. $-(2x - 4)$
      $-1(2x - 4)$ ▶ Just as $-x = -1x$, $-(2x - 4) = -1(2x - 4)$.
      $-1(2x) - (-1)(4)$ ▶ Use the Distributive Property.
      $-2x + 4$ ▶ Note: When a negative sign immediately precedes the parentheses, remove the parentheses and change the sign of *each* term inside the parentheses.

   C. $(2y - 6)2$
      $(2y)(2) - (6)(2)$ ▶ Use the Distributive Property $(b + c)a = ba + ca$.
      $4y - 12$

   D. $5(3x + 7y - z)$
      $5(3x) + 5(7y) - 5(z)$ ▶ Use the Distributive Property.
      $15x + 35y - 5z$

Problem 4   Simplify.
   A. $7(4 + 2y)$ B. $-(5x - 12)$ C. $(3a - 1)5$ D. $-3(6a^2 - 8a + 9)$

   Solution   See page A6.

4 Simplify general variable expressions

When simplifying variable expressions, use the Distributive Property to remove parentheses and brackets used as grouping symbols.

Example 5 Simplify: $4(x - y) - 2(-3x + 6y)$

Solution $4(x - y) - 2(-3x + 6y)$
$4x - 4y + 6x - 12y$ ▶ Use the Distributive Property to remove parentheses.
$10x - 16y$ ▶ Combine like terms.

Problem 5 Simplify: $7(x - 2y) - 3(-x - 2y)$

Solution See page A7.

Example 6 Simplify: $2x - 3[2x - 3(x + 7)]$

Solution $2x - 3[2x - 3(x + 7)]$
$2x - 3[2x - 3x - 21]$ ▶ Use the Distributive Property to remove the inner grouping symbols.

$2x - 3[-x - 21]$ ▶ Combine like terms inside the grouping symbols.
$2x + 3x + 63$ ▶ Use the Distributive Property to remove the brackets.
$5x + 63$ ▶ Combine like terms.

Problem 6 Simplify: $3y - 2[x - 4(2 - 3y)]$

Solution See page A7.

EXERCISES 2.2

1 Simplify.

1. $6x + 8x$

2. $12x + 13x$

3. $9a - 4a$

4. $12a - 3a$

5. $4y + (-10y)$

6. $8y + (-6y)$

7. $-3b - 7$

8. $-12y - 3$

9. $-12a + 17a$

10. $-3a + 12a$

11. $5ab - 7ab$

12. $9ab - 3ab$

13. $-12xy + 17xy$

14. $-15xy + 3xy$

15. $-3ab + 3ab$

16. $-7ab + 7ab$

17. $-\frac{1}{2}x - \frac{1}{3}x$

18. $-\frac{2}{5}y + \frac{3}{10}y$

19. $\frac{3}{8}x^2 - \frac{5}{12}x^2$ **20.** $\frac{2}{3}y^2 - \frac{4}{9}y^2$ **21.** $3x + 5x + 3x$

22. $8x + 5x + 7x$ **23.** $5a - 3a + 5a$ **24.** $10a - 17a + 3a$

25. $-5x^2 - 12x^2 + 3x^2$ **26.** $-y^2 - 8y^2 + 7y^2$

27. $7x + (-8x) + 3y$ **28.** $8y + (-10x) + 8x$

29. $7x - 3y + 10x$ **30.** $8y + 8x - 8y$

31. $3a + (-7b) - 5a + b$ **32.** $-5b + 7a - 7b + 12a$

33. $3x + (-8y) - 10x + 4x$ **34.** $3y + (-12x) - 7y + 2y$

35. $x^2 - 7x + (-5x^2) + 5x$ **36.** $3x^2 + 5x - 10x^2 - 10x$

2   Simplify.

37. $4(3x)$ **38.** $12(5x)$ **39.** $-3(7a)$

40. $-2(5a)$ **41.** $-2(-3y)$ **42.** $-5(-6y)$

43. $(4x)2$ **44.** $(6x)12$ **45.** $(3a)(-2)$

46. $(7a)(-4)$ **47.** $(-3b)(-4)$ **48.** $(-12b)(-9)$

49. $-5(3x^2)$ **50.** $-8(7x^2)$ **51.** $\frac{1}{3}(3x^2)$

52. $\frac{1}{6}(6x^2)$ **53.** $\frac{1}{5}(5a)$ **54.** $\frac{1}{8}(8x)$

55. $-\frac{1}{2}(-2x)$ **56.** $-\frac{1}{4}(-4a)$ **57.** $-\frac{1}{7}(-7n)$

58. $-\frac{1}{9}(-9b)$ **59.** $(3x)\left(\frac{1}{3}\right)$ **60.** $(12x)\left(\frac{1}{12}\right)$

61. $(-6y)\left(-\frac{1}{6}\right)$ **62.** $(-10n)\left(-\frac{1}{10}\right)$ **63.** $\frac{1}{3}(9x)$

64. $\frac{1}{7}(14x)$ **65.** $-\frac{1}{5}(10x)$ **66.** $-\frac{1}{8}(16x)$

67. $-\frac{2}{3}(12a^2)$ **68.** $-\frac{5}{8}(24a^2)$ **69.** $-\frac{1}{2}(-16y)$

70. $-\frac{3}{4}(-8y)$ **71.** $(16y)\left(\frac{1}{4}\right)$ **72.** $(33y)\left(\frac{1}{11}\right)$

73. $(-6x)\left(\frac{1}{3}\right)$ **74.** $(-10x)\left(\frac{1}{5}\right)$ **75.** $(-8a)\left(-\frac{3}{4}\right)$

3   Simplify.

76. $-(x + 2)$ 77. $-(x + 7)$ 78. $2(4x - 3)$

79. $5(2x - 7)$ 80. $-2(a + 7)$ 81. $-5(a + 16)$

82. $-3(2y - 8)$ 83. $-5(3y - 7)$ 84. $(5 - 3b)7$

85. $(10 - 7b)2$ 86. $-3(3 - 5x)$ 87. $-5(7 - 10x)$

88. $3(5x^2 + 2x)$ 89. $6(3x^2 + 2x)$ 90. $-2(-y + 9)$

91. $-5(-2x + 7)$ 92. $(-3x - 6)5$

93. $(-2x + 7)7$ 94. $2(-3x^2 - 14)$

95. $5(-6x^2 - 3)$ 96. $-3(2y^2 - 7)$

97. $-8(3y^2 - 12)$ 98. $3(x^2 - y^2)$

99. $5(x^2 + y^2)$ 100. $-2(x^2 - 3y^2)$

101. $-4(x^2 - 5y^2)$ 102. $-(6a^2 - 7b^2)$

103. $3(x^2 + 2x - 6)$ 104. $4(x^2 - 3x + 5)$

105. $-2(y^2 - 2y + 4)$ 106. $-3(y^2 - 3y - 7)$

107. $2(-a^2 - 2a + 3)$ 108. $4(-3a^2 - 5a + 7)$

109. $-5(-2x^2 - 3x + 7)$ 110. $-3(-4x^2 + 3x - 4)$

111. $3(2x^2 + xy - 3y^2)$ 112. $5(2x^2 - 4xy - y^2)$

113. $-(3a^2 + 5a - 4)$ 114. $-(8b^2 - 6b + 9)$

4   Simplify.

115. $4x - 2(3x + 8)$ 116. $6a - (5a + 7)$

117. $9 - 3(4y + 6)$ 118. $10 - (11x - 3)$

119. $5n - (7 - 2n)$ 120. $8 - (12 + 4y)$

121. $3(x + 2) - 5(x - 7)$ 122. $2(x - 4) - 4(x + 2)$

123. $12(y - 2) + 3(7 - 3y)$ 124. $6(2y - 7) - 3(3 - 2y)$

125. $3(a - b) - 4(a + b)$ 126. $2(a + 2b) - (a - 3b)$

127. $4[x - 2(x - 3)]$ 128. $2[x + 2(x + 7)]$

129. $-2[3x + 2(4 - x)]$ 130. $-5[2x + 3(5 - x)]$

131. $-3[2x - (x + 7)]$ 132. $-2[3x - (5x - 2)]$

133. $2x - 3[x - 2(4 - x)]$ 134. $-7x + 3[x - 7(3 - 2x)]$

135. $-5x - 2[2x - 4(x + 7)] - 6$ 136. $4a - 2[2b - (b - 2a)] + 3b$

137. $2x + 3(x - 2y) + 5(3x - 7y)$ **138.** $5y - 2(y - 3x) + 2(7x - y)$

SUPPLEMENTAL EXERCISES 2.2

Simplify.

139. $C - 0.7C$ **140.** $\frac{1}{3}(3x + y) - \frac{2}{3}(6x - y)$ **141.** $-\frac{1}{4}[2x + 2(y - 6y)]$

Complete.

142. A number that has no reciprocal is _____.

143. The additive inverse of $a - b$ is _____.

S E C T I O N **2.3**

Translating Verbal Expressions Into Variable Expressions

1 Translate a verbal expression into a variable expression given the variable

One of the major skills required in applied mathematics is the ability to translate a verbal expression into a variable expression. This requires recognizing the verbal phrases that translate into mathematical operations. Here is a partial list of the phrases used to indicate the different mathematical operations.

Addition	added to	6 added to y	$y + 6$
	more than	8 more than x	$x + 8$
	the sum of	the sum of x and z	$x + z$
	increased by	t increased by 9	$t + 9$
	the total of	the total of 5 and y	$5 + y$
Subtraction	minus	x minus 2	$x - 2$
	less than	7 less than t	$t - 7$
	decreased by	m decreased by 3	$m - 3$
	the difference between	the difference between y and 4	$y - 4$
Multiplication	times	10 times t	$10t$
	of	one-half of x	$\frac{1}{2}x$
	the product of	the product of y and z	yz
	multiplied by	y multiplied by 11	$11y$
	twice	twice n	$2n$

Division	divided by	x divided by 12	$\dfrac{x}{12}$
	the quotient of	the quotient of y and z	$\dfrac{y}{z}$
	the ratio of	the ratio of t to 9	$\dfrac{t}{9}$
Power	the square of	the square of x	x^2
	the cube of	the cube of a	a^3

Example 1   Translate into a variable expression.

A. the total of 5 times b and c
B. the quotient of 8 less than n and 14
C. 13 more than the sum of 7 and the square of x

Solution   A. the <u>total</u> of 5 <u>times</u> b and c ▶ Identify the words that indicate
the mathematical operations.
$5b + c$ ▶ Use the identified operations to
write the variable expression.

B. the <u>quotient</u> of 8 <u>less than</u> n and 14   ▶ Identify the words that indicate
the mathematical operations.
$\dfrac{n - 8}{14}$ ▶ Use the identified operations to
write the variable expression.

C. 13 <u>more than</u> the <u>sum</u> of 7
and the <u>square</u> of x

$(7 + x^2) + 13$

Problem 1   Translate into a variable expression.

A. 18 less than the cube of x
B. y decreased by the sum of z and 9
C. the difference between the square of q and the sum of r and t

Solution   See page A7.

2   ## Translate a verbal expression into a variable expression by assigning the variable

In most applications that involve translating phrases into variable expressions, the variable to be used is not given. To translate these phrases, a variable must be assigned to an unknown quantity before the variable expression can be written.

Example 2   Translate "a number multiplied by the total of six and the cube of the number" into a variable expression.

Solution   the unknown number: n ▶ Assign a variable to one of the unknown quantities.

the cube of the number: n^3
the total of six and the
   cube of the number: $6 + n^3$

$n(6 + n^3)$

▶ Use the assigned variable to write an expression for any other unknown quantity.

▶ Use the assigned variable to write the variable expression.

Problem 2   Translate "a number added to the product of five and the square of the number" into a variable expression.

Solution   See page A7.

Example 3   Translate "the quotient of twice a number and the difference between the number and twenty" into a variable expression.

Solution   the unknown number: n
twice the number: $2n$
the difference between the number and twenty: $n - 20$

$$\frac{2n}{n - 20}$$

Problem 3   Translate "the product of three and the sum of seven and twice a number" into a variable expression.

Solution   See page A7.

3   Translate a verbal expression into a variable expression and then simplify the resulting expression

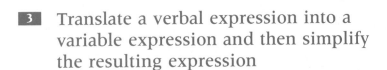

After translating a verbal expression into a variable expression, simplify the variable expression by using the Properties of the Real Numbers.

Example 4   Translate and simplify "the total of four times an unknown number and twice the difference between the number and eight."

Solution   the unknown number: n ▶ Assign a variable to one of the unknown quantities.

four times the unknown number: $4n$ ▶ Use the assigned variable to write an expression for any other unknown quantity.

twice the difference between the number and eight: $2(n - 8)$

$4n + 2(n - 8)$ ▶ Use the assigned variable to write the variable expression.

$4n + 2n - 16$ ▶ Simplify the variable expression.
$6n - 16$

Problem 4   Translate and simplify "a number minus the difference between twice the number and seventeen."

Solution   See page A7.

Example 5   Translate and simplify "the difference between five-eighths of a number and two-thirds of the same number."

Solution   the unknown number: n ▶ Assign a variable to one of the unknown quantities.

five-eighths of the number: $\frac{5}{8}n$ ▶ Use the assigned variable to write an expression for any other unknown quantity.

two-thirds of the number: $\frac{2}{3}n$

$\frac{5}{8}n - \frac{2}{3}n$ ▶ Use the assigned variable to write the variable expression.

$\frac{15}{24}n - \frac{16}{24}n$ ▶ Simplify the variable expression.

$-\frac{1}{24}n$

Problem 5   Translate and simplify "the sum of three-fourths of a number and one-fifth of the same number."

Solution   See page A7.

EXERCISES 2.3

1   Translate into a variable expression.

1. d less than 19

2. the sum of 6 and c

3. r decreased by 12

4. w increased by 55

5. a multiplied by 28

6. y added to 16

7. 5 times the difference between n and 7

8. 30 less than the square of b

9. y decreased by the product of 3 and y

10. the sum of four-fifths of m and 18

11. the product of -6 and b

12. 9 increased by the quotient of t and 5

13. 4 divided by the difference between p and 6

14. the product of 7 and the total of r and 8

15. the quotient of 9 less than x and twice x

16. the product of a and the sum of a and 13

17. 21 less than the product of s and

18. 14 more than one-half of the square of z

19. the ratio of 8 more than d to d

20. the total of 9 times the cube of m and the square of m

21. three-eighths of the sum of t and 15

22. s decreased by the quotient of s and 2

23. w increased by the quotient of 7 and w

24. the difference between the square of c and the total of c and 14

25. d increased by the difference between 16 times d and 3

26. the product of 8 and the total of b and 5

2   Translate into a variable expression.

27. a number divided by nineteen

28. thirteen minus a number

29. forty more than a number

30. three-sevenths of a number

31. the square of the difference between a number and ninety

32. the quotient of twice a number and five

33. the sum of four-ninths of a number and twenty

34. eight less than the product of fifteen and a number

35. the product of a number and ten more than the number

36. six less than the total of a number and the cube of the number

37. fourteen added to the product of seven and a number

38. the quotient of three and the total of four and a number

39. the quotient of twelve and the sum of a number and two

40. eleven increased by one-half of a number

41. the ratio of two and the sum of a number and one

42. a number multiplied by the difference between twice the number and nine

43. the difference between sixty and the quotient of a number and fifty

44. the product of nine less than a number and the number

45. the sum of the square of a number and three times the number

46. the quotient of seven more than twice a number and the number

47. the sum of three more than a number and the cube of the number

48. a number decreased by the difference between the cube of the number and ten

49. the square of a number decreased by one-fourth of the number

50. four less than seven times the square of a number

51. twice a number decreased by the quotient of seven and the number

52. eighty decreased by the product of thirteen and a number

53. the cube of a number decreased by the product of twelve and the number

54. the quotient of five and the sum of a number and nineteen

3   Translate into a variable expression. Then simplify.

55. a number increased by the total of the number and ten

56. a number added to the product of five and the number

57. a number decreased by the difference between nine and the number

58. eight more than the sum of a number and eleven

59. the difference between one-fifth of a number and three-eighths of the number

60. a number minus the sum of the number and fourteen

61. four more than the total of a number and nine

62. the sum of one-eighth of a number and one-twelfth of the number

63. twice the sum of three times a number and forty

64. the sum of a number divided by two and the number

65. seven times the product of five and a number

66. sixteen multiplied by one-fourth of a number

67. the total of seventeen times a number and twice the number

68. the difference between nine times a number and twice the number

69. a number plus the product of the number and twelve

70. nineteen more than the difference between a number and five

71. three times the sum of the square of a number and four

72. a number subtracted from the product of the number and seven

73. three-fourths of the sum of sixteen times a number and four

74. the difference between fourteen times a number and the product of the number and seven

75. sixteen decreased by the sum of a number and nine

76. eleven less than the difference between eight and a number

77. five more than the quotient of four times a number and two

78. twenty minus the sum of four-ninths of a number and three

79. six times the total of a number and eight

80. four times the sum of a number and twenty

81. seven minus the sum of a number and two

82. three less than the sum of a number and ten

83. one-third of the sum of a
number and six times the number

84. twice the quotient of four times a
number and eight

85. the total of eight increased
by the cube of a number and twice
the cube of the number

86. the sum of five more than the
square of a number and twice
the square of the number

87. twelve more than a number added
to the difference between the
number and six

88. a number plus four added to the
difference between three and
twice the number

89. the sum of a number and nine
added to the difference between
the number and twenty

90. seven increased by a number
added to twice the difference
between the number and two

91. fourteen minus the product of three
less than a number and ten

92. a number plus the product of
the number minus five and seven

SUPPLEMENTAL EXERCISES 2.3

Complete each statement with the word *even* or *odd*.

93. If k is an odd integer, then $k + 1$ is an
_____ integer.

94. If k is an odd integer, then $k - 2$ is an
_____ integer.

95. If n is an integer, then $2n$ is an _____
integer.

96. If m and n are even integers, then $m - n$ is
an _____ integer.

97. If m and n are even integers, then mn is an
_____ integer.

98. If m and n are odd integers, then $m + n$ is
an _____ integer.

99. If m and n are odd integers, then $m - n$ is
an _____ integer.

100. If m and n are odd integers, then mn is an
_____ integer.

101. If m is an even integer and n is an odd
integer, then $m - n$ is an _____ integer.

102. If m is an even integer and n is an odd
integer, then $m + n$ is an _____ integer.

Write a variable expression.

103. An employee is paid $640 per week plus $24 for each hour of overtime
worked. Write a variable expression for the employee's weekly pay.

104. An auto repair bill is $92 for parts and $25 for each hour of labor. Write a variable expression for the amount of the repair bill.

105. In a coin bank, there are 10 more dimes than quarters. Write a variable expression for the number of dimes in the bank in terms of the number of quarters.

106. A 5¢ stamp in a stamp collection is 25 years older than an 8¢ stamp in the collection. Write a variable expression for the age of the 5¢ stamp in terms of the age of the 8¢ stamp.

Calculators and Computers

 Evaluating Variable Expressions

Evaluating variable expressions with your calculator will at times require the use of the $\boxed{+/-}$ key, the $\boxed{x^2}$ key, and the parentheses keys. The $\boxed{+/-}$ key changes the sign of the number currently in the display. The parentheses keys are used as grouping symbols, and the $\boxed{x^2}$ key is used to square the number in the display.

For the examples below, $a = 2$, $b = -3$, and $c = 4$.

Evaluate $a + bc$.

Replace a, b, and c by their values. $2 + (-3)(4)$

Use the Order of Operations Agreement (multiply before adding), and enter the following on your calculator:

$$3\ \boxed{+/-}\ \boxed{\times}\ 4\ \boxed{+}\ 2\ \boxed{=}$$

The answer in the display should be -10.

Evaluate $ac^2 - b$.

Replace a, b, and c by their values. $(2)(4)^2 - (-3)$

Use the Order of Operations Agreement (exponents, multiply, subtract), and enter the following on your calculator:

$$4\ \boxed{x^2}\ \boxed{\times}\ 2\ \boxed{-}\ 3\ \boxed{+/-}\ \boxed{=}$$

The answer in the display should be 35.

Evaluate $5a^2 - 6bc$.

This example is a little more complicated and requires using the parentheses keys. Because of the Order of Operations Agreement, the expression $5a^2 - 6bc$ is evaluated as if there were parentheses around each of the terms: $(5a^2) - (6bc)$. (Do all multiplications before subtraction.)

Replace a, b, and c by their values. $5(2)^2 - 6(-3)(4)$

Use the Order of Operations Agreement, and enter the following on your calculator:

$$(\;\; 2 \;\; x^2 \;\; \times \;\; 5 \;\;) \;\; - \;\; (\;\; 6 \;\; \times \;\; 3 \;\; +/- \;\; \times \;\; 4 \;\;) \;\; =$$

The answer in the display should be 92.

Something Extra

Dimensional Analysis

Recall that $2 \cdot 2 = 2^2$ and $2 \cdot 2 \cdot 2 = 2^3$. These properties hold true for variables.

$$x \cdot x = x^2 \qquad\qquad x \cdot x \cdot x = x^3$$
$$m \cdot m = m^2 \qquad\qquad m \cdot m \cdot m = m^3$$

Quantities such as 4 m, 12¢, and 3 h are number quantities written with units. In these examples, the units are meters, cents, and hours, respectively. In calculations that involve quantities, the units are operated on algebraically.

The area of a rectangle is found by multiplying the length times the width. If a rectangle measures 3 m by 5 m, then the area is given by (3 m)(5 m).

Simplify this expression algebraically.   (3 m)(5 m)
$(3 \cdot 5)(m \cdot m)$
15 m^2

The area of the rectangle is 15 m^2 (square meters).

The volume of a box is found by multiplying the length times the width times the height. If a box measures 10 cm by 5 cm by 3 cm, then the volume is given by (10 cm)(5 cm)(3 cm).

Simplify this expression algebraically.   (10 cm)(5 cm)(3 cm)
$(10 \cdot 5 \cdot 3)(cm \cdot cm \cdot cm)$
150 cm^3

The volume of the box is 150 cm^3 (cubic centimeters).

Solve.

1. Find the area of a triangle with a base of 3 cm and a height of 7 cm. (The area of a triangle is found by multiplying one-half times the base times the height.)

2. Find the area of a rectangle with a length of 9 cm and a width of 12 cm.

3. Find the perimeter of a square that has sides measuring 8 m.

4. A rectangle measures 5 m by 10 m. Find the perimeter.

5. Find the volume of a rectangular solid with a length of 4 m, a width of 3 m, and a height of 2 m.

6. Each side of a cube measures 25 mm. Find the volume.

7. Three packages of fertilizer are purchased. Each package weighs 30 kg. Find the total weight of the fertilizer purchased.

8. Each of 18 students in a chemistry lab uses 12 ml of acid for an experiment. Find the total amount of acid used by the students.

Chapter Summary

Key Words

A *variable* is a letter that is used to stand for a quantity that is unknown.

A *variable expression* is an expression that contains one or more variables.

The *terms* of a variable expression are the addends of the expression.

A *variable term* is composed of a numerical coefficient and a variable part.

Like terms of a variable expression are the terms with the same variable part.

The *additive inverse* of a number is the opposite of the number.

The *multiplicative inverse* of a number is the reciprocal of the number.

Essential Rules

The Commutative Property of Addition If a and b are real numbers, then $a + b = b + a$.

The Commutative Property of Multiplication If a and b are real numbers, then $ab = ba$.

The Associative Property of Addition	If a, b, and c are real numbers, then $(a + b) + c = a + (b + c)$.
The Associative Property of Multiplication	If a, b, and c are real numbers, then $(ab)c = a(bc)$.
The Addition Property of Zero	If a is a real number, then $a + 0 = 0 + a = a$.
The Multiplication Property of Zero	If a is a real number, then $a \cdot 0 = 0 \cdot a = 0$.
The Multiplication Property of One	If a is a real number, then $1 \cdot a = a \cdot 1 = a$.
The Inverse Property of Addition	If a is a real number, then $a + (-a) = (-a) + a = 0$.
The Inverse Property of Multiplication	If a is a real number and $a \neq 0$, then $a \cdot \dfrac{1}{a} = \dfrac{1}{a} \cdot a = 1$.
The Distributive Property	If a, b, and c are real numbers, then $a(b + c) = ab + ac$.

Chapter Review

1. Simplify: $-7y^2 + 6y^2 - (-2y^2)$

2. Simplify: $(12x)\left(\frac{1}{4}\right)$

3. Simplify: $\frac{2}{3}(-15a)$

4. Simplify: $-2(2x - 4)$

5. Simplify: $5(2x + 4) - 3(x - 6)$

6. Evaluate $a^2 - 3b$ when $a = 2$ and $b = -4$.

7. Complete the statement by using the Inverse Property of Addition.
$-9 + ? = 0$

8. Simplify: $-4(-9y)$

9. Simplify: $-2(-3y + 9)$

10. Simplify: $3[2x - 3(x - 2y)] + 3y$

11. Simplify: $-4(2x^2 - 3y^2)$

12. Simplify: $3x - 5x + 7x$

13. Evaluate $b^2 - 3ab$ when $a = 3$ and $b = -2$.

14. Simplify: $\frac{1}{5}(10x)$

15. Simplify: $5(3 - 7b)$

16. Simplify: $2x + 3[4 - (3x - 7)]$

17. Identify the property that justifies the statement.
$-4(3) = 3(-4)$

18. Simplify: $3(8 - 2x)$

19. Simplify: $-2x^2 - (-3x^2) + 4x^2$

20. Simplify: $-3x - 2(2x - 7)$

21. Simplify: $-3(3y^2 - 3y - 7)$

22. Simplify: $-2[x - 2(x - y)] + 5y$

23. Evaluate $\frac{-2ab}{2b - a}$ when $a = -4$ and $b = 6$.

24. Simplify: $(-3)(-12y)$

25. Simplify: $4(3x - 2) - 7(x + 5)$

26. Simplify: $(16x)\left(\frac{1}{8}\right)$

27. Simplify: $-3(2x^2 - 7y^2)$

28. Evaluate $3(a - c) - 2ab$ when $a = 2$, $b = 3$, and $c = -4$.

29. Simplify: $2x - 3(x - 2)$

30. Simplify: $2a - (-3b) - 7a - 5b$

31. Simplify: $-5(2x^2 - 3x + 6)$

32. Simplify: $3x - 7y - 12x$

33. Simplify: $\frac{1}{2}(12a)$

34. Simplify: $2x + 3[x - 2(4 - 2x)]$

35. Simplify: $3x + (-12y) - 5x - (-7y)$

36. Simplify: $\left(-\frac{5}{6}\right)(-36b)$

37. Complete the statement by using the Distributive Property.
$(6 + 3)7 = 42 + ?$ 84

38. Simplify: $4x^2 + 9x - 6x^2 - 5x$

39. Simplify: $-\frac{3}{8}(16x^2)$

40. Simplify: $-3[2x - (7x - 9)]$

41. Simplify: $-(8a^2 - 3b^2)$

42. Identify the property that justifies the statement.
$-32(0) = 0$

43. Translate "b decreased by the product of 7 and b" into a variable expression.

44. Translate "the sum of a number and twice the square of the number" into a variable expression.

45. Translate "three less than the quotient of six and a number" into a variable expression.

46. Translate "the difference between eight and the quotient of a number and twelve" into a variable expression.

47. Translate "10 divided by the difference between y and 2" into a variable expression.

48. Translate and simplify "twelve more than the product of three plus a number and five."

49. Translate and simplify "eight times the quotient of twice a number and sixteen."

50. Translate and simplify "the product of four and the sum of two and five times a number."

Chapter Test

1. Simplify: $(9y)4$

2. Simplify: $7x + 5y - 3x - 8y$

3. Simplify: $8n - (6 - 2n)$

4. Evaluate $3ab - (2a)^2$ when $a = -2$ and $b = -3$.

5. Identify the property that justifies the statement.
$\frac{1}{4}(1) = \frac{1}{4}$

6. Simplify: $-4(-x + 10)$

7. Simplify: $\frac{2}{3}x^2 - \frac{7}{12}x^2$

8. Simplify: $(-10x)\left(-\frac{2}{5}\right)$

9. Simplify: $(-4y^2 + 8)6$

10. Complete the statement by using the Inverse Property of Addition.
$-19 + ? = 0$

11. Evaluate $\frac{-3ab}{2a + b}$ when $a = -1$ and $b = 4$.

12. Simplify: $5(x + y) - 8(x - y)$

13. Simplify: $6b - 9b + 4b$

14. Simplify: $13(6a)$

15. Simplify: $3(x^2 - 5x + 4)$

16. Evaluate $4(b - a) + bc$ when $a = 2$, $b = -3$, and $c = 4$.

17. Simplify: $6x - 3(y - 7x) + 2(5x - y)$

18. Translate "the quotient of 8 more than n and 17" into a variable expression.

19. Translate "the difference between the sum of a and b and the square of b" into a variable expression.

20. Translate "a number increased by the product of five and the cube of the number" into a variable expression.

21. Translate "the sum of the square of a number and the product of the number and eleven" into a variable expression.

22. Translate "six less than twice a number" into a variable expression.

23. Translate and simplify "twenty times the sum of a number and nine."

24. Translate and simplify "two more than a number added to the difference between the number and three."

25. Translate and simplify "a number minus the product of one-fourth and twice the number."

Cumulative Review

1. Add: $-4 + 7 + (-10)$

2. Subtract: $-16 - (-25) - 4$

3. Multiply: $(-2)(3)(-4)$

4. Divide: $(-60) \div 12$

5. Write $1\frac{1}{4}$ as a decimal.

6. Simplify: $\frac{7}{12} - \frac{11}{16} - \left(-\frac{1}{3}\right)$

7. Simplify: $\frac{5}{12} \div \left(2\frac{1}{2}\right)$

8. Simplify: $\left(-\frac{9}{16}\right)\left(\frac{8}{27}\right)\left(-\frac{3}{2}\right)$

9. Simplify: $-3^2 \cdot \left(-\frac{2}{3}\right)^3$

10. Simplify: $-2^5 \div (3 - 5)^2 - (-3)$

11. Simplify: $\left(-\frac{3}{4}\right)^2 - \left(\frac{3}{8} - \frac{11}{12}\right)$

12. Evaluate $a - 3b^2$ when $a = 4$ and $b = -2$.

13. Simplify: $-2x^2 - (-3x^2) + 4x^2$

14. Simplify: $8a - 12b - 9a$

15. Simplify: $\frac{1}{3}(9a)$

16. Simplify: $\left(-\frac{5}{8}\right)(-32b)$

17. Simplify: $5(4 - 2x)$

18. Simplify: $-3(-2y + 7)$

19. Simplify: $-2(3x^2 - 4y^2)$

20. Simplify: $-4(2y^2 - 5y - 8)$

21. Simplify: $-4x - 3(2x - 5)$

22. Simplify: $3(4x - 1) - 7(x + 2)$

23. Simplify: $3x + 2[x - 4(2 - x)]$

24. Simplify: $3[4x - 2(x - 4y)] + 5y$

25. Translate "the sum of one-half of *b* and *b*" into a variable expression.

26. Translate "the quotient of eight and the difference between *n* and 3" into a variable expression.

27. Translate "the difference between six and the product of a number and twelve" into a variable expression.

28. Translate and simplify "the sum of a number and two more than the number."

29. Translate and simplify "the total of five and the difference between a number and seven."

30. Translate and simplify "nine more than the product of five plus a number and ten."

3

Solving Equations

Mersenne Primes

A prime number that can be written in the form $2^n - 1$, where n is also prime, is called a Mersenne prime. The table below shows some Mersenne primes.

$$3 = 2^2 - 1$$
$$7 = 2^3 - 1$$
$$31 = 2^5 - 1$$
$$127 = 2^7 - 1$$

You might notice that not every prime number is a Mersenne prime. For example, 5 is a prime number but not a Mersenne prime. Also, not all numbers in the form $2^n - 1$, where n is prime, yield a prime number. For example, $2^{11} - 1 = 2047$ is not a prime number.

The search for Mersenne primes has been quite extensive, especially since the advent of the computer. One reason for the extensive research into large prime numbers (not only Mersenne primes) involves cryptology.

Cryptology is the study of making or breaking secret codes. One method of making a code that is difficult to break is called public key cryptology. For this method to work, it is necessary to use very large prime numbers. To keep anyone from breaking the code, each prime should have at least 200 digits.

Today, the largest known Mersenne prime is $2^{216091} - 1$. This number has 65,050 digits in its representation.

Another Mersenne prime got special recognition in a postage-meter stamp. It is the number $2^{11213} - 1$. This number has 3276 digits in its representation.

Introduction to Equations

1 Determine if a given number is a solution of an equation

An **equation** expresses the equality of two mathematical expressions. The expressions can be either numerical or variable expressions.

$$\left.\begin{array}{l} 9 + 3 = 12 \\ 3x - 2 = 10 \\ y^2 + 4 = 2y - 1 \\ z = 2 \end{array}\right\} \text{Equations}$$

The equation at the right is true if the variable is replaced by 5.

$x + 8 = 13$
$5 + 8 = 13$ A true equation

The equation $x + 8 = 13$ is false if the variable is replaced by 7.

$7 + 8 = 13$ A false equation

A **solution** of an equation is a number that, when substituted for the variable, results in a true equation. 5 is a solution of the equation $x + 8 = 13$. 7 is not a solution of the equation $x + 8 = 13$.

Example 1   Is -3 a solution of the equation $4x + 16 = x^2 - 5$?

Solution $4x + 16 = x^2 - 5$

$4(-3) + 16$	$(-3)^2 - 5$
$-12 + 16$	$9 - 5$

       $4 = 4$

▶ Replace the variable by the given number, -3.
▶ Evaluate the numerical expressions using the Order of Operations Agreement.
▶ Compare the results. If the results are equal, the given number is a solution. If the results are not equal, the given number is not a solution.

Yes, -3 is a solution of the equation $4x + 16 = x^2 - 5$.

Problem 1   Is $\frac{1}{4}$ a solution of $5 - 4x = 8x + 2$?

Solution   See page A8.

3.1 - 3.4

4.1 - 4.5

Example 2   Is -4 a solution of $4 + 5x = x^2 - 2x$?

Solution $4 + 5x = x^2 - 2x$

$4 + 5(-4)$	$(-4)^2 - 2(-4)$
$4 + (-20)$	$16 - (-8)$

▶ Replace the variable by the given number, -4.

$$-16 \neq 24$$

▶ Evaluate the numerical expressions using the Order of Operations Agreement.

No, -4 is not a solution of the equation $4 + 5x = x^2 - 2x$.

▶ Compare the results. If the results are equal, the given number is a solution. If the results are not equal, the given number is not a solution.

Problem 2   Is 5 a solution of $10x - x^2 = 3x - 10$?

Solution   See page A8.

2   Solve equations of the form $x + a = b$

To **solve** an equation means to find a solution of the equation. The simplest equation to solve is an equation of the form **variable = constant** because the constant is the solution.

If $x = 5$, then 5 is the solution of the equation because $5 = 5$ is a true equation.

The solution of the equation shown at the right is 7.

$$x + 2 = 9 \qquad 7 + 2 = 9$$

Note that if 4 is added to each side of the equation, the solution is still 7.

$$x + 2 + 4 = 9 + 4$$
$$x + 6 = 13 \qquad 7 + 6 = 13$$

If -5 is added to each side of the equation, the solution is still 7.

$$x + 2 + (-5) = 9 + (-5)$$
$$x - 3 = 4 \qquad 7 - 3 = 4$$

This illustrates the Addition Property of Equations.

Addition Property of Equations

> The same number or variable term can be added to each side of an equation without changing the solution of the equation.

This property is used in solving equations. Note the effect of adding, to each side of the equation $x + 2 = 9$, the opposite of the constant term 2. After simplifying each side of the equation, the equation is in the form variable = constant. The solution is the constant.

$$x + 2 = 9$$
$$x + 2 + (-2) = 9 + (-2)$$
$$x + 0 = 7$$
$$x = 7$$

variable	=	constant

The solution is 7.

In solving an equation, the goal is to rewrite the given equation in the form *variable = constant*   The Addition Property of Equations can be used to rewrite an equation in this form. The Addition Property of Equations is used to **remove a term** from one side of an equation **by adding the opposite of that term** to each side of the equation.

Example 3   Solve and check: $y - 6 = 9$

Solution   The goal is to rewrite the equation in the form *variable = constant*.

$$y - 6 = 9$$
$$y - 6 + 6 = 9 + 6$$ ▶ Add the opposite of the constant term -6 to each side of the equation (the Addition Property of Equations).
$$y + 0 = 15$$ ▶ Simplify using the Addition Property of Inverses.
$$y = 15$$ ▶ Simplify using the Addition Property of Zero. Now the equation is in the form *variable = constant*.

Check $y - 6 = 9$
$$\overline{15 - 6 \mid 9}$$
$$9 = 9$$ ▶ This is a true equation. The solution checks.

The solution is 15.   ▶ Write the solution.

Problem 3   Solve and check: $x - \dfrac{1}{3} = -\dfrac{3}{4}$

Solution   See page A8.

Because subtraction is defined in terms of addition, the Addition Property of Equations allows the same number to be subtracted from each side of an equation without changing the solution of the equation.

Solve: $y + \dfrac{1}{2} = \dfrac{5}{4}$

The goal is to rewrite the equation in the form *variable = constant*.

$$y + \dfrac{1}{2} = \dfrac{5}{4}$$

Add the opposite of the constant term $\dfrac{1}{2}$ to each side of the equation. This is equivalent to subtracting $\dfrac{1}{2}$ from each side of the equation.

$$y + \dfrac{1}{2} - \dfrac{1}{2} = \dfrac{5}{4} - \dfrac{1}{2}$$
$$y + 0 = \dfrac{5}{4} - \dfrac{2}{4}$$

$$y = \dfrac{3}{4}$$

The solution $\dfrac{3}{4}$ checks.

The solution is $\dfrac{3}{4}$.

Example 4   Solve: $\frac{1}{2} = x + \frac{2}{3}$

Solution $\frac{1}{2} = x + \frac{2}{3}$

$\frac{1}{2} - \frac{2}{3} = x + \frac{2}{3} - \frac{2}{3}$ ▶ Subtract $\frac{2}{3}$ from each side of the equation.

$-\frac{1}{6} = x$ ▶ Simplify each side of the equation.

The solution is $-\frac{1}{6}$.

Problem 4   Solve: $-8 = 5 + x$

Solution   See page A8.

Note from the solution to Example 4 above that an equation can be rewritten in the form *constant = variable*. Whether the equation is written in the form *variable = constant* or in the form *constant = variable*, the solution is the constant.

3   Solve equations of the form $ax = b$

The solution of the equation shown at the right is 3.	$2x = 6$	$2 \cdot 3 = 6$
Note that if each side of the equation is multiplied by 5, the solution is still 3.	$5 \cdot 2x = 5 \cdot 6$ $10x = 30$	$10 \cdot 3 = 30$
If each side is multiplied by -4, the solution is still 3.	$(-4) \cdot 2x = (-4) \cdot 6$ $-8x = -24$	$-8 \cdot 3 = -24$

This illustrates the Multiplication Property of Equations.

Multiplication Property of Equations

> Each side of an equation can be multiplied by the same nonzero number without changing the solution of the equation.

This property is used in solving equations. Note the effect of multiplying each side of the equation $2x = 6$ by the reciprocal of the coefficient 2. After simplifying each side of the equation, the equation is in the form *variable = constant*. The solution is the constant.

$2x = 6$

$\frac{1}{2} \cdot 2x = \frac{1}{2} \cdot 6$

$1x = 3$

$x = 3$

variable	=	constant

The solution is 3.

In solving an equation, the goal is to rewrite the given equation in the form variable = constant. The Multiplication Property of Equations can be used to rewrite an equation in this form. The Multiplication Property of Equations is used to **remove a coefficient** from a variable term in an equation **by multiplying each side of the equation by the reciprocal of the coefficient.**

Example 5   Solve: $\frac{3x}{4} = -9$

Solution $\dfrac{3x}{4} = -9$ ▶ $\dfrac{3x}{4} = \dfrac{3}{4}x$

$\dfrac{4}{3} \cdot \dfrac{3}{4}x = \dfrac{4}{3}(-9)$ ▶ Multiply each side of the equation by the reciprocal of the coefficient $\frac{3}{4}$ (the Multiplication Property of Equations).

$1x = -12$ ▶ Simplify using the Inverse Property of Multiplication.
$x = -12$ ▶ Simplify using the Multiplication Property of One. Now the equation is in the form *variable = constant*.

The solution is -12.   ▶ Write the solution.

Problem 5   Solve: $-\dfrac{2x}{5} = 6$

Solution   See page A8.

Because division is defined in terms of multiplication, the Multiplication Property of Equations allows each side of an equation to be divided by the same number without changing the solution of the equation.

Solve: $8x = 16$

The goal is to rewrite the equation in the form *variable = constant*. $8x = 16$

Multiply each side of the equation by the reciprocal of 8. This is equivalent to dividing each side by 8. $\dfrac{8x}{8} = \dfrac{16}{8}$

$x = 2$

The solution 2 checks.   The solution is 2.

When using the Multiplication Property of Equations to solve an equation, multiply each side of the equation by the reciprocal of the coefficient when the coefficient is a fraction. Divide each side of the equation by the coefficient when the coefficient is an integer or a decimal.

Example 6   Solve and check: $4x = 6$

Solution $4x = 6$

$\dfrac{4x}{4} = \dfrac{6}{4}$ ▶ Divide each side of the equation by 4, the coefficient of x.

$x = \dfrac{3}{2}$ ▶ Simplify each side of the equation.

Check $\dfrac{4x = 6}{\quad}$

$4\left(\dfrac{3}{2}\right) \,\Big|\, 6$

$6 = 6$ ▶ This is a true equation. The solution checks.

The solution is $\dfrac{3}{2}$.

Problem 6   Solve and check: $6x = 10$

Solution   See page A8.

Example 7   Solve: $5x - 9x = 12$

Solution $5x - 9x = 12$

$-4x = 12$ ▶ Combine like terms.

$\dfrac{-4x}{-4} = \dfrac{12}{-4}$ ▶ Divide each side of the equation by -4.

$x = -3$

The solution is -3.

Problem 7   Solve: $4x - 8x = 16$

Solution   See page A8.

EXERCISES 3.1

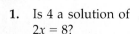

1. Is 4 a solution of
$2x = 8$?

2. Is 3 a solution of
$y + 4 = 7$?

3. Is -1 a solution of
$2b - 1 = 3$?

4. Is -2 a solution of
$3a - 4 = 10$?

5. Is 1 a solution of
$4 - 2m = 3$?

6. Is 2 a solution of
$7 - 3n = 2$?

7. Is 5 a solution of
$2x + 5 = 3x$?

8. Is 4 a solution of
$3y - 4 = 2y$?

9. Is 0 a solution of
$4a + 5 = 3a + 5$?

10. Is 0 a solution of
$4 - 3b = 4 - 5b$?

11. Is -2 a solution of
$4 - 2n = n + 10$?

12. Is -3 a solution of
$5 - m = 2 - 2m$?

13. Is 3 a solution of
$z^2 + 1 = 4 + 3z$?

14. Is 2 a solution of
$2x^2 - 1 = 4x - 1$?

15. Is -1 a solution of
$y^2 - 1 = 4y + 3$?

16. Is -2 a solution of
$m^2 - 4 = m + 3$?

17. Is 5 a solution of
$x^2 + 2x + 1 = (x + 1)^2$?

18. Is -6 a solution of
$(n - 2)^2 = n^2 - 4n + 4$?

19. Is 4 a solution of
$x(x + 1) = x^2 + 5$?

20. Is 3 a solution of
$2a(a - 1) = 3a + 3$?

21. Is $-\frac{1}{4}$ a solution of
$8t + 1 = -1$?

22. Is $\frac{1}{2}$ a solution of
$4y + 1 = 3$?

23. Is $\frac{2}{5}$ a solution of
$5m + 1 = 10m - 3$?

24. Is $\frac{3}{4}$ a solution of
$8x - 1 = 12x + 3$?

25. Is $\frac{1}{3}$ a solution of
$2n + 2 = 5n - 1$?

26. Is $\frac{1}{4}$ a solution of
$5x + 4 = x + 5$?

27. Is $-\frac{1}{3}$ a solution of
$3b(b - 1) = 2b + 2$?

28. Is 2.1 a solution of
$x^2 - 4x = x + 1.89$?

29. Is 1.5 a solution of
$c^2 - 3c = 4c - 8.25$?

30. Is -1.8 a solution of
$a(a + 1) = 2.6 - 2a$?

31. Is 4.2 a solution of
$m(2m + 1) = (m + 1)^2$?

32. Is 2.3 a solution of
$d^2 - 2d + 5 = d + 3.39$?

33. Is 0.08 a solution of
$9 - 2y = 5 + 48y$?

2   Solve and check.

34. $x + 5 = 7$

35. $y + 3 = 9$

36. $b - 4 = 11$

37. $z - 6 = 10$

38. $2 + a = 8$

39. $5 + x = 12$

40. $m + 9 = 3$

41. $t + 12 = 10$

42. $n - 5 = -2$

43. $x - 6 = -5$

44. $b + 7 = 7$

45. $y - 5 = -5$

46. $a - 3 = -5$

47. $x - 6 = -3$

48. $z + 9 = 2$

49. $n + 11 = 1$

50. $10 + m = 3$

51. $8 + x = 5$

52. $9 + x = -3$

53. $10 + y = -4$

54. $b - 5 = -3$

55. $t - 6 = -4$

56. $4 + x = 10$

57. $9 + a = 20$

58. $2 = x + 7$

59. $-8 = n + 1$

60. $4 = m - 11$

61. $-6 = y - 5$

62. $12 = 3 + w$

63. $-9 = 5 + x$

64. $4 = -10 + b$

65. $-7 = -2 + x$

66. $13 = -6 + a$

67. $m + \dfrac{2}{3} = -\dfrac{1}{3}$

68. $c + \dfrac{3}{4} = -\dfrac{1}{4}$

69. $x - \dfrac{1}{2} = \dfrac{1}{2}$

70. $x - \dfrac{2}{5} = \dfrac{3}{5}$

71. $\dfrac{5}{8} + y = \dfrac{1}{8}$

72. $\dfrac{4}{9} + a = -\dfrac{2}{9}$

73. $m + \dfrac{1}{2} = -\dfrac{1}{4}$

74. $b + \dfrac{1}{6} = -\dfrac{1}{3}$

75. $x + \dfrac{2}{3} = \dfrac{3}{4}$

76. $n + \dfrac{2}{5} = \dfrac{2}{3}$

77. $-\dfrac{5}{6} = x - \dfrac{1}{4}$

78. $-\dfrac{1}{4} = c - \dfrac{2}{3}$

79. $a - \dfrac{2}{7} = \dfrac{1}{14}$

80. $y + \dfrac{3}{8} = -\dfrac{1}{2}$

81. $t - \dfrac{4}{15} = \dfrac{2}{3}$

82. $-\dfrac{1}{21} = m + \dfrac{2}{3}$

83. $\dfrac{5}{9} = b - \dfrac{1}{3}$

84. $\dfrac{5}{12} = n + \dfrac{3}{4}$

85. $d + 1.3619 = 2.0148$

86. $w + 2.932 = 4.801$

87. $-0.813 + x = -1.096$

88. $-1.926 + t = -1.042$

89. $6.149 = -3.108 + z$

90. $5.237 = -2.014 + x$

3   Solve and check.

91. $5x = 15$

92. $4y = 28$

93. $3b = -12$

94. $2a = -14$

95. $-3x = 6$

96. $-5m = 20$

97. $-3x = -27$

98. $-6n = -30$

99. $20 = 4c$

100. $18 = 2t$

101. $-32 = 8w$

102. $-56 = 7x$

103. $8d = 0$

104. $-5x = 0$

105. $36 = 9z$

106. $35 = -5x$

107. $-64 = 8a$

108. $-32 = -4y$

109. $-42 = 6t$

110. $-12m = -144$

111. $-9c = 81$

112. $-54 = 6c$

113. $49 = -7t$

114. $\dfrac{x}{3} = 2$

115. $\dfrac{x}{4} = 3$

116. $-\dfrac{y}{2} = 5$

117. $-\dfrac{b}{3} = 6$

118. $\dfrac{n}{7} = -4$

119. $-\dfrac{a}{9} = 5$

120. $\dfrac{t}{6} = -3$

121. $\dfrac{2}{5}x = 12$

122. $-\dfrac{4}{3}c = -8$

123. $\dfrac{5}{6}y = -20$

124. $\dfrac{3}{4}y = 9$

125. $\dfrac{2}{5}x = 6$

126. $-\dfrac{2}{3}d = 8$

127. $-\dfrac{3}{5}m = 12$

128. $\dfrac{2n}{3} = 2$

129. $\dfrac{5x}{6} = -10$

130. $\dfrac{-3z}{8} = 9$

131. $\dfrac{-4x}{5} = -12$

132. $-6 = -\dfrac{2}{3}y$

133. $-15 = -\dfrac{3}{5}x$

134. $\dfrac{2}{5}a = 3$

135. $\dfrac{3x}{4} = 2$

136. $\dfrac{3}{4}c = \dfrac{3}{5}$

137. $\dfrac{2}{9} = \dfrac{2}{3}y$

138. $-\dfrac{6}{7} = -\dfrac{3}{4}b$

139. $-\dfrac{2}{5}m = -\dfrac{6}{7}$

140. $5x + 2x = 14$

141. $3n + 2n = 20$

142. $7d - 4d = 9$

143. $10y - 3y = 21$

144. $2x - 5x = 9$

145. $\dfrac{x}{1.4} = 3.2$

146. $\dfrac{z}{2.9} = -7.8$

147. $3.4a = 7.004$

148. $2.3m = 2.415$

149. $-3.7x = 7.77$

150. $\dfrac{n}{2.6} = 9.08$

151. $\dfrac{x}{5.7} = 2.06$

152. $-1.6m = 5.44$

153. $4.3c = 4.687$

SUPPLEMENTAL EXERCISES 3.1

Solve.

154. $\dfrac{3}{2}x - \dfrac{7}{5}x = 3$

155. $\dfrac{3}{2}n - \dfrac{4}{3}n = 6$

156. $\dfrac{3}{4}t - \dfrac{5}{8}t = -9$

157. $\dfrac{2}{5}d - \dfrac{5}{8}d = -18$

158. $10 = \dfrac{4}{9}c - \dfrac{2}{3}c$

159. $-22 = -\dfrac{5}{8}b + \dfrac{1}{6}b$

160. $\dfrac{n + n}{2} = -8$

161. $\dfrac{2m + m}{5} = -9$

162. $\dfrac{3y - 8y}{7} = 15$

163. $\dfrac{4t - 7t}{6} = -11$

164. $\dfrac{1}{\dfrac{1}{x}} = 5$

165. $\dfrac{1}{\dfrac{1}{y}} = -7$

166. $\dfrac{1}{\dfrac{1}{x}} + 8 = -19$

167. $\dfrac{1}{\dfrac{1}{y}} + 18 = 11$

168. $\dfrac{4}{\dfrac{3}{b}} = 8$

169. $\dfrac{6}{\dfrac{5}{a}} = -12$

170. $\dfrac{5}{\dfrac{7}{a}} - \dfrac{3}{\dfrac{7}{a}} = 6$

171. $-\dfrac{3}{\dfrac{4}{a}} + \dfrac{2}{\dfrac{5}{a}} = 7$

S E C T I O N   **3.2**

General Equations—Part 1

1 Solve equations of the form $ax + b = c$

In solving an equation of the form $ax + b = c$, the goal is to rewrite the equation in the form *variable* = *constant*. This requires the application of both the Addition and Multiplication Properties of Equations.

To solve the equation $\frac{2}{5}x - 3 = -7$,

add 3 to each side of the equation.

$$\frac{2}{5}x - 3 = -7$$

$$\frac{2}{5}x - 3 + 3 = -7 + 3$$

Simplify.

$$\frac{2}{5}x = -4$$

Multiply each side of the equation by the reciprocal of the coefficient $\frac{2}{5}$.

$$\frac{5}{2} \cdot \frac{2}{5}x = \frac{5}{2}(-4)$$

Simplify.
Now the equation is in the form
variable = *constant*.

$$x = -10$$

Check
$$\frac{2}{5}x - 3 = -7$$
$$\begin{array}{c|c} \frac{2}{5}(-10) - 3 & -7 \\ -4 - 3 & -7 \\ -7 = -7 \end{array}$$

Write the solution.

The solution is -10.

Example 1 Solve: $3x - 7 = -5$

Solution $3x - 7 = -5$

$3x - 7 + 7 = -5 + 7$ ▶ Add 7 to each side of the equation.

$3x = 2$ ▶ Simplify.

$\dfrac{3x}{3} = \dfrac{2}{3}$ ▶ Divide each side of the equation by 3.

$x = \dfrac{2}{3}$ ▶ Simplify. Now the equation is in the form *variable* = *constant*.

The solution is $\frac{2}{3}$. ▶ Write the solution.

Problem 1   Solve: $5x + 7 = 10$

Solution   See page A9.

Example 2   Solve: $5 = 9 - 2x$

Solution
$$5 = 9 - 2x$$
$$5 - 9 = 9 - 9 - 2x \qquad \blacktriangleright \text{Subtract 9 from each side of the equation.}$$
$$-4 = -2x \qquad \blacktriangleright \text{Simplify.}$$

$$\frac{-4}{-2} = \frac{-2x}{-2} \qquad \blacktriangleright \text{Divide each side of the equation by } -2.$$

$$2 = x \qquad \blacktriangleright \text{Simplify.}$$

The solution is 2.   ▶ Write the solution.

Problem 2   Solve: $11 = 11 + 3x$

Solution   See page A9.

2   Application problems

Example 3   The pressure at a certain depth in the ocean can be approximated by the equation $P = 15 + \frac{1}{2}D$, where P is the pressure in pounds per square inch, and D is the depth in feet. Use this equation to find the depth when the pressure is 35 pounds per square inch.

Strategy   To find the depth, replace P with the given value and solve for D.

Solution
$$P = 15 + \frac{1}{2}D$$

$$35 = 15 + \frac{1}{2}D$$

$$35 - 15 = 15 - 15 + \frac{1}{2}D$$

$$20 = \frac{1}{2}D$$

$$2 \cdot 20 = 2 \cdot \frac{1}{2}D$$

$$40 = D$$

The depth is 40 ft.

Problem 3 To determine the cost of production, an economist uses the equation $T = U \cdot N + F$, where T is the total cost, U is the unit cost, N is the number of units made, and F is the fixed cost. Use this equation to find the number of units made during a month when the total cost was $8000, the unit cost was $15, and the fixed costs were $2000.

Solution See page A9.

EXERCISE 3.2

1 Solve and check.

1. $3x + 1 = 10$

2. $4y + 3 = 11$

3. $2a - 5 = 7$

4. $5m - 6 = 9$

5. $5 = 4x + 9$

6. $2 = 5b + 12$

7. $13 = 9 + 4z$

8. $7 - c = 9$

9. $2 - x = 11$

10. $4 - 3w = -2$

11. $5 - 6x = -13$

12. $8 - 3t = 2$

13. $-5d + 3 = -12$

14. $-8x - 3 = -19$

15. $-7n - 4 = -25$

16. $-12x + 30 = -6$

17. $-13 = -11y + 9$

18. $2 = 7 - 5a$

19. $3 = 11 - 4n$

20. $-35 = -6b + 1$

21. $-8x + 3 = -29$

22. $-3m - 21 = 0$

23. $-5x - 30 = 0$

24. $-4y + 15 = 15$

25. $7x - 3 = 3$

26. $8y + 3 = 7$

27. $6a + 5 = 9$

28. $3m + 4 = 11$

29. $14 = 9x + 1$

30. $4 = 5b + 6$

31. $11 = 15 + 4n$

32. $4 = 2 - 3c$

33. $9 - 4x = 6$

34. $3t - 2 = 0$

35. $9x - 4 = 0$

36. $7 - 8z = 0$

37. $1 - 3x = 0$

38. $9d + 10 = 7$

39. $12w + 11 = 5$

40. $6y - 5 = -7$

41. $8b - 3 = -9$

42. $7 + 12x = 3$

43. $9 + 14x = 7$

44. $5 - 6m = 2$

45. $7 - 9a = 4$

46. $9 = -12c + 5$

47. $10 = -18x + 7$

48. $2y + \dfrac{1}{3} = \dfrac{7}{3}$

49. $4a + \dfrac{3}{4} = \dfrac{19}{4}$

50. $2n - \dfrac{3}{4} = \dfrac{13}{4}$

51. $3x - \dfrac{5}{6} = \dfrac{13}{6}$

52. $5y + \dfrac{3}{7} = \dfrac{3}{7}$

53. $9x + \dfrac{4}{5} = \dfrac{4}{5}$

54. $8 = 7d - 1$

55. $8 = 10x - 5$

56. $4 = 7 - 2w$

57. $7 = 9 - 5a$

58. $8t + 13 = 3$

59. $12x + 19 = 3$

60. $-6y + 5 = 13$

61. $-4x + 3 = 9$

62. $\dfrac{1}{2}a - 3 = 1$

63. $\dfrac{1}{3}m - 1 = 5$

64. $\dfrac{2}{5}y + 4 = 6$

65. $\dfrac{3}{4}n + 7 = 13$

66. $-\dfrac{2}{3}x + 1 = 7$

67. $-\dfrac{3}{8}b + 4 = 10$

68. $\dfrac{x}{4} - 6 = 1$

69. $\dfrac{y}{5} - 2 = 3$

70. $\dfrac{2x}{3} - 1 = 5$

71. $\dfrac{3c}{7} - 1 = 8$

72. $4 - \dfrac{3}{4}z = -2$

73. $3 - \dfrac{4}{5}w = -9$

74. $5 + \dfrac{2}{3}y = 3$

75. $17 + \dfrac{5}{8}x = 7$

76. $17 = 7 - \dfrac{5}{6}t$

77. $9 = 3 - \dfrac{2x}{7}$

78. $3 = \dfrac{3a}{4} + 1$

79. $7 = \dfrac{2x}{5} + 4$

80. $5 - \dfrac{4c}{7} = 8$

81. $7 - \dfrac{5}{9}y = 9$

82. $6a + 3 + 2a = 11$

83. $5y + 9 + 2y = 23$

84. $7x - 4 - 2x = 6$

85. $11z - 3 - 7z = 9$

86. $2x - 6x + 1 = 9$

87. $b - 8b + 1 = -6$

88. $3 = 7x + 9 - 4x$

89. $-1 = 5m + 7 - m$

90. $8 = 4n - 6 + 3n$

91. $0.15y + 0.025 = -0.074$

92. $1.2x - 3.44 = 1.3$

93. $3.5 = 3.5 + 0.076x$

94. $-6.5 = 4.3y - 3.06$

Solve.

95. If $2x - 3 = 7$, evaluate $3x + 4$.

96. If $3x + 5 = -4$, evaluate $2x - 5$.

97. If $4 - 5x = -1$, evaluate $x^2 - 3x + 1$.

98. If $2 - 3x = 11$, evaluate $x^2 + 2x - 3$.

99. If $5x + 3 - 2x = 12$, evaluate $4 - 5x$.

100. If $2x - 4 - 7x = 16$, evaluate $x^2 + 1$.

2   Solve.

The distance s that an object will fall in t seconds is given by $s = 16t^2 + vt$, where v is the initial velocity of the object.

101. Find the initial velocity of an object that falls 80 ft in 2 s.

102. Find the initial velocity of an object that falls 144 ft in 3 s.

A company uses the equation $V = C - 6000t$ to determine the depreciated value V, after t years, of a milling machine that originally cost C dollars. Equations such as this are used in accounting for straight-line depreciation.

103. A milling machine originally cost $50,000. In how many years will the depreciated value be $38,000?

104. A milling machine originally cost $78,000. In how many years will the depreciated value be $48,000?

Anthropologists can approximate the height of a primate by the size of its humerus (the bone extending from the shoulder to the elbow) by using the equation $H = 1.2L + 27.8$, where L is the length of the humerus and H is the height of the primate.

105. An anthropoligist estimates the height of a primate to be 66 in. What is the approximate length of the humerus of this primate? Round to the nearest tenth.

106. The height of a primate is estimated to be 62 in. Find the approximate length of the humerus of this primate.

The world record time for a 1-mile race can be approximated by the equation $t = 17.08 - 0.0067y$, where t is the time and y is the year of the race.

107. Approximate the year in which the first "4-minute mile" was run. (The actual year was 1954.)

108. In 1985, the world record for a 1-mile race was 3.77 min. For what year does the equation predict this record time?

A telephone company estimates that the number N of phone calls made per day between two cities of population P_1 and P_2 that are d miles apart is given by the equation $N = \frac{2.51P_1P_2}{d^2}$.

109. Estimate the population P_2, given that P_1 is 48,000, the number of phone calls is 1,100,000, and the distance between the cities is 75 mi. Round to the nearest thousand.

110. Estimate the population P_1, given that P_2 is 125,000, the number of phone calls is 2,500,000, and the distance between the cities is 50 mi. Round to the nearest thousand.

SUPPLEMENTAL EXERCISES 3.2

Solve.

111. $\frac{2}{3}x + 7 + \frac{1}{6}x = 12$

112. $\frac{3}{4}x - 8 + \frac{5}{8}x = 3$

Solve for x.

113. $32{,}166 \div x = 518$ remainder 50

114. $22{,}683 \div x = 482$ remainder 29

SECTION **3.3**

General Equations—Part II

1 Solve equations of the form
$ax + b = cx + d$

In solving an equation of the form $ax + b = cx + d$, the goal is to rewrite the equation in the form *variable* = *constant*. Begin by rewriting the equation so that there is only one variable term in the equation. Then rewrite the equation so that there is only one constant term.

Solve: $4x - 5 = 6x + 11$	$4x - 5 = 6x + 11$
Subtract $6x$ from each side of the equation.	$4x - 6x - 5 = 6x - 6x + 11$
Simplify. Now there is only one variable term in the equation.	$-2x - 5 = 11$
Add 5 to each side of the equation.	$-2x - 5 + 5 = 11 + 5$
Simplify. Now there is only one constant term in the equation.	$-2x = 16$
Divide each side of the equation by -2.	$\dfrac{-2x}{-2} = \dfrac{16}{-2}$
Simplify. Now the equation is in the form *variable* = *constant*.	$x = -8$

Check $4x - 5 = 6x + 11$

$4(-8) - 5$	$6(-8) + 11$
$-32 - 5$	$-48 + 11$

$-37 = -37$

Write the solution.   The solution is -8.

Example 1   Solve: $4x - 3 = 8x - 7$

Solution

$$4x - 3 = 8x - 7$$
$$4x - 8x - 3 = 8x - 8x - 7$$

▶ Subtract $8x$ from each side of the equation.

$$-4x - 3 = -7$$

▶ Simplify. Now there is only one variable term in the equation.

$$-4x - 3 + 3 = -7 + 3$$

▶ Add 3 to each side of the equation.

$$-4x = -4$$

▶ Simplify. Now there is only one constant term in the equation.

$$\frac{-4x}{-4} = \frac{-4}{-4}$$

▶ Divide each side of the equation by -4.

$$x = 1$$

▶ Simplify. Now the equation is in the form *variable = constant*.

The solution is 1.

▶ Write the solution.

Problem 1   Solve: $5x + 4 = 6 + 10x$

Solution   See page A9.

Example 2   Solve: $3x + 4 - 5x = 2 - 4x$

Solution

$$3x + 4 - 5x = 2 - 4x$$
$$-2x + 4 = 2 - 4x$$

▶ Combine like terms.

$$-2x + 4x + 4 = 2 - 4x + 4x$$

▶ Add $4x$ to each side of the equation.

$$2x + 4 = 2$$

▶ Simplify.

$$2x + 4 - 4 = 2 - 4$$

▶ Subtract 4 from each side of the equation.

$$2x = -2$$

▶ Simplify.

$$\frac{2x}{2} = \frac{-2}{2}$$

▶ Divide each side of the equation by 2.

$$x = -1$$

▶ Simplify.

The solution is -1.

▶ Write the solution.

Problem 2   Solve: $5x - 10 - 3x = 6 - 4x$

Solution   See page A9.

2 Solve equations containing parentheses

When an equation contains parentheses, one of the steps in solving the equation requires the use of the Distributive Property. The Distributive Property is used to remove parentheses from a variable expression.

$a(b + c) = ab + ac$

Solve: $4 + 5(2x - 3) = 3(4x - 1)$

Use the Distributive Property to remove parentheses.

$$4 + 5(2x - 3) = 3(4x - 1)$$
$$4 + 10x - 15 = 12x - 3$$

Simplify.

$$10x - 11 = 12x - 3$$

Subtract $12x$ from each side of the equation.

$$10x - 12x - 11 = 12x - 12x - 3$$

Simplify.
Now there is only one variable term in the equation.

$$-2x - 11 = -3$$

Add 11 to each side of the equation.

$$-2x - 11 + 11 = -3 + 11$$

Simplify.
Now there is only one constant term in the equation.

$$-2x = 8$$

Divide each side of the equation by -2.

$$\frac{-2x}{-2} = \frac{8}{-2}$$

Simplify.
Now the equation is in the form *variable = constant*.

$$x = -4$$

Check

$$4 + 5(2x - 3) = 3(4x - 1)$$

$4 + 5[2(-4) - 3]$	$3[4(-4) - 1]$
$4 + 5(-8 - 3)$	$3(-16 - 1)$
$4 + 5(-11)$	$3(-17)$
$4 - 55$	-51
$-51 = -51$	

Write the solution.

The solution is -4.

Example 3   Solve: $3x - 4(2 - x) = 3(x - 2) - 4$

Solution $3x - 4(2 - x) = 3(x - 2) - 4$
$3x - 8 + 4x = 3x - 6 - 4$ ▶ Use the Distributive Property to remove parentheses.

$7x - 8 = 3x - 10$ ▶ Simplify.
$7x - 3x - 8 = 3x - 3x - 10$ ▶ Subtract $3x$ from each side of the equation.

$4x - 8 = -10$
$4x - 8 + 8 = -10 + 8$ ▶ Add 8 to each side of the equation.
$4x = -2$

$\dfrac{4x}{4} = \dfrac{-2}{4}$ ▶ Divide each side of the equation by 4.

$x = -\dfrac{1}{2}$ ▶ The equation is in the form *variable = constant*.

The solution is $-\dfrac{1}{2}$.

Problem 3   Solve: $5x - 4(3 - 2x) = 2(3x - 2) + 6$

Solution   See page A10.

Example 4   Solve: $3[2 - 4(2x - 1)] = 4x - 10$

Solution $3[2 - 4(2x - 1)] = 4x - 10$
$3[2 - 8x + 4] = 4x - 10$ ▶ Use the Distributive Property to remove the parentheses.

$3[6 - 8x] = 4x - 10$ ▶ Simplify inside the brackets.
$18 - 24x = 4x - 10$ ▶ Use the Distributive Property to remove the brackets.

$18 - 24x - 4x = 4x - 4x - 10$ ▶ Subtract $4x$ from each side of the equation.

$18 - 28x = -10$
$18 - 18 - 28x = -10 - 18$ ▶ Subtract 18 from each side of the equation.

$-28x = -28$

$\dfrac{-28x}{-28} = \dfrac{-28}{-28}$ ▶ Divide each side of the equation by -28.

$x = 1$
The solution is 1.

Problem 4   Solve: $-2[3x - 5(2x - 3)] = 3x - 8$

Solution   See page A10.

3 Application problems

A lever system is shown below. It consists of a lever, or bar; a fulcrum; and two forces, F_1 and F_2. The distance d represents the length of the lever, x represents the distance from F_1 to the fulcrum, and $d - x$ represents the distance from F_2 to the fulcrum.

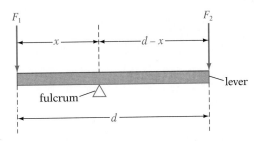

A principle of physics states that when the lever system balances,

$$F_1 x = F_2(d - x).$$

Example 5 A lever is 10 ft long. A force of 100 lb is applied to one end of the lever, and a force of 400 lb is applied to the other end. What is the location of the fulcrum when the system balances?

Strategy To find the location of the fulcrum when the system balances, replace the variables F_1, F_2, and d in the lever system equation with the given values, and solve for x.

Solution
$$F_1 x = F_2(d - x)$$
$$100x = 400(10 - x)$$
$$100x = 4000 - 400x$$
$$100x + 400x = 4000 - 400x + 400x$$
$$500x = 4000$$

$$\frac{500x}{500} = \frac{4000}{500}$$

$$x = 8$$

The fulcrum is 8 ft from the 100-lb force.

Problem 5 A lever is 8 ft long. A force of 80 lb is applied to one end of the lever, and a force of 560 lb is applied to the other end. Where is the fulcrum located when the system balances?

Solution See page A10.

Example 6   A lever is 25 ft long. At a distance of 12 ft from the fulcrum, a force of 26 lb is applied. How large a force must be applied to the other end of the lever so that the system will balance?

Strategy   To find the force when the system balances, replace the variables F_1, x, and d in the lever system equation with the given values, and solve for F_2.

Solution

$$F_1x = F_2(d - x)$$
$$26(12) = F_2(25 - 12)$$
$$312 = F_2(13)$$
$$312 = 13F_2$$

$$\frac{312}{13} = \frac{13F_2}{13}$$

$$24 = F_2$$

A 24-lb force must be applied to the other end.

Problem 6   A lever is 14 ft long. At a distance of 6 ft from the fulcrum, a force of 40 lb is applied. How large a force must be applied to the other end of the lever so that the system will balance?

Solution   See page A10.

EXERCISES 3.3

1   Solve and check.

1. $8x + 5 = 4x + 13$
2. $6y + 2 = y + 17$
3. $7m + 4 = 6m + 7$
4. $11n + 3 = 10n + 11$
5. $5x - 4 = 2x + 5$
6. $9a - 10 = 3a + 2$
7. $12y - 4 = 9y - 7$
8. $13b - 1 = 4b - 19$
9. $15x - 2 = 4x - 13$
10. $7a - 5 = 2a - 20$
11. $3x + 1 = 11 - 2x$
12. $n - 2 = 6 - 3n$
13. $2x - 3 = -11 - 2x$
14. $4y - 2 = -16 - 3y$
15. $2b + 3 = 5b + 12$
16. $m + 4 = 3m + 8$
17. $4x - 7 = 5x + 1$
18. $6d - 2 = 7d + 5$
19. $4y - 8 = y - 8$
20. $5a + 7 = 2a + 7$
21. $6 - 5x = 8 - 3x$
22. $10 - 4n = 16 - n$
23. $5 + 7x = 11 + 9x$
24. $3 - 2y = 15 + 4y$
25. $2x - 4 = 6x$
26. $2b - 10 = 7b$
27. $8m = 3m + 20$

28. $9y = 5y + 16$

29. $-3x - 4 = 2x + 6$

30. $-5a - 3 = 2a + 18$

31. $-8n + 3 = -5n - 6$

32. $-10x + 4 = -x - 14$

33. $-x - 4 = -3x - 16$

34. $8 - 4x = 18 - 5x$

35. $6 - 10a = 8 - 9a$

36. $5 - 7m = 2 - 6m$

37. $8b + 5 = 5b + 7$

38. $6y - 1 = 2y + 2$

39. $7x - 8 = x - 3$

40. $10x - 3 = 3x - 1$

41. $5n + 3 = 2n + 1$

42. $8a - 2 = 4a - 5$

43. $8.7y = 3.9y + 9.6$

44. $4.5x - 5.4 = 2.7x$

45. $5.6x = 7.2x - 6.4$

Solve.

46. If $5x = 3x - 8$, evaluate $4x + 2$.

47. If $7x + 3 = 5x - 7$, evaluate $3x - 2$.

48. If $2 - 6a = 5 - 3a$, evaluate $4a^2 - 2a + 1$.

49. If $1 - 5c = 4 - 4c$, evaluate $3c^2 - 4c + 2$.

50. If $2y + 3 = 5 - 4y$, evaluate $6y - 7$.

51. If $3z + 1 = 1 - 5z$, evaluate $3z^2 - 7z + 8$.

2   Solve and check.

52. $5x + 2(x + 1) = 23$

53. $6y + 2(2y + 3) = 16$

54. $9n - 3(2n - 1) = 15$

55. $12x - 2(4x - 6) = 28$

56. $7a - (3a - 4) = 12$

57. $9m - 4(2m - 3) = 11$

58. $5(3 - 2y) + 4y = 3$

59. $4(1 - 3x) + 7x = 9$

60. $10x + 1 = 2(3x + 5) - 1$

61. $5y - 3 = 7 + 4(y - 2)$

62. $4 - 3a = 7 - 2(2a + 5)$

63. $9 - 5x = 12 - (6x + 7)$

64. $3y - 7 = 5(2y - 3) + 4$

65. $2a - 5 = 4(3a + 1) - 2$

66. $5 - (9 - 6x) = 2x - 2$

67. $7 - (5 - 8x) = 4x + 3$

68. $3[2 - 4(y - 1)] = 3(2y + 8)$

69. $5[2 - (2x - 4)] = 2(5 - 3x)$

70. $3a + 2[2 + 3(a - 1)] = 2(3a + 4)$

71. $5 + 3[1 + 2(2x - 3)] = 6(x + 5)$

72. $-2[4 - (3b + 2)] = 5 - 2(3b + 6)$

73. $-4[x - 2(2x - 3)] + 1 = 2x - 3$

74. $0.3x - 2(1.6x) - 8 = 3(1.9x - 4.1)$

75. $0.56 - 0.4(2.1y + 3) = 0.2(2y + 6.1)$

Solve.

76. If $4 - 3a = 7 - 2(2a + 5)$, evaluate $a^2 + 7a$.

77. If $9 - 5x = 12 - (6x + 7)$, evaluate $x^2 - 3x - 2$.

78. If $2z - 5 = 3(4z + 5)$, evaluate $\dfrac{z^2}{z - 2}$.

79. If $3n - 7 = 5(2n + 7)$, evaluate $\dfrac{n^2}{2n - 6}$.

3   Solve.

Use the lever system equation $F_1x = F_2(d - x)$.

80. An adult and a child are on a see-saw 14 ft long. The adult weighs 175 lb and the child weighs 70 lb. How many feet from the child must the fulcrum be placed so that the see-saw balances?

81. Two children are sitting 8 ft apart on a see-saw. One child weighs 60 lb and the second child weighs 50 lb. The fulcrum is 3.5 ft from the child weighing 60 lb. Is the see-saw balanced?

82. A 50-lb weight is applied to the left end of a see-saw 10 ft long. The fulcrum is 4 ft from the 50-lb weight. A weight of 30 lb is applied to the right end of the see-saw. Is the 30-lb weight adequate to balance the see-saw?

83. In preparation for a stunt, two acrobats are standing on a plank 18 ft long. One acrobat weighs 128 lb and the second acrobat weighs 160 lb. How far from the 128-lb acrobat must the fulcrum be placed so that the acrobats are balanced on the plank?

84. A lever 10 ft long is used to move a 100-lb rock. The fulcrum is placed 2 ft from the rock. What force must be applied to the other end of the lever to move the rock?

85. A screwdriver 9 in. long is used as a lever to open a can of paint. The tip of the screwdriver is placed under the lip of the can with the fulcrum 0.15 in. from the lip. A force of 30 lb is applied to the other end of the screwdriver. Find the force on the lip of the can.

To determine the breakeven point, or the number of units that must be sold so that no profit or loss occurs, an economist uses the equation $Px = Cx + F$, where P is the selling price per unit, x is the number of units sold, C is the cost to make each unit, and F is the fixed cost.

86. An economist has determined that the selling price per unit for a television is $250. The cost to make one television is $165, and the fixed cost is $29,750. Find the breakeven point.

87. A business analyst has determined that the selling price per unit for a cellular telephone is $675. The cost to make one cellular telephone is $485, and the fixed cost is $36,100. Find the breakeven point.

88. A manufacturing engineer determines that the cost per unit for a compact disc is $3.35 and that the fixed cost is $6180. The selling price for the compact disc is $8.50. Find the breakeven point.

89. The manufacture of a softball bat requires two steps: (1) cut the rough shape and (2) sand the bat to its final form. The cost to rough-shape the bat is $.45, and the cost to sand the bat to final form is $1.05. Find the breakeven point if the selling price per bat is $7 and the fixed cost is $16,500.

SUPPLEMENTAL EXERCISES 3.3

Solve. If the equation has no solution, write "No solution."

90. $2(5x - 3) + 6 - 10x = -8$

91. $3(2x - 1) - (6x - 4) = -9$

92. $4(3 - x) - 3(4 - x) = -15$

93. $7(3x + 6) - 4(3 + 5x) = 13 + x$

94. $\frac{1}{2}(6a - 8) + 5 = -\frac{1}{4}(12 - 4a) - 3$

95. $\frac{1}{5}(25 - 10b) + 4 = \frac{1}{3}(9b - 15) - 6$

96. $3[4(w + 2) - (w + 1)] = 5(2 + w)$

97. $2(n + 7) = 4[3(n - 5) - 6(1 + n)]$

98. $4[2(b - 3) + 5] = 8[3(b - 2) - (b - 4)]$

99. $5[m + 2(3 - m)] = 3[2(4 - m) - 5]$

100. $\dfrac{5(2x - 3) - 3(3x - 5)}{4} = -6$

101. $\dfrac{2(5x - 6) - 3(x - 4)}{7} = x + 2$

102. $3[2 - 2(4x + 2) + 15x] = 4(3 + 7x) - (7x + 18)$

103. $x\left[3 - \dfrac{2}{x}(x + 1)\right] = 5(3x + 1) - x\left[2 + \dfrac{7}{x}(2x + 1)\right]$

SECTION **3.4**
Translating Sentences into Equations

1 Translate a sentence into an equation and solve

An equation states that two mathematical expressions are equal. Therefore, to translate a sentence into an equation requires recognizing the words or phrases that mean "equals." Besides "equals," some of these phrases are "is," "is equal to," "amounts to," and "represents."

Once the sentence is translated into an equation, the equation can be solved by rewriting the equation in the form *variable = constant.*

Example 1   Translate "five less than twice a number is the number minus nine" into an equation and solve.

Solution   the unknown number: n ▶ Assign a variable to the unknown quantity.

| five less than twice a number | is | the number minus nine |

▶ Find two verbal expressions for the same value.

$$2n - 5 = n - 9$$

▶ Write a mathematical expression for each verbal expression. Write the equals sign.

$$2n - n - 5 = n - n - 9$$
$$n - 5 = -9$$
$$n - 5 + 5 = -9 + 5$$
$$n = -4$$

▶ Solve the equation.

The number is -4.

Problem 1   Translate "three more than a number is equal to the difference between seven and the number" into an equation and solve.

Solution   See page A10.

Example 2   Translate "two times the sum of a number and eight equals the sum of four times the number and six" into an equation and solve.

Solution   the unknown number: n ▶ Assign a variable to the unknown quantity.

| two times the sum of a number and eight | equals | the sum of four times the number and six |

▶ Find two verbal expressions for the same value.

$$2(n + 8) = 4n + 6$$

▶ Write a mathematical expression for each verbal expression. Write the equals sign.

$$2n + 16 = 4n + 6$$
$$2n - 4n + 16 = 4n - 4n + 6$$
$$-2n + 16 = 6$$
$$-2n + 16 - 16 = 6 - 16$$
$$-2n = -10$$
$$\frac{-2n}{-2} = \frac{-10}{-2}$$
$$n = 5$$

▶ Solve the equation.

The number is 5.

Problem 2   Translate "nine less than twice a number is five times the sum of the number and twelve" into an equation and solve.

Solution   See page A11.

2   ## Application problems

Example 3   The temperature of the sun on the Kelvin scale is 6500 K. This is 454° more than the temperature on the Fahrenheit scale. Find the Fahrenheit temperature.

Strategy   To find the Fahrenheit temperature, write and solve an equation using F to represent the Fahrenheit temperature.

Solution

| 6500 | is | 454° more than the Fahrenheit temperature |

$$6500 = F + 454$$
$$6500 - 454 = F + 454 - 454$$
$$6046 = F$$

The temperature of the sun is 6046°F.

Problem 3   A molecule of octane gas has eight carbon atoms. This represents twice the number of carbon atoms in a butane gas molecule. Find the number of carbon atoms in a butane molecule.

Solution   See page A11.

Example 4   A board 10 ft long is cut into two pieces. Three times the length of the shorter piece is twice the length of the longer piece. Find the length of each piece.

Strategy   To find the length of each piece, write and solve an equation using x to represent the length of the shorter piece and $10 - x$ to represent the length of the longer piece.

Solution

| three times the shorter piece | is | twice the longer piece |

$$3x = 2(10 - x)$$
$$3x = 20 - 2x$$
$$3x + 2x = 20 - 2x + 2x$$
$$5x = 20$$

$$\frac{5x}{5} = \frac{20}{5}$$

$$x = 4$$ ▶ The length of the shorter piece is 4 ft.

$$10 - x = 10 - 4 = 6$$ ▶ Substitute the value of x into the variable expression for the longer piece and evaluate.

The shorter piece is 4 ft. The longer piece is 6 ft.

Problem 4   A company manufactures 160 bicycles per day. Four times the number of 3-speed bicycles made equals 30 less than the number of 10-speed bicycles made. Find the number of 10-speed bicycles manufactured each day.

Solution   See page A11.

EXERCISES 3.4

1   Translate into an equation and solve.

1. The difference between a number and fifteen is seven. Find the number.

2. The sum of five and a number is three. Find the number.

3. The product of seven and a number is negative twenty-one. Find the number.

4. The quotient of a number and four is two. Find the number.

5. Four less than three times a number is five. Find the number.

6. The difference between five and twice a number is one. Find the number.

7. Four times the sum of twice a number and three is twelve. Find the number.

8. Twenty-one is three times the difference between four times a number and five. Find the number.

9. Twelve is six times the difference between a number and three. Find the number.

10. The difference between six times a number and four times the number is negative fourteen. Find the number.

11. The product of a number and four added to seven is equal to three. Find the number.

12. The total of six times a number and four times the number is ninety. Find the number.

13. Twenty-two is two less than six times a number. Find the number.

14. Negative fifteen is three more than twice a number. Find the number.

15. Three times a number is equal to four more than the product of the number and two. Find the number.

16. A number is equal to four more than five times the number. Find the number.

17. Seven more than four times a number is three more than two times the number. Find the number.

18. The difference between three times a number and four is five times the number. Find the number.

19. Eight less than five times a number is four more than eight times the number. Find the number.

20. The sum of a number and six is four less than six times the number. Find the number.

21. Twice the difference between a number and twenty-five is three times the number. Find the number.

22. Four times a number is three times the difference between thirty-five and the number. Find the number.

23. The sum of two numbers is twenty. Three times the smaller is equal to two times the larger. Find the two numbers.

24. The sum of two numbers is fifteen. One less than three times the smaller is equal to the larger. Find the two numbers.

25. The sum of two numbers is twenty-four. Four less than three times the smaller is twelve less than twice the larger. Find the two numbers.

26. The sum of two numbers is fourteen. The difference between two times the smaller and the larger is one. Find the two numbers.

27. The sum of two numbers is eighteen. The total of three times the smaller and twice the larger is forty-four. Find the two numbers.

28. The sum of two numbers is two. The difference between eight and twice the smaller number is two less than four times the larger. Find the two numbers.

2 Write an equation and solve.

29. The retail price of a compact disc player is $238. This price is $96 more than the cost of the compact disc player. Find the cost of the compact disc player.

30. As a result of depreciation, the value of a car now is $9600. This is three-fifths of its original value. Find the original value of the car.

31. The operating speed of a personal computer is 8 megahertz. This is one-fourth the speed of a newer model. Find the speed of the newer personal computer.

32. One measure of computer speed is mips (*millions of instructions per second*). One computer has a rating of 10 mips, which is two-thirds the speed of a second computer. Find the mips rating of the second computer.

33. A team scored 26 points during a football game. This team scored twice as many field goals (3 points each) as it did touchdowns (7 points each). Find the number of field goals scored by the team.

34. An office furniture store maintains an inventory of twice as many chairs as desks. There are currently 36 desks and chairs in stock. How many chairs are in stock?

35. A university employs a total of 600 teaching assistants and research assistants. There are three times as many teaching assistants as research assistants. Find the number of research assistants employed by the university.

36. A basketball team scored 105 points during one game. There were as many two-point baskets as free throws (1 point each), and the number of three-point baskets was five less than the number of free throws. Find the number of three-point baskets.

37. A 24-lb soil supplement contains iron, potassium, and a mulch. There is five times as much mulch as iron and twice as much potassium as iron. Find the amount of mulch in the soil supplement.

38. A real estate agent sold two homes and received commissions totaling $6000. The agent's commission on one home was one and one-half times the commission on the second home. Find the agent's commission for each home.

39. The purchase price of a new big-screen TV, including finance charges, was $3276. A down payment of $450 was made. The remainder was paid in 24 equal monthly installments. Find the monthly payment.

40. The purchase price of a new computer system, including finance charges, was $6350. A down payment of $350 was made. The remainder was paid in 24 equal monthly installments. Find the monthly payment.

41. The cost to replace a water pump in a sports car was $600. This included $375 for the water pump and $45 per hour for labor. How many hours of labor were required to replace the water pump?

42. The cost of electricity in a certain city is $.08 for each of the first 300 kWh (kilowatt-hours) and $.13 for each kWh over 300 kWh. Find the number of kilowatt-hours used by a family that receives a $51.95 electric bill.

43. An investor deposited $5000 into two accounts. Two times the smaller deposit is $1000 more than the larger deposit. Find the amount deposited into each account.

44. The length of a rectangular cement patio is 3 ft less than three times the width. The sum of the length and width of the patio is 21 ft. Find the length of the patio.

45. Greek architects considered a rectangle whose length was approximately 1.6 times the width to be most visually appealing. Find the length and width of a rectangle constructed in this manner if the sum of the length and width is 130 ft.

46. The sum of the measures of the angles of a triangle is always 180°. If one angle in a triangle is twice the smallest angle, and the third angle is three times the smallest angle, find the measure of each angle of the triangle.

47. A computer screen consists of tiny dots of light called pixels. In a certain graphics mode, there are 640 horizontal pixels. This is 40 more than three times the number of vertical pixels. Find the number of vertical pixels.

48. The velocity of a falling object equals the sum of its initial velocity and the product of 32 and the time it falls, in seconds. Given an initial velocity of 24 ft/s, how long has an object been falling when the velocity is 88 ft/s?

49. A wire 12 ft long is cut into two pieces. Each piece is bent into the shape of a square. The perimeter of the larger square is twice the perimeter of the smaller square. Find the perimeter of the larger square.

50. Five thousand dollars is divided between two scholarships. Three times the smaller scholarship is twice the larger. Find the amount of the larger scholarship.

51. A carpenter is building a wood door frame. The height of the frame is 1 ft less than three times the width. What is the width of the largest door frame that can be constructed from a board 19 ft long?

52. A 20-ft board is cut into two pieces. Twice the length of the shorter piece is 4 ft longer than the length of the longer piece. Find the length of the shorter piece.

SUPPLEMENTAL EXERCISES 3.4

Solve.

53. The amount of liquid in a container triples every minute. The container becomes completely filled at 3:40 P.M. What fractional part of the container is filled at 3:39 P.M.?

54. A cyclist traveling at a constant speed completes $\frac{3}{5}$ of a trip in $1\frac{1}{2}$ h. In how many additional hours will the cyclist complete the entire trip?

55. The charges for a long-distance telephone call are $1.21 for the first three minutes and $.42 for each additional minute or fraction of a minute. If the charges for a call were $6.25, how many minutes did the phone call last?

56. Four employees are paid at four consecutive levels on a wage scale. The difference between any two consecutive levels is $160 per month. The average of the four employee's monthly wages is $1440. What is the monthly wage of the highest-paid employee?

57. A coin bank contains nickels, dimes, and quarters. There are 14 nickels in the bank, $\frac{1}{6}$ of the coins are dimes, and $\frac{3}{5}$ of the coins are quarters. How many coins are in the coin bank?

58. During one day at an office, one-half of the amount of money in the petty cash drawer was used in the morning, and one-third of the remaining money was used in the afternoon, leaving $5 in the petty cash drawer at the end of the day. How much money was in the petty cash drawer at the start of the day?

Calculators and Computers

 Solving a First-Degree Equation

Solving equations is a learned skill and requires practice. To provide you with additional practice, the program SOLVE A FIRST-DEGREE EQUATION on the Math ACE Disk will allow you to practice solving the following three types of equations:

1) $ax + b = c$

2) $ax + b = cx + d$

3) Equations with parentheses

1) After you select the type of equation you want to practice, a problem will be displayed on the screen.

2) Using paper and pencil, solve the problem.

3) When you are ready, press the RETURN key, and the complete solution will be displayed.

4) Compare your solution with the displayed solution.

5) All answers are rounded to the nearest hundredth.

6) When you finish a problem, you may continue practicing the type of problem you have selected, return to the main menu and select a different type, or quit the program.

Something Extra

Density

Mass is a measure of the amount of matter in an object. Mass determines how hard or how easy it is to change an object's motion. It is easier to stop a baseball travelling at 50 mph than to stop a car travelling at the same speed because the car has a larger mass than the baseball.

The **density** of an object is its mass per unit of volume. A given volume of lead is heavier than an equal volume of aluminum because lead is denser than aluminum. The density of water is 1 g/cm^3, which means that 1 cubic centimeter of water has a mass of 1 gram. The density of aluminum is 2.7 g/cm^3. Therefore, a cubic centimeter of aluminum contains 2.7 times as much mass as a cubic centimeter of water.

The formula used to calculate density is $D = \dfrac{M}{V}$, where D is the density of an object, M is the mass, and V is the volume.

A sample of metal has a mass of 200 g and a volume of 20 cm^3. Find the density of the metal.

$$D = \frac{M}{V}$$

$$= \frac{200 \text{ g}}{20 \text{ cm}^3}$$

$$= \frac{10 \text{ g}}{\text{cm}^3}$$

$= 10$ g/cm^3 (This is read "10 grams per cubic centimeter.")

The density is 10 g/cm^3.

The density of copper is 8.90 g/cm^3. What is the mass of a piece of copper whose volume is 63.5 cm^3?

$$D = \frac{M}{V}$$

$$8.90 \text{ g/cm}^3 = \frac{M}{63.5 \text{ cm}^3}$$

$$(63.5 \text{ cm}^3)\left(\frac{8.90 \text{ g}}{\text{cm}^3}\right) = (63.5 \text{ cm}^3)\left(\frac{M}{63.5 \text{ cm}^3}\right)$$

$$565.15 \text{ g} = M$$

The mass is 565.15 g.

Solve.

1. A sample of metal has a mass of 150 g and a volume of 10 cm^3. Find the density of the metal.

2. The mass of a block of wood is 11 g. Its volume is 10 cm^3. Find the density of the block of wood.

3. Find the density of a sulfuric acid solution that has a mass of 65 g and a volume of 50 cm^3.

4. The density of aluminum is 2700 kg/m^3. Find the mass of a piece of aluminum that has a volume of 0.05 m^3.

5. The density of mercury is 13.6 g/cm^3. Find the mass of 50 cm^3 of mercury.

6. The density of silver is 10.5 g/cm^3. Find the volume of a piece of silver that has a mass of 210 g.

7. The density of sulfur is 2 g/cm^3. What is the mass of sulfur that fills a rectangular container that is 10 cm long, 5 cm wide, and 3 cm high?

Chapter Summary

Key Words

An *equation* expresses the equality of two mathematical expressions.

A *solution* of an equation is a number that, when substituted for the variable, results in a true equation.

To *solve* an equation means to find a solution of the equation. The goal is to rewrite the equation in the form *variable = constant*.

Essential Rules

Addition Property of Equations   The same number or variable term can be added to each side of an equation without changing the solution of the equation.

Multiplication Property of Equations   Each side of an equation can be multiplied by the same nonzero number without changing the solution of the equation.

Chapter Review

1.   Is 3 a solution of $5x - 2 = 4x + 5$?

2.   Solve: $x - 4 = 16$

3.   Solve: $8x = -56$

4.   Solve: $5x - 6 = 29$

5.   Solve: $5x + 3 = 10x - 17$

6.   Is $\frac{3}{4}$ a solution of $8x - 1 = 5 - 12x$?

7.   Solve:
$3(5x + 2) + 2 = 10x + 5[x - (3x - 1)]$

8.   Solve: $x - 2 = 4x - 47$

9.   Solve: $-4x - 2 = 10$

10.   Solve: $12y - 1 = 3y + 2$

11.   Is -5 a solution of $4x + 10 = 5 + 3x$?

12.   Solve: $6x + 3(2x - 1) = -27$

13.   Solve: $7 - [4 + 2(x - 3)] = 11(x + 2)$

14.   Solve: $-5x = 45$

15.   Solve: $3x + 7 + 4x = 42$

16.   Solve: $4.6 = 2.1 + x$

17.   Solve: $\frac{x}{7} = -7$

18.   Solve: $14 + 6x = 17$

19.   Solve: $8a - 3 = 5a - 6$

20.   Solve: $x + 3 = 24$

21.   Solve: $-4[x + 3(x - 5)] = 3(8x + 20)$

22.   Solve: $x + 5(3x - 20) = 10(x - 4)$

23.   Is 2 a solution of $x^2 + 4x + 1 = 3x + 7$?

24.   Solve: $\frac{3}{5}a = 12$

25. Solve: $a - \dfrac{1}{6} = \dfrac{2}{3}$

26. Solve: $32 = 9x - 4 - 3x$

27. Solve: $-6x + 16 = -2x$

28. Solve: $12 + x = 19$

29. Solve: $9b - 3b = 24$

30. Solve: $5 + 2(x + 1) = 13$

31. Solve: $14x + 7x + 8 = -10$

32. Solve: $-7x - 5 = 4x + 50$

33. Find the measure of the third angle of a triangle if the first angle is 20° and the second angle is 50°. Use the equation $A + B + C = 180°$, where A, B, and C are the measures of the angles of a triangle.

34. A lever is 12 ft long. At a distance of 2 ft from the fulcrum, a force of 120 lb is applied. How large a force must be applied to the other end of the lever so that the system will balance? Use the lever system equation $F_1x = F_2(d - x)$.

35. Find the time it takes for the velocity of a falling object to increase from 4 ft/s to 100 ft/s. Use the equation $v = v_0 + 32t$, where v is the final velocity of a falling object, v_0 is the initial velocity, and t is the time it takes for the object to fall.

36. Translate "four less than the product of five and a number is sixteen" into an equation and solve.

37. A piano wire is 35 in. long. A note can be produced by dividing this wire into two parts so that three times the length of the shorter piece is twice the length of the longer piece. Find the length of the shorter piece.

38. The sum of two numbers is twenty-one. Three times the smaller number is two less than twice the larger number. Find the two numbers.

39. Translate "the product of six and three more than a number is ten less than twice the number" into an equation and solve.

40. A lever is 8 ft long. A force of 25 lb is applied to one end of the lever and a force of 15 lb is applied to the other end. Find the location of the fulcrum when the system balances. Use the lever system equation $F_1x = F_2(d - x)$.

41. Find the length of a rectangle when the perimeter is 84 ft and the width is 18 ft. Use the equation $P = 2L + 2W$, where P is the perimeter of a rectangle, L is the length, and W is the width.

42. The sum of two numbers is thirty-six. The difference between the larger number and eight equals the total of four and three times the smaller number. Find the two numbers.

43. Find the discount on a mattress set if the sale price is $220.75 and the regular price is $300. Use the equation $S = R - D$, where S is the sale price and D is the discount.

44. Translate "the quotient of a number and nine is twenty-four less than the number" into an equation and solve.

45. The Empire State Building is 1472 ft tall. This is 514 ft less than twice the height of the Eiffel Tower. Find the height of the Eiffel Tower.

46. The pressure at a certain depth in the ocean can be approximated by the equation $P = 15 + \frac{1}{2}D$, where P is the pressure in pounds per square inch and D is the depth in feet. Use this equation to find the depth when the pressure is 55 pounds per square inch.

47. Translate "fifteen is equal to the total of two-thirds of a number and three" into an equation and solve.

48. A rectangular mirror has a width of 24 in. This is 6 in. less than one-half the length of the mirror. Find the length of the mirror.

49. A board 10 ft long is cut into two pieces. Four times the length of the shorter piece is 2 ft less than two times the longer piece. Find the length of the longer piece.

50. An optical engineer's consulting fee was $600. This included $80 for supplies and $65 for each hour of consultation. Find the number of hours of consultation.

C hapter Test

1. Solve: $\frac{3}{4}x = -9$

2. Solve: $6 - 5x = 5x + 11$

3. Solve: $3x - 5 = -14$

4. Is -2 a solution of $x^2 - 3x = 2x - 6$?

5. Solve: $x + \frac{1}{2} = \frac{5}{8}$

6. Solve: $5x - 2(4x - 3) = 6x + 9$

7. Is $\frac{2}{3}$ a solution of $6x - 7 = 3 - 9x$?

8. Solve: $7 - 4x = -13$

9. Solve: $x - 3 = -8$

10. Solve: $11 - 4x = 2x + 8$

11. Solve: $-\dfrac{3}{8}x = 5$

12. Solve: $3x - 2 = 5x + 8$

13. Solve: $6 - 2(5x - 8) = 3x - 4$

14. Solve: $6x - 3(2 - 3x) = 4(2x - 7)$

15. Solve: $3(2x - 5) = 8x - 9$

16. Solve: $9 - 3(2x - 5) = 12 + 5x$

17. A financial manager has determined that the cost per unit for a calculator is $15 and that the fixed costs per month are $2000. Find the number of calculators produced during a month in which the total cost was $5000. Use the equation $T = U \cdot N + F$, where T is the total cost, U is the cost per unit, N is the number of units produced, and F is the fixed cost.

18. Translate "the sum of six times a number and thirteen is five less than the product of three and the number" into an equation and solve.

19. A chemist mixes 100 g of water at 80°C with 50 g of water at 20°C. Use the equation $m_1 \cdot (T_1 - T) = m_2 \cdot (T - T_2)$ to find the final temperature of the water after mixing. In this equation, m_1 is the quantity of water at the hotter temperature, T_1 is the temperature of the hotter water, m_2 is the quantity of water at the cooler temperature, T_2 is the temperature of the cooler water, and T is the final temperature of the water after mixing.

20. Translate "the sum of five times a number and six equals the product of the number plus twelve and three" into an equation and solve.

21. Translate "the difference between three times a number and fifteen is twenty-seven" into an equation and solve.

22. The sum of two numbers is 18. The difference between four times the smaller number and seven is equal to the sum of two times the larger number and five. Find the two numbers.

23. A train travels between two cities in 26 h. This is 5 h more than the product of three and the time required for a plane to fly between the two cities. Find the number of hours required for the plane to fly between the two cities.

24. A business manager has determined that the cost per unit for a camera is $90 and that the fixed costs per month are $3500. Find the number of cameras produced during a month in which the total cost was $21,500. Use the equation $T = U \cdot N + F$, where T is the total cost, U is the cost per unit, N is the number of units produced, and F is the fixed cost.

25. A board 18 ft long is cut into two pieces. Two feet less than the product of five and the length of the shorter piece is equal to the difference between three times the length of the longer piece and eight. Find the length of each piece.

$$2 - 5x = 3 \times y$$

Cumulative Review

1. Subtract: $-6 - (-20) - 8$

2. Multiply: $(-2)(-6)(-4)$

3. Simplify: $-\frac{5}{6} - \left(-\frac{7}{16}\right)$

4. Simplify: $-2\frac{1}{3} \div 1\frac{1}{6}$

5. Simplify: $-4^2 \cdot \left(-\frac{3}{2}\right)^3$

6. Simplify: $25 - 3 \cdot \dfrac{(5-2)^2}{2^3+1} - (-2)$

7. Evaluate $3(a - c) - 2ab$, when $a = 2$, $b = 3$, and $c = -4$.

8. Simplify: $3x - 8x + (-12x)$

9. Simplify: $2a - (-3b) - 7a - 5b$

10. Simplify: $(16x)\left(\frac{1}{8}\right)$

11. Simplify: $-4(-9y)$

12. Simplify: $-2(-x^2 - 3x + 2)$

13. Simplify: $-2(x - 3) + 2(4 - x)$

14. Simplify: $-3[2x - 4(x - 3)] + 2$

15. Is -3 a solution of $x^2 + 6x + 9 = x + 3$?

16. Is $\frac{1}{2}$ a solution of $3 - 8x = 12x - 2$?

17. Solve: $x - 4 = -9$

18. Solve: $\frac{3}{5}x = -15$

19. Solve: $7x - 8 = -29$

20. Solve: $13 - 9x = -14$

21. Solve: $8x - 3(4x - 5) = -2x - 11$

22. Solve: $\frac{3}{8}x = -\frac{3}{4}$

23. Solve: $5x - 8 = 12x + 13$

24. Solve: $3(x - 7) = 5x - 12$

25. A business manager has determined that the cost per unit for a camera is $70 and that the fixed costs per month are $3500. Find the number of cameras produced during a month in which the total cost was $21,000. Use the equation $T = U \cdot N + F$, where T is the total cost, U is the cost per unit, N is the number of units produced, and F is the fixed cost.

26. A chemist mixes 300 g of water at 75°C with 100 g of water at 15°C. Use the equation $m_1 \cdot (T_1 - T) = m_2 \cdot (T - T_2)$ to find the final temperature of the water. In this equation, m_1 is the quantity of water at the hotter temperature, T_1 is the temperature of the hotter water, m_2 is the quantity of water at the cooler temperature, T_2 is the temperature of the cooler water, and T is the final temperature of the water after mixing.

27. Translate "the difference between twelve and the product of five and a number is negative eighteen" into an equation and solve.

28. Translate "the sum of eight times a number and twelve is equal to the product of four and the number" into an equation and solve.

29. The area of the cement foundation of a house is 2000 ft². This is 200 ft² more than three times the area of the garage. Find the area of the garage.

30. A board 16 ft long is cut into two pieces. Four feet more than the product of three and the length of the shorter piece is equal to three feet less than twice the length of the longer piece. Find the length of each piece.

4

Solving Equations: Applications

Objectives

- Write percents as fractions and as decimals
- Write fractions and decimals as percents
- The basic percent equation
- Application problems
- Markup problems
- Discount problems
- Investment problems
- Value mixture problems
- Percent mixture problems
- Uniform motion problems
- Perimeter problems
- Problems involving the angles of a triangle
- Consecutive integer problems
- Coin and stamp problems
- Age problems

Word Problems

Word problems have been challenging students of mathematics for a long time. Here are two types of problems you may have seen before:

A number added to $\frac{1}{7}$ of the number is 19. What is the number?

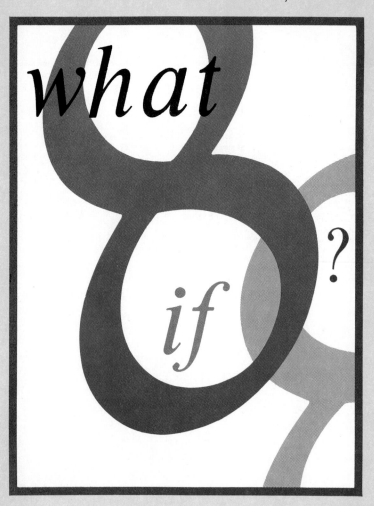

A dog is chasing a rabbit that has a head start of 150 feet. The dog jumps 9 feet every time the rabbit jumps 7 feet. In how many jumps will the dog catch up with the rabbit?

What is unusual about these problems is their age. The first one is about 4000 years old and occurred as Problem 1 in the Rhind Papyrus. The second problem is about 1500 years old and comes from a Latin algebra book written in 450 A.D.

These examples illustrate that word problems have been around for a long time. The long history of word problems also reflects the importance that each generation has placed on solving these problems. It is through word problems that the initial steps of applying mathematics are taken.

The answer to the first problem is $16\frac{5}{8}$.

The answer to the second problem is 75 jumps.

Introduction to Percent

1 Write percents as fractions and as decimals

"A population growth rate of 3%," "a manufacturer's discount of 25%," and "an 8% increase in pay" are typical examples of the many ways in which percent is used in applied problems. **Percent** means "parts of 100." Thus 27% means 27 parts of 100.

In applied problems involving a percent, it is usually necessary either to rewrite the percent as a fraction or a decimal or to rewrite a fraction or a decimal as a percent.

To write 27% as a fraction, remove the percent sign and multiply by $\frac{1}{100}$.

$$27\% = 27\left(\frac{1}{100}\right) = \frac{27}{100}$$

To write a percent as a decimal, remove the percent sign and multiply by 0.01.

To write 33% as a decimal, remove the percent sign and multiply by 0.01.

$$33\% = 33(0.01) = 0.33$$

Move the decimal point two places to the left. Then remove the percent sign.

Write 100% as a decimal.

$$100\% = 100(0.01) = 1$$

Example 1   Write 130% as a fraction and as a decimal.

Solution $130\% = 130\left(\frac{1}{100}\right) = \frac{130}{100} = 1\frac{3}{10}$ ▶ To write a percent as a fraction, remove the percent sign and multiply by $\frac{1}{100}$.

$130\% = 130(0.01) = 1.30$ ▶ To write a percent as a decimal, remove the percent sign and multiply by 0.01.

Problem 1   Write 125% as a fraction and as a decimal.

Solution   See page A12.

Example 2 Write $33\frac{1}{3}\%$ as a fraction.

Solution $33\frac{1}{3}\% = 33\frac{1}{3}\left(\frac{1}{100}\right) = \frac{100}{3}\left(\frac{1}{100}\right)$ ▶ Write the mixed number $33\frac{1}{3}$ as the improper fraction $\frac{100}{3}$.

$= \frac{1}{3}$

Problem 2 Write $16\frac{2}{3}\%$ as a fraction.

Solution See page A12.

Example 3 Write 0.25% as a decimal.

Solution $0.25\% = 0.25(0.01) = 0.0025$ ▶ Remove the percent sign and multiply by 0.01.

Problem 3 Write 6.08% as a decimal.

Solution See page A12.

2 # Write fractions and decimals as percents

A fraction or decimal can be written as a percent by multiplying by 100%. Recall that 100% = 1.

To write $\frac{5}{8}$ as a percent, multiply by 100%.

$$\frac{5}{8} = \frac{5}{8}(100\%) = \frac{500}{8}\% = 62.5\% \text{ or } 62\frac{1}{2}\%$$

To write 0.82 as a percent, multiply by 100%.

$$0.82 \quad = \quad 0.82(100\%) \quad = \quad 82\%$$

Move the decimal point two places to the right. Then write the percent sign.

Example 4 Write 0.027 as a percent.

Solution $0.027 = 0.027(100\%) = 2.7\%$ ▶ To write a decimal as a percent, multiply by 100%.

Problem 4 Write 0.043 as a percent.

Solution See page A12.

Example 5 Write 1.34 as a percent.

Solution $1.34 = 1.34(100\%) = 134\%$ ▶ Multiply the decimal by 100%.

Problem 5 Write 2.57 as a percent.

Solution See page A12.

Example 6 Write $\frac{5}{6}$ as a percent. Round to the nearest tenth of a percent.

Solution $\frac{5}{6} = \frac{5}{6}(100\%) = \frac{500}{6}\% \approx 83.3\%$ ▶ To write a fraction as a percent, multiply by 100%.

Problem 6 Write $\frac{5}{9}$ as a percent. Round to the nearest tenth of a percent.

Solution See page A12.

Example 7 Write $\frac{7}{16}$ as a percent. Write the remainder in fractional form.

Solution $\frac{7}{16} = \frac{7}{16}(100\%) = \frac{700}{16}\% = 43\frac{3}{4}\%$ ▶ Multiply the fraction by 100%.

Problem 7 Write $\frac{9}{16}$ as a percent. Write the remainder in fractional form.

Solution See page A12.

EXERCISES 4.1

1 Write as a fraction and as a decimal.

1. 75%	**2.** 40%	**3.** 50%	**4.** 10%
5. 64%	**6.** 88%	**7.** 175%	**8.** 160%
9. 19%	**10.** 87%	**11.** 5%	**12.** 2%
13. 450%	**14.** 380%	**15.** 8%	**16.** 4%

Write as a fraction.

17. $11\frac{1}{9}\%$	**18.** $4\frac{2}{7}\%$	**19.** $12\frac{1}{2}\%$	**20.** $37\frac{1}{2}\%$
21. $31\frac{1}{4}\%$	**22.** $66\frac{2}{3}\%$	**23.** $\frac{1}{4}\%$	**24.** $\frac{1}{2}\%$
25. $5\frac{3}{4}\%$	**26.** $68\frac{3}{4}\%$	**27.** $6\frac{1}{4}\%$	**28.** $83\frac{1}{3}\%$

Write as a decimal.

29. 7.3%	**30.** 9.1%	**31.** 15.8%	**32.** 16.7%

33. 0.3%   **34.** 0.9%   **35.** 9.15%   **36.** 121.2%

37. 18.23%   **38.** 62.14%   **39.** 0.15%   **40.** 0.27%

2   Write as a percent.

41. 0.15   **42.** 0.37   **43.** 0.05   **44.** 0.02

45. 0.175   **46.** 0.125   **47.** 1.15   **48.** 1.36

49. 0.62   **50.** 0.96   **51.** 3.165   **52.** 2.142

53. 0.008   **54.** 0.004   **55.** 0.065   **56.** 0.083

Write as a percent. Round to the nearest tenth of a percent.

57. $\dfrac{27}{50}$ **58.** $\dfrac{83}{100}$ **59.** $\dfrac{1}{3}$ **60.** $\dfrac{3}{8}$

61. $\dfrac{5}{11}$ **62.** $\dfrac{4}{9}$ **63.** $\dfrac{7}{8}$ **64.** $\dfrac{9}{20}$

65. $1\dfrac{2}{3}$ **66.** $2\dfrac{1}{2}$ **67.** $1\dfrac{2}{7}$ **68.** $1\dfrac{11}{12}$

Write as a percent. Write the remainder in fractional form.

69. $\dfrac{17}{50}$ **70.** $\dfrac{17}{25}$ **71.** $\dfrac{3}{8}$ **72.** $\dfrac{7}{16}$

73. $\dfrac{5}{14}$ **74.** $\dfrac{3}{19}$ **75.** $\dfrac{3}{16}$ **76.** $\dfrac{4}{7}$

77. $1\dfrac{1}{4}$ **78.** $2\dfrac{5}{8}$ **79.** $1\dfrac{5}{9}$ **80.** $1\dfrac{13}{16}$

SUPPLEMENTAL EXERCISES 4.1

Rewrite each fraction as a fraction with a denominator of 100. Then write the fraction as a percent.

81. $\dfrac{7}{50}$ **82.** $\dfrac{19}{25}$ **83.** $\dfrac{11}{20}$

84. $\dfrac{9}{10}$ **85.** $\dfrac{4}{5}$ **86.** $\dfrac{3}{4}$

Solve.

87. Let x represent the price of a car. If the sales tax is 6% of the price, express the total of the price of the car and the sales tax in terms of x.

88. Let x represent the price of a suit. If the suit is on sale at a discount rate of 30%, express the price of the suit after the discount in terms of x.

S E C T I O N **4.2**

The Percent Equation

1 The basic percent equation

The solution of a problem that involves a percent requires solving the basic percent equation shown below.

The Basic Percent Equation

Percent · Base = Amount
$$P \quad \cdot \quad B \quad = \quad A$$

To translate a problem involving a percent into an equation, remember that the word *of* translates to "multiply" and the word *is* translates to "=". The base usually follows the word *of*.

20% of what number is 30?

Given: $P = 20\% = 0.20$
         $A = 30$
Unknown: Base

$$PB = A$$
$$(0.20)B = 30$$
$$\frac{0.20B}{0.20} = \frac{30}{0.20}$$
$$B = 150$$

The number is 150.

Find 25% of 200.

Given: $P = 25\% = 0.25$
         $B = 200$
Unknown: Amount

$$PB = A$$
$$0.25(200) = A$$
$$50 = A$$

25% of 200 is 50.

In most cases, the percent is written as a decimal before solving the basic percent equation. However, some percents are more easily written as a fraction. For example,

$$33\frac{1}{3}\% = \frac{1}{3} \qquad\qquad 66\frac{2}{3}\% = \frac{2}{3}$$

$$16\frac{2}{3}\% = \frac{1}{6} \qquad\qquad 83\frac{1}{3}\% = \frac{5}{6}$$

Example 1   12 is $33\frac{1}{3}$% of what number?

Solution $12 = \frac{1}{3}B$ ▶ $33\frac{1}{3}\% = \frac{1}{3}$

$3 \cdot 12 = 3 \cdot \frac{1}{3}B$

$36 = B$

The number is 36.

Problem 1   27 is what percent of 60?

Solution   See page A12.

Example 2   20 is $83\frac{1}{3}$% of what number?

Solution $20 = \frac{5}{6}B$ ▶ $83\frac{1}{3}\% = \frac{5}{6}$

$\frac{6}{5} \cdot 20 = \frac{6}{5} \cdot \frac{5}{6}B$

$24 = B$

The number is 24.

Problem 2   What percent of 50 is 12?

Solution   See page A12.

2   Application problems

The key to solving a percent problem is identifying the percent, the base, and the amount. The base usually follows the word *of*.

Example 3   A fundraiser has raised $3460. Their goal is to raise $5000. What percent of the goal has been raised?

Strategy   To find the percent of the goal raised, solve the basic percent equation using $B = 5000$ and $A = 3460$. The percent is unknown.

Solution $PB = A$

$P(5000) = 3460$

$5000P = 3460$

$\dfrac{5000P}{5000} = \dfrac{3460}{5000}$

$P = 0.692$

69.2% of the goal has been raised.

Problem 3   A student correctly answered 72 of the 80 questions on an exam. What percent of the questions were answered correctly?

Solution   See page A12.

Example 4   In a survey of 1500 registered voters, 27% responded that they believed that the economy in this country was improving. How many people believed the country's economy was improving?

Strategy   To find the number of people, solve the basic percent equation using $B = 1500$ and $P = 27\% = 0.27$. The amount is unknown.

Solution
$$PB = A$$
$$0.27(1500) = A$$
$$405 = A$$

Of those surveyed, 405 people believed the country's economy was improving.

Problem 4   An engineer estimates that 32% of the gasoline used by a car is used efficiently. Using this estimate, determine how many gallons out of 15 gal of gasoline are used efficiently?

Solution   See page A12.

EXERCISES 4.2

1   Solve.

1. 12 is what percent of 50?
2. What percent of 125 is 50?
3. Find 18% of 40.
4. What is 25% of 60?
5. 12% of what is 48?
6. 45% of what is 9?
7. What is $33\frac{1}{3}\%$ of 27?
8. Find $16\frac{2}{3}\%$ of 30.
9. What percent of 12 is 3?
10. 10 is what percent of 15?
11. 60% of what is 3?
12. 75% of what is 6?
13. 12 is what percent of 6?
14. 20 is what percent of 16?
15. $5\frac{1}{4}\%$ of what is 21?
16. $37\frac{1}{2}\%$ of what is 15?
17. Find 15.4% of 50.
18. What is 18.5% of 46?
19. 1 is 0.5% of what?
20. 3 is 1.5% of what?

21. $\frac{3}{4}$% of what is 3?

22. $\frac{1}{2}$% of what is 3?

23. Find 125% of 16.

24. What is 250% of 12?

25. 16.4 is what percent of 20.4? Round to the nearest percent.

26. Find 18.3% of 625. Round to the nearest tenth.

2   Solve.

27. A large grocery store chain expects to make a profit of 1.8% of total sales. What is the expected profit for a day in which total sales are $1,000,000?

28. A university consists of three colleges: business, engineering, and fine arts. There are 2900 students in the business college, 1500 students in the engineering college, and 1000 students in the fine arts college. What percent of the total number of students in the university are in the fine arts college? Round to the nearest percent.

29. Approximately 21% of air is oxygen. Using this estimate, determine how many liters of oxygen are in a room containing 21,600 L of air.

30. A baseball playoff series was increased from 5 games to 7 games. What percent increase does this represent?

31. The cost of a 60-second television commercial during the 1967 Super Bowl was $80,000. In 1983, a 60-second commercial for the Super Bowl cost $800,000. What percent of the 1967 cost is the 1983 cost?

32. A ski vacation package regularly costs $850 for one week, but people who make an early reservation receive a 15% discount. Find the amount of the early-reservation discount.

33. A football stadium increased its 60,000-seat capacity by 15%. How many seats were added to the stadium?

34. To override a presidential veto, at least $66\frac{2}{3}$% of the Senate must vote to override the veto. There are 100 senators in the Senate. What is the minimum number of votes needed to override a veto?

35. To receive a B grade in a history course, a student must correctly answer 75 of the 90 questions on an exam. What percent of the questions must a student answer correctly to receive a B grade?

36. An airline knowingly overbooks flights by selling 18% more tickets than there are seats available. How many tickets would this airline sell for an airplane that has 150 seats?

37. A survey of 250 people was conducted to determine soft drink preferences. 100 people preferred cola flavor, 60 people preferred lemon/lime flavor, 50 people preferred orange flavor, and 40 people preferred cherry flavor. What percent of the people surveyed preferred a drink that was not cola flavored?

38. In a recent city election, 16,400 out of 80,000 registered voters voted. What percent of the people voted in the election?

39. There are approximately 8760 h in one year. It is estimated that children aged 2 to 5 watch 25 h of television each week. What percent of the year does a child aged 2 to 5 spend watching TV? Round to the nearest hundredth of a percent.

SUPPLEMENTAL EXERCISES 4.2

Solve.

40. Your bill for dinner at a restaurant is $62.80. You add a 15% tip to the bill. What is the total payment?

41. Because of a decrease in demand for 3-speed bicycles, a bicycle dealer reduced the order for these models from 16 per month to 6 per month. What percent decrease does this represent?

42. A clerical typist was earning a wage of $7.50 an hour before a 6% increase in pay. What is the typist's new hourly wage?

43. The total sales for December for a hobby shop were $25,000. For January, total sales showed an 11% decrease from December's sales. What were the total sales for January?

44. Your bill for dinner, including a 7.25% sales tax, was $62.74. You want to leave a 15% tip on the cost of the dinner before the sales tax. Find the amount of the tip to the nearest dollar.

45. The total cost for a dinner was $54.86. This included a 15% tip calculated on the cost of the dinner after a 6% sales tax. Find the cost of dinner.

46. A two-year labor agreement for a carpenter calls for an 8% increase in pay the first year and a 9% increase in pay the second year. If the carpenter is earning $22.50 per hour this year, what will be the carpenter's wage in two years?

47. A retailer decides to increase the original price of each item in the store by 10%. After the price increase, the retailer notices a significant drop in sales and so decides to reduce the current price of each item in the store by 10%. Are the prices back to the original prices? If not, are the prices lower or higher than the original prices?

SECTION 4.3

Markup and Discount

1 Markup problems

Cost is the price that a business pays for a product. **Selling price** is the price for which a business sells a product to a customer. The difference between selling price and cost is called **markup.** Markup is added to a retailer's cost to cover the expenses of operating a business. Markup is usually expressed as a percent of the retailer's cost. This percent is called the **markup rate.**

The basic markup equations used by a business are

$$\text{Selling price} = \text{Cost} + \text{Markup}$$
$$S = C + M$$

$$\text{Markup} = \text{Markup rate} \cdot \text{Cost}$$
$$M = r \cdot C$$

By substituting $r \cdot C$ for M in the first equation, selling price can also be written as:

$$S = C + M$$
$$S = C + (r \cdot C)$$
$$S = C + rC$$

Example 1 The manager of a clothing store buys a suit for $90 and sells the suit for $126. Find the markup rate.

Strategy Given: $C = \$90$
$S = \$126$
Unknown markup rate: r
Use the equation $S = C + rC$.

Solution

$$S = C + rC$$
$$126 = 90 + 90r \qquad \blacktriangleright \text{Substitute the values of } C \text{ and } S \text{ into the equation.}$$
$$126 - 90 = 90 - 90 + 90r \qquad \blacktriangleright \text{Subtract 90 from each side of the equation.}$$
$$36 = 90r$$
$$\frac{36}{90} = \frac{90r}{90} \qquad \blacktriangleright \text{Divide each side of the equation by 90.}$$
$$0.4 = r \qquad \blacktriangleright \text{The decimal must be changed to a percent.}$$

The markup rate is 40%.

Problem 1   The cost to the manager of a sporting goods store for a tennis racket is $60. The selling price of the racket is $90. Find the markup rate.

Solution   See page A13.

Example 2   The manager of a furniture store uses a markup rate of 45% on all items. The selling price of a chair is $232. Find the cost of the chair.

Strategy   Given: $r = 45\% = 0.45$
$S = \$232$
Unknown cost: C
Use the equation $S = C + rC$.

Solution $S = C + rC$
$232 = C + 0.45C$
$232 = 1.45C$ ▶ $C + 0.45C = 1C + 0.45C = (1 + 0.45)C$
$160 = C$

The cost of the chair is $160.

Problem 2   A hardware store employee uses a markup rate of 40% on all items. The selling price of a lawnmower is $133. Find the cost.

Solution   See page A13.

2   Discount problems

Discount is the amount by which a retailer reduces the regular price of a product for a promotional sale. Discount is usually expressed as a percent of the regular price. This percent is called the **discount rate.**

The basic discount equations used by a business are:

$$\begin{array}{ccccc} \text{Sale price} & = & \text{Regular price} & - & \text{Discount} \\ S & = & R & - & D \end{array}$$

$$\begin{array}{ccccc} \text{Discount} & = & \text{Discount rate} & \cdot & \text{Regular price} \\ D & = & r & \cdot & R \end{array}$$

By substituting $r \cdot R$ for D in the first equation, sale price can also be written as:

$$S = R - D$$
$$S = R - (r \cdot R)$$
$$S = R - rR$$

Example 3   In a garden supply store, the regular price of a 100-ft garden hose is $48. During an "after-summer sale," the hose is being sold for $36. Find the discount rate.

Strategy   Given: $R = \$48$
$S = \$36$
Unknown discount rate: r
Use the equation $S = R - rR$.

Solution

$$S = R - rR$$
$$36 = 48 - 48r$$ ▶ Substitute the values of R and S into the equation.
$$36 - 48 = 48 - 48 - 48r$$ ▶ Subtract 48 from each side of the equation.
$$-12 = -48r$$
$$\frac{-12}{-48} = \frac{-48r}{-48}$$ ▶ Divide each side of the equation by -48.
$$0.25 = r$$ ▶ The decimal must be changed to a percent.

The discount rate is 25%.

Problem 3   A case of motor oil that regularly sells for $29.80 is on sale for $22.35. What is the discount rate?

Solution   See page A13.

Example 4   The sale price for a chemical sprayer is $27.30. This price is 35% off the regular price. Find the regular price.

Strategy   Given: $S = \$27.30$
$r = 35\% = 0.35$
Unknown regular price: R
Use the equation $S = R - rR$.

Solution

$$S = R - rR$$
$$27.30 = R - 0.35R$$
$$27.30 = 0.65R$$ ▶ $R - 0.35R = 1R - 0.35R = (1 - 0.35)R$
$$42 = R$$

The regular price is $42.00.

Problem 4   The sale price for a telephone is $43.50. This price is 25% off the regular price. Find the regular price.

Solution   See page A13.

EXERCISES 4.3

1   Solve.

1. A computer software retailer uses a markup rate of 40%. Find the selling price of a computer game that costs the retailer $25.

2. A car dealer advertises a 5% markup over cost. Find the selling price of a car that costs the dealer $12,000.

3. The pro in a golf shop purchases a one-iron for $40. The selling price of the one-iron is $75. Find the markup rate.

4. A jeweler purchases a diamond ring for $350. The selling price of the ring is $700. Find the markup rate.

5. A leather jacket costs a clothing store manager $140. Find the selling price of the leather jacket if the markup rate is 40%.

6. The cost to a landscape architect for a 25-gal tree is $65. Find the selling price of the tree if the markup rate used by the architect is 30%.

7. A digitally recorded compact disc costs the manager of a music store $8.50. The selling price of the disc is $11.90. Find the markup rate.

8. A grocer purchases a can of fruit juice for $.68. The selling price of the fruit juice is $.85. Find the markup rate.

9. A tire dealer uses a markup rate of 55% on steel-belted tires. Find the selling price of a steel-belted tire that costs the dealer $45.

10. A cobbler uses a markup rate of 40% on rubber heels for shoes. Find the selling price of a rubber heel that costs the cobbler $3.50.

11. The manager of an electronics store adds $50 to the cost of every 17-inch television, regardless of the cost of the set. Find the markup rate on a television that costs the manager $215. Round to the nearest tenth of a percent.

12. A department store manager uses a markup rate of 40% on items that cost over $100 and a markup rate of 50% on items that cost less than $100. Find the selling price of a ceramic bowl that costs the department store $86.

2   Solve.

13. A tennis racket that regularly sells for $55 is on sale for 25% off the regular price. Find the sale price.

14. A fax machine that regularly sells for $975 is on sale for $33\frac{1}{3}$% off the regular price. Find the sale price.

15. A car stereo system that regularly sells for $425 is on sale for $318.75. Find the discount rate.

16. During a year-end clearance sale, a car dealer offers $2500 off the regular price of a car. Find the discount rate for a car that regularly sells for $12,500.

17. A college bookstore sells a used book at a discount of 30% off the regular price of a new book. Find the price of a used book that costs $35 when it is new.

18. An airline is offering a 35% discount on round-trip air fares. Find the sale price of a round-trip ticket that normally costs $385.

19. A gold bracelet that regularly sells for $1250 is on sale for $750. Find the discount rate.

20. A pair of skis that regularly sells for $325 is on sale for $250. Find the discount rate. Round to the nearest percent.

21. A supplier of electrical equipment offers a 5% discount for a purchase that is paid for within 30 days. A transformer regularly sells for $230. Find the discount price of a transformer that is paid for 10 days after the purchase.

22. A clothing wholesaler offers a discount of 10% per shirt when 10 to 20 shirts are purchased and a discount of 15% per shirt when 21 to 50 shirts are purchased. A shirt regularly sells for $17. Find the sale price per shirt when 35 shirts are purchased.

23. A service station offers a discount of $10 per tire when two tires are purchased and a discount of $25 per tire when four tires are purchased. Find the discount rate when a customer buys four tires that regularly sell for $95 each. Round to the nearest percent.

24. A department store offers a discount of $3 per dinner plate when five or fewer plates are purchased and a discount of $5 per plate when more than five plates are purchased. Find the discount rate when a customer buys three dinner plates that regularly sell for $18 each. Round to the nearest percent.

SUPPLEMENTAL EXERCISES 4.3

Solve.

25. A pair of shoes that now sells for $63 has been marked up 40%. Find the markup on the pair of shoes.

26. The sale price of a typewriter is 25% off the regular price. The discount is $70. Find the sale price.

27. A refrigerator selling for $770 has a markup of $220. Find the markup rate.

28. The sale price of a word processor is $765 after a discount of $135. Find the discount rate.

29. The manager of a camera store uses a markup rate of 30%. Find the cost of a camera selling for $299.

30. The sale price of a television was $180. Find the regular price if the sale price was computed by taking $\frac{1}{3}$ off the regular price followed by an additional 25% discount on the reduced price.

31. A customer buys four tires, three at the regular price and one for 20% off the regular price. The four tires cost $209. What was the regular price of a tire?

32. A lamp, originally priced at under $100, was on sale for 25% off the regular price. When the regular price, a whole number of dollars, was discounted, the discounted price was also a whole number of dollars. Find the largest possible number of dollars in the regular price of the lamp.

SECTION 4.4
Investment Problems

1   Investment problems

The annual simple interest that an investment earns is given by the equation $I = Pr$, where I is the simple interest, P is the principal, or the amount invested, and r is the simple interest rate.

The annual simple interest rate on a $4500 investment is 8%. Find the annual simple interest earned on the investment.

Given: $P = \$4500$ $I = Pr$
        $r = 8\% = 0.08$ $I = 4500(0.08)$
Unknown interest: I $I = 360$

The annual simple interest is $360.

Solve: An investor has a total of $10,000 to deposit into two simple interest accounts. On one account, the annual simple interest rate is 7%. On the second account, the annual simple interest rate is 8%. How much should be invested in each account so that the total annual interest earned is $785?

STRATEGY for solving a problem involving money deposited in two simple interest accounts

■ For each amount invested, write a numerical or variable expression for the principal, the interest rate, and the interest earned. The results can be recorded in a table.

The sum of the amounts invested is $10,000.

Amount invested at 7%: x
Amount invested at 8%: $10,000 - x$

	Principal, P	·	Interest rate, r	=	Interest earned, I
Amount at 7%	x	·	0.07	=	$0.07x$
Amount at 8%	$10,000 - x$	·	0.08	=	$0.08(10,000 - x)$

■ Determine how the amounts of interest earned on each amount are related. For example, the total interest earned by both accounts may be known, or it may be known that the interest earned on one account is equal to the interest earned by the other account.

The sum of the interest earned by the two investments equals the total annual interest earned ($785).

$$0.07x + 0.08(10,000 - x) = 785$$
$$0.07x + 800 - 0.08x = 785$$
$$-0.01x + 800 = 785$$
$$-0.01x = -15$$
$$x = 1500$$

$10,000 - x = 10,000 - 1500 = 8500$ ▶ Substitute the value of x into the variable expression for the amount invested at 8%.

The amount invested at 7% is $1500. The amount invested at 8% is $8500.

Example 1   An investment counselor invested 75% of a client's money into a 9% annual simple interest money market fund. The remainder was invested in 6% annual simple interest government securities. Find the amount invested in each if the total annual interest earned is $3300.

Strategy   ■ Amount invested: x
Amount invested at 9%: $0.75x$
Amount invested at 6%: $0.25x$

	Principal	·	Rate	=	Interest
Amount at 9%	$0.75x$	·	0.09	=	$0.09(0.75x)$
Amount at 6%	$0.25x$	·	0.06	=	$0.06(0.25x)$

■ The sum of the interest earned by the two investments equals the total annual interest earned ($3300).

Solution $0.09(0.75x) + 0.06(0.25x) = 3300$
$0.0675x + 0.015x = 3300$
$0.0825x = 3300$
$x = 40,000$ ▶ The amount invested is $40,000.

$0.75x = 0.75(40,000) = 30,000$ ▶ Find the amount invested at 9%.

$0.25x = 0.25(40,000) = 10,000$ ▶ Find the amount invested at 6%.

The amount invested at 9% is $30,000.
The amount invested at 6% is $10,000.

Problem 1   An investment of $2500 is made at an annual simple interest rate of 7%. How much additional money must be invested at 10% so that the total interest earned will be 9% of the total investment?

Solution   See pages A13 and A14.

EXERCISES 4.4

1   Solve.

1. An investment of $2500 is made at an annual simple interest rate of 7%. How much additional money must be invested at an annual simple interest rate of 11% so that the total interest earned is 9% of the total investment?

2. A total of $6000 is invested into two simple interest accounts. The annual simple interest rate on one account is 9%. The annual simple interest rate on the second account is 6%. How much should be invested in each account so that both accounts earn the same amount of interest?

3. An engineer invested a portion of $15,000 in a 7% annual simple interest account and the remainder in a 6.5% annual simple interest government bond. The two investments earn $1020 in interest annually. How much was invested in each account?

4. An investment club invested part of $20,000 in preferred stock that earns 8% annual simple interest and the remainder in a municipal bond that earns 7% annual simple interest. The amount of interest earned each year is $1520. How much was invested in each account?

5. A grocer deposited an amount of money into a high-yield mutual fund that earns 13% annual simple interest. A second deposit, $2500 more than the first, was placed in a certificate of deposit earning 7% annual simple interest. In one year, the total interest earned on both investments was $475. How much money was invested in the mutual fund?

6. A deposit was made into a 7% annual simple interest account. Another deposit, $1500 less than the first, was placed in a certificate of deposit earning 9% annual simple interest. The total interest earned on both investments for one year was $505. How much money was deposited in the certificate of deposit?

7. A corporation gave a university $300,000 to support product safety research. The university deposited some of the money in a 10% simple interest account and the remainder in an 8.5% annual simple interest account. How much was deposited in each account if the annual interest is $28,500?

8. A financial consultant invested part of a client's $30,000 in municipal bonds that earn 6.5% annual simple interest and the remainder of the money in 8.5% corporate bonds. How much is invested in each account if the total annual interest earned is $2190?

9. To provide for retirement income, an electrician purchases a $5000 bond that earns 7.5% annual simple interest. How much additional money must be invested in bonds that earn 8% annual simple interest so that the total annual interest earned from the two investments is $615?

10. The portfolio manager for an investment group invested $40,000 in a certificate of deposit that earns 7.25% annual simple interest. How much additional money must be invested in certificates that earn an annual simple interest rate of 8.5% so that the total annual interest earned from the two investments is $5025?

11. A charity deposited a total of $54,000 into two simple interest accounts. The annual simple interest rate on one account is 8%. The annual simple interest rate on the second account is 12%. How much was invested in each account if the total interest earned is 9% of the total investment?

12. A college sports foundation deposited a total of $24,000 into two simple interest accounts. The annual simple interest rate on one account is 7%. The annual simple interest rate on the second account is 11%. How much is invested in each account if the total annual interest earned is 10% of the total investment?

13. An investment banker invested 55% of the bank's available cash in an account that earns 8.25% annual simple interest. The remainder of the cash was placed in an account that earns 10% annual simple interest. The interest earned in one year was $58,743.75. Find the total amount invested.

14. A financial planner recommended that 40% of a client's cash account be invested in preferred stock earning 9% annual simple interest. The remainder of the client's cash was placed in Treasury bonds earning 7% annual simple interest. The total annual interest earned from the two investments was $2496. Find the total amount invested.

15. The manager of a mutual fund placed 30% of the fund's available cash in a 6% annual simple interest account, 25% in 8% corporate bonds, and the remainder in a money market fund earning 7.5% annual simple interest. The total annual interest earned from the investments was $35,875. Find the total amount invested.

16. The manager of a trust invested 30% of a client's cash in government bonds that earn 6.5% annual simple interest, 30% in utility stocks that earn 7% annual simple interest, and the remainder in an account that earns 8% annual simple interest. The total interest earned from the investments was $5437.50. Find the total amount invested.

SUPPLEMENTAL EXERCISES 4.4

Solve.

17. A sales representative invests in a stock paying 9% dividends. A research consultant invests $5000 more than the sales representative in bonds paying 8% annual simple interest. The research consultant's income from the investment is equal to the sales representative's. Find the amount of the research consultant's investment.

18.   A financial manager invested 20% of a client's money in bonds paying 9% annual simple interest, 35% in an 8% simple interest account, and the remainder in 9.5% corporate bonds. Find the amount invested in each if the total annual interest earned is $5325.

19.   A plant manager invested $3000 more in stocks than in bonds. The stocks paid 8% annual simple interest, and the bonds paid 9.5% annual simple interest. Both investments yielded the same income. Find the total annual interest received on both investments.

20.   A bank offers a customer a 2-year certificate of deposit (CD) which earns 8% compound annual interest. This means that the interest earned each year is added to the principal before the interest for the next year is calculated. Find the value in 2 years of a nurse's investment of $2500 in this CD.

21.   A bank offers a customer a 3-year certificate of deposit (CD) which earns 8.5% compound annual interest. This means that the interest earned each year is added to the principal before the interest for the next year is calculated. Find the value in 3 years of an accountant's investment of $3000 in this CD.

SECTION 4.5
Mixture Problems

1   Value mixture problems

A value mixture problem involves combining two ingredients that have different prices into a single blend. For example, a coffee merchant may blend two types of coffee into a single blend, or a candy manufacturer may combine two types of candy to sell as a "variety pack."

The solution of a value mixture problem is based on the equation $V = AC$, where V is the value of an ingredient, A is the amount of the ingredient, and C is the cost per unit of the ingredient.

Find the value of 5 oz of a gold alloy that costs $185 per ounce.

Given: $A = 5$ oz $V = AC$
       $C = \$185$ $V = 5\,(185)$
Unknown value: V $V = 925$

The value of the 5 oz of gold alloy is $925.

Solve: A coffee merchant wants to make 9 lb of a blend of coffee costing $6 per pound. The blend is made using a $7 grade and a $4 grade of coffee. How many pounds of each of these grades should be used?

STRATEGY *for solving a value mixture problem*

- For each ingredient in the mixture, write a numerical or variable expression for the amount of the ingredient used, the unit cost of the ingredient, and the value of the amount used. For the blend, write a numerical or variable expression for the amount, the unit cost of the blend, and the value of the amount. The results can be recorded in a table.

The sum of the amounts is 9 lb.

Amount of $7 coffee: x
Amount of $4 coffee: $9 - x$

	Amount, A	$\cdot$	Unit cost, C	$=$	Value, V
$7 grade	x	$\cdot$	$7	$=$	$7x$
$4 grade	$9 - x$	$\cdot$	$4	$=$	$4(9 - x)$
$6 blend	9	$\cdot$	$6	$=$	$6(9)$

- Determine how the values of the ingredients are related. Use the fact that the sum of the values of all ingredients is equal to the value of the blend.

The sum of the values of the $7 grade and the $4 grade is equal to the value of the $6 blend.

$$7x + 4(9 - x) = 6(9)$$
$$7x + 36 - 4x = 54$$
$$3x + 36 = 54$$
$$3x = 18$$
$$x = 6$$

$9 - x = 9 - 6 = 3$ ▶ Substitute the value of x into the variable expression for the amount of the $4 grade.

The merchant must use 6 lb of the $7 coffee and 3 lb of the $4 coffee.

Example 1   How many ounces of a silver alloy that costs $6 an ounce must be mixed with 10 oz of a silver alloy that costs $8 an ounce to make a mixture that costs $6.50 an ounce?

Strategy   ■ Ounces of $6 alloy: x

	Amount	Cost	Value
$6 alloy	x	$6	$6x$
$8 alloy	10	$8	8(10)
$6.50 mixture	$10 + x$	$6.50	$6.50(10 + x)$

■ The sum of the values before mixing equals the value after mixing.

Solution $6x + 8(10) = 6.50(10 + x)$
$6x + 80 = 65 + 6.5x$
$-0.5x + 80 = 65$
$-0.5x = -15$
$x = 30$

30 oz of the $6 silver alloy must be used.

Problem 1   A gardener has 20 lb of a lawn fertilizer that costs $.90 per pound. How many pounds of a fertilizer that costs $.75 per pound should be mixed with this 20 lb of lawn fertilizer to produce a mixture that costs $.85 per pound?

Solution   See page A14.

2 Percent mixture problems

The amount of a substance in a solution can be given as a percent of the total solution. For example, in a 5% saltwater solution, 5% of the total solution is salt. The remaining 95% is water.

The solution of a percent mixture problem is based on the equation $Q = Ar$, where Q is the quantity of a substance in the solution, r is the percent of concentration, and A is the amount of solution.

A 500-ml bottle contains a 3% solution of hydrogen peroxide. Find the amount of hydrogen peroxide in the solution.

Given: $A = 500$ $Q = Ar$
             $r = 3\% = 0.03$ $Q = 500(0.03)$
Unknown amount: Q $Q = 15$

The bottle contains 15 ml of hydrogen peroxide.

Solve: How many gallons of a 15% salt solution must be mixed with 4 gal of a 20% salt solution to make a 17% salt solution?

STRATEGY *for solving a percent mixture problem*

■ For each solution, write a numerical or variable expression for the amount of solution, the percent of concentration, and the quantity of the substance in the solution. The results can be recorded in a table.

The unknown quantity of 15% solution: x

	Amount of solution, A	·	Percent of concentration, r	=	Quantity of substance, Q
15% solution	x	·	0.15	=	$0.15x$
20% solution	4	·	0.20	=	$0.20(4)$
17% solution	$x + 4$	·	0.17	=	$0.17(x + 4)$

■ Determine how the quantities of the substance in each solution are related. Use the fact that the sum of the quantities of the substances being mixed is equal to the quantity of the substance after mixing.

The sum of the quantities of salt in the 15% solution and the 20% solution is equal to the quantity of salt in the 17% solution.

$$0.15x + 0.20(4) = 0.17(x + 4)$$
$$0.15x + 0.8 = 0.17x + 0.68$$
$$-0.02x + 0.8 = 0.68$$
$$-0.02x = -0.12$$
$$x = 6$$

6 gal of the 15% solution are required.

Example 2   A chemist wishes to make 3 L of a 7% acid solution by mixing a 9% acid solution and a 4% acid solution. How many liters of each solution should the chemist use?

Strategy   ▪ Liters of 9% solution: x
          Liters of 4% solution: $3 - x$

	Amount	Percent	Quantity
9%	x	0.09	$0.09x$
4%	$3 - x$	0.04	$0.04(3 - x)$
7%	3	0.07	$0.07(3)$

▪ The sum of the quantities before mixing is equal to the quantity after mixing.

Solution
$$0.09x + 0.04(3 - x) = 0.07(3)$$
$$0.09x + 0.12 - 0.04x = 0.21$$
$$0.05x + 0.12 = 0.21$$
$$0.05x = 0.09$$
$$x = 1.8$$

▶ 1.8 L of the 9% solution are needed.

$$3 - x = 3 - 1.8 = 1.2$$

▶ Find the amount of the 4% solution needed.

The chemist needs 1.8 L of the 9% solution and 1.2 L of the 4% solution.

Problem 2   A pharmacist dilutes 6 L of a 10% solution by adding water. How many liters of water are added to make an 8% solution?

Solution   See pages A14 and A15.

EXERCISES 4.5

1   Solve.   *Help*

1. A high-protein diet supplement that costs $6.75 per pound is mixed with a vitamin supplement that costs $3.25 per pound. How many pounds of each should be used to make 5 lb of a mixture that costs $4.65 per pound?

2. A 20-oz alloy of platinum that costs $220 per ounce is mixed with an alloy that costs $400 per ounce. How many ounces of the $400 alloy should be used to make an alloy that costs $300 per ounce?

3. Find the cost per pound of a coffee mixture made from 8 lb of coffee that costs $9.20 per pound and 12 lb of coffee that costs $5.50 per pound.

4. How many pounds of tea that cost $4.20 per pound must be mixed with 12 lb of tea that cost $2.25 per pound to make a mixture that costs $3.40 per pound?

5. A goldsmith combined an alloy that costs $4.30 per ounce with an alloy that costs $1.80 per ounce. How many ounces of each were used to make a mixture of 200 oz costing $2.50 per ounce?

6. How many liters of a solvent that costs $80 per liter must be mixed with 6 L of a solvent that costs $25 per liter to make a solvent that costs $36 per liter?

7. Find the cost per pound of a trail mix made from 40 lb of raisins that cost $4.40 per pound and 100 lb of granola that cost $2.30 per pound.

8. Find the cost per ounce of a mixture of 200 oz of a cologne that costs $5.50 per ounce and 500 oz of a cologne that costs $2.00 per ounce.

9. How many kilograms of hard candy that cost $7.50 per kilogram must be mixed with 24 kg of jelly beans that cost $3.25 per kilogram to make a mixture that sells for $4.50 per kilogram?

10. A grocery store offers a cheese and fruit sampler that combines cheddar cheese that costs $8 per kilogram with kiwis that cost $3 per kilogram. How many kilograms of each were used to make a 5-kg mixture that costs $4.50 per kilogram?

11. A ground meat mixture is formed by combining meat that costs $2.20 per pound with meat that costs $4.20 per pound. How many pounds of each were used to make a 50-lb mixture that costs $3.00 per pound?

12. A lumber company combined oak wood chips that cost $3.10 per pound with pine wood chips that cost $2.50 per pound. How many pounds of each were used to make an 80-lb mixture costing $2.65 per pound?

13. How many kilograms of soil supplement that costs $7.00 per kilogram must be mixed with 20 kg of aluminum nitrate that costs $3.50 per kilogram to make a fertilizer that costs $4.50 per kilogram?

14. A caterer made an ice cream punch by combining fruit juice that costs $2.25 per gallon with ice cream that costs $3.25 per gallon. How many gallons of each were used to make 100 gal of punch costing $2.50 per gallon?

15. The manager of a specialty food store combined almonds that cost $4.50 per pound with walnuts that cost $2.50 per pound. How many pounds of each were used to make a 100-lb mixture that cost $3.24 per pound?

16. Find the cost per gallon of a carbonated fruit drink made from 12 gal of fruit juice that costs $4.00 per gallon and 30 gal of carbonated water that costs $2.25 per gallon.

17. Find the cost per pound of a sugar-coated breakfast cereal made from 40 lb of sugar that costs $1.00 per pound and 120 lb of corn flakes that cost $.60 per pound.

18. Find the cost per ounce of a gold alloy made from 25 oz of pure gold that costs $482 per ounce and 40 oz of an alloy that costs $300 per ounce.

19. How many pounds of lima beans that cost $.90 per pound must be mixed with 16 lb of corn that costs $.50 per pound to make a mixture of vegetables that costs $.65 per pound?

20. How many liters of a blue dye that costs $1.60 per liter must be mixed with 18 L of anil that costs $2.50 per liter to make a mixture that costs $1.90 per liter?

2   Solve.

21. A chemist wants to make 50 ml of a 16% acid solution by mixing a 13% acid solution and an 18% acid solution. How many milliliters of each solution should the chemist use?

22. How many pounds of coffee that is 40% java beans must be mixed with 80 lb of coffee that is 30% java beans to make a coffee blend that is 32% java beans?

23. Thirty ounces of pure silver are added to 50 oz of a silver alloy that is 20% silver. What is the percent concentration of silver in the resulting alloy?

24. Two hundred liters of punch that contains 35% fruit juice is mixed with 300 L of a second punch. The resulting fruit punch is 20% fruit juice. Find the percent concentration of fruit juice in the second punch.

25. The manager of a garden shop mixes grass seed that is 60% rye grass with 70 lb of grass seed that is 80% rye grass to make a mixture that is 74% rye grass. How much of the 60% mixture is used?

26. Ten grams of sugar are added to a 40-g serving of a breakfast cereal that is 30% sugar. What is the percent concentration of sugar in the resulting mixture?

27. A dermatologist mixes 50 g of a cream that is 0.5% hydrocortisone with 150 g of a second hydrocortisone cream. The resulting mixture is 0.68% hydrocortisone. Find the percent concentration of hydrocortisone in the second hydrocortisone cream.

28. A carpet manufacturer blends two fibers, one 20% wool and the second 50% wool. How many pounds of each fiber should be woven together to produce 600 lb of a fabric that is 28% wool?

29. A hair dye is made by blending a 7% hydrogen peroxide solution and a 4% hydrogen peroxide solution. How many milliliters of each are used to make a 300-ml solution that is 5% hydrogen peroxide?

30. How many grams of pure salt must be added to 40 g of a 20% salt solution to make a solution that is 36% salt?

31. How many ounces of pure water must be added to 50 oz of a 15% saline solution to make a saline solution that is 10% salt?

32. A paint that contains 21% green dye is mixed with a paint that contains 15% green dye. How many gallons of each must be used to make 60 gal of paint that is 19% green dye?

33. A goldsmith mixes 8 oz of a 30% alloy with 12 oz of a 25% gold alloy. What is the percent concentration of the resulting alloy?

34. A physicist mixes 40 L of liquid oxygen with 50 L of liquid air that is 64% oxygen. What is the percent concentration of oxygen in the resulting mixture?

35. How many ounces of pure bran flakes must be added to 50 oz of cereal that is 40% bran flakes to produce a mixture that is 50% bran flakes?

36. How many milliliters of pure chocolate must be added to 150 ml of chocolate topping that is 50% chocolate to make a topping that is 75% chocolate?

37. A tea that is 20% jasmine is blended with a tea that is 15% jasmine. How many pounds of each tea are used to make 5 lb of tea that is 18% jasmine?

38. A clothing manufacturer has some pure silk thread and some thread that is 85% silk. How many kilograms of each must be woven together to make 75 kg of cloth that is 96% silk?

39. How many ounces of dried apricots must be added to 18 oz of a snack mix that contains 20% dried apricots to make a mixture that is 25% dried apricots?

40. A recipe for a rice dish calls for 12 oz of a rice mixture that is 20% wild rice and 8 oz of pure wild rice. What is the percent concentration of wild rice in the 20-oz mixture?

SUPPLEMENTAL EXERCISES 4.5

Solve.

41. Find the cost per ounce of a mixture of 30 oz of an alloy that costs $4.50 per ounce, 40 oz of an alloy that costs $3.50 per ounce, and 30 oz of an alloy that costs $3.00 per ounce.

42. A grocer combined walnuts that cost $1.60 per pound and cashews that cost $2.50 per pound with 20 lb of peanuts that cost $1.00 per pound. Find the amount of walnuts and the amount of cashews used to make the 50-lb mixture costing $1.72 per pound.

43. How many ounces of water evaporated from 50 oz of a 12% salt solution to produce a 15% salt solution?

44. A chemist mixed pure acid with water to make 10 L of a 30% acid solution. How much pure acid and how much water did the chemist use?

45. How many grams of pure water must be added to 50 g of pure acid to make a solution that is 40% acid?

46. A radiator contains 15 gal of a 20% antifreeze solution. How many gallons must be drained from the radiator and replaced by pure antifreeze so that the radiator will contain 15 gal of a 40% antifreeze solution?

SECTION 4.6

Uniform Motion Problems

1 Uniform motion problems

A train that travels constantly in a straight line at 50 mph is in *uniform motion*. **Uniform motion** means the speed of an object does not change.

The solution of a uniform motion problem is based on the equation $d = rt$, where d is the distance traveled, r is the rate of travel, and t is the time spent traveling.

Solve: A car leaves a town traveling at 35 mph. Two hours later, a second car leaves the same town, on the same road, traveling at 55 mph. In how many hours will the second car be passing the first car?

STRATEGY *for solving a uniform motion problem*

■ For each object, write a numerical or variable expression for the distance, rate, and time. The results can be recorded in a table.

The first car traveled 2 h longer than the second car.

Unknown time for the second car: t
Time for the first car: $t + 2$

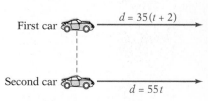

	Rate, r	·	Time, t	=	Distance, d
First car	35	·	$t + 2$	=	$35(t + 2)$
Second car	55	·	t	=	$55t$

■ Determine how the distances traveled by each object are related. For example, the total distance traveled by both objects may be known, or it may be known that the two objects traveled the same distance.

The two cars travel the same distance. The distance traveled by the first car equals the distance traveled by the second car.

$$35(t + 2) = 55t$$
$$35t + 70 = 55t$$
$$70 = 20t$$
$$3.5 = t$$

The second car will be passing the first car in 3.5 h.

Example 1   Two cars, one traveling 10 mph faster than the second car, start at the same time from the same point and travel in opposite directions. In 3 h, they are 288 mi apart. Find the rate of the second car.

Strategy   ▪ Rate of second car: r
Rate of first car: $r + 10$

	Rate	Time	Distance
First car	$r + 10$	3	$3(r + 10)$
Second car	r	3	$3r$

▪ The total distance traveled by the two cars is 288 mi. The distance traveled by the first car plus the distance traveled by the second car is 288 mi.

Solution $3(r + 10) + 3r = 288$
$3r + 30 + 3r = 288$
$6r + 30 = 288$
$6r = 258$
$r = 43$
The second car is traveling 43 mph.

Problem 1   Two trains, one traveling at twice the speed of the other, start at the same time from stations that are 306 mi apart and travel toward each other. In 3 h, the trains pass each other. Find the rate of each train.

Solution   See page A15.

Example 2   A bicycling club rides out into the country at a speed of 16 mph and returns over the same road at 12 mph. How far does the club ride out into the country if it travels a total of 7 h?

Strategy   ▪ Time spent riding out: t
Time spent riding back: $7 - t$

	Rate	Time	Distance
Out	16	t	$16t$
Back	12	$7 - t$	$12(7 - t)$

▪ The distance out equals the distance back.

Solution $16t = 12(7 - t)$
$16t = 84 - 12t$
$28t = 84$
$t = 3$ ▶ The time is 3 h. Find the distance.
The distance out $= 16t = 16(4) = 48$ mi.
The club rides 48 mi into the country.

Problem 2   On a survey mission, a pilot flew out to a parcel of land and back in 7 h. The rate out was 120 mph. The rate back was 90 mph. How far was the parcel of land?

Solution   See pages A15 and A16.

EXERCISES 4.6

1   Solve.

1. Two small planes start from the same point and fly in opposite directions. The first plane is flying 25 mph slower than the second plane. In 2 h, the planes are 470 mi apart. Find the rate of each plane.

2. Two cyclists start from the same point and ride in opposite directions. One cyclist rides twice as fast as the other. In 3 h, they are 81 mi apart. Find the rate of each cyclist.

3. A long-distance runner started on a course running at an average speed of 6 mph. One half-hour later, a second runner began the same course at an average speed of 7 mph. How long after the second runner started will the second runner overtake the first runner?

4. A motorboat leaves a harbor and travels at an average speed of 9 mph toward a small island. Two hours later a cabin cruiser leaves the same harbor and travels at an average speed of 18 mph toward the same island. In how many hours after the cabin cruiser leaves will the cabin cruiser be alongside the motorboat?

5. On a 130-mi trip, a car traveled at an average speed of 55 mph and then reduced its speed to 40 mph for the remainder of the trip. The trip took a total of 2.5 h. For how long did the car travel at 40 mph?

6. A motorboat leaves a harbor and travels at an average speed of 18 mph to an island. The average speed on the return trip was 12 mph. How far was the island from the harbor if the total trip took 5 h?

7. As part of flight training, a student pilot was required to fly to an airport and then return. The average speed on the way to the airport was 100 mph, and the average speed returning was 150 mph. Find the distance between the two airports if the total flying time was 5 h.

8. A family drove to a resort at an average speed of 25 mph and later returned over the same road at an average speed of 40 mph. Find the distance to the resort if the total driving time was 13 h.

9. Three campers left their campsite by canoe and paddled downstream at an average rate of 10 mph. They then turned around and paddled back upstream at an average rate of 5 mph to return to their campsite. How long did it take the campers to canoe downstream if the total trip took 1 h?

10. Running at an average rate of 8 m/s, a sprinter ran to the end of a track and then jogged back to the starting point at an average rate of 6 m/s. The sprinter took 35 s to run to the end of the track and jog back. Find the length of the track.

11. A jet plane traveling at 570 mph overtakes a propeller-driven plane that has had a 2-h head start. The propeller-driven plane is traveling at 190 mph. How far from the starting point does the jet overtake the propeller-driven plane?

12. A car traveling at 56 mph overtakes a cyclist who, riding at 14 mph, has had a 3-h head start. How far from the starting point does the car overtake the cyclist?

13. A 605-mi, 5-h plane trip was flown at two speeds. For the first part of the trip, the average speed was 115 mph. For the remainder of the trip, the average speed was 125 mph. How long did the plane fly at each speed?

14. On a 220-mi trip, a car traveled at an average speed of 50 mph and then reduced its average speed to 35 mph for the remainder of the trip. The trip took a total of 5 h. How long did the car travel at each speed?

15. After a sailboat had been on the water for 3 h, a change in wind direction reduced the average speed of the boat by 5 mph. The entire distance sailed was 51 mi. The total time spent sailing was 6 h. How far did the sailboat travel in the first 3 h?

16. A bus traveled on a level road for 2 h at an average speed that was 20 mph faster than its average speed on a winding road. The time spent on the winding road was 3 h. Find the average speed on the winding road if the total trip was 210 mi.

17. A passenger train leaves a train depot 1 h after a freight train leaves the same depot. The freight train is traveling 15 mph slower than the passenger train. Find the rate of each train if the passenger train overtakes the freight train in 3 h.

18. An executive drove from home at an average speed of 40 mph to an airport where a helicopter was waiting. The executive boarded the helicopter and flew to the corporate offices at an average speed of 60 mph. The entire distance was 150 mi. The entire trip took 3 h. Find the distance from the airport to the corporate offices.

19. A bus traveling at a rate of 60 mph overtakes a car traveling at a rate of 45 mph. If the car had a 1-h head start, how far from the starting point does the bus overtake the car?

20. A car and a cyclist start at 10 A.M. from the same point, headed in the same direction. The average speed of the car is 5 mph more than three times the average speed of the cyclist. In 1.5 h, the car is 46.5 mi ahead of the cyclist. Find the rate of the cyclist.

21. A cyclist and a jogger set out at 11 A.M. from the same point, headed in the same direction. The average speed of the cyclist is twice the average speed of the jogger. In 1 h, the cyclist is 7 mi ahead of the jogger. Find the rate of the cyclist.

22. A car and a bus set out at 2 P.M. from the same point, headed in the same direction. The average speed of the car is 30 mph slower than twice the average speed of the bus. In 2 h, the car is 30 mi ahead of the bus. Find the rate of the car.

23. Two joggers start at the same time from opposite ends of a 12-mi course. One jogger is running at a rate of 5 mph, and the other is running at a rate of 7 mph. How long after they begin will they meet?

24. Two cyclists start at the same time from opposite ends of a course that is 51 mi long. One cyclist is riding at a rate of 16 mph, and the second cyclist is riding at a rate of 18 mph. How long after they begin will they meet?

SUPPLEMENTAL EXERCISES 4.6

Solve.

25. At 10 A.M., two campers left their campsite by canoe and paddled downstream at an average speed of 12 mph. They then turned around and paddled back upstream at an average rate of 4 mph. The total trip took 1 h. At what time did the campers turn around downstream?

26. At 7 A.M., two joggers start from opposite ends of an 8-mi course. One jogger is running at a rate of 4 mph, and the other is running at a rate of 6 mph. At what time will the joggers meet?

27. A truck leaves a depot at 11 A.M. and travels at a speed of 45 mph. At noon, a van leaves the same place and travels the same route at a speed of 65 mph. At what time does the van overtake the truck?

28. A bicyclist rides for 2 h at a speed of 10 mph and then returns at a speed of 20 mph. Find the cyclist's average speed for the trip.

29. A car travels a 1-mile track at an average speed of 30 mph. At what average speed must the car travel the next mile so that the average speed for the 2 mi is 60 mph?

SECTION 4.7

Geometry Problems

1 Perimeter problems

The **perimeter** of a geometric figure is a measure of the distance around the figure. The equations for the perimeters of a rectangle and a triangle are shown below.

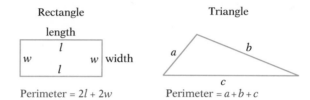

Rectangle

Perimeter = $2l + 2w$

Triangle

Perimeter = $a + b + c$

Solve: The perimeter of a rectangle is 32 ft. The length of the rectangle is 1 ft more than twice the width. Find the width of the rectangle.

STRATEGY *for solving a perimeter problem*

■ Let a variable represent the measure of one of the unknown sides of the figure. Express the measures of the remaining sides in terms of that variable.

Width: w
Length: $2w + 1$

■ Determine which perimeter equation to use.

Use the equation for the perimeter of a rectangle.

$$2l + 2w = P$$
$$2(2w + 1) + 2w = 32$$
$$4w + 2 + 2w = 32$$
$$6w + 2 = 32$$
$$6w = 30$$
$$w = 5$$

The width is 5 ft.

Example 1   The perimeter of a triangle is 22 ft. Two sides of the triangle are equal. The third side is 2 ft less than the length of one of the equal sides. Find the measures of the three sides of the triangle.

Strategy   ■ Each equal side: x
     The third side: $x - 2$
   ■ Use the equation for the perimeter of a triangle.

Solution
$$a + b + c = P$$
$$x + x + (x - 2) = 22$$
$$3x - 2 = 22$$
$$3x = 24$$
$$x = 8$$

$x - 2 = 8 - 2 = 6$ ▶ Substitute the value of x into the variable expression for the length of the third side.

Each of the equal sides measures 8 ft. The third side measures 6 ft.

Problem 1   The perimeter of a rectangle is 42 m. The length of the rectangle is 3 m more than the width. Find the measure of the width.

Solution   See page A16.

2   Problems involving the angles of a triangle

In a triangle, the sum of the measures of all the angles is 180°.

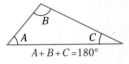

$A + B + C = 180°$

Two special types of triangles are shown at the right. A **right triangle** has one right angle (90°).

Right triangle

An **isosceles triangle** has two equal angles.

Equal angles

Isosceles triangle

Solve: In a right triangle, the measure of one angle is four times the measure of the smallest angle. Find the measure of the smallest angle.

STRATEGY for solving a problem involving the angles of a triangle

▬ Let a variable represent one of the unknown angles. Express the other angles in terms of that variable.

Measure of smallest angle: x
Measure of second angle: $4x$
Measure of right angle: $90°$

▬ Use the equation $\angle A + \angle B + \angle C = 180°$.

$$x + 4x + 90 = 180$$
$$5x + 90 = 180$$
$$5x = 90$$
$$x = 18$$

The measure of the smallest angle is 18°.

Example 2   In an isosceles triangle, the measure of one angle is 40° more than twice the measure of one of the equal angles. Find the measure of one of the equal angles.

Strategy   ▬ Measure of one of the equal angles: x
Measure of the second equal angle: x
Measure of the third angle: $2x + 40$
▬ Use the equation $\angle A + \angle B + \angle C = 180°$.

Solution $$x + x + (2x + 40) = 180$$
$$4x + 40 = 180$$
$$4x = 140$$
$$x = 35$$

The measure of one of the equal angles is 35°.

Problem 2   In a triangle, the measure of one angle is twice the measure of the second angle. The measure of the third angle is 8° less than the measure of the second angle. Find the measure of each angle.

Solution   See page A16.

EXERCISES 4.7

1 Solve.

1. The perimeter of a rectangle is 150 ft. The length of the rectangle is twice the width. Find the length and width of the rectangle.

2. The perimeter of a rectangle is 58 m. The width of the rectangle is 5 m less than the length. Find the length and width of the rectangle.

3. The width of a rectangle is 40% of the length. The perimeter of the rectangle is 364 ft. Find the length and width of the rectangle.

4. The width of a rectangle is 20% of the length. The perimeter is 240 cm. Find the length and width of the rectangle.

5. The perimeter of a triangle is 39 ft. One side of the triangle is 1 ft longer than the second side. The third side is 2 ft longer than the second side. Find the length of each side.

6. In an isosceles triangle, two sides are equal. The third side is 2 m less than one of the equal sides. The perimeter is 16 m. Find the length of each side.

7. The perimeter of a triangle is 30 ft. The length of the first side is 1 ft less than twice the length of the second side. The length of the third side is 1 ft more than twice the length of the second side. Find the length of each side.

8. The perimeter of a rectangle is 36 m. The length of the rectangle is 3 m less than twice the width. Find the length and width of the rectangle.

9. In an isosceles triangle, two sides are equal. The length of the third side is 25% of the length of one of the equal sides. Find the length of each side when the perimeter is 135 ft.

10. The perimeter of a triangle is 130 cm. One side is twice the second side. The third side is 30 cm more than the second side. Find the length of each side.

11. The perimeter of a rectangle is 52.6 m. The width of the rectangle is 0.7 m less than the length. Find the length and width of the rectangle.

12. In an isosceles triangle, two sides are equal. The length of one of the equal sides is 2.5 times the length of the third side. The perimeter is 10.8 m. Find the length of each side.

2 Solve.

13. In an isosceles triangle, one angle is three times the measure of one of the equal angles. Find the measure of each angle.

14. In an equiangular triangle, all three angles are equal. Find the measures of the equal angles.

15. One angle of a right triangle is 6° less than twice the measure of the smallest angle. Find the measure of each angle.

16. In an isosceles right triangle, two angles are equal, and the third angle is 90°. Find the measures of the equal angles.

17. In an isosceles triangle, one angle is 10° less than three times the measure of one of the equal angles. Find the measure of each angle.

18. In an isosceles triangle, one angle is 16° more than twice the measure of one of the equal angles. Find the measure of each angle.

19. In a triangle, one angle is 15° more than the measure of the second angle. The third angle is 30° more than the measure of the second angle. Find the measure of each angle.

20. In a triangle, one angle is twice the measure of the second angle. The third angle is three times the measure of the second angle. Find the measure of each angle.

21. One angle of a triangle is three times the measure of the third angle. The second angle is 5° more than the measure of the third angle. Find the measure of each angle.

22. One angle of a triangle is five times the measure of the second angle. The third angle is six times the measure of the first angle. Find the measure of each angle.

23. The first angle of a triangle is three times the measure of the second angle. The third angle is 47° more than the measure of the first angle. Find the measure of each angle.

24. The first angle of a triangle is twice the measure of the second angle. The third angle is 5° less than the measure of the first angle. Find the measure of each angle.

SUPPLEMENTAL EXERCISES 4.7

Solve.

25. A rectangle and an equilateral triangle have the same perimeter. The length of the rectangle is three times the width. Each side of the triangle is 8 cm. Find the length and width of the rectangle.

26. An equilateral triangle and a rectangle have the same perimeter. The length of the rectangle is 3 cm less than twice the width. Each side of the triangle is 10 cm. Find the length and width of the rectangle.

27. The length of a rectangle is 1 cm more than twice the width. If the length of the rectangle is decreased by 2 cm and the width is decreased by 1 cm, the perimeter is 20 cm. Find the length and width of the original rectangle.

28. The length of a rectangle is 2 cm more than twice the width. If the length of the rectangle is decreased by 3 cm and the width is decreased by 2 cm, the perimeter is 24 cm. Find the length and width of the original rectangle.

29. The length of a rectangle is $14x$. The perimeter is $50x$. Find the width of the rectangle in terms of the variable x.

30. The width of a rectangle is $8x$. The perimeter is $48x$. Find the length of the rectangle in terms of the variable x.

SECTION 4.8
Puzzle Problems

1   Consecutive integer problems

Recall that the integers are the numbers . . ., -3, -2, -1, 0, 1, 2, 3, 4, ...

An **even integer** is an integer that is divisible by 2. Examples of even integers are -8, 0, and 22.

An **odd integer** is an integer that is not divisible by 2. Examples of odd integers are -17, 1, and 39.

Consecutive integers are integers that follow one another in order. Examples of consecutive integers are shown at the right. (Assume the variable n represents an integer.)	11, 12, 13 -8, -7, -6 n, $n + 1$, $n + 2$
Examples of **consecutive even integers** are shown at the right. (Assume the variable n represents an even integer.)	24, 26, 28 -10, -8, -6 n, $n + 2$, $n + 4$
Examples of **consecutive odd integers** are shown at the right. (Assume the variable n represents an odd integer.)	19, 21, 23 -1, 1, 3 n, $n + 2$, $n + 4$

Solve: The sum of three consecutive odd integers is 51. Find the integers.

STRATEGY *for solving a consecutive integer problem*

■ Let a variable represent one of the integers. Express each of the other integers in terms of that variable. Remember that consecutive integers differ by 1. Consecutive even or consecutive odd integers differ by 2.

First odd integer: n
Second odd integer: $n + 2$
Third odd integer: $n + 4$

■ Determine the relationship among the integers.

The sum of the three odd integers is 51.

$$n + (n + 2) + (n + 4) = 51$$
$$3n + 6 = 51$$
$$3n = 45$$
$$n = 15$$

$n + 2 = 15 + 2 = 17$ ▶ Substitute the value of n into the variable
$n + 4 = 15 + 4 = 19$ expressions for the second and third integers.

The three consecutive odd integers are 15, 17, and 19.

Example 1   Find three consecutive even integers such that three times the second is six more than the sum of the first and third.

Strategy   ■ First even integer: n
Second even integer: $n + 2$
Third even integer: $n + 4$
■ Three times the second equals six more than the sum of the first and third.

Solution $3(n + 2) = n + (n + 4) + 6$
$3n + 6 = 2n + 10$
$n + 6 = 10$
$n = 4$ ▶ The first even integer is 4.

$n + 2 = 4 + 2 = 6$ ▶ Substitute the value of n into the variable
$n + 4 = 4 + 4 = 8$ expressions for the second and third integers.

The three consecutive even integers are 4, 6, and 8.

Problem 1   Find three consecutive integers whose sum is -12.

Solution   See page A17.

2 Coin and stamp problems

In solving problems that deal with coins or stamps of different values, it is necessary to represent the value of the coins or stamps in the same unit of money. Frequently, the unit of money is cents.

The value of 3 quarters in cents is $3 \cdot 25$, or 75 cents.
The value of 4 nickels in cents is $4 \cdot 5$, or 20 cents.
The value of d dimes in cents is $d \cdot 10$, or $10d$ cents.

Solve: A coin bank contains $1.20 in dimes and quarters. In all, there are nine coins in the bank. Find the number of quarters in the bank.

STRATEGY *for solving a coin problem*

■ For each denomination of coin, write a numerical or variable expression for the number of coins, the value of the coin in cents, and the total value of the coins in cents. The results can be recorded in a table.

The total number of coins is 9.

Number of quarters: x
Number of dimes: $9 - x$

Coin	Number of coins	·	Value of coin in cents	=	Total value in cents
Quarter	x	·	25	=	$25x$
Dime	$9 - x$	·	10	=	$10(9 - x)$

■ Determine the relationship between the total values of the different coins. Use the fact that the sum of the total values of each denomination of coin is equal to the total value of all the coins.

The sum of the total values of the different denominations of coins is equal to the total value of all the coins (120 cents).

$$25x + 10(9 - x) = 120$$
$$25x + 90 - 10x = 120$$
$$15x + 90 = 120$$
$$15x = 30$$
$$x = 2$$

There are 2 quarters in the bank.

Example 2 A collection of stamps consists of 3¢ stamps and 8¢ stamps. The number of 8¢ stamps is five more than three times the number of 3¢ stamps. The total value of all the stamps is $1.75. Find the number of each type of stamp in the collection.

Strategy ■ Number of 3¢ stamps: x
Number of 8¢ stamps: $3x + 5$

Stamp	Number	Value	Total Value
3¢	x	3	$3x$
8¢	$3x + 5$	8	$8(3x + 5)$

■ The sum of the total values of the different types of stamps equals the total value of all the stamps (175 cents).

Solution
$$3x + 8(3x + 5) = 175$$
$$3x + 24x + 40 = 175$$
$$27x + 40 = 175$$
$$27x = 135$$
$$x = 5$$

$3x + 5 = 3(5) + 5 = 15 + 5 = 20$ ▶ Find the number of 8¢ stamps.

There are five 3¢ stamps and twenty 8¢ stamps in the collection.

Problem 2 A coin bank contains nickels, dimes, and quarters. There are five times as many nickels as dimes and six more quarters than dimes. The total value of all the coins is $6.30. Find the number of each kind of coin in the bank.

Solution See page A17.

3 Age problems

In solving a problem involving the ages of different objects, it is necessary to represent each object's present age and its age either in the past or in the future. For example, if a coin is 20 years old, its age x years ago can be represented by the variable expression $20 - x$; its age x years from now can be represented by the variable expression $20 + x$.

Solve: A painting is 40 years old, and a sculpture is 20 years old. How many years ago was the painting three times as old as the sculpture was then?

STRATEGY *for solving an age problem*

■ Represent the ages in terms of numerical or variable expressions. To represent a past age, subtract from the present age. To represent a future age, add to the present age. The results can be recorded in a table.

The number of years ago: x

	Present age	Past age
Painting	40	$40 - x$
Sculpture	20	$20 - x$

■ Determine the relationship among the ages.

At a past age, the painting was three times as old as the sculpture was then.

$$40 - x = 3(20 - x)$$
$$40 - x = 60 - 3x$$
$$40 + 2x = 60$$
$$2x = 20$$
$$x = 10$$

Ten years ago, the painting was three times as old as the sculpture.

Example 3   A stamp collector has a 3¢ stamp that is 25 years older than a 5¢ stamp. In 16 years, the 3¢ stamp will be twice as old as the 5¢ stamp will be then. How old is the 5¢ stamp now?

Strategy   ■ Present age of the 5¢ stamp: x

	Present	Future
3¢ stamp	$x + 25$	$x + 41$
5¢ stamp	x	$x + 16$

■ At a future age, the 3¢ stamp will be twice as old as the 5¢ stamp.

Solution $2(x + 16) = x + 41$
$2x + 32 = x + 41$
$x + 32 = 41$
$x = 9$

The 5¢ stamp is 9 years old.

Problem 3 A half-dollar is now 35 years old. A dime is 25 years old. How many years ago was the half-dollar twice as old as the dime?

Solution See page A18.

EXERCISES 4.8

1 Solve.

1. The sum of three consecutive integers is 54. Find the integers.

2. The sum of three consecutive integers is 75. Find the integers.

3. The sum of three consecutive even integers is 84. Find the integers.

4. The sum of three consecutive even integers is 48. Find the integers.

5. The sum of three consecutive odd integers is 57. Find the integers.

6. The sum of three consecutive odd integers is 81. Find the integers.

7. Find two consecutive even integers such that five times the first is equal to four times the second.

8. Find two consecutive even integers such that six times the first equals three times the second.

9. Nine times the first of two consecutive odd integers equals seven times the second. Find the integers.

10. Five times the first of two consecutive odd integers is three times the second. Find the integers.

11. Find three consecutive integers whose sum is negative twenty-four.

12. Find three consecutive even integers whose sum is negative twelve.

13. Three times the smallest of three consecutive even integers is two more than twice the largest. Find the integers.

14. Twice the smallest of three consecutive odd integers is five more than the largest. Find the integers.

15. Find three consecutive even integers such that three times the middle integer is six more than the sum of the first and third.

16. Find three consecutive odd integers such that four times the middle integer is equal to two less than the sum of the first and third.

2   Solve.

17. A bank contains 27 coins in dimes and quarters. The coins have a total value of $4.95. Find the number of dimes and quarters in the bank.

18. A coin purse contains 18 coins in nickels and dimes. The coins have a total value of $1.15. Find the number of nickels and dimes in the coin purse.

19. A business executive bought 40 stamps for $9.60. The purchase included 25¢ stamps and 20¢ stamps. How many of each type of stamp were bought?

20. A postal clerk sold some 15¢ stamps and some 25¢ stamps. Altogether, 15 stamps were sold for a total cost of $3.15. How many of each type of stamp were sold?

21. A drawer contains 15¢ stamps and 18¢ stamps. The number of 15¢ stamps is four less than three times the number of 18¢ stamps. The total value of all the stamps is $1.29. How many 15¢ stamps are in the drawer?

22. The total value of the dimes and quarters in a bank is $6.05. There are six more quarters than dimes. Find the number of each type of coin in the bank.

23. A child's piggy bank contains 44 coins in quarters and dimes. The coins have a total value of $8.60. Find the number of quarters in the bank.

24. A coin bank contains nickels and dimes. The number of dimes is 10 less than twice the number of nickels. The total value of all the coins is $2.75. Find the number of each type of coin in the bank.

25. A total of 26 bills are in a cash box. Some of the bills are one-dollar bills, and the rest are five-dollar bills. The total amount of cash in the box is $50. Find the number of each type of bill in the cash box.

26. A bank teller cashed a check for $200 using twenty-dollar bills and ten-dollar bills. In all, twelve bills were handed to the customer. Find the number of twenty-dollar bills and the number of ten-dollar bills.

27. A coin bank contains pennies, nickels, and dimes. There are six times as many nickels as pennies and four times as many dimes as pennies. The total amount of money in the bank is $7.81. Find the number of pennies in the bank.

28. A coin bank contains pennies, nickels, and quarters. There are seven times as many nickels as pennies and three times as many quarters as pennies. The total amount of money in the bank is $5.55. Find the number of pennies in the bank.

29. A collection of stamps consists of 22¢ stamps and 40¢ stamps. The number of 22¢ stamps is three more than four times the number of 40¢ stamps. The total value of the stamps is $8.34. Find the number of 22¢ stamps in the collection.

30. A collection of stamps consists of 2¢ stamps, 8¢ stamps, and 14¢ stamps. The number of 2¢ stamps is five more than twice the number of 8¢ stamps. The number of 14¢ stamps is three times the number of 8¢ stamps. The total value of the stamps is $2.26. Find the number of each type of stamp in the collection.

31. A collection of stamps consists of 3¢ stamps, 7¢ stamps, and 12¢ stamps. The number of 3¢ stamps is five less than the number of 7¢ stamps. The number of 12¢ stamps is one-half the number of 7¢ stamps. The total value of all the stamps is $2.73. Find the number of each type of stamp in the collection.

32. A collection of stamps consists of 2¢ stamps, 5¢ stamps, and 7¢ stamps. There are nine more 2¢ stamps than 5¢ stamps and twice as many 7¢ stamps as 5¢ stamps. The total value of the stamps is $1.44. Find the number of each type of stamp in the collection.

33. A collection of stamps consists of 6¢ stamps, 8¢ stamps, and 15¢ stamps. The number of 6¢ stamps is three times the number of 8¢ stamps. There are six more 15¢ stamps than there are 6¢ stamps. The total value of all the stamps is $5.16. Find the number of each type of stamp.

34. A child's piggy bank contains nickels, dimes, and quarters. There are twice as many nickels as dimes and four more quarters than nickels. The total value of all the coins is $9.40. Find the number of each type of coin.

3   Solve.

35. An Oriental rug is 52 years old and a Persian rug is 16 years old. How many years ago was the Oriental rug four times as old as the Persian rug was then?

36. A log cabin quilt is 24 years old and a friendship quilt is 6 years old. In how many years will the log cabin quilt be three times as old as the friendship quilt will be then?

37. A wool tapestry is 32 years older than a linen tapestry. Twenty years ago, the wool tapestry was twice as old as the linen tapestry was then. Find the present age of each.

38. A silver coin is 28 years older than a bronze coin. In 6 years, the silver coin will be twice as old as the bronze coin will be then. Find the present age of each coin.

39. A pitcher is 30 years old, and a vase is 22 years old. How many years ago was the pitcher twice as old as the vase was then?

40. A marble bust is 25 years old, and a terra-cotta bust is 85 years old. In how many years will the terra-cotta bust be three times as old as the marble bust will be then?

41. A pewter bowl is 8 years old, and a silver bowl is 22 years old. In how many years will the silver bowl be twice the age the pewter bowl will be then?

42. A kerosene lamp is 95 years old, and an electric lamp is 55 years old. How many years ago was the kerosene lamp twice the age the electric lamp was then?

43. A mosaic is 74 years older than an engraving. Thirty years ago, the mosaic was three times as old as the engraving was then. Find the present age of each.

44. A limestone statue is 56 years older than a marble statue. In 12 years, the limestone statue will be three times as old as the marble statue will be then. Find the present age of each statue.

45. A medallion is 60 years old, and a medal is 8 years old. In how many years will the medallion be twice the age the medal will be then?

46. A pen drawing is 15 years older than a pencil drawing. Ten years ago, the pen drawing was four times the age the pencil drawing was then. What is the present age of the pen drawing?

47. The sum of the ages of a china plate and a glass plate is 16 years. Four years ago, the china plate was three times the age the glass plate was then. Find the present age of each plate.

48. The sum of the ages of a wood plaque and a bronze plaque is 20 years. Four years ago, the bronze plaque was one-half the age the wood plaque was then. Find the present age of each plaque.

49. The sum of the ages of two ships is 12 years. Two years ago, the age of the older ship was three times the age the newer ship was then. Find the present age of each ship.

50. The sum of the ages of two children is 16 years. Four years ago, the age of the older child was three times the age the younger child was then. Find the present age of each child.

51. The sum of the ages of a 5¢ stamp and an 8¢ stamp is 24 years. Six years from now, the age of the 5¢ stamp will equal the age of the 8¢ stamp four years ago. Find the present age of each stamp.

52. The sum of the ages of a dime and a quarter is 32 years. Four years from now, the age of the dime will equal the age of the quarter four years ago. Find the present age of each coin.

SUPPLEMENTAL EXERCISES 4.8

Solve.

53. Find four consecutive even integers whose sum is -36.

54. Find four consecutive odd integers whose sum is -48.

55. The three children in a family were born at three-year intervals. Five years ago, the sum of their ages was 18. Find the present ages of the three children.

56. The four children in a family were born at two-year intervals. Three years ago, the sum of their ages was 20. Find the present ages of the children.

57. The age of an oil painting next year will be twice the age of a watercolor last year. The sum of their present ages is 60. Find the present age of each.

58. The age of a gold coin next year will be three times the age of a silver coin last year. The sum of their present ages is 100. Find the present age of each.

59. A coin bank contains only dimes and quarters. The number of quarters in the bank is two less than twice the number of dimes. There are 34 coins in the bank. How much money is in the bank?

60. A postal clerk sold twenty stamps to a customer. The number of 13¢ stamps purchased was two more than twice the number of 22¢ stamps purchased. If the customer bought only 13¢ stamps and 22¢ stamps, how much money did the clerk collect from the customer?

61. Find three consecutive odd integers such that the sum of the first and third is twice the second.

62. Find four consecutive integers such that the sum of the first and fourth equals the sum of the second and third.

Calculators and Computers

 The EEX Key on a Calculator

Many application problems require the use of very large or very small numbers. The EEX key, or on some calculators the EXP key, is used for entering these numbers. For example, the speed of light is approximately 29,800,000,000 cm/s. This number cannot be directly entered on a calculator. The EEX key (which means exponent) is used.

To understand this key, consider the following table:

$$5489 = 548.9 \times 10^1$$
$$5489 = 54.89 \times 10^2$$
$$5489 = 5.489 \times 10^3$$

The number 5489 can be represented in various forms. Note that each time the decimal point is moved to the *left*, it is necessary to *multiply* by a power of ten so that the value of the number is not changed.

Now study the following table:

$$0.004638 = 0.04638 \div 10^1$$
$$0.004638 = 0.4638 \div 10^2$$
$$0.004638 = 4.638 \div 10^3$$

The number 0.004638 can be represented in various forms. Note that each time the decimal point is moved to the *right*, it is necessary to *divide* by a power of ten. It is customary to express dividing by a power of ten by using multiplication and a negative exponent. Thus the above table could be written as follows:

$$0.004638 = 0.04638 \times 10^{-1}$$
$$0.004638 = 0.4638 \times 10^{-2}$$
$$0.004638 = 4.638 \times 10^{-3}$$

To enter the speed of light on a calculator, rewrite the number by moving the decimal point 10 places to the left and then multiplying by 10^{10}.

Now enter the number.   2.98 EEX 10

The EEX key is used to enter the exponent on 10.

The wavelength of an x ray is approximately 0.00000000537 cm. To enter this number on your calculator, rewrite the number by moving the decimal point 9 places to the right and then dividing by 10^9 (or multiplying by 10^{-9}).

$$0.00000000537 = 5.37 \div 10^9 = 5.37 \times 10^{-9}$$

Now enter the number.　　　5.37 $\boxed{\text{EEX}}$ 9 $\boxed{+/-}$

The $\boxed{+/-}$ key is used to change the sign of the exponent and thus make it negative.

S omething Extra

Acceleration

Uniform motion was the topic of Section 4.6. Uniform motion means that the speed or direction of an object does not change.

A more complete description of the motion of an object is given by its velocity. **Velocity** describes both the speed of an object and its direction of motion. For example, the velocity of an airplane might be given as 150 mph, eastward.

In uniform motion, both the speed and the direction of a moving object remain unchanged. Uniform motion is motion at constant velocity.

Accelerated motion is motion with changing velocity. Accelerated motion involves a change in direction, a change in speed (either increasing or decreasing), or a change in both speed and direction. If we discuss only acceleration that involves increasing or decreasing speed, then acceleration is the rate at which the speed of a moving object is changing.

Suppose that the speed of a car increases steadily during the first 5 s of operation from zero to 20 m/s (meters per second). The change in its speed is $20 - 0 = 20$ m/s. Its acceleration is $20 \div 5 = 4$ m/s² (meters per second per second). During each of the five seconds, the car increased its speed by 4 m/s.

The speed of a car is steadily decreasing when the car is coming to a stop. Its acceleration is calculated in the same way, but the acceleration is negative.

The acceleration of a moving object can be calculated by dividing the change in the speed of the object during a given time by the time:

$$a = \frac{v - v_0}{t},$$

where a is the acceleration, v_0 is the object's velocity at the start, and v is the final velocity acquired by the object after being uniformly accelerated for time t.

Solve.

1. The speed of a car increases steadily from 20 m/s to 30 m/s in 5 s. What is the acceleration of the car?

2. During the first 5 s of take-off, the speed of an airplane increases steadily from zero to 15 m/s. Find the acceleration of the airplane.

3. An airplane's speed decreases steadily from 200 m/s to 150 m/s in 10 s. Find the acceleration of the airplane.

4. The speed of a car decreases steadily from 40 m/s to 20 m/s in 10 s. What is the acceleration of the car?

5. A car traveling at 18 m/s accelerates at the rate of 0.5 m/s^2. Find the speed of the car after 10 s.

6. An airplane traveling at 50 m/s accelerates at the rate of 2 m/s^2. Find the airplane's speed after 5 s.

7. An airplane flying at 45 m/s accelerates at the rate of 1 m/s^2. What is the speed of the airplane after 10 s?

8. A car traveling at 20 m/s accelerates at the rate of 0.25 m/s^2. What is the speed of the car after 8 s?

Chapter Summary

Key Words

Percent means parts of 100.

Cost is the price that a business pays for a product.

Selling price is the price for which a business sells a product to a customer.

Markup is the difference between selling price and cost.

Discount is the amount by which a retailer reduces the regular price of a product.

An object in *uniform motion* moves at a constant speed and in a straight line.

The *perimeter* of a geometric figure is a measure of the distance around the figure.

A *right angle* is an angle that measures 90°.

A *right triangle* has one right angle.

An *isosceles triangle* has two equal angles and two equal sides.

Consecutive integers are integers that follow one another in order.

Essential Rules

Basic Percent Equation

Amount = Percent · Base
$$A = PB$$

Basic Markup Equations

Selling price = Cost + Markup
$$S = C + M$$

Markup = Markup rate · Cost
$$M = rC$$

Basic Discount Equations

Sale price = Regular price − Discount
$$S = R - D$$

Discount = Discount rate · Regular price
$$D = rR$$

Annual Simple Interest Equation

Simple interest = Principal · Simple interest rate
$$I = Pr$$

Value Mixture Equation

Value = Amount · Unit cost
$$V = AC$$

Percent Mixture Equation

Quantity = Amount · Percent of concentration
$$Q = Ar$$

Uniform Motion Equation

Distance = Rate · Time
$$d = rt$$

Perimeter of a rectangle

$$P = 2l + 2w$$

Perimeter of a triangle

$$P = a + b + c$$

Triangle Equation

$$\angle A + \angle B + \angle C = 180°$$

Chapter Review

1. Write 55% as a fraction.

2. Write $2\frac{8}{9}$ as a percent. Write the remainder in fractional form.

3. 7 is 28% of what number?

4. Write $79\frac{1}{2}$% as a fraction.

5. 0.5 is what percent of 3?

6. What is $66\frac{2}{3}$% of 24?

7. Write 240% as a decimal.

8. Write 1.59 as a percent.

9. 8 is what percent of 200?

10. Write 342% as a decimal.

11. Write $\frac{5}{8}$ as a percent.

12. What is $\frac{1}{2}$% of 3000?

13. Write 7% as a decimal.

14. Write $2\frac{7}{9}$ as a percent. Round to the nearest tenth of a percent.

15. 18 is 72% of what number?

16. Write 6.2% as a decimal.

17. Write $\frac{3}{13}$ as a percent. Round to the nearest tenth of a percent.

18. 60 is 48% of what number?

19. Write 0.672 as a percent.

20. Write $\frac{16}{23}$ as a percent. Write the remainder in fractional form.

21. 27 is what percent of 40?

22. Write 0.002 as a percent.

23. Write $\frac{19}{35}$ as a percent. Write the remainder in fractional form.

24. What is 81% of 500?

25. A ceiling fan that regularly sells for $60 is on sale for $40. Find the discount rate.

26. A total of $15,000 is deposited into two simple interest accounts. The annual simple interest rate on one account is 6%. The annual simple interest rate on the second account is 7%. How much should be invested in each account so that the total interest earned is $970?

27. Find the cost per pound of a meatloaf mixture made from 3 lb of ground beef costing $1.99 per pound and 1 lb of ground turkey costing $1.39 per pound.

28. A motorcyclist and a bicyclist set out at 8 A.M. from the same point, headed in the same direction. The speed of the motorcyclist is three times the speed of the bicyclist. In 2 h, the motorcyclist is 60 mi ahead of the bicyclist. Find the rate of the motorcyclist.

29. In an isosceles triangle, the measure of one angle is 25° less than half the measure of one of the equal angles. Find the measure of each angle.

30. In 1987, the value of a Yogi Berra rookie card was $75. By 1990, the value of the card had increased by $300. What percent increase does this represent?

31. The manager of a sporting goods store buys indoor/outdoor basketballs for $8.50 and sells them for $14.45. Find the markup rate.

32. A dairy owner mixes 5 gal of cream that is 30% butterfat with 8 gal of milk that is 4% butterfat. Find the percent concentration of butterfat in the resulting mixture.

33. The length of a rectangle is four times the width. The perimeter is 200 ft. Find the length and width of the rectangle.

34. A person's weight on the moon is $16\frac{2}{3}$% of the person's weight on the earth. If an astronaut weighs 180 lb on the earth, how much would the astronaut weigh on the moon?

35. An engineering consultant invested $14,000 in an individual retirement account paying 8.15% annual simple interest. How much additional money must be deposited into an account paying 12% annual simple interest so that the total interest earned is 9.25% of the total investment?

36. Find two consecutive integers such that five times the first integer is 15 more than three times the second integer.

37. A pharmacist has 15 L of an 80% alcohol solution. How many liters of pure water should be added to the alcohol solution to make an alcohol solution that is 75% alcohol?

38. An antique telephone is 85 years old, and an antique typewriter is 160 years old. How many years ago was the typewriter twice the age the telephone was then?

39. One angle of a triangle is 15° more than the measure of the second angle. The third angle is 15° less than the measure of the second angle. Find the measure of each angle.

40. A ticket seller at a baseball card show had $145 in one-dollar bills and five-dollar bills. In all, there were 53 bills. Find the number of one-dollar bills held by the ticket seller.

41. In a certain state, the sales tax is $7\frac{1}{4}\%$. The sales tax on a chemistry text-book is $2.61. Find the cost of the textbook before the tax is added.

42. The sale price for a carpet sweeper is $26.56, which is 17% off the regular price. Find the regular price.

43. The owner of a health food store combined cranberry juice that cost $1.79 per quart with apple juice that cost $1.19 per quart. How many quarts of each were used to make 10 qt of a cranapple juice mixture costing $1.61 per quart?

44. The manager of a telescope supply company buys a 16-in. mirror blank for $340 and uses a 75% markup rate. Find the selling price of the mirror blank.

45. A furniture store uses a markup rate of 60%. The store sells a solid oak curio cabinet for $1074. Find the cost of the curio cabinet.

46. The largest swimming pool in the world is located in Casablanca, Morocco, and is 480 m long. Two swimmers start at the same time from opposite ends of the pool and start swimming toward each other. One swimmer's rate is 65 meters per minute. The other swimmer's rate is 55 meters per minute. In how many minutes after they begin will they meet?

47. The perimeter of a triangle is 35 in. The second side is 4 in. longer than the first side. The third side is 1 in. shorter than twice the first side. Find the measure of each side.

48. The sum of three consecutive odd integers is -45. Find the integers.

49. A grandparent is eight times the age of a child. One year from now, the grandparent will be seven times the age the child will be then. Find the present ages of the grandparent and the child.

50. A restaurant diner left a tip of $2.25 in dimes and quarters. There were two more quarters than dimes. Find the number of dimes and the number of quarters left for a tip.

C hapter Test

1. Write 60% as a fraction and as a decimal.

2. 20 is what percent of 16?

3. Write 0.375 as a percent. $\frac{375}{1000}$

4. Write $62\frac{1}{2}\%$ as a fraction.

5. Write $\frac{7}{8}$ as a percent. Write the remainder in fractional form.

6. Find 16% of 40.

7. Write 80% as a fraction and as a decimal.

8. Write $\frac{2}{25}$ as a percent.

9. 30% of what is 12?

10. Write $16\frac{2}{3}\%$ as a fraction.

11. Find $83\frac{1}{3}\%$ of 24.

12. Write 0.075 as a percent.

13. The manager of a sports shop uses a markup rate of 50%. The selling price for a set of golf clubs is $300. Find the cost of the golf clubs.

14. A portable typewriter that regularly sells for $100 is on sale for $80. Find the discount rate.

15. How many gallons of a 15% acid solution must be mixed with 5 gal of a 20% acid solution to make a 16% acid solution?

16. The perimeter of a rectangle is 38 m. The length of the rectangle is 1 m less than three times the width. Find the length and width of the rectangle.

17. Find three consecutive odd integers such that three times the first integer is one less than the sum of the second and third integers.

18. The age of a 20¢ stamp is 5 years, and the age of a 5¢ stamp is 35 years. In how many years will the 5¢ stamp be three times the age the 20¢ stamp will be then?

19. The value of a personal computer today is $2400. This is 80% of the computer's value last year. Find the value of the computer last year.

20. A total of $7000 is deposited into two simple interest accounts. On one account, the annual simple interest rate is 10%, and on the second account, the annual simple interest rate is 15%. How much should be invested in each account so that the total annual interest earned is $800?

21. A coffee merchant wants to make 12 lb of a blend of coffee costing $6 per pound. The blend is made using a $7 grade and a $4 grade of coffee. How many pounds of each of these grades should be used?

22. Two planes start at the same time from the same point and fly in opposite directions. The first plane is flying 100 mph faster than the second plane. In 3 h, the two planes are 1050 mi apart. Find the rate of each plane.

23. In a triangle, the first angle is 15° more than the second angle. The third angle is three times the second angle. Find the measure of each angle.

24. A coin bank contains 50 coins in nickels and quarters. The total amount of money in the bank is $9.50. Find the number of nickels and the number of quarters in the bank.

25. A club treasurer deposited $2400 in two simple interest accounts. The annual simple interest rate on one account is 6.75%. The annual simple interest rate on the other account is 9.45%. How much should be deposited in each account so that the same interest is earned on each account?

$$\frac{2d}{100}$$

Cumulative Review

1. Simplify: $-2 + (-8) - (-16)$

2. Simplify: $\left(-\frac{2}{3}\right)^3\left(-\frac{3}{4}\right)^2$

3. Simplify: $\frac{5}{6} - \left(\frac{2}{3}\right)^2 \div \left(\frac{1}{2} - \frac{1}{3}\right)$

4. Evaluate $-|-18|$.

5. Evaluate $b^2 - (a - b)^2$ when $a = 4$ and $b = -1$.

6. Simplify: $5x - 3y - (-4x) + 7y$

7. Simplify: $-4(3 - 2x - 5x^3)$

8. Simplify: $-2[x - 3(x - 1) - 5]$

9. Simplify: $-3x^2 - (-5x^2) + 4x^2$

10. Is 2 a solution of $4 - 2x - x^2 = 2 - 4x$?

11. Solve: $9 - x = 12$

12. Solve: $-\frac{4}{5}x = 12$

13. Solve: $8 - 5x = -7$

14. Solve: $-6x - 4(3 - 2x) = 4x + 8$

15. Write 40% as a fraction.

16. Write $83\frac{1}{3}\%$ as a fraction.

17. Write 0.025 as a percent.

18. Write $\frac{3}{25}$ as a percent.

19. Find $16\frac{2}{3}\%$ of 18.

20. 40% of what is 18?

21. The sum of two numbers is fifteen. The sum of five times the smaller number and eight equals five less than the product of three and the larger number. Find the two numbers.

22. An auto repair bill was $213. This includes $88 for parts and $25 for each hour of labor. Find the number of hours of labor.

23. A survey of 250 librarians showed that 50 of the libraries had a particular reference book on their shelves. What percent of the libraries had the reference book?

24. A deposit of $4000 is made into an account that earns 11% annual simple interest. How much additional money must be deposited into an account that pays 14% annual simple interest so that the total interest earned is 12% of the total investment?

25. The manager of a department store buys a chain necklace for $8 and sells it for $14. Find the markup rate.

26. How many grams of a gold alloy that costs $4 a gram must be mixed with 30 g of a gold alloy that costs $7 a gram to make an alloy costing $5 a gram?

27. How many ounces of pure water must be added to 70 oz of a 10% salt solution to make a 7% salt solution?

28. In an isosceles triangle, two angles are equal. The third angle is 8° less than twice the measure of one of the equal angles. Find the measure of one of the equal angles.

29. Three times the second of three consecutive even integers is 14 more than the sum of the first and third integers. Find the middle even integer.

30. A coin bank contains dimes and quarters. The number of quarters is five less than four times the number of dimes. The total amount in the bank is $6.45. Find the number of dimes in the bank.

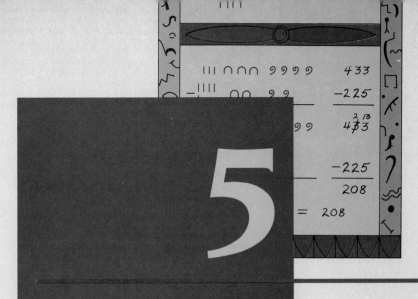

5

Polynomials

Objectives

- Add polynomials
- Subtract polynomials
- Multiply monomials
- Simplify powers of monomials
- Multiply a polynomial by a monomial
- Multiply two polynomials
- Multiply two binomials
- Multiply binomials that have special products
- Application problems
- Integer exponents
- Divide a polynomial by a monomial
- Divide polynomials

Early Egyptian Arithmetic Operations

The early Egyptian arithmetic processes are recorded on the Rhind Papyrus, but the underlying principles are not included. Scholars of today can only guess how these early developments were discovered.

Egyptian hieroglyphics used a base-ten system of numbers in which a vertical line represented 1, a heel bone, ∩, represented 10, and a scroll, 𝟿, represented 100.

The symbols at the right represent the number 237. There are 7 vertical lines, 3 heel bones, and 2 scrolls. Thus the symbols at the right represent 7 + 30 + 200, or 237.

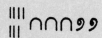

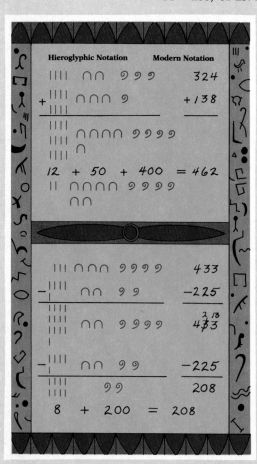

Addition in hieroglyphic notation does not require memorization of addition facts. Addition is done just by counting symbols.

Addition is a simple grouping operation.

Write down the total of each kind of symbol.

Group the 10 straight lines into one heel bone.

Subtraction in the hieroglyphic system is similar to making change. For example, what change do you get from a 1.00 bill when buying a $.55 item?

5 cannot be subtracted from 3, so a 10 is "borrowed," and 10 ones are added.

Note that no zero is provided in this number system. That place value symbol is just not used. As shown at the right, the heel bone is not used because there are no 10's necessary in 208.

Addition and Subtraction of Polynomials

1   Add polynomials

A **monomial** is a number, a variable, or a product of numbers and variables. A **polynomial** is a variable expression in which the terms are monomials.

A polynomial of *one* term is a **monomial.**	$5x^3$ is a monomial.
A polynomial of *two* terms is a **binomial.**	$5y^2 - 3x$ is a binomial.
A polynomial of *three* terms is a **trinomial.**	$6xy - 2r^2s + 4r$ is a trinomial.

The terms of a polynomial in one variable are usually arranged so that the exponents of the variable decrease from left to right. This is called **descending order.**

$$4x^3 - 3x^2 + 6x - 1$$
$$5y^4 - 2y^3 + y^2 - 7y + 8$$

The **degree of a polynomial** in one variable is its largest exponent. The degree of $4x^3 - 3x^2 + 6x - 1$ is 3. The degree of $5y^4 - 2y^3 + y^2 - 7y + 8$ is 4.

Polynomials can be added, using either a vertical or a horizontal format, by combining like terms.

Example 1   Simplify $(2x^2 + x - 1) + (3x^3 + 4x^2 - 5)$. Use a vertical format.

Solution

$$\begin{array}{r} 2x^2 + x - 1 \\ 3x^3 + 4x^2 \quad\ - 5 \\ \hline 3x^3 + 6x^2 + x - 6 \end{array}$$

▶ Arrange the terms of each polynomial in descending order with like terms in the same column.

▶ Combine the terms in each column.

Problem 1   Simplify $(2x^2 + 4x - 3) + (5x^2 - 6x)$. Use a vertical format.

Solution   See page A18.

Example 2   Simplify $(3x^3 - 7x + 2) + (7x^2 + 2x - 7)$. Use a horizontal format.

Solution $(3x^3 - 7x + 2) + (7x^2 + 2x - 7)$
$3x^3 + 7x^2 + (-7x + 2x) + (2 - 7)$

$3x^3 + 7x^2 - 5x - 5$

▶ Use the Commutative and Associative Properties of Addition to rearrange and group like terms.

▶ Combine like terms, and write the polynomial in descending order.

Problem 2   Simplify $(-4x^2 - 3xy + 2y^2) + (3x^2 - 4y^2)$. Use a horizontal format.

Solution   See page A18.

2   Subtract polynomials

The **opposite** of the polynomial $x^2 - 2x + 3$ is $-(x^2 - 2x + 3)$.

To simplify the opposite of a polynomial, remove the parentheses and change the sign of every term inside the parentheses.

$$-(x^2 - 2x + 3) = -x^2 + 2x - 3$$

Polynomials can be subtracted using either a vertical or horizontal format. To subtract, add the opposite of the second polynomial to the first.

Example 3   Simplify $(-3x^2 - 7) - (-8x^2 + 3x - 4)$. Use a vertical format.

Solution
$$\begin{array}{l} -3x^2 \qquad - 7 \\ \underline{-8x^2 + 3x - 4} \end{array}$$

▶ Arrange the terms of each polynomial in descending order with like terms in the same column.

$$\begin{array}{l} -3x^2 \qquad - 7 \\ \underline{8x^2 - 3x + 4} \\ 5x^2 - 3x - 3 \end{array}$$

▶ Write the opposite of the second polynomial.
▶ Combine the terms in each column.

Problem 3   Simplify $(8y^2 - 4xy + x^2) - (2y^2 - xy + 5x^2)$. Use a vertical format.

Solution   See page A18.

Example 4   Simplify $(5x^2 - 3x + 4) - (-3x^3 - 2x + 8)$. Use a horizontal format.

Solution
$(5x^2 - 3x + 4) - (-3x^3 - 2x + 8)$
$(5x^2 - 3x + 4) + (3x^3 + 2x - 8)$

▶ Rewrite subtraction as addition of the opposite.

$3x^3 + 5x^2 - x - 4$

▶ Combine like terms, and write the polynomial in descending order.

Problem 4   Simplify $(-3a^2 - 4a + 2) - (5a^3 + 2a - 6)$.

Solution   See page A18.

EXERCISES 5.1

1   Simplify. Use a vertical format.

1. $(x^2 + 7x) + (-3x^2 - 4x)$

2. $(3y^2 - 2y) + (5y^2 + 6y)$

3. $(y^2 + 4y) + (-4y - 8)$

4. $(3x^2 + 9x) + (6x - 24)$

5. $(2x^2 + 6x + 12) + (3x^2 + x + 8)$

6. $(x^2 + x + 5) + (3x^2 - 10x + 4)$

7. $(x^3 - 7x + 4) + (2x^2 + x - 10)$

8. $(3y^3 + y^2 + 1) + (-4y^3 - 6y - 3)$

9. $(2a^3 - 7a + 1) + (-3a^2 - 4a + 1)$

10. $(5r^3 - 6r^2 + 3r) + (r^2 - 2r - 3)$

Simplify. Use a horizontal format.

11. $(4x^2 + 2x) + (x^2 + 6x)$

12. $(-3y^2 + y) + (4y^2 + 6y)$

13. $(4x^2 - 5xy) + (3x^2 + 6xy - 4y^2)$

14. $(2x^2 - 4y^2) + (6x^2 - 2xy + 4y^2)$

15. $(2a^2 - 7a + 10) + (a^2 + 4a + 7)$

16. $(-6x^2 + 7x + 3) + (3x^2 + x + 3)$

17. $(5x^3 + 7x - 7) + (10x^2 - 8x + 3)$

18. $(3y^3 + 4y + 9) + (2y^2 + 4y - 21)$

19. $(2r^2 - 5r + 7) + (3r^3 - 6r)$

20. $(3y^3 + 4y + 14) + (-4y^2 + 21)$

21. $(3x^2 + 7x + 10) + (-2x^3 + 3x + 1)$

22. $(7x^3 + 4x - 1) + (2x^2 - 6x + 2)$

2   Simplify. Use a vertical format.

23. $(x^2 - 6x) - (x^2 - 10x)$

24. $(y^2 + 4y) - (y^2 + 10y)$

25. $(2y^2 - 4y) - (-y^2 + 2)$

26. $(-3a^2 - 2a) - (4a^2 - 4)$

27. $(x^2 - 2x + 1) - (x^2 + 5x + 8)$

28. $(3x^2 + 2x - 2) - (5x^2 - 5x + 6)$

29. $(4x^3 + 5x + 2) - (-3x^2 + 2x + 1)$

30. $(5y^2 - y + 2) - (-2y^3 + 3y - 3)$

31. $(2y^3 + 6y - 2) - (y^3 + y^2 + 4)$

32. $(-2x^2 - x + 4) - (-x^3 + 3x - 2)$

Simplify. Use a horizontal format.

33. $(y^2 - 10xy) - (2y^2 + 3xy)$

34. $(x^2 - 3xy) - (-2x^2 + xy)$

35. $(3x^2 + x - 3) - (x^2 + 4x - 2)$

36. $(5y^2 - 2y + 1) - (-3y^2 - y - 2)$

37. $(-2x^3 + x - 1) - (-x^2 + x - 3)$

38. $(2x^2 + 5x - 3) - (3x^3 + 2x - 5)$

39. $(4a^3 - 2a + 1) - (a^3 - 2a + 3)$

40. $(b^2 - 8b + 7) - (4b^3 - 7b - 8)$

41. $(4y^3 - y - 1) - (2y^2 - 3y + 3)$

42. $(3x^2 - 2x - 3) - (2x^3 - 2x^2 + 4)$

SUPPLEMENTAL EXERCISES 5.1

State whether the polynomial is a monomial, a binomial, or a trinomial.

43. $8x^4 - 6x^2$

44. $4a^2b^2 + 9ab + 10$

45. $7x^3y^4$

Simplify.

46. $\left(\frac{2}{3}a^2 + \frac{1}{2}a - \frac{3}{4}\right) - \left(\frac{5}{3}a^2 + \frac{1}{2}a + \frac{1}{4}\right)$

47. $\left(\frac{3}{5}x^2 + \frac{1}{6}x - \frac{5}{8}\right) + \left(\frac{2}{5}x^2 + \frac{5}{6}x - \frac{3}{8}\right)$

State whether or not the expression is a monomial.

48. $3\sqrt{x}$

49. $\frac{4}{x}$

50. x^2y^2

State whether or not the expression is a polynomial.

51. $\frac{1}{5}x^3 + \frac{1}{2}x$

52. $\frac{1}{5x^2} + \frac{1}{2x}$

53. $x + \sqrt{5}$

Solve.

54. What polynomial must be added to $3x^2 - 4x - 2$ so that the sum is $-x^2 + 2x + 1$?

55. What polynomial must be added to $-2x^3 + 4x - 7$ so that the sum is $x^2 - x - 1$?

56. What polynomial must be subtracted from $6x^2 - 4x - 2$ so that the difference is $2x^2 + 2x - 5$?

57. What polynomial must be subtracted from $2x^3 - x^2 + 4x - 2$ so that the difference is $x^3 + 2x - 8$?

S E C T I O N **5.2**

Multiplication of Monomials

1 Multiply monomials

Recall that in the exponential expression x^5, x is the base and 5 is the exponent. The exponent indicates the number of times the base occurs as a factor.

The product of exponential expressions with the *same* base can be simplified by writing each expression in factored form and writing the result with an exponent.

$$x^3 \cdot x^2 = \overbrace{(x \cdot x \cdot x)}^{3 \text{ factors}} \cdot \overbrace{(x \cdot x)}^{2 \text{ factors}}$$
$$\underbrace{}_{5 \text{ factors}}$$
$$= x \cdot x \cdot x \cdot x \cdot x$$
$$= x^5$$

Adding the exponents results in the same product.

$$x^3 \cdot x^2 = x^{3+2} = x^5$$

Rule for Multiplying Exponential Expressions

If m and n are integers, then $x^m \cdot x^n = x^{m+n}$.

For example, in the expression $a^2 \cdot a^6 \cdot a$, the bases are the same. The expression can be simplified by adding the exponents. Recall that $a = a^1$.

$$a^2 \cdot a^6 \cdot a = a^{2+6+1} = a^9$$

Example 1 Simplify: $(2xy)(3x^2y)$

Solution $(2xy)(3x^2y) =$
$(2 \cdot 3)(x \cdot x^2)(y \cdot y) =$ ▶ Use the Commutative and Associative Properties of Multiplication to rearrange and group factors.

$6x^{1+2}y^{1+1} =$ ▶ Multiply variables with the same base by adding the exponents.

$6x^3y^2$

Problem 1 Simplify: $(3x^2)(6x^3)$

Solution See page A18.

Example 2   Simplify: $(2x^2y)(-5xy^4)$

Solution $(2x^2y)(-5xy^4) =$
$[2(-5)](x^2 \cdot x)(y \cdot y^4) =$ ▶ Use the Properties of Multiplication to rearrange and group factors.

$-10x^3y^5$ ▶ Multiply variables with the same base by adding the exponents.

Problem 2   Simplify: $(-3xy^2)(-4x^2y^3)$

Solution   See page A18.

2   Simplify powers of monomials

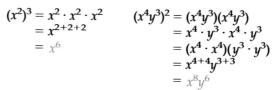

A power of a monomial can be simplified by rewriting the expression in factored form and then using the Rule for Multiplying Exponential Expressions.

$$(x^2)^3 = x^2 \cdot x^2 \cdot x^2$$
$$= x^{2+2+2}$$
$$= x^6$$

$$(x^4y^3)^2 = (x^4y^3)(x^4y^3)$$
$$= x^4 \cdot y^3 \cdot x^4 \cdot y^3$$
$$= (x^4 \cdot x^4)(y^3 \cdot y^3)$$
$$= x^{4+4}y^{3+3}$$
$$= x^8y^6$$

Note that multiplying each exponent inside the parentheses by the exponent outside the parentheses gives the same result.

$$(x^2)^3 = x^{2 \cdot 3} = x^6$$

$$(x^4y^3)^2 = x^{4 \cdot 2}y^{3 \cdot 2} = x^8y^6$$

Rule for Simplifying Powers of Exponential Expressions

If m and n are integers, then $(x^m)^n = x^{mn}$.

Rule for Simplifying Powers of Products

If m, n, and p are integers, then $(x^m y^n)^p = x^{mp}y^{np}$.

To simplify $(x^5)^2$, multiply the exponents.

$$(x^5)^2 = x^{5 \cdot 2} = x^{10}$$

To simplify $(3a^2b)^3$, multiply each exponent inside the parentheses by the exponent outside the parentheses.

$$(3a^2b)^3 = 3^{1 \cdot 3}a^{2 \cdot 3}b^{1 \cdot 3}$$
$$= 3^3a^6b^3$$
$$= 27a^6b^3$$

Example 3   Simplify: $(-2x)(-3xy^2)^3$

Solution $(-2x)(-3xy^2)^3 =$
$(-2x)(-3)^3x^3y^6 =$

► Multiply each exponent in $-3xy^2$ by the exponent out-side the parentheses (3).

$(-2x)(-27)x^3y^6 =$
$[-2(-27)](x \cdot x^3)y^6 =$

► Simplify $(-3)^3$.
► Use the Properties of Multiplication to rearrange and group factors.

$54x^4y^6$

► Multiply variables with the same base by adding the exponents.

Problem 3   Simplify: $(3x)(2x^2y)^3$
Solution   See page A18.

EXERCISES 5.2

1   Simplify.

1. $(x)(2x)$
2. $(-3y)(y)$
3. $(3x)(4x)$
4. $(7y^3)(7y^2)$
5. $(-2a^3)(-3a^4)$
6. $(5a^6)(-2a^5)$
7. $(x^2y)(xy^4)$
8. $(x^2y^4)(xy^7)$
9. $(-2x^4)(5x^5y)$
10. $(-3a^3)(2a^2b^4)$
11. $(x^2y^4)(x^5y^4)$
12. $(a^2b^4)(ab^3)$
13. $(2xy)(-3x^2y^4)$
14. $(-3a^2b)(-2ab^3)$
15. $(x^2yz)(x^2y^4)$
16. $(-ab^2c)(a^2b^5)$
17. $(a^2b^3)(ab^2c^4)$
18. $(x^2y^3z)(x^3y^4)$
19. $(-a^2b^2)(a^3b^6)$
20. $(xy^4)(-xy^3)$
21. $(-6a^3)(a^2b)$
22. $(2a^2b^3)(-4ab^2)$
23. $(-5y^4z)(-8y^6z^5)$
24. $(3x^2y)(-4xy^2)$
25. $(10ab^2)(-2ab)$
26. $(x^2y)(yz)(xyz)$
27. $(xy^2z)(x^2y)(z^2y^2)$
28. $(-2x^2y^3)(3xy)(-5x^3y^4)$
29. $(4a^2b)(-3a^3b^4)(a^5b^2)$
30. $(3ab^2)(-2abc)(4ac^2)$

2   Simplify.

31. $(2^2)^3$
32. $(3^2)^2$
33. $(-2)^2$
34. $(-3)^3$
35. $(-2^2)^3$
36. $(-2^3)^3$
37. $(x^3)^3$
38. $(y^4)^2$
39. $(x^7)^2$
40. $(y^5)^3$
41. $(-x^2)^2$
42. $(-x^2)^3$
43. $(2x)^2$
44. $(3y)^3$
45. $(-2x^2)^3$
46. $(-3y^3)^2$
47. $(x^2y^3)^2$
48. $(x^3y^4)^5$
49. $(3x^2y)^2$
50. $(-2ab^3)^4$

51. $(a^2)(3a^2)^3$

52. $(b^2)(2a^3)^4$

53. $(-2x)(2x^3)^2$

54. $(2y)(-3y^4)^3$

55. $(x^2y)(x^2y)^3$

56. $(a^3b)(ab)^3$

57. $(ab^2)^2(ab)^2$

58. $(x^2y)^2(x^3y)^3$

59. $(-2x)(-2x^3y)^3$

60. $(-3y)(-4x^2y^3)^3$

61. $(-2x)(-3xy^2)^2$

62. $(-3y)(-2x^2y)^3$

63. $(ab^2)(-2a^2b)^3$

64. $(a^2b^2)(-3ab^4)^2$

65. $(-2a^3)(3a^2b)^3$

66. $(-3b^2)(2ab^2)^3$

67. $(-3ab)^2(-2ab)^3$

68. $(-3a^2b)^3(-3ab)^3$

SUPPLEMENTAL EXERCISES 5.2

Simplify.

69. $(6x)(2x^2) + (4x^2)(5x)$

70. $(2a^7)(7a^2) - (6a^3)(5a^6)$

71. $(3a^2b^2)(2ab) - (9ab^2)(a^2b)$

72. $(3x^2y^2)^2 - (2xy)^4$

73. $(5xy^3)(3x^4y^2) - (2x^3y)(x^2y^4)$

74. $a^2(ab^2)^3 - a^3(ab^3)^2$

75. $4a^2(2ab)^3 - 5b^2(a^5b)$

76. $9x^3(3x^2y)^2 - x(x^3y)^2$

77. $-2xy(x^2y)^3 - 3x^5(xy^2)^2$

78. $5a^2b(ab^2)^2 + b^3(2a^2b)^2$

79. $a^n \cdot a^n$

80. $(a^n)^2$

81. $(a^2)^n$

82. $a^2 \cdot a^n$

Solve.

83. The length of a rectangle is $4ab$. The width is $2ab$. Find the perimeter of the rectangle in terms of ab.

SECTION 5.3
Multiplication of Polynomials

1   Multiply a polynomial by a monomial

To multiply a polynomial by a monomial, use the Distributive Property and the Rule for Multiplying Exponential Expressions.

To simplify $-2x(x^2 - 4x - 3)$, use the Distributive Property and the Rule for Multiplying Exponential Expressions.

$-2x(x^2 - 4x - 3) =$
$-2x(x^2) - (-2x)(4x) - (-2x)(3) =$

$-2x^3 + 8x^2 + 6x$

Example 1   Simplify.

A. $(5x + 4)(-2x)$ B. $x^3(2x^2 - 3x + 2)$

Solution   A. $(5x + 4)(-2x) = 5x(-2x) + 4(-2x)$ ▶ Use the Distributive Property.
$$= -10x^2 - 8x$$ ▶ Use the Rule for Multiplying Exponential Expressions.

B. $x^3(2x^2 - 3x + 2) =$
$2x^5 - 3x^4 + 2x^3$

Problem 1   Simplify.

A. $(-2y + 3)(-4y)$ B. $-a^2(3a^2 + 2a - 7)$

Solution   See page A19.

2   Multiply two polynomials

Multiplication of two polynomials requires the repeated application of the Distributive Property.

$$(y - 2)(y^2 + 3y + 1) = (y - 2)(y^2) + (y - 2)(3y) + (y - 2)(1)$$
$$= y^3 - 2y^2 + 3y^2 - 6y + y - 2$$
$$= y^3 + y^2 - 5y - 2$$

A more convenient method of multiplying two polynomials is to use a vertical format similar to that used for multiplication of whole numbers.

$$y^2 + 3y + 1$$
$$y - 2$$

Multiply each term in the trinomial by -2. $-2y^2 - 6y - 2$
Multiply each term in the trinomial by y. $y^3 + \quad 3y^2 + \quad y$
Like terms must be in the same column. $\overline{y^3 + \quad y^2 - 5y - 2}$
Add the terms in each column.

Example 2   Simplify: $(2b^3 - b + 1)(2b + 3)$

Solution $2b^3 - \ b + 1$
$$2b + 3$$
$$\overline{\quad 6b^3 \qquad\quad - 3b + 3}$$ ▶ Multiply each term of $2b^3 - b + 1$ by 3.
$$4b^4 \qquad\quad - 2b^2 + 2b$$ ▶ Multiply each term of $2b^3 - b + 1$ by 2b. Arrange the terms in descending order.
$$\overline{4b^4 + 6b^3 - 2b^2 - \ b + 3}$$ ▶ Add the terms in each column.

Problem 2   Simplify: $(2y^3 + 2y^2 - 3)(3y - 1)$

Solution   See page A19.

Example 3   Simplify: $(4a^3 - 5a - 2)(3a - 2)$

Solution

$$
\begin{array}{r}
4a^3 - 5a - 2 \\
3a - 2 \\
\hline
-8a^3 \qquad\qquad + 10a + 4 \\
12a^4 \qquad - 15a^2 - 6a \\
\hline
12a^4 - 8a^3 - 15a^2 + 4a + 4
\end{array}
$$

▶ Multiply each term of $4a^3 - 5a - 2$ by -2.
▶ Multiply each term of $4a^3 - 5a - 2$ by $3a$.
▶ Add the terms in each column.

Problem 3   Simplify: $(3x^3 - 2x^2 + x - 3)(2x + 5)$

Solution   See page A19.

3   Multiply two binomials

It is often necessary to find the product of two binomials. The product can be found using a method called **FOIL**, which is based on the Distributive Property. The letters of FOIL stand for **First, Outer, Inner,** and **Last.**

Simplify $(2x + 3)(x + 5)$.

Multiply the First terms. $(2x + 3)(x + 5)$ $2x \cdot x = 2x^2$

Multiply the Outer terms. $(2x + 3)(x + 5)$ $2x \cdot 5 = 10x$

Multiply the Inner terms. $(2x + 3)(x + 5)$ $3 \cdot x = 3x$

Multiply the Last terms. $(2x + 3)(x + 5)$ $3 \cdot 5 = 15$

$$\qquad\qquad\qquad\qquad\qquad \textbf{F}\quad\textbf{O}\quad\textbf{I}\quad\textbf{L}$$

Add the products. $(2x + 3)(x + 5)$ $= 2x^2 + 10x + 3x + 15$
Combine like terms. $= 2x^2 + 13x + 15$

Example 4   Simplify: $(4x - 3)(3x - 2)$

Solution $(4x - 3)(3x - 2) =$
$4x(3x) + 4x(-2) + (-3)(3x) + (-3)(-2) =$ ▶ Use the FOIL method.
$12x^2 - 8x - 9x + 6 =$
$12x^2 - 17x + 6$ ▶ Combine like terms.

Problem 4   Simplify: $(4y - 5)(3y - 3)$

Solution   See page A19.

Example 5   Simplify: $(3x - 2y)(x + 4y)$

Solution $(3x - 2y)(x + 4y) =$
$3x(x) + 3x(4y) + (-2y)(x) + (-2y)(4y) =$ ► Use the FOIL method.
$3x^2 + 12xy - 2xy - 8y^2 =$
$3x^2 + 10xy - 8y^2$ ► Combine like terms.

Problem 5   Simplify: $(3a + 2b)(3a - 5b)$

Solution   See page A19.

4 Multiply binomials that have special products

The expression $(a + b)(a - b)$ is the product of the sum and difference of two terms. The first binomial in the expression is a sum; the second is a difference. The two terms are a and b. The first term in each binomial is a. The second term in each binomial is b.

The expression $(a + b)^2$ is the square of a binomial. The first term in the binomial is a. The second term in the binomial is b.

Using FOIL, a pattern for the product of the sum and difference of two terms and for the square of a binomial can be found.

The Sum and Difference of Two Terms

$$(a + b)(a - b) = a^2 - ab + ab - b^2$$
$$= a^2 - b^2$$

Square of first term ┘
Square of second term ┘

The Square of a Binomial

$$(a + b)^2 = (a + b)(a + b) = a^2 + ab + ab + b^2$$
$$= a^2 + 2ab + b^2$$

Square of first term ┘
Twice the product of the two terms ┘
Square of last term ┘

Example 6   Simplify: $(2x + 3)(2x - 3)$

Solution $(2x + 3)(2x - 3) =$ ▶ $(2x + 3)(2x - 3)$ is the sum and difference of two terms.
$(2x)^2 - 3^2 =$ ▶ Square the first term. Square the second term.
$4x^2 - 9$ ▶ Simplify.

Problem 6   Simplify: $(2a + 5c)(2a - 5c)$

Solution   See page A19.

Example 7   Simplify: $(3x - 2)^2$

Solution $(3x - 2)^2 =$ ▶ $(3x - 2)^2$ is the square of a binomial.
$(3x)^2 + 2(3x)(-2) + (-2)^2 =$ ▶ Square the first term. Find twice the product of the two terms. Square the last term.

$9x^2 - 12x + 4$ ▶ Simplify.

Problem 7   Simplify: $(3x + 2y)^2$

Solution   See page A19.

5 Application problems

Example 8   The radius of a circle is $x - 3$. Use the equation $A = \pi r^2$, where r is the radius, to find the area of the circle in terms of the variable x. Leave the answer in terms of π.

Strategy   To find the area, replace the variable r in the equation $A = \pi r^2$ with the given value and solve for A.

Solution $A = \pi r^2$
$A = \pi(x - 3)^2$ ▶ This is the square of a binomial.
$A = \pi(x^2 - 6x + 9)$
$A = \pi x^2 - 6\pi x + 9\pi$

The area is $\pi x^2 - 6\pi x + 9\pi$.

Problem 8   The length of a rectangle is $x + 8$. The width is $x - 5$. Find the area of the rectangle in terms of the variable x.

Solution   See page A19.

Example 9   The length of a side of a square is $3x + 5$. Use the formula $A = s^2$, where s is the length of a side of a square and A is the area, to find the area of the square in terms of the variable x.

Strategy   To find the area of the square, replace the variable s in the equation $A = s^2$ with the given value and simplify.

Solution $A = s^2$
$A = (3x + 5)^2$ ▶ This is the square of a binomial.
$A = 9x^2 + 30x + 25$

The area is $9x^2 + 30x + 25$.

Problem 9   The base of a triangle is $x + 3$, and the height is $4x - 6$. Use the formula $A = \frac{1}{2} bh$, where b is the base, h is the height, and A is the area of the triangle, to find the area of the triangle in terms of the variable x.

Solution   See page A19.

EXERCISES 5.3

1   Simplify.

1. $x(x - 2)$

2. $y(3 - y)$

3. $-x(x + 7)$

4. $-y(7 - y)$

5. $3a^2(a - 2)$

6. $4b^2(b + 8)$

7. $-5x^2(x^2 - x)$

8. $-6y^2(y + 2y^2)$

9. $-x^3(3x^2 - 7)$

10. $-y^4(2y^2 - y^6)$

11. $2x(6x^2 - 3x)$

12. $3y(4y - y^2)$

13. $(2x - 4)3x$

14. $(3y - 2)y$

15. $(3x + 4)x$

16. $(2x + 1)2x$

17. $-xy(x^2 - y^2)$

18. $-x^2y(2xy - y^2)$

19. $x(2x^3 - 3x + 2)$

20. $y(-3y^2 - 2y + 6)$

21. $-a(-2a^2 - 3a - 2)$

22. $-b(5b^2 + 7b - 35)$

23. $x^2(3x^4 - 3x^2 - 2)$

24. $y^3(-4y^3 - 6y + 7)$

25. $2y^2(-3y^2 - 6y + 7)$

26. $4x^2(3x^2 - 2x + 6)$

27. $(a^2 + 3a - 4)(-2a)$

28. $(b^3 - 2b + 2)(-5b)$

29. $-3y^2(-2y^2 + y - 2)$

30. $-5x^2(3x^2 - 3x - 7)$

31. $xy(x^2 - 3xy + y^2)$ **32.** $ab(2a^2 - 4ab - 6b^2)$

2   Simplify.

33. $(x^2 + 3x + 2)(x + 1)$

34. $(x^2 - 2x + 7)(x - 2)$

35. $(a^2 - 3a + 4)(a - 3)$

36. $(x^2 - 3x + 5)(2x - 3)$

37. $(-2b^2 - 3b + 4)(b - 5)$

38. $(-a^2 + 3a - 2)(2a - 1)$

39. $(-2x^2 + 7x - 2)(3x - 5)$

40. $(-a^2 - 2a + 3)(2a - 1)$

41. $(x^2 + 5)(x - 3)$

42. $(y^2 - 2y)(2y + 5)$

43. $(x^3 - 3x + 2)(x - 4)$

44. $(y^3 + 4y^2 - 8)(2y - 1)$

45. $(5y^2 + 8y - 2)(3y - 8)$

46. $(3y^2 + 3y - 5)(4y - 3)$

47. $(5a^3 - 15a + 2)(a - 4)$

48. $(3b^3 - 5b^2 + 7)(6b - 1)$

49. $(y^3 + 2y^2 - 3y + 1)(y + 2)$

50. $(2a^3 - 3a^2 + 2a - 1)(2a - 3)$

3   Simplify.

51. $(x + 1)(x + 3)$

52. $(y + 2)(y + 5)$

53. $(a - 3)(a + 4)$

54. $(b - 6)(b + 3)$

55. $(y + 3)(y - 8)$

56. $(x + 10)(x - 5)$

57. $(y - 7)(y - 3)$

58. $(a - 8)(a - 9)$

59. $(2x + 1)(x + 7)$

60. $(y + 2)(5y + 1)$

61. $(3x - 1)(x + 4)$

62. $(7x - 2)(x + 4)$

63. $(4x - 3)(x - 7)$

64. $(2x - 3)(4x - 7)$

65. $(3y - 8)(y + 2)$

66. $(5y - 9)(y + 5)$

67. $(3x + 7)(3x + 11)$

68. $(5a + 6)(6a + 5)$

69. $(7a - 16)(3a - 5)$

70. $(5a - 12)(3a - 7)$

71. $(3b + 13)(5b - 6)$

72. $(x + y)(2x + y)$ **73.** $(2a + b)(a + 3b)$ **74.** $(3x - 4y)(x - 2y)$

75. $(2a - b)(3a + 2b)$ **76.** $(5a - 3b)(2a + 4b)$ **77.** $(2x + y)(x - 2y)$

78. $(3x - 7y)(3x + 5y)$ **79.** $(2x + 3y)(5x + 7y)$ **80.** $(5x + 3y)(7x + 2y)$

81. $(3a - 2b)(2a - 7b)$ **82.** $(5a - b)(7a - b)$ **83.** $(a - 9b)(2a + 7b)$

84. $(2a + 5b)(7a - 2b)$ **85.** $(10a - 3b)(10a - 7b)$ **86.** $(12a - 5b)(3a - 4b)$

87. $(5x + 12y)(3x + 4y)$ **88.** $(11x + 2y)(3x + 7y)$ **89.** $(2x - 15y)(7x + 4y)$

90. $(5x + 2y)(2x - 5y)$ **91.** $(8x - 3y)(7x - 5y)$ **92.** $(2x - 9y)(8x - 3y)$

---

4   Simplify.

93. $(y - 5)(y + 5)$ **94.** $(y + 6)(y - 6)$ **95.** $(2x + 3)(2x - 3)$

96. $(4x - 7)(4x + 7)$ **97.** $(x + 1)^2$ **98.** $(y - 3)^2$

99. $(3a - 5)^2$ **100.** $(6x - 5)^2$ **101.** $(3x - 7)(3x + 7)$

102. $(9x - 2)(9x + 2)$ **103.** $(2a + b)^2$ **104.** $(x + 3y)^2$

105. $(x - 2y)^2$ **106.** $(2x - 3y)^2$ **107.** $(4 - 3y)(4 + 3y)$

108. $(4x - 9y)(4x + 9y)$ **109.** $(5x + 2y)^2$ **110.** $(2a - 9b)^2$

---

5   Solve.

111.   The length of a rectangle is $5x$. The width is $2x - 7$. Find the area of the rectangle in terms of the variable x.

112.   The width of a rectangle is $x - 6$. The length is $2x + 3$. Find the area of the rectangle in terms of the variable x.

113.   The length of a rectangle is $4x - 1$. The width is $2x + 9$. Find the area of the rectangle in terms of the variable x.

114.   The length of a side of a square is $2x + 1$. Use the equation $A = s^2$, where s is the length of the side of a square, to find the area of the square in terms of the variable x.

115.   The length of a side of a square is $x + 7$. Use the equation $A = s^2$, where s is the length of the side of a square, to find the area of the square in terms of the variable x.

116.   The length of a side of a square is $2x - 3$. Use the equation $A = s^2$, where s is the length of the side of a square, to find the area of the square in terms of the variable x.

117.   The base of a triangle is $4x$, and the height is $2x + 5$. Use the equation $A = \frac{1}{2}bh$, where b is the base of the triangle and h is the height, to find the area of the triangle in terms of the variable x.

118.   The base of a triangle is $2x + 6$, and the height is $x - 8$. Use the equation $A = \frac{1}{2}bh$, where b is the base of the triangle and h is the height, to find the area of the triangle in terms of the variable x.

119.   The radius of a circle is $x + 4$. Use the equation $A = \pi r^2$, where r is the radius, to find the area of the circle in terms of the variable x. Leave the answer in terms of π.

120.   The radius of a circle is $x - 3$. Use the equation $A = \pi r^2$, where r is the radius, to find the area of the circle in terms of the variable x. Leave the answer in terms of π.

SUPPLEMENTAL EXERCISES 5.3

Simplify.

121. $(a + b)^2 - (a - b)^2$

122. $(x + 3y)^2 + (x + 3y)(x - 3y)$

123. $(3a^2 - 4a + 2)^2$

124. $(x + 4)^3$

125. $3x^2(2x^3 + 4x - 1) - 6x^3(x^2 - 2)$

126. $(3b + 2)(b - 6) + (4 + 2b)(3 - b)$

127. $x^n(x^n + 1)$

128. $(x^n + 1)(x^n - 1)$

129. $(x^n + 1)(x^n + 1)$

130. $(x^n - 1)^2$

Solve.

131.   What polynomial when divided by $2x - 1$ has a quotient of $x + 6$?

132.   Subtract the product of $4x - y$ and $x + 2y$ from $6x^2 - 7xy$.

133.   Find $(4n^3)^2$ if $2n - 3 = 4n - 7$.

SECTION **5.4**

Division of Polynomials

1 Integer exponents

The quotient of two exponential expressions with the *same* base can be simplified by writing each expression in factored form, dividing by the common factors, and then writing the result with an exponent.

$$\frac{x^5}{x^2} = \frac{\overset{1}{\cancel{x}} \cdot \overset{1}{\cancel{x}} \cdot x \cdot x \cdot x}{\underset{1}{\cancel{x}} \cdot \underset{1}{\cancel{x}}} = x^3$$

Note that subtracting the exponents results in the same quotient.

$$\frac{x^5}{x^2} = x^{5-2} = x^3$$

To divide two monomials with the same base, subtract the exponents of the like bases.

Simplify: $\dfrac{a^7}{a^3}$

The bases are the same.
Subtract the exponents.

$$\frac{a^7}{a^3} = a^{7-3} = a^4$$

Simplify: $\dfrac{r^8 s^6}{r^7 s}$

Subtract the exponents of the like bases.

$$\frac{r^8 s^6}{r^7 s} = r^{8-7} s^{6-1} = r s^5$$

Recall that for any number a, $a \neq 0$, $\dfrac{a}{a} = 1$. This property is true for exponential expressions as well. For example, for $x \neq 0$, $\dfrac{x^4}{x^4} = 1$.

This expression also can be simplified using the rule for dividing exponential expressions with the same base.

$$\frac{x^4}{x^4} = x^{4-4} = x^0$$

Because $\dfrac{x^4}{x^4} = 1$ and $\dfrac{x^4}{x^4} = x^{4-4} = x^0$, the following definition of zero as an exponent is used.

Zero as an Exponent

If $x \neq 0$, then $x^0 = 1$. The expression 0^0 is not defined.

Simplify: $(12a^3)^0$, $a \neq 0$

Any nonzero expression to the zero power is 1. $(12a^3)^0 = 1$

Simplify: $-(xy^4)^0$, $x \neq 0$, $y \neq 0$

Any nonzero expression to the zero power is 1. $-(xy^4)^0 = -(1) = -1$
Because the negative sign is outside the paren-
theses, the answer is -1.

The meaning of a negative exponent can be developed by examining the quotient $\dfrac{x^4}{x^6}$.

The expression can be simplified by
writing the numerator and denominator
in factored form, dividing by the common
factors, and then writing the result with
an exponent.

$$\frac{x^4}{x^6} = \frac{\overset{1}{\cancel{x}} \cdot \overset{1}{\cancel{x}} \cdot \overset{1}{\cancel{x}} \cdot \overset{1}{\cancel{x}}}{\underset{1}{\cancel{x}} \cdot \underset{1}{\cancel{x}} \cdot \underset{1}{\cancel{x}} \cdot \underset{1}{\cancel{x}} \cdot x \cdot x} = \frac{1}{x^2}$$

Now simplify the same expression by
subtracting the exponents of the like
bases.

$$\frac{x^4}{x^6} = x^{4-6} = x^{-2}$$

Because $\dfrac{x^4}{x^6} = \dfrac{1}{x^2}$ and $\dfrac{x^4}{x^6} = x^{-2}$, the following definition of a negative
exponent is used.

Definition of Negative Exponents

If n is a positive integer and $x \neq 0$, then $x^{-n} = \dfrac{1}{x^n}$ and $\dfrac{1}{x^{-n}} = x^n$.

To evaluate 2^{-4}, write the expression $2^{-4} = \dfrac{1}{2^4} = \dfrac{1}{16}$
with a positive exponent. Then simplify.

Now that negative exponents have been defined, the Rule for Dividing Exponential Expressions can be stated.

Rule for Dividing Exponential Expressions

If m and n are integers and $x \neq 0$, then $\dfrac{x^m}{x^n} = x^{m-n}$.

Example 1   Write $\dfrac{3^{-3}}{3^2}$ with a positive exponent. Then evaluate.

Solution $\dfrac{3^{-3}}{3^2} = 3^{-3-2}$ ▶ 3^{-3} and 3^2 have the same base. Subtract the exponents.

   $= 3^{-5}$

   $= \dfrac{1}{3^5}$ ▶ Use the Definition of Negative Exponents to write the expression with a positive exponent.

   $= \dfrac{1}{243}$ ▶ Evaluate.

Problem 1   Write $\dfrac{2^{-2}}{2^3}$ with a positive exponent. Then evaluate.

Solution   See page A20.

The rules for simplifying exponential expressions and powers of exponential expressions are true for all integers. These rules are restated here.

Rules of Exponents

If m, n, and p are integers, then

$$x^m \cdot x^n = x^{m+n} \qquad\qquad (x^m)^n = x^{mn} \qquad\qquad (x^m y^n)^p = x^{mp} y^{np}$$

$$\dfrac{x^m}{x^n} = x^{m-n},\ x \neq 0 \qquad\qquad \left(\dfrac{x^m}{y^n}\right)^p = \dfrac{x^{mp}}{y^{np}},\ y \neq 0 \qquad\qquad x^{-n} = \dfrac{1}{x^n},\ x \neq 0$$

$$x^0 = 1,\ x \neq 0$$

An exponential expression is in simplest form when it is written with only positive exponents.

Example 2   Simplify: $\dfrac{x^{-4}y^6}{xy^2}$

Solution $\dfrac{x^{-4}y^6}{xy^2} = x^{-5}y^4$ ▶ Divide variables with the same base by subtracting the exponents.

   $= \dfrac{y^4}{x^5}$ ▶ Write the expression with only positive exponents.

Problem 2   Simplify: $\dfrac{b^8}{a^{-5}b^6}$

Solution   See page A20.

Example 3   Simplify.

A. $\dfrac{-35a^6b^{-2}}{25a^{-2}b^5}$ B. $(-2x)(3x^{-2})^{-3}$

Solution   A. $\dfrac{-35a^6b^{-2}}{25a^{-2}b^5} = -\dfrac{35a^6b^{-2}}{25a^{-2}b^5}$

▶ A negative sign is placed in front of a fraction.

$$= -\dfrac{\overset{1}{\cancel{5}} \cdot 7a^{6-(-2)}b^{-2-5}}{\underset{1}{\cancel{5}} \cdot 5}$$

▶ Factor the coefficients. Divide by the common factors. Divide variables with the same base by subtracting the exponents.

$$= -\dfrac{7a^8b^{-7}}{5}$$

$$= -\dfrac{7a^8}{5b^7}$$

▶ Write the expression with only positive exponents.

B. $(-2x)(3x^{-2})^{-3} = (-2x)(3^{-3}x^6)$

▶ Use the Rule for Simplifying Powers of Products.

$$= \dfrac{-2x \cdot x^6}{3^3}$$

▶ Write the expression with positive exponents.

$$= -\dfrac{2x^7}{27}$$

▶ Use the Rule for Multiplying Exponential Expressions, and simplify the numerical exponential expression.

Problem 3   Simplify.

A. $\dfrac{12x^{-8}y^4}{-16xy^{-3}}$ B. $(-3ab)(2a^3b^{-2})^{-3}$

Solution   See page A20.

Example 4   Simplify: $\dfrac{(2r^2t^{-1})^{-3}}{(r^{-3}t^4)^2}$

Solution $\dfrac{(2r^2t^{-1})^{-3}}{(r^{-3}t^4)^2} = \dfrac{2^{-3}r^{-6}t^3}{r^{-6}t^8}$ ▶ Use the Rule for Simplifying Powers of Products.

$= 2^{-3}r^0t^{-5}$ ▶ Divide variables with the same base by subtracting the exponents.

$= \dfrac{r^0}{2^3t^5}$ ▶ Write the expression with only positive exponents.

$= \dfrac{1}{8t^5}$ ▶ Simplify. $(r^0 = 1,\ 2^3 = 8)$

Problem 4   Simplify: $\dfrac{(6a^{-2}b^3)^{-1}}{(4a^3b^{-2})^{-2}}$

Solution   See page A20.

2   Divide a polynomial by a monomial

Note that $\dfrac{8+4}{2}$ can be simplified by first

adding the terms in the numerator and then dividing the result. It can also be simplified by first dividing each term in the numerator by the denominator and then adding the result.

$\dfrac{8+4}{2} = \dfrac{12}{2} = 6$

$\dfrac{8+4}{2} = \dfrac{8}{2} + \dfrac{4}{2} = 4 + 2 = 6$

To divide a polynomial by a monomial, divide each term in the numerator by the denominator, and write the sum of the quotients.

$\dfrac{a+b}{c} = \dfrac{a}{c} + \dfrac{b}{c}$

Simplify: $\dfrac{6x^2 + 4x}{2x}$

Divide each term of the polynomial $6x^2 + 4x$ by the monomial $2x$. Simplify.

$\dfrac{6x^2 + 4x}{2x} = \dfrac{6x^2}{2x} + \dfrac{4x}{2x}$

$= 3x + 2$

Example 5   Simplify: $\dfrac{6x^3 - 3x^2 + 9x}{3x}$

Solution $\dfrac{6x^3 - 3x^2 + 9x}{3x} = \dfrac{6x^3}{3x} - \dfrac{3x^2}{3x} + \dfrac{9x}{3x}$ ▶ Divide each term of the polynomial by the monomial $3x$.

$= 2x^2 - x + 3$ ▶ Simplify each expression.

Problem 5   Simplify: $\dfrac{4x^3y + 8x^2y^2 - 4xy^3}{2xy}$

Solution   See page A20.

Example 6   Simplify: $\dfrac{12x^2y - 6xy + 4x^2}{2xy}$

Solution $\dfrac{12x^2y - 6xy + 4x^2}{2xy} =$

$\dfrac{12x^2y}{2xy} - \dfrac{6xy}{2xy} + \dfrac{4x^2}{2xy} =$ ▶ Divide each term of the polynomial by the monomial $2xy$.

$6x - 3 + \dfrac{2x}{y}$ ▶ Simplify each expression.

Problem 6   Simplify: $\dfrac{24x^2y^2 - 18xy + 6y}{6xy}$

Solution   See page A20.

3 ## Divide polynomials

To divide polynomials, use a method similar to that used for division of whole numbers. The same equation used to check division of whole numbers is used to check polynomial division.

Dividend = (Quotient × Divisor) + Remainder

Simplify: $(x^2 - 5x + 8) \div (x - 3)$

Step 1

$$x - 3\overline{)x^2 - 5x + 8}$$

$$\begin{array}{r} x \\ \underline{x^2 - 3x} \\ -2x + 8 \end{array}$$

Think: $x\overline{)x^2} = \dfrac{x^2}{x} = x$

Multiply: $x(x - 3) = x^2 - 3x$

Subtract: $(x^2 - 5x) - (x^2 - 3x) = -2x$

Step 2

$$x - 3\overline{)x^2 - 5x + 8}$$

$$\begin{array}{r} x - 2 \\ \underline{x^2 - 3x} \\ -2x + 8 \\ \underline{-2x + 6} \\ 2 \end{array}$$

Think: $x\overline{)-2x} = \dfrac{-2x}{x} = -2$

Multiply: $-2(x - 3) = -2x + 6$

Subtract: $(-2x + 8) - (-2x + 6) = 2$

The remainder is 2.

Check: $(x - 2)(x - 3) + 2 = x^2 - 3x - 2x + 6 + 2 = x^2 - 5x + 8$

$(x^2 - 5x + 8) \div (x - 3) = x - 2 + \dfrac{2}{x - 3}$

Example 7   Simplify: $(6x + 2x^3 + 26) \div (x + 2)$

Solution

$$x + 2\overline{)2x^3 + 0x^2 + 6x + 26}$$

$$\begin{array}{r} 2x^2 - 4x + 14 \\ \underline{2x^3 + 4x^2} \\ -4x^2 + 6x \\ \underline{-4x^2 - 8x} \\ 14x + 26 \\ \underline{14x + 28} \\ -2 \end{array}$$

▶ Arrange the terms in descending order.
There is no term of x^2 in $2x^3 + 6x + 26$.
Insert $0x^2$ for the missing term so
that like terms will be in columns.

$(6x + 2x^3 + 26) \div (x + 2) = 2x^2 - 4x + 14 - \dfrac{2}{x + 2}$

Problem 7   Simplify: $(x^3 - 2x - 4) \div (x - 2)$

Solution   See page A20.

EXERCISES 5.4

1   Write with a positive or zero exponent. Then evaluate.

1. 5^{-2}

2. 3^{-3}

3. $\dfrac{1}{8^{-2}}$

4. $\dfrac{1}{12^{-1}}$

5. $\dfrac{3^{-2}}{3}$ **6.** $\dfrac{5^{-3}}{5}$ **7.** $\dfrac{2^3}{2^3}$ **8.** $\dfrac{3^{-2}}{3^{-2}}$

Simplify.

9. x^{-2} **10.** y^{-10} **11.** $\dfrac{1}{a^{-6}}$ **12.** $\dfrac{1}{b^{-4}}$

13. $x^2 y^{-3}$ **14.** $a^{-2}b$ **15.** $x^{-1}y^{-2}$ **16.** $x^{-3}y^{-4}$

17. $x^{-3}x^4$ **18.** $x \cdot x^{-2}$ **19.** $\dfrac{a^{-3}}{a^5}$ **20.** $\dfrac{a^{-10}}{a^{10}}$

21. $\dfrac{x^{-2}y}{x}$ **22.** $\dfrac{x^4 y^{-3}}{x^2}$ **23.** $\dfrac{a^2 b^{-3}}{a^{-2}}$ **24.** $\dfrac{x^4 y^{-5}}{y^{-2}}$

25. $\dfrac{a^{-2}b}{b^0}$ **26.** $\dfrac{a^2 b^{-4}}{b^0}$ **27.** $(y^{-4})^3$ **28.** $(x^{-2})^{-3}$

29. $(ab^{-1})^0$ **30.** $(a^2 b)^0$ **31.** $(x^{-2}y^2)^2$ **32.** $(x^{-3}y^{-1})^2$

33. $(x^2 y^{-1})^{-2}$ **34.** $(x^{-2}y^{-3})^{-4}$ **35.** $(-2xy^{-2})^3$ **36.** $(-3x^{-1}y^2)^2$

37. $(4x^2 y^{-3})^2$ **38.** $(5xy^{-3})^{-2}$ **39.** $(2x^{-1})(x^{-3})$ **40.** $(-2x^{-5})x^7$

41. $(-5a^2)(a^{-5})^2$ **42.** $\dfrac{a^{-1}b^{-1}}{ab}$ **43.** $\dfrac{a^{-3}b^{-4}}{a^2 b^2}$ **44.** $\dfrac{3x^{-2}y^2}{6xy^2}$

45. $\dfrac{2x^{-2}y}{8xy}$ **46.** $\dfrac{3x^{-2}y}{xy}$ **47.** $\dfrac{16a^{-4}y}{3ay}$ **48.** $\dfrac{3x^{-2}y}{xy^2}$

49. $\dfrac{2x^{-1}y^4}{x^2 y^3}$ **50.** $\dfrac{2x^{-1}y^{-4}}{4xy^2}$ **51.** $\dfrac{3a^{-1}b^{-5}}{6a^{-3}b^4}$ **52.** $\dfrac{12x^3 y^{-6}}{4x^{-2}y}$

53. $\dfrac{-8y^2}{4y^{-1}}$ **54.** $\dfrac{12x^4}{3x^{-2}}$ **55.** $\dfrac{5x^2}{15x}$ **56.** $\dfrac{-16x}{4x^3}$

57. $\dfrac{27y}{-12y^3}$ **58.** $\dfrac{(x^{-1}y)^2}{xy^2}$ **59.** $\dfrac{(x^{-2}y)^2}{x^2 y^3}$ **60.** $\dfrac{(6b)^3}{(-3b^2)^2}$

61. $\dfrac{(-3a^2)^3}{(9a)^2}$ **62.** $\dfrac{-36a^4 b^7}{60a^5 b^9}$ **63.** $\dfrac{-16xy^4}{96x^4 y^4}$ **64.** $\dfrac{-(8a^2 b^4)^3}{64a^3 b^8}$

65. $(2a^{-3})(a^7b^{-1})^3$

66. $(3ab^{-2})(2a^{-1}b)^{-3}$

67. $\dfrac{(a^{-2}y^3)^{-3}}{a^2y^{-5}}$

68. $\dfrac{(x^{-3}y^{-2})^2}{x^6y^{-8}}$

69. $\dfrac{-20a^3b^4}{-45ab^7}$

70. $\dfrac{(-2ab^2)^3}{-8ab^7}$

71. $\dfrac{(-3x^3y)^2}{-12xy^5}$

72. $\dfrac{(3a^{-1}b^4)^2}{9a^{-5}b^{-3}}$

73. $\dfrac{(2x^{-2}y)^{-3}}{x^{-4}y^{-5}}$

74. $\dfrac{(2a^{-2}b^3)^{-2}}{(4a^2b^{-4})^{-1}}$

75. $\dfrac{(3^{-1}r^4s^{-3})^{-2}}{(6r^2t^{-2}s^{-1})^2}$

76. $\dfrac{(4^{-1}x^{-2}y^5)^{-2}}{(2x^3y^{-6})^3}$

2 Simplify.

77. $\dfrac{2x + 2}{2}$

78. $\dfrac{5y + 5}{5}$ *y+1*

79. $\dfrac{10a - 25}{5}$ *2a-5*

80. $\dfrac{16b - 40}{8}$ *2b-5*

81. $\dfrac{3a^2 + 2a}{a}$ *3a+2*

82. $\dfrac{6y^2 + 4y}{y}$

83. $\dfrac{4b^3 - 3b}{b}$

84. $\dfrac{12x^2 - 7x}{x}$

85. $\dfrac{3x^2 - 6x}{3x}$

86. $\dfrac{10y^2 - 6y}{2y}$

87. $\dfrac{5x^2 - 10x}{-5x}$

88. $\dfrac{3y^2 - 27y}{-3y}$

89. $\dfrac{x^3 + 3x^2 - 5x}{x}$

90. $\dfrac{a^3 - 5a^2 + 7a}{a}$

91. $\dfrac{x^6 - 3x^4 - x^2}{x^2}$

92. $\dfrac{a^8 - 5a^5 - 3a^3}{a^2}$

93. $\dfrac{5x^2y^2 + 10xy}{5xy}$

94. $\dfrac{8x^2y^2 - 24xy}{8xy}$

95. $\dfrac{9y^6 - 15y^3}{-3y^3}$

96. $\dfrac{4x^4 - 6x^2}{-2x^2}$

97. $\dfrac{3x^2 - 2x + 1}{x}$

98. $\dfrac{8y^2 + 2y - 3}{y}$

99. $\dfrac{-3x^2 + 7x - 6}{x}$

100. $\dfrac{2y^2 - 6y + 9}{y}$

101. $\dfrac{16a^2b - 20ab + 24ab^2}{4ab}$

102. $\dfrac{22a^2b + 11ab - 33ab^2}{11ab}$

103. $\dfrac{9x^2y + 6xy - 3xy^2}{xy}$

3   Simplify.

104. $(x^2 + 2x + 1) \div (x + 1)$

105. $(x^2 + 10x + 25) \div (x + 5)$

106. $(a^2 - 6a + 9) \div (a - 3)$

107. $(b^2 - 14b + 49) \div (b - 7)$

108. $(x^2 - x - 6) \div (x - 3)$

109. $(y^2 + 2y - 35) \div (y + 7)$

110. $(5x + 2x^2 + 2) \div (x + 2)$

111. $(21 - 13y + 2y^2) \div (y - 3)$

112. $(4x^2 - 16) \div (2x + 4)$

113. $(2y^2 + 7) \div (y - 3)$

114. $(x^2 + 1) \div (x - 1)$

115. $(x^2 + 4) \div (x + 2)$

116. $(6x^2 - 7x) \div (3x - 2)$

117. $(6y^2 + 2y) \div (2y + 4)$

118. $(5x^2 + 7x) \div (x - 1)$

119. $(6x^2 - 5) \div (x + 2)$

120. $(a^2 + 5a + 10) \div (a + 2)$

121. $(b^2 - 8b - 9) \div (b - 3)$

122. $(2y^2 - 9y + 8) \div (2y + 3)$

123. $(3x^2 + 5x - 4) \div (x - 4)$

124. $(3 + 8x + 4x^2) \div (2x - 1)$

125. $(21y + 10y^2 + 10) \div (2y + 3)$

126. $(15a^2 - 8a - 8) \div (3a + 2)$

127. $(12a^2 - 25a - 7) \div (3a - 7)$

128. $(5 - 23x + 12x^2) \div (4x - 1)$

129. $(25a + 6a^2 + 24) \div (3a - 1)$

130. $(x^3 + 3x^2 + 5x + 3) \div (x + 1)$

131. $(x^3 - 6x^2 + 7x - 2) \div (x - 1)$

132. $(y^3 + 6y^2 + 4y - 5) \div (y + 3)$

133. $(4a^3 + 8a^2 + 5a + 9) \div (2a + 3)$

134. $(6a^3 - 5a^2 + 5) \div (3a + 2)$

135. $(4b^3 - b - 5) \div (2b + 3)$

SUPPLEMENTAL EXERCISES 5.4

Evaluate.

136. $4^{-2} + 2^{-4}$

137. $8^{-2} + 2^{-5}$

138. $9^{-2} + 3^{-3}$

Write in decimal notation.

139. 5^{-3}

140. 2^{-4}

141. 25^{-2}

Simplify using division of polynomials.

142. $\left(\frac{4x^2}{x}\right) + \left(\frac{5x^3y}{x^2y}\right)$

143. $\left(\frac{a^2b^4c^5}{ab^2c^4}\right) + \left(\frac{a^3b^6c^3}{a^2b^4c^2}\right)$

144. $\left(\frac{9x^2y^4}{3xy^2}\right) - \left(\frac{12x^5y^6}{6x^4y^4}\right)$

145. $\left(\frac{6x^2 + 9x}{3x}\right) + \left(\frac{8xy^2 + 4y^2}{4y^2}\right)$

146. $\left(\frac{6x^4yz^3}{2x^2y^3}\right)\left(\frac{2x^2z^3}{4y^2z}\right) \div \left(\frac{6x^2y^3}{x^4y^2z}\right)$

147. $\left(\frac{x^3yz^2}{3x^3y^2z^2}\right)\left(\frac{2x^3y^2z^3}{2y^2z}\right) \div \left(\frac{6y^2z^3}{x^3y^4z}\right)$

148. $\left(\frac{2x^2y^2z}{2x^3y^3}\right) \div \left(\frac{4x^2yz^3}{2y^2z}\right) \div \left(\frac{3x^5y^2}{x^3y^2z}\right)$

149. $\left(\frac{5x^2yz^3}{3x^4yz^3}\right) \div \left(\frac{10x^2y^5z^4}{2y^3z}\right) \div \left(\frac{5y^4z^2}{x^2y^6z}\right)$

Complete.

150. $\left(\frac{2}{3}\right)^{-3} = ($_____$)^3$

151. $\left(\frac{4}{5}\right)^{-4} = ($_____$)^4$

152. If $m = n$ and $a \neq 0$, then $\frac{a^m}{a^n} = $ _____.

153. If $m = n + 1$ and $a \neq 0$, then $\frac{a^m}{a^n} = $ _____.

Solve.

154. $(-4.8)^x = 1$

155. $(-39.7)^x = 1$

Solve.

156. The product of a monomial and $4b$ is $12a^2b$. Find the monomial.

157. The product of a monomial and $6x$ is $24xy^2$. Find the monomial.

158. The quotient of a polynomial and $2x + 1$ is $2x - 4 + \frac{7}{2x + 1}$. Find the polynomial.

159. The quotient of a polynomial and $x - 3$ is $x^2 - x + 8 + \frac{22}{x - 3}$. Find the polynomial.

alculators and Computers

Evaluating Polynomials

One way to evaluate a polynomial is first to express the polynomial in a form that suggests a sequence of steps on the calculator. To illustrate this method, consider the polynomial $4x^2 - 5x + 2$. First the polynomial is rewritten as:

$$4x^2 - 5x + 2 = (4x - 5)x + 2$$

To evaluate the polynomial, work through the rewritten expression from left to right, substituting the appropriate value for x.

Here are some examples.

Evaluate $5x^2 - 2x + 4$ when $x = 3$.

Rewrite the polynomial. $5x^2 - 2x + 4 = (5x - 2)x + 4$

Replace x in the rewritten expression by the given value. $(5 \cdot 3 - 2) \cdot 3 + 4$

Work through the expression
from left to right. $\boxed{(}\ \boxed{5}\ \boxed{\times}\ \boxed{3}\ \boxed{-}\ \boxed{2}\ \boxed{)}\ \boxed{\times}\ \boxed{3}\ \boxed{+}\ \boxed{4}\ \boxed{=}$

The result in the display should be 43.

Evaluate $2x^3 - 4x^2 + 7x - 12$ when $x = 4$.

Rewrite the polynomial. $2x^3 - 4x^2 + 7x - 12 = [(2x - 4)x + 7]x - 12$

Replace x in the rewritten expression by the $[(2 \cdot 4 - 4) \cdot 4 + 7] \cdot 4 - 12$
given value.

Work through the expression from left to right.

$\boxed{(}\ \boxed{(}\ \boxed{2}\ \boxed{\times}\ \boxed{4}\ \boxed{-}\ \boxed{4}\ \boxed{)}\ \boxed{\times}\ \boxed{4}\ \boxed{+}\ \boxed{7}\ \boxed{)}\ \boxed{\times}\ \boxed{4}\ \boxed{-}\ \boxed{12}\ \boxed{=}$

The result in the display should be 80.

Evaluate $4x^2 - 3x + 5$ when $x = -2$.

Rewrite the polynomial. $4x^2 - 3x + 5 = (4x - 3)x + 5$

Replace x in the expression by the given value. $[4 \cdot (-2) - 3] \cdot (-2) + 5$

Work through the expression
from left to right. $\boxed{(}\ \boxed{4}\ \boxed{\times}\ \boxed{2}\ \boxed{+/-}\ \boxed{-}\ \boxed{3}\ \boxed{)}\ \boxed{\times}\ \boxed{2}\ \boxed{+/-}\ \boxed{+}\ \boxed{5}\ \boxed{=}$

The result in the display should be 27.

Here are some practice exercises.
Evaluate for the given value.

1. $2x^2 - 3x + 7; x = 4$

2. $3x^2 + 7x - 12; x = -3$

3. $3x^3 - 2x^2 + 6x - 8; x = 3$

4. $2x^3 + 4x^2 - x - 2; x = -2$

5. $x^4 - 3x^3 + 6x^2 + 5x - 1; x = 2$

6. $2x^3 - 4x + 8; x = 2$
   (*Hint:* $2x^3 - 4x + 8 =$
   $2x^3 + 0x^2 - 4x + 8$)

Something Extra

Scientific Notation

Very large numbers and very small numbers are encountered in the fields of science and engineering. For example, the charge of an electron is 0.000000000000000000160 coulomb. These numbers can be written more easily in scientific notation. In scientific notation, a number is expressed as the product of two factors, one a number between 1 and 10, and the other a power of ten. Some examples are given below.

Number	*Number Written in Scientific Notation*
40	$4.0 \cdot 10^1$
400	$4.00 \cdot 10^2$
609	$6.09 \cdot 10^2$
5000	$5.000 \cdot 10^3$
0.8	$8 \cdot 10^{-1}$
0.073	$7.3 \cdot 10^{-2}$
0.00192	$1.92 \cdot 10^{-3}$

As illustrated below, the rules for calculating with numbers in scientific notation are the same as those for calculating with algebraic expressions. The power of ten corresponds to the variable, and the number between 1 and 10 corresponds to the coefficient of the variable.

	Algebraic Expressions	*Scientific Notation*
Addition	$6x^3 + 5x^3 = 11x^3$	$6 \cdot 10^3 + 5 \cdot 10^3 = 11 \cdot 10^3$
Subtraction	$8x^{-4} - 3x^{-4} = 5x^{-4}$	$8 \cdot 10^{-4} - 3 \cdot 10^{-4} = 5 \cdot 10^{-4}$
Multiplication	$(4x^{-3})(2x^5) = 8x^2$	$(4 \cdot 10^{-3})(2 \cdot 10^5) = 8 \cdot 10^2$
Division	$\dfrac{6x^5}{3x^{-2}} = 2x^7$	$\dfrac{6 \cdot 10^5}{3 \cdot 10^{-2}} = 2 \cdot 10^7$

Note that in addition and subtraction of algebraic expressions, the variable parts must be the same, and in addition and subtraction of numbers written in scientific notation, the powers of ten must be the same.

Simplify.

1. $1.3 \cdot 10^7 + 8.4 \cdot 10^7$

2. $7.19 \cdot 10^{-6} + 1.56 \cdot 10^{-6}$

3. $5.0 \cdot 10^{-14} - 2.7 \cdot 10^{-14}$

4. $4.62 \cdot 10^{12} - 3.08 \cdot 10^{12}$

5. $(3.0 \cdot 10^5)(1.1 \cdot 10^{-8})$

6. $(2.4 \cdot 10^{-9})(1.6 \cdot 10^3)$

7. $\dfrac{7.2 \cdot 10^{13}}{2.4 \cdot 10^{-3}}$

8. $\dfrac{5.4 \cdot 10^{-2}}{1.8 \cdot 10^{-4}}$

Chapter Summary

Key Words

A *monomial* is a number, a variable, or a product of numbers and variables.

A *polynomial* is a variable expression in which the terms are monomials.

A polynomial of one term is a *monomial*.

A polynomial of two terms is a *binomial*.

A polynomial of three terms is a *trinomial*.

The *degree of a polynomial* in one variable is the largest exponent on a variable.

Essential Rules

Rule for Multiplying Exponential Expressions	If m and n are integers, then $x^m \cdot x^n = x^{m+n}$.
Rule for Simplifying Powers of Exponential Expressions	If m and n are integers, then $(x^m)^n = x^{mn}$.

Rule for Simplifying Powers of Products	If m, n, and p are integers, then $(x^m y^n)^p = x^{mp} y^{np}$.
FOIL method of finding the product of two binomials	Add the products of the First terms, the Outer terms, the Inner terms, and the Last terms.
The Sum and Difference of Two Terms	$(a + b)(a - b) = a^2 - b^2$
The Square of a Binomial	$(a + b)^2 = a^2 + 2ab + b^2$ $(a - b)^2 = a^2 - 2ab + b^2$
Definition of x^0	$x^0 = 1,\ x \neq 0$
Definition of Negative Exponents	If n is a positive integer and $x \neq 0$, then $x^{-n} = \dfrac{1}{x^n}$ and $\dfrac{1}{x^{-n}} = x^n$.
Rule for Dividing Exponential Expressions	If m and n are integers and $x \neq 0$, then $\dfrac{x^m}{x^n} = x^{m-n}$.
Rule for Simplifying Powers of Quotients	If m, n, and p are integers and $y \neq 0$, then $\left(\dfrac{x^m}{y^n}\right)^p = \dfrac{x^{mp}}{y^{np}}$.

Chapter Review

1. Simplify: $(12y^2 + 17y - 4) + (9y^2 - 13y + 3)$

2. Simplify: $(5xy^2)(-4x^2y^3)$

3. Simplify: $-2x(4x^2 + 7x - 9)$

4. Simplify: $(5a - 7)(2a + 9)$

5. Simplify: $\dfrac{36x^2 - 42x + 60}{6}$

6. Simplify: $(5x^2 - 2x - 1) - (3x^2 - 5x + 7)$

7. Simplify: $(-3^2)^3$

8. Simplify: $(x^2 - 5x + 2)(x - 1)$

9. Simplify: $(a + 7)(a - 7)$

10. Evaluate $\dfrac{6^2}{6^{-2}}$.

11. Simplify: $(x^2 + x - 42) \div (x + 7)$

12. Simplify: $(2x^3 + 7x^2 + x) + (2x^2 - 4x - 12)$

13. Simplify: $(6a^2b^5)(3a^6b)$

14. Simplify: $x^2y(3x^2 - 2x + 12)$

15. Simplify: $(2b - 3)(4b + 5)$

16. Simplify: $\dfrac{16y^2 - 32y}{-4y}$

17. Simplify: $(13y^3 - 7y - 2) - (12y^2 - 2y - 1)$

18. Simplify: $(2^3)^2$

19. Simplify: $(3y^2 + 4y - 7)(2y + 3)$

20. Simplify: $(2b - 9)(2b + 9)$

21. Simplify: $(a^{-2}b^3c)^2$

22. Simplify: $(6y^2 - 35y + 36) \div (3y - 4)$

23. Simplify: $(6x^2 - 9x + 7) + (-6x^2 + 9x + 7)$

24. Simplify: $(xy^5z^3)(x^3y^3z)$

25. Simplify: $(6y^2 - 2y + 9)(-2y^3)$

26. Simplify: $(6x - 12)(3x - 2)$

27. Simplify: $\dfrac{a^5 - 12a^4 + 3a^3}{a^2}$

28. Simplify: $(8a^2 - a) - (15a^2 - 4)$

29. Simplify: $(-3x^2y^3)^2$

30. Simplify: $(4a^2 - 3)(3a - 2)$

31. Simplify: $(5y - 7)^2$

32. Simplify: $(-3x^{-2}y^{-3})^{-2}$

33. Simplify: $(x^2 + 17x + 64) \div (x + 12)$

34. Simplify: $(5a^2 + 6a - 11) + (5a^2 + 6a - 11)$

35. Simplify: $(a^2b^7c^6)(ab^3c)(a^3bc^2)$

36. Simplify: $2ab^3(4a^2 - 2ab + 3b^2)$

37. Simplify: $(3x + 4y)(2x - 5y)$

38. Simplify: $\dfrac{12b^7 + 36b^5 - 3b^3}{3b^3}$

39. Simplify: $(b^2 - 11b + 19) - (5b^2 + 2b - 9)$

40. Simplify: $(5a^7b^6)^2(4ab)$

41. Simplify: $(6b^3 - 2b^2 - 5)(2b^2 - 1)$

42. Simplify: $(6 - 5x)(6 + 5x)$

43. Simplify: $\dfrac{6x^{-2}y^4}{(3xy)^2}$

44. Simplify: $(a^3 + a^2 + 18) \div (a + 3)$

45. Simplify: $(4b^3 - 7b^2 + 10) + (2b^2 - 9b - 3)$

46. Simplify: $(2a^{12}b^3)(-9b^2c^6)(3ac)$

47. Simplify: $-9x^2(2x^2 + 3x - 7)$

48. Simplify: $(10y - 3)(3y - 10)$

49. Simplify: $\dfrac{15x^8 + 10x^4 + 2x^2}{5x^3}$

50. Simplify: $(6y^2 + 2y + 7) - (8y^2 + y + 12)$

51. Simplify: $(6x^4y^7z^2)^2(-2x^3y^2z^6)^2$

52. Simplify: $(-3x^3 - 2x^2 + x - 9)(4x + 3)$

53. Simplify: $(8a + 1)^2$

54. Simplify: $\dfrac{(4a^{-2}b^{-8})^2}{(2a^{-1}b^{-2})^4}$

55. Simplify: $(b^3 - 2b^2 - 33b - 7) \div (b - 7)$

56. The length of a rectangle is $3x$. The width is $4x - 7$. Find the area of the rectangle in terms of the variable x.

57. The length of a side of a square is $5x + 4$. Use the equation $A = s^2$, where s is the length of the side of a square, to find the area of the square in terms of the variable x.

58. The base of a triangle is $3x - 2$, and the height is $6x + 4$. Use the equation $A = \dfrac{1}{2}bh$, where b is the base of the triangle and h is the height, to find the area of the triangle in terms of the variable x.

59. The radius of a circle is $x - 6$. Use the equation $A = \pi r^2$, where r is the radius, to find the area of the circle in terms of the variable x. Leave the answer in terms of π.

60. The width of a rectangle is $3x - 8$. The length is $5x + 4$. Find the area of the rectangle in terms of the variable x.

Chapter Test

1. Simplify:
   $(3x^3 - 2x^2 - 4) + (8x^2 - 8x + 7)$

2. Simplify: $(-2x^3 + x^2 - 7)(2x - 3)$

3. Simplify: $2x(2x^2 - 3x)$

4. Simplify: $(-2a^2b)^3$

5.   Simplify: $\dfrac{12x^2}{-3x^{-4}}$

6.   Simplify: $(2ab^{-3})(3a^{-2}b^4)$

7.   Simplify:
     $(3a^2 - 2a - 7) - (5a^3 + 2a - 10)$

8.   Simplify: $(a - 2b)(a + 5b)$

9.   Simplify: $\dfrac{16x^5 - 8x^3 + 20x}{4x}$

10. Simplify: $(4x^2 - 7) \div (2x - 3)$

11. Simplify: $(-2xy^2)(3x^2y^4)$

12. Simplify: $-3y^2(-2y^2 + 3y - 6)$

13. Simplify: $\dfrac{(3xy^3)^3}{3x^4y^3}$

14. Simplify: $(2x - 5)^2$

15. Simplify: $(2x - 7y)(5x - 4y)$

16. Simplify: $(x - 3)(x^2 - 4x + 5)$

17. Simplify: $(a^2b^{-3})^2$

18. Simplify: $(x^2 + 6x - 7) \div (x - 1)$

19. Simplify: $(4y - 3)(4y + 3)$

20. Simplify:
     $(3y^3 - 5y + 8) - (-2y^2 + 5y + 8)$

21. Simplify: $(-3a^3b^2)^2$

22. Simplify: $(2a - 7)(5a^2 - 2a + 3)$

23. Simplify: $(3b + 2)^2$

24. Simplify: $\dfrac{(-2a^2b^3)^2}{8a^4b^8}$

25. Simplify: $(8x^2 + 4x - 3) \div (2x - 3)$

26. Simplify: $(a^2b^5)(ab^2)$

27. Simplify: $(a - 3b)(a + 4b)$

28. Simplify: $\dfrac{12x^3 - 3x^2 + 9}{3x^2}$

29. The length of a side of a square is $2x + 3$. Use the equation $A = s^2$, where s is the length of the side of a square, to find the area of the square in terms of the variable x.

30. The radius of a circle is $x - 5$. Use the equation $A = \pi r^2$, where r is the radius, to find the area of the circle in terms of the variable x. Leave the answer in terms of π.

Cumulative Review

1. Simplify: $\frac{3}{16} - \left(-\frac{3}{8}\right) - \frac{5}{9}$

2. Simplify: $-5^2 \cdot \left(\frac{2}{3}\right)^3 \cdot \left(-\frac{3}{8}\right)$

3. Simplify: $\left(-\frac{1}{2}\right)^2 \div \left(\frac{5}{8} - \frac{5}{6}\right) + 2$

4. Find the opposite of -87.

5. Write $\frac{31}{40}$ as a decimal.

6. Evaluate $\frac{b - (a - b)^2}{b^2}$ when $a = 3$ and $b = -2$.

7. Simplify: $-3x - (-xy) + 2x - 5xy$

8. Simplify: $(16x)\left(-\frac{3}{4}\right)$

9. Simplify: $-2[3x - 4(3 - 2x) + 2]$

10. Complete the statement by using the Inverse Property of Addition.
$-8 + ? = 0$

11. Solve: $12 = -\frac{2}{3}x$

12. Solve: $3x - 7 = 2x + 9$

13. Solve: $3 - 4(2 - x) = 3x + 7$

14. Solve: $-\frac{4}{5}x = 16 - x$

15. 38.4 is what percent of 160?

16. Simplify:
$(5b^3 - 4b^2 - 7) - (3b^2 - 8b + 3)$

17. Simplify: $(3x - 4)(5x^2 - 2x + 1)$

18. Simplify: $(4b - 3)(5b - 8)$

19. Simplify: $(5b + 3)^2$

20. Simplify: $\frac{(-3a^3b^2)^3}{12a^4b^{-2}}$

21. Simplify: $\frac{-15y^2 + 12y - 3}{-3y}$

22. Simplify: $(a^2 - 3a - 28) \div (a + 4)$

23. Simplify: $(-3x^{-4}y)(-3x^{-2}y)$

24. Translate and simplify "the product of five and the difference between a number and twelve."

25. Translate "the difference between eight times a number and twice the number is eighteen" into an equation and solve.

26. The width of a rectangle is 40% of the length. The perimeter of the rectangle is 42 m. Find the length and width of the rectangle.

27. A calculator costs a retailer $24. Find the selling price when the markup rate is 80%.

28. Fifty ounces of pure orange juice are added to 200 oz of a fruit punch that is 10% orange juice. What is the percent concentration of orange juice in the resulting mixture?

29. A car traveling at 50 mph overtakes a cyclist who, riding at 10 mph, has had a 2-h head start. How far from the starting point does the car overtake the cyclist?

30. The length of a side of a square is $3x + 2$. Use the equation $A = s^2$, where s is the length of the side of a square, to find the area of the square in terms of the variable x.

6

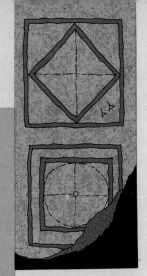

Factoring

Objectives

- Factor a monomial from a polynomial
- Factor by grouping
- Factor trinomials of the form $x^2 + bx + c$
- Factor completely
- Factor trinomials of the form $ax^2 + bx + c$ using trial factors
- Factor trinomials of the form $ax^2 + bx + c$ by grouping
- Factor the difference of two squares and perfect square trinomials
- Factor completely
- Solve equations by factoring
- Application problems

Algebra from Geometry

The early Babylonians made substantial progress in both algebra and geometry. Often the progress they made in algebra was based on geometric concepts.

Here are some geometric proofs of algebraic identities.

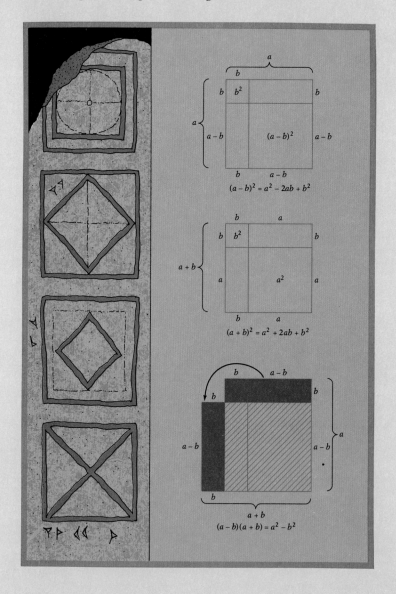

$$(a-b)^2 = a^2 - 2ab + b^2$$

$$(a+b)^2 = a^2 + 2ab + b^2$$

$$(a-b)(a+b) = a^2 - b^2$$

Common Factors

1 Factor a monomial from a polynomial

The **greatest common factor (GCF)** of two or more integers is the greatest integer that is a factor of all the integers.

$24 = 2 \cdot 2 \cdot 2 \cdot 3$
$60 = 2 \cdot 2 \cdot 3 \cdot 5$
$\text{GCF} = 2 \cdot 2 \cdot 3 = 12$

The GCF of two or more monomials is the product of the GCF of the coefficients and the common variable factors.

$6x^3y = 2 \cdot 3 \cdot x \cdot x \cdot x \cdot y$
$8x^2y^2 = 2 \cdot 2 \cdot 2 \cdot x \cdot x \cdot y \cdot y$
$\text{GCF} = 2 \cdot x \cdot x \cdot y = 2x^2y$

Note that the exponent of each variable in the GCF is the same as the *smallest* exponent of that variable in either of the monomials.

The GCF of $6x^3y$ and $8x^2y^2$ is $2x^2y$.

Example 1   Find the GCF of $12a^4b$ and $18a^2b^2c$.

Solution $12a^4b = 2 \cdot 2 \cdot 3 \cdot a^4 \cdot b$ ▶ Factor each monomial.
$18a^2b^2c = 2 \cdot 3 \cdot 3 \cdot a^2 \cdot b^2 \cdot c$

$\text{GCF} = 2 \cdot 3 \cdot a^2 \cdot b = 6a^2b$ ▶ The common variable factors are a^2 and b.
c is not a common variable factor.

Problem 1   Find the GCF of $4x^6y$ and $18x^2y^6$.

Solution   See page A20.

The Distributive Property is used to multiply factors of a polynomial. To **factor** a polynomial means to write the polynomial as a product of other polynomials.

Multiply

Factors		Polynomial
$2x(x + 5)$	$=$	$2x^2 + 10x$

Factor

In the example above, $2x$ is the GCF of the terms $2x^2$ and $10x$. It is a **common monomial factor** of the terms. $x + 5$ is a **binomial factor** of $2x^2 + 10x$.

213

Example 2   Factor.   A. $5x^3 - 35x^2 + 10x$ B. $16x^2y + 8x^4y^2 - 12x^4y^5$

Solution   A. $5x^3 = 5 \cdot x^3$
$35x^2 = 5 \cdot 7 \cdot x^2$
$10x = 2 \cdot 5 \cdot x$
The GCF is $5x$.

▶ Find the GCF of the terms of the polynomial.

$\dfrac{5x^3}{5x} = x^2, \dfrac{-35x^2}{5x} = -7x, \dfrac{10x}{5x} = 2$

▶ Divide each term of the polynomial by the GCF.

$5x^3 - 35x^2 + 10x =$
$5x(x^2) + 5x(-7x) + 5x(2) =$

▶ Use the quotients to rewrite the polynomial, expressing each term as a product with the GCF as one of the factors.

$5x(x^2 - 7x + 2)$

▶ Use the Distributive Property to write the polynomial as a product of factors.

B. $16x^2y = 2 \cdot 2 \cdot 2 \cdot 2 \cdot x^2 \cdot y$
$8x^4y^2 = 2 \cdot 2 \cdot 2 \cdot x^4 \cdot y^2$
$12x^4y^5 = 2 \cdot 2 \cdot 3 \cdot x^4 \cdot y^5$
The GCF is $4x^2y$.

▶ Find the GCF of the terms of the polynomial.

$16x^2y + 8x^4y^2 - 12x^4y^5 =$
$4x^2y(4) + 4x^2y(2x^2y) + 4x^2y(-3x^2y^4) =$

▶ Rewrite the polynomial, expressing each term as a product with the GCF as one of the factors.

$4x^2y(4 + 2x^2y - 3x^2y^4)$

▶ Use the Distributive Property to write the polynomial as a product of factors.

Problem 2   Factor.   A. $14a^2 - 21a^4b$ B. $6x^4y^2 - 9x^3y^2 + 12x^2y^4$

Solution   See page A21. $7a^2$
$2a^2(2) - 3a^2b$ $7a(2 - 3a^2b)$

2   Factor by grouping

In the examples below, the binomials in parentheses are called **binomial factors**.

$$2a(a + b) \text{ and } 3xy(x - y)$$

The Distributive Property is used to factor a common binomial factor from an expression.

In the expression at the right, the common binomial factor is $x - 3$. The Distributive Property is used to write the expression as a product of factors.

$x(x - 3) + 4(x - 3) =$

$(x - 3)(x + 4)$

Example 3   Factor: $y(x + 2) + 3(x + 2)$

Solution $y(x + 2) + 3(x + 2) =$ ▶ The common binomial factor is $x + 2$.
$(x + 2)(y + 3)$

Problem 3   Factor: $a(b - 7) + b(b - 7)$

Solution   See page A21. $(b - 7)(a + b)$

Sometimes a binomial factor must be rewritten before a common binomial factor can be found.

Factor: $a(a - b) + 5(b - a)$

$a - b$ and $b - a$ are different binomials.
Note that $(b - a) = (-a + b) = -(a - b)$.
Rewrite $(b - a)$ as $-(a - b)$ so that there is a common factor.

$$a(a - b) + 5(b - a) = a(a - b) + 5[-(a - b)]$$
$$= a(a - b) - 5(a - b)$$
$$= (a - b)(a - 5)$$

Example 4   Factor: $2x(x - 5) + y(5 - x)$

Solution $2x(x - 5) + y(5 - x) =$
$2x(x - 5) - y(x - 5) =$ ▶ Rewrite $5 - x$ as $-(x - 5)$ so that there is a common factor.

$(x - 5)(2x - y)$ ▶ Write the expression as a product of factors.

Problem 4   Factor: $3y(5x - 2) + 4(2 - 5x)$

Solution   See page A21. $(5x - 2)(3y + 4)$

Some polynomials can be factored by grouping the terms so that a common binomial factor is found.

Factor: $2x^3 - 3x^2 + 4x - 6$

$2x^3 - 3x^2 + 4x - 6 =$

Group the first two terms and the last two terms. $(2x^3 - 3x^2) + (4x - 6) =$

Factor out the GCF from each group. $x^2(2x - 3) + 2(2x - 3) =$

Write the expression as a product of factors. $(2x - 3)(x^2 + 2)$

Example 5   Factor: $3y^3 - 4y^2 - 6y + 8$

Solution $3y^3 - 4y^2 - 6y + 8 =$
$(3y^3 - 4y^2) - (6y - 8) =$ ▶ Group the first two terms and the last two terms.
Note that $-6y + 8 = -(6y - 8)$.

$y^2(3y - 4) - 2(3y - 4) =$ ▶ Factor out the GCF from each group.
$(3y - 4)(y^2 - 2)$ ▶ Write the expression as a product of factors.

Problem 5   Factor: $(y^5 - 5y^3) + (4y^2 - 20)$

Solution   See page A21.

$y^3(y^2 - 5) \quad 4(y^2 - 5)$

$(y^3 + 4)(y^2 - 5)$

EXERCISES 6.1

1 Find the greatest common factor.

1. x^7, x^3
2. y^6, y^{12}
3. x^2y^4, xy^6

4. a^5b^3, a^3b^8
5. $x^2y^4z^6, xy^8z^2$
6. ab^2c^3, a^3b^2c

7. $a^3b^2c^3, ab^4c^3$
8. x^3y^2z, x^4yz^5
9. $3x^4, 12x^2$

10. $12x, 30x^2$
11. $16a^3, 18a$
12. $8y^3, 12y^6$

13. $14a^3, 49a^7$
14. $12y^2, 27y^4$
15. $3x^2y^2, 5ab^2$

16. $8x^2y^3, 7ab^4$
17. $9a^2b^4, 24a^4b^2$
18. $15a^4b^2, 9ab^5$

19. $ab^3, 4a^2b, 12a^2b^3$
20. $12x^2y, x^4y, 16x$
21. $2x^2y, 4xy, 8x$

22. $16x^2, 8x^4y^2, 12xy$
23. $3x^2y^2, 6x, 9x^3y^3$
24. $4a^2b^3, 8a^3, 12ab^4$

Factor.

25. $5a + 5$ $5(a+1)$
26. $7b - 7$ $7(b-1)$
27. $16 - 8a^2$ $8(2-a^2)$

28. $12 + 12y^2$
29. $8x + 12$
30. $16a - 24$

31. $30a - 6$
32. $20b + 5$
33. $7x^2 - 3x$

34. $12y^2 - 5y$
35. $3a^2 + 5a^5$
36. $9x - 5x^2$

37. $14y^2 + 11y$
38. $6b^3 - 5b^2$
39. $2x^4 - 4x$

40. $3y^4 - 9y$
41. $10x^4 - 12x^2$
42. $12a^5 - 32a^2$

43. $8a^8 - 4a^5$
44. $16y^4 - 8y^7$
45. $x^2y^2 - xy$

46. $a^2b^2 + ab$
47. $3x^2y^4 - 6xy$
48. $12a^2b^5 - 9ab$

49. $x^2y - xy^3$
50. $a^2b + a^4b^2$
51. $2a^5b + 3xy^3$

52. $5x^2y - 7ab^3$
53. $6a^2b^3 - 12b^2$ **6**
54. $8x^2y^3 - 4x^2$

55. $6a^2bc + 4ab^2c$

56. $10x^2yz^2 + 15xy^3z$

57. $18x^2y^2 - 9a^2b^2$

58. $9a^2x - 27a^3x^3$

59. $6x^3y^3 - 12x^6y^6$

60. $3a^2b^2 - 12a^5b^5$

61. $x^3 - 3x^2 - x$

62. $a^3 + 4a^2 + 8a$

63. $2x^2 + 8x - 12$

64. $a^3 - 3a^2 + 5a$

65. $b^3 - 5b^2 - 7b$

66. $5x^2 - 15x + 35$

67. $8y^2 - 12y + 32$

68. $3x^3 + 6x^2 + 9x$

69. $5y^3 - 20y^2 + 10y$

70. $2x^4 - 4x^3 + 6x^2$

71. $3y^4 - 9y^3 - 6y^2$

72. $2x^3 + 6x^2 - 14x$

73. $3y^3 - 9y^2 + 24y$

74. $2y^5 - 3y^4 + 7y^3$

75. $6a^5 - 3a^3 - 2a^2$

76. $x^3y - 3x^2y^2 + 7xy^3$

77. $2a^2b - 5a^2b^2 + 7ab^2$

78. $5y^3 + 10y^2 - 25y$

79. $4b^5 + 6b^3 - 12b$

80. $3a^2b^2 - 9ab^2 + 15b^2$

81. $8x^2y^2 - 4x^2y + x^2$

82. $x^4y^4 - 3x^3y^3 + 6x^2y^2$

83. $4x^5y^5 - 8x^4y^4 + x^3y^3$

84. $16x^2y - 8x^3y^4 - 48x^2y^2$

2   Factor.

85. $x(a + b) + 2(a + b)$

86. $a(x + y) + 4(x + y)$

87. $x(b + 2) - y(b + 2)$

88. $a(y - 4) - b(y - 4)$

89. $a(x - 2) + 5(2 - x)$

90. $a(x - 7) + b(7 - x)$

91. $b(y - 3) + 3(3 - y)$

92. $c(a - 2) - b(2 - a)$

93. $a(x - y) - 2(y - x)$

94. $3(a - b) - x(b - a)$

95. $x^3 + 4x^2 + 3x + 12$

96. $x^3 - 4x^2 - 3x + 12$

97. $2y^3 + 4y^2 + 3y + 6$

98. $3y^3 - 12y^2 + y - 4$

99. $ab + 3b - 2a - 6$

100. $yz + 6z - 3y - 18$

101. $x^2a - 2x^2 - 3a + 6$

102. $x^2y + 4x^2 + 3y + 12$

103. $3ax - 3bx - 2ay + 2by$

104. $8 + 2c + 4a^2 + a^2c$

105. $x^2 - 3x + 4ax - 12a$

106. $t^2 + 4t - st - 4s$

107. $xy - 5y - 2x + 10$

108. $2y^2 - 10y + 7xy - 35x$

109. $21x^2 + 6xy - 49x - 14y$

110. $4a^2 + 5ab - 10b - 8a$

111. $2ra + a^2 - 2r - a$ **112.** $2ab - 3b^2 - 3b + 2a$

113. $4x^2 + 3xy - 12y - 16x$ **114.** $8s + 12r - 6s^2 - 9rs$

115. $10xy^2 - 15xy + 6y - 9$ **116.** $10a^2b - 15ab - 4a + 6$

SUPPLEMENTAL EXERCISES 6.1

Factor by grouping.

117. **a.** $\left(2x^2 + 6x\right) + \left(5x + 15\right)$ $2x(x+3)$ **b.** $2x^2 + 5x + 6x + 15$
                                $(2x+5)(x+3)$

118. **a.** $3x^2 + 3xy - xy - y^2$ **b.** $3x^2 - xy + 3xy - y^2$

119. **a.** $2a^2 - 2ab - 3ab + 3b^2$ **b.** $2a^2 - 3ab - 2ab + 3b^2$

Compare your answers to parts **a** and **b** of Exercises 117–119 to answer Exercise 120.

120. Do different groupings of the terms in a polynomial affect the binomial factoring?

A whole number is a perfect number if it is the sum of all of its factors less than itself. For example, 6 is a perfect number because all the factors of 6 that are less than 6 are 1, 2, and 3, and $1 + 2 + 3 = 6$.

121. Find the one perfect number between 20 and 30.

122. Find the one perfect number between 490 and 500.

Solve.

123. In the equation $P = 2l + 2w$, what is the effect on P when the quantity $l + w$ doubles?

SECTION 6.2

Factoring Polynomials of the Form $x^2 + bx + c$

1   Factor trinomials of the form $x^2 + bx + c$

Trinomials of the form $x^2 + bx + c$, where b and c are integers, are shown at the right.

$x^2 + 9x + 14,$ $b = 9,$ $c = 14$
$x^2 - x - 12,$ $b = -1,$ $c = -12$
$x^2 - 2x - 15,$ $b = -2,$ $c = -15$

To factor a trinomial of this form means to express the trinomial as the product of two binomials. Some trinomials expressed as the product of binomials (factored form) are shown below.

$$\begin{array}{rcl}
\underline{\text{Trinomial}} & & \underline{\text{Factored Form}} \\
x^2 + 9x + 14 & = & (x + 2)(x + 7) \\
x^2 - x - 12 & = & (x + 3)(x - 4) \\
x^2 - 2x - 15 & = & (x + 3)(x - 5)
\end{array}$$

The method by which the factors of a trinomial are found is based on FOIL. Consider the following binomial products, noting the relationship between the constant terms of the binomials and the terms of the trinomials.

Signs in the binomials are the same

$$(x + 6)(x + 2) = x^2 + 2x + 6x + (6)(2) = x^2 + 8x + 12$$

Sum of 6 and 2 —————
Product of 6 and 2 —————

$$(x - 3)(x - 4) = x^2 - 4x - 3x + (-3)(-4) = x^2 - 7x + 12$$

Sum of -3 and -4 —————
Product of -3 and -4 —————

Signs in the binomials are opposite

$$(x + 3)(x - 5) = x^2 - 5x + 3x + (3)(-5) = x^2 - 2x - 15$$

Sum of 3 and -5 —————
Product of 3 and -5 —————

$$(x - 4)(x + 6) = x^2 + 6x - 4x + (-4)(6) = x^2 + 2x - 24$$

Sum of -4 and 6 —————
Product of -4 and 6 —————

Points to Remember in Factoring $x^2 + bx + c$

1. In the trinomial, the coefficient of x is the sum of the constant terms of the binomials.

2. In the trinomial, the constant term is the product of the constant terms of the binomials.

3. When the constant term of the trinomial is positive, the constant terms of the binomials have the same sign as the coefficient of x in the trinomial.

4. When the constant term of the trinomial is negative, the constant terms of the binomials have opposite signs.

Success at factoring a trinomial depends on remembering the four points listed above. For example, to factor

$$x^2 - 2x - 24,$$

find two numbers whose sum is -2 and whose product is -24 [Points 1 and 2]. Because the constant term of the trinomial is negative (-24), the numbers will have opposite signs [Point 4].

A systematic method of finding these numbers involves listing the factors of the constant term of the trinomial and the sum of those factors.

Factors of -24	Sum of the Factors
1, -24	$1 + (-24) = -23$
-1, 24	$-1 + 24 = 23$
2, -12	$2 + (-12) = -10$
-2, 12	$-2 + 12 = 10$
3, -8	$3 + (-8) = -5$
-3, 8	$-3 + 8 = 5$
4, -6	**$4 + (-6) = -2$**
-4, 6	$-4 + 6 = 2$

4 and -6 are two numbers whose sum is -2 and whose product is -24. Write the binomial factors of the trinomial.

$$x^2 - 2x - 24 = (x + 4)(x - 6)$$

Check: $(x + 4)(x - 6) = x^2 - 6x + 4x - 24 = x^2 - 2x - 24$

By the Commutative Property of Multiplication, the binomial factors can also be written as

$$x^2 - 2x - 24 = (x - 6)(x + 4).$$

Example 1   Factor: $x^2 + 18x + 32$

Solution

Factors of 32	Sum
1, 32	33
2, 16	18
4, 8	12

▶ Try only positive factors of 32 [Point 3].

▶ Once the correct pair is found, the other factors need not be tried.

$x^2 + 18x + 32 = (x + 2)(x + 16)$ ▶ Write the factors of the trinomial.

Check $(x + 2)(x + 16) =$
$x^2 + 16x + 2x + 32 =$
$x^2 + 18x + 32$

Problem 1   Factor: $x^2 - 8x + 15$

Solution   See page A21.

15

-35 $x - 3$ $x - 5$

Example 2   Factor: $x^2 - 6x - 16$

Solution

Factors of -16	Sum
1, -16	-15
-1, 16	15
2, -8	-6
-2, 8	6
4, -4	0

▶ The factors must be of opposite signs [Point 4].

$x^2 - 6x - 16 = (x + 2)(x - 8)$ ▶ Write the factors of the trinomial.

Check $(x + 2)(x - 8) =$
$x^2 - 8x + 2x - 16 =$
$x^2 - 6x - 16$

Problem 2   Factor: $x^2 + 3x - 18$

Solution   See page A21.

Not all trinomials can be factored when using only integers. Consider $x^2 - 6x - 8$.

Factors of -8	Sum
1, -8	-7
-1, 8	7
2, -4	-2
-2, 4	2

Because none of the pairs of factors of -8 have a sum of -6, the trinomial is not factorable. The trinomial is said to be **nonfactorable over the integers.**

2   Factor completely

A polynomial is factored completely when it is written as a product of factors that are nonfactorable over the integers.

Example 3 Factor: $3x^3 + 15x^2 + 18x$

Solution The GCF of $3x^3$, $15x^2$, and $18x$ is $3x$.

▶ Find the GCF of the terms of the polynomial.

$3x^3 + 15x^2 + 18x =$
$3x(x^2) + 3x(5x) + 3x(6) =$
$3x(x^2 + 5x + 6)$

▶ Factor out the GCF.
▶ Write the polynomial as a product of factors.
▶ Factor the trinomial $x^2 + 5x + 6$. Try only positive factors of 6.

Factors of 6	Sum
1, 6	7
2, 3	5

$3x^3 + 15x^2 + 18x = 3x(x + 2)(x + 3)$

Check $3x(x + 2)(x + 3) = 3x(x^2 + 3x + 2x + 6)$
$= 3x(x^2 + 5x + 6)$
$= 3x^3 + 15x^2 + 18x$

Problem 3 Factor: $3a^2b - 18ab - 81b$

Solution See page A21.

Example 4 Factor: $x^2 + 9xy + 20y^2$

Solution

Factors of 20	Sum
1, 20	21
2, 10	12
4, 5	9

▶ Try only positive factors of 20.

$x^2 + 9xy + 20y^2 = (x + 4y)(x + 5y)$

Check $(x + 4y)(x + 5y) = x^2 + 5xy + 4xy + 20y^2$
$= x^2 + 9xy + 20y^2$

Problem 4 Factor: $4x^2 - 40xy + 84y^2$

Solution See page A21.

EXERCISES 6.2

1 Factor.

1. $x^2 + 3x + 2$

2. $x^2 + 5x + 6$

3. $x^2 - x - 2$

4. $x^2 + x - 6$

5. $a^2 + a - 12$ *handwritten: -12 1 -12 $4-3$ $(a-3)a+4$*

6. $a^2 - 2a - 35$

7. $a^2 - 3a + 2$

8. $a^2 - 5a + 4$

9. $a^2 + a - 2$

10. $a^2 - 2a - 3$

11. $b^2 - 6b + 9$

12. $b^2 + 8b + 16$

13. $b^2 + 7b - 8$

14. $y^2 - y - 6$

15. $y^2 + 6y - 55$

16. $z^2 - 4z - 45$

17. $y^2 - 5y + 6$

18. $y^2 - 8y + 15$

19. $z^2 - 14z + 45$ *handwritten: $(x+10)(x+10)$*

20. $z^2 - 14z + 49$

21. $z^2 - 12z - 160$

22. $p^2 + 2p - 35$

23. $p^2 + 12p + 27$ *handwritten: 100*

24. $p^2 - 6p + 8$

25. $x^2 + 20x + 100$

26. $x^2 + 18x + 81$

27. $b^2 + 9b + 20$ *handwritten: -42*

28. $b^2 + 13b + 40$

29. $x^2 - 11x - 42$ *handwritten: 2*

30. $x^2 + 9x - 70$

31. $b^2 - b - 20$ *handwritten: -11*

32. $b^2 + 3b - 40$

33. $y^2 - 14y - 51$

34. $y^2 - y - 72$

35. $p^2 - 4p - 21$

36. $p^2 + 16p + 39$

37. $y^2 - 8y + 32$

38. $y^2 - 9y + 81$

39. $x^2 - 20x + 75$

40. $p^2 + 24p + 63$

41. $x^2 - 15x + 56$

42. $x^2 + 21x + 38$

43. $x^2 + x - 56$

44. $x^2 + 5x - 36$

45. $a^2 - 21a - 72$

46. $a^2 - 7a - 44$

47. $a^2 - 15a + 36$

48. $a^2 - 21a + 54$

49. $z^2 - 9z - 136$

50. $z^2 + 14z - 147$

51. $c^2 - c - 90$

52. $c^2 - 3c - 180$

53. $z^2 + 15z + 44$

54. $p^2 + 24p + 135$

55. $c^2 + 19c + 34$

56. $c^2 + 11c + 18$

57. $x^2 - 4x - 96$

58. $x^2 + 10x - 75$

59. $x^2 - 22x + 112$

60. $x^2 + 21x - 100$

61. $b^2 + 8b - 105$

62. $b^2 - 22b + 72$

63. $a^2 - 9a - 36$

64. $a^2 + 42a - 135$

65. $b^2 - 23b + 102$

66. $b^2 - 25b + 126$

67. $a^2 + 27a + 72$

68. $z^2 + 24z + 144$

69. $x^2 + 25x + 156$

70. $x^2 - 29x + 100$

71. $x^2 - 10x - 96$

72. $x^2 + 9x - 112$

2   Factor.

$2\left(x^2 + 3x + 2\right)$

$2((x+2)(x+1))$ 21

73. $2x^2 + 6x + 4$

74. $3x^2 + 15x + 18$

75. $3a^2 + 3a - 18$

76. $4x^2 - 4x - 8$

77. $ab^2 + 2ab - 15a$

78. $ab^2 + 7ab - 8a$

79. $xy^2 - 5xy + 6x$

80. $xy^2 + 8xy + 15x$

81. $z^3 - 7z^2 + 12z$

82. $2a^3 + 6a^2 + 4a$

83. $3y^3 - 15y^2 + 18y$

84. $4y^3 + 12y^2 - 72y$

85. $3x^2 + 3x - 36$

86. $2x^3 - 2x^2 - 4x$

87. $5z^2 - 15z - 140$

88. $6z^2 + 12z - 90$

89. $2a^3 + 8a^2 - 64a$

90. $3a^3 - 9a^2 - 54a$

91. $x^2 - 5xy + 6y^2$

92. $x^2 + 4xy - 21y^2$

93. $a^2 - 9ab + 20b^2$

94. $a^2 - 15ab + 50b^2$

95. $x^2 - 3xy - 28y^2$

96. $s^2 + 2st - 48t^2$

97. $y^2 - 15yz - 41z^2$

98. $y^2 + 85yz + 36z^2$

99. $z^4 - 12z^3 + 35z^2$

100. $z^4 + 2z^3 - 80z^2$

101. $b^4 - 22b^3 + 120b^2$

102. $b^4 - 3b^3 - 10b^2$

103. $2y^4 - 26y^3 - 96y^2$

104. $3y^4 + 54y^3 + 135y^2$

105. $x^4 + 7x^3 - 8x^2$

106. $x^4 - 11x^3 - 12x^2$

107. $4x^2y + 20xy - 56y$

108. $3x^2y - 6xy - 45y$

109. $8y^2 - 32y + 24$

110. $10y^2 - 100y + 90$

111. $c^3 + 13c^2 + 30c$

112. $c^3 + 18c^2 - 40c$

113. $3x^3 - 36x^2 + 81x$

114. $4x^3 + 4x^2 - 24x$

115. $x^2 - 8xy + 15y^2$

116. $y^2 - 7xy - 8x^2$

117. $a^2 - 13ab + 42b^2$

118. $y^2 + 4yz - 21z^2$

119. $y^2 + 8yz + 7z^2$

120. $y^2 - 16yz + 15z^2$

121. $3x^2y + 60xy - 63y$

122. $4x^2y - 68xy - 72y$

123. $3x^3 + 3x^2 - 36x$

124. $4x^3 + 12x^2 - 160x$

125. $4z^3 + 32z^2 - 132z$

126. $5z^3 - 50z^2 - 120z$

127. $4x^3 + 8x^2 - 12x$

128. $5x^3 + 30x^2 + 40x$

129. $5p^2 + 25p - 420$

130. $4p^2 - 28p - 480$

131. $p^4 + 9p^3 - 36p^2$

132. $\boxed{p^4 + p^3 - 56p^2}$ $p^2\left(p^2 + p - 56\right)$

133. $a^2 - 8ab - 33b^2$

134. $x^2 + 4xy - 60y^2$

$(p-7\ \ p+8)$ $\overset{8\ \ -7}{}$

135. $15ab^2 + 45ab - 60a$

136. $20a^2b - 100ab + 120b$

$p^2(p-7)(p+8)$

SUPPLEMENTAL EXERCISES 6.2

Factor.

137. $20 + c^2 + 9c$

138. $x^2y - 54y - 3xy$

139. $45a^2 + a^2b^2 - 14a^2b$

140. $12p^2 - 96p + 3p^3$

Find all integers k such that the trinomial can be factored over the integers.

141. $x^2 + kx + 35$

142. $x^2 + kx + 18$

143. $x^2 - kx + 21$

144. $x^2 - kx + 14$

SECTION 6.3

Factoring Polynomials of the Form $ax^2 + bx + c$

1 Factor trinomials of the form $ax^2 + bx + c$ using trial factors

Trinomials of the form $ax^2 + bx + c$, where a, b, and c are integers and $a \neq 0$, are shown at the right.

$3x^2 - x + 4$, $a = 3$, $b = -1$, $c = 4$
$4x^2 + 5x - 8$, $a = 4$, $b = 5$, $c = -8$

These trinomials differ from those in the previous section in that the coefficient of x^2 is not 1. There are various methods of factoring these trinomials. The method described in this objective is factoring trinomials using trial factors.

Factoring polynomials of the form $ax^2 + bx + c$ by trial and error may require testing many trial factors. To reduce the number of trial factors, remember the following points.

Points to Remember in Factoring $ax^2 + bx + c$

1. If the terms of the trinomial do not have a common factor, then the terms of a binomial factor cannot have a common factor.
2. When the constant term of the trinomial is positive, the constant terms of the binomials have the same sign as the coefficient of x in the trinomial.
3. When the constant term of the trinomial is negative, the constant terms of the binomials have opposite signs.

Factor: $10x^2 - x - 3$

The terms of the trinomial do not have a common factor; therefore, a binomial factor will not have a common factor.

Because the constant term, c, of the trinomial is negative (-3), the binomial factors will have opposite signs (Point 3).

Find the factors of a (10) and the factors of c (-3).	Factors of 10	Factors of -3
	1, 10	1, -3
	2, 5	-1, 3

Using these factors, write trial factors, and use the Outer and Inner products of FOIL to check the middle term.	Trial Factors	Middle Term
	$(x + 1)(10x - 3)$	$-3x + 10x = 7x$
	$(x - 1)(10x + 3)$	$3x - 10x = -7x$
	$\mathbf{(2x + 1)(5x - 3)}$	$-6x + 5x = -x$
	$(2x - 1)(5x + 3)$	$6x - 5x = x$
	$(10x + 1)(x - 3)$	$-30x + x = -29x$
	$(10x - 1)(x + 3)$	$30x - x = 29x$
	$(5x + 1)(2x - 3)$	$-15x + 2x = -13x$
	$(5x - 1)(2x + 3)$	$15x - 2x = 13x$

From the list of trial factors, $10x^2 - x - 3 = (2x + 1)(5x - 3)$

Check: $(2x + 1)(5x - 3) = 10x^2 - 6x + 5x - 3$
$$= 10x^2 - x - 3$$

All the trial factors for this trinomial were listed in this example. However, once the correct binomial factors are found, it is not necessary to continue checking the remaining trial factors.

Factor: $4x^2 - 27x + 18$

The terms of the trinomial do not have a common factor; therefore, a binomial factor will not have a common factor.

Because the constant term, c, of the trinomial is positive (18), the binomial factors will have the same sign as the coefficient of x. Because the coefficient of x is -27, both signs will be negative (Point 2).

Find the factors of a (4) and the negative factors of c (18).	**Factors of 4** 1, 4 2, 2	**Factors of 18** $-1, -18$ $-2, -9$ $-3, -6$

Using these factors, write trial factors, and use the Outer and Inner products of FOIL to check the middle term.	**Trial Factors** $(x - 1)(4x - 18)$ $(x - 2)(4x - 9)$ $(x - 3)(4x - 6)$ $(2x - 1)(2x - 18)$ $(2x - 2)(2x - 9)$ $(2x - 3)(2x - 6)$ $(4x - 1)(x - 18)$ $(4x - 2)(x - 9)$ $\mathbf{(4x - 3)(x - 6)}$	**Middle Term** Common factor $-9x - 8x = -17x$ Common factor Common factor Common factor Common factor $-72x - x = -73x$ Common factor $\mathbf{-24x - 3x = -27x}$

The correct factors have been found. The remaining trial factors need not be checked.

$$4x^2 - 27x + 18 = (4x - 3)(x - 6)$$

This example illustrates that many of the trial factors may have common factors and thus need not be tried. For the remainder of this chapter, the trial factors with a common factor will not be listed.

Example 1   Factor: $3x^2 + 20x + 12$

Solution

Factors of 3	**Factors of 12**
1, 3	1, 12
	2, 6
	3, 4

▶ Because 20 is positive, only the positive factors of 12 need be tried.

Trial Factors	**Middle Term**
$(x + 3)(3x + 4)$	$4x + 9x = 13x$
$(3x + 1)(x + 12)$	$36x + x = 37x$
$(3x + 2)(x + 6)$	$18x + 2x = 20x$

▶ Write the trial factors. Use FOIL to check the middle term.

$3x^2 + 20x + 12 = (3x + 2)(x + 6)$

Problem 1 Factor: $6x^2 - 11x + 5$

Solution See page A22.

Example 2 Factor: $6x^2 - 5x - 6$

Solution

Factors of 6	Factors of -6
1, 6	1, -6
2, 3	-1, 6
	2, -3
	-2, 3

▶ Find the factors of a (6) and the factors of c (-6).

Trial Factors	Middle Term
$(x - 6)(6x + 1)$	$x - 36x = -35x$
$(x + 6)(6x - 1)$	$-x + 36x = 35x$
$(2x - 3)(3x + 2)$	$4x - 9x = -5x$

▶ Write the trial factors. Use FOIL to check the middle term.

$$6x^2 - 5x - 6 = (2x - 3)(3x + 2)$$

Check $(2x - 3)(3x + 2) = 6x^2 + 4x - 9x - 6$
$$= 6x^2 - 5x - 6$$

Problem 2 Factor: $8x^2 + 14x - 15$

Solution See page A22.

Example 3 Factor : $15 - 2x - x^2$

Solution

Factors of 15	Factors of -1
1, 15	1, -1
3, 5	

▶ The terms have no common factor. The coefficient of x^2 is -1.

Trial Factors	Middle Term
$(1 + x)(15 - x)$	$-x + 15x = 14x$
$(1 - x)(15 + x)$	$x - 15x = -14x$
$(3 + x)(5 - x)$	$-3x + 5x = 2x$
$(3 - x)(5 + x)$	$3x - 5x = -2x$

▶ Write trial factors. Determine the middle term of the trinomial.

$$15 - 2x - x^2 = (3 - x)(5 + x)$$

Check $(3 - x)(5 + x) = 15 + 3x - 5x - x^2$
$$= 15 - 2x - x^2$$

Problem 3 Factor: $24 - 10y + y^2$

Solution See page A22.

The first step in factoring a trinomial is to determine if there is a common factor. If there is a common factor, factor out the GCF of the terms.

Example 4   Factor: $3x^3 - 23x^2 + 14x$

Solution   The GCF of $3x^3$, $23x^2$, and $14x$ is x. ▶ Find the GCF of the terms of the polynomial.

$3x^3 - 23x^2 + 14x =$
$x(3x^2 - 23x + 14)$ ▶ Factor out the GCF.

Factors of 3	Factors of 14
1, 3	−1, −14
	−2, −7

▶ Factor the trinomial $3x^2 - 23x + 14$.

Trial Factors	Middle Term
$(x - 1)(3x - 14)$	$-14x - 3x = -17x$
$(x - 14)(3x - 1)$	$-x - 42x = -43x$
$(x - 2)(3x - 7)$	$-7x - 6x = -13x$
$(x - 7)(3x - 2)$	$-2x - 21x = -23x$

$3x^3 - 23x^2 + 14x = x(x - 7)(3x - 2)$

Check $x(x - 7)(3x - 2) =$
$x(3x^2 - 2x - 21x + 14) =$
$x(3x^2 - 23x + 14) =$
$3x^3 - 23x^2 + 14x$

Problem 4   Factor: $4a^2b^2 - 30a^2b + 14a^2$

Solution   See page A22.

2 Factor trinomials of the form $ax^2 + bx + c$ by grouping

In the previous objective, trinomials of the form $ax^2 + bx + c$ were factored using trial factors. In this objective, factoring by grouping is used.

To factor $ax^2 + bx + c$, first find two factors of $a \cdot c$ whose sum is b. Use the two factors to rewrite the middle term of the trinomial as the sum of two terms. Then use factoring by grouping to write the factorization of the trinomial.

Factor: $2x^2 + 13x + 15$

$a = 2$, $c = 15$, $a \cdot c = 2 \cdot 15 = 30$
Find two positive factors of 30
whose sum is 13.

The factors are 3 and 10.

Positive Factors of 30	Sum
1, 30	31
2, 15	17
3, 10	13
5, 6	11

Use the factors 3 and 10 to
rewrite $13x$ as $3x + 10x$.

$2x^2 + \quad 13x \quad + 15 =$
$2x^2 + 3x + 10x + 15 =$

Factor by grouping.

$(2x^2 + 3x) + (10x + 15) =$
$x(2x + 3) + 5(2x + 3) =$
$(2x + 3)(x + 5)$

Check: $(2x + 3)(x + 5) = 2x^2 + 10x + 3x + 15$
$\qquad\qquad\qquad\quad = 2x^2 + 13x + 15$

Factor: $6x^2 - 11x - 10$

$a = 6$, $c = -10$, $a \cdot c = 6(-10) = -60$
Find two factors of -60 whose
sum is -11.

Factors of -60	Sum
1, -60	-59
-1, $\ 60$	59
2, -30	-28
-2, $\ 30$	28
3, -20	-17
-3, $\ 20$	17
4, -15	-11

The required sum has been found.
The remaining factors need not be
checked.
The factors are 4 and -15.

Use the factors 4 and -15 to
rewrite $-11x$ as $4x - 15x$.

$6x^2 - \quad 11x \quad - 10 =$
$6x^2 + 4x - 15x - 10 =$

Factor by grouping.
Note: $-15x - 10 = -(15x + 10)$

$(6x^2 + 4x) - (15x + 10) =$
$2x(3x + 2) - 5(3x + 2) =$
$(3x + 2)(2x - 5)$

Check: $(3x + 2)(2x - 5) = 6x^2 - 15x + 4x - 10$
$\qquad\qquad\qquad\qquad = 6x^2 - 11x - 10$

Factor: $3x^2 - 2x - 4$

$a = 3$, $c = -4$, $a \cdot c = 3(-4) = -12$
Find two factors of -12 whose sum is -2.

Factors of -12	Sum
1, -12	-11
-1, 12	11
2, -6	-4
-2, 6	4
3, -4	-1
-3, 4	1

No integer factors of -12 have a sum of -2. Therefore, $3x^2 - 2x - 4$ is non-factorable over the integers

Example 5   Factor: $2x^2 + 19x - 10$

Solution $-1(20) = -20$, $-1 + 20 = 19$ ▶ $a \cdot c = 2(-10) = -20$
Find two factors of -20 whose sum is 19.

$2x^2 + 19x - 10 =$
$2x^2 - x + 20x - 10 =$ ▶ Rewrite $19x$ as $-x + 20x$.
$(2x^2 - x) + (20x - 10) =$ ▶ Factor by grouping.
$x(2x - 1) + 10(2x - 1) =$
$(2x - 1)(x + 10)$

$(a + 7.)(2a - 1)$

Problem 5   Factor: $2a^2 + 13a - 7$

Solution   See page A22.

$2a^2 + 13a - 7$

$14 \qquad -1 \ 14$

13

$(2a^2 - 1a) + (14a - 7)$

$a(2a - 1) \qquad 7(2a - 1)$

Example 6   Factor: $8y^2 - 10y - 3$

$2(-12) = -24$, $2 - 12 = -10$ ▶ $a \cdot c = 8(-3) = -24$
Find two factors of -24 whose sum is -10.

$8y^2 - 10y - 3 =$
$8y^2 + 2y - 12y - 3 =$ ▶ Rewrite $-10y$ as $2y - 12y$.
$(8y^2 + 2y) - (12y + 3) =$ ▶ Factor by grouping.
$2y(4y + 1) - 3(4y + 1) =$
$(4y + 1)(2y - 3)$

Problem 6   Factor: $4a^2 - 11a - 3$

Solution   See page A22.

Remember that the first step in factoring a trinomial is to determine if there is a common factor. If there is a common factor, factor out the GCF of the terms.

Example 7   Factor: $24x^2y - 76xy + 40y$

Solution $24x^2y - 76xy + 40y =$
$4y(6x^2 - 19x + 10)$

▶ The terms of the polynomial have a common factor, $4y$. Factor out the GCF.

$-4(-15) = 60,\ -4 - 15 = -19$

▶ Factor $6x^2 - 19x + 10$.
$a \cdot c = 6(10) = 60$
Find two negative factors of 60 whose sum is -19.

$6x^2 - 19x + 10 =$
$6x^2 - 4x - 15x + 10 =$
$(6x^2 - 4x) - (15x - 10) =$
$2x(3x - 2) - 5(3x - 2) =$
$(3x - 2)(2x - 5)$

▶ Rewrite $-19x$ as $-4x - 15x$.
▶ Factor by grouping.

$24x^2y - 76xy + 40y =$
$4y(6x^2 - 19x + 10) =$
$4y(3x - 2)(2x - 5)$

▶ Write the complete factorization of the given polynomial.

Problem 7   Factor: $15x^3 + 40x^2 - 80x$

Solution   See page A23.

EXERCISES 6.3

1   Factor by the method of using trial factors.

1. $2x^2 + 3x + 1$

2. $5x^2 + 6x + 1$

3. $2y^2 + 7y + 3$

4. $3y^2 + 7y + 2$

5. $2a^2 - 3a + 1$

6. $3a^2 - 4a + 1$

7. $2b^2 - 11b + 5$ $(2b - 1)(b - 5)$

8. $3b^2 - 13b + 4$

9. $2x^2 + x - 1$ $2x + 1$ $x + 1$

10. $4x^2 - 3x - 1$

11. $2x^2 - 5x - 3$

12. $3x^2 + 5x - 2$

13. $6z^2 - 7z + 3$

14. $9z^2 + 3z + 2$

15. $6t^2 - 11t + 4$

16. $10t^2 + 11t + 3$

17. $8x^2 + 33x + 4$

18. $7x^2 + 50x + 7$

19. $3b^2 - 16b + 16$

20. $6b^2 - 19b + 15$

21. $2z^2 - 27z - 14$

22. $4z^2 + 5z - 6$

23. $3p^2 + 22p - 16$

24. $7p^2 + 19p + 10$

25. $6x^2 - 17x + 12$

26. $15x^2 - 19x + 6$

27. $5b^2 + 33b - 14$

28. $8x^2 - 30x + 25$

29. $6a^2 + 7a - 24$

30. $14a^2 + 15a - 9$

31. $18t^2 - 9t - 5$

32. $12t^2 + 28t - 5$

33. $6b^2 + 71b - 12$

34. $8b^2 + 65b + 8$

35. $9x^2 + 12x + 4$

36. $25x^2 - 30x + 9$

37. $15a^2 + 26a - 21$

38. $6a^2 + 23a + 21$

39. $8y^2 - 26y + 15$

40. $18y^2 - 27y + 4$

41. $8z^2 + 2z - 15$

42. $10z^2 + 3z - 4$

43. $3x^2 + 14x - 5$

44. $15a^2 - 22a + 8$

45. $12x^2 + 25x + 12$

46. $10b^2 + 43b - 9$

47. $12z^2 + 32z - 35$

48. $16y^2 - 46y + 15$

49. $3a^2b - 16ab + 16b$

50. $2a^2b - ab - 21b$

51. $3z^2 + 95z + 10$

52. $8z^2 - 36z + 1$

53. $3x^2 + xy - 2y^2$

54. $6x^2 + 10xy + 4y^2$

55. $28 + 3z - z^2$

56. $15 - 2z - z^2$

57. $8 - 7x - x^2$

58. $12 + 11x - x^2$

59. $9x^2 + 33x - 60$

60. $16x^2 - 16x - 12$

61. $24x^2 - 52x + 24$

62. $60x^2 + 95x + 20$

63. $35a^4 + 9a^3 - 2a^2$

64. $15a^4 + 26a^3 + 7a^2$

65. $15b^2 - 115b + 70$

66. $25b^2 + 35b - 30$

67. $15a^2 + 11ab - 14b^2$

68. $15a^2 - 31ab + 10b^2$

69. $33z - 8z^2 - z^3$

70. $24z + 10z^2 - z^3$

71. $10x^3 + 12x^2 + 2x$

72. $9x^3 - 39x^2 + 12x$

73. $10y^3 - 44y^2 + 16y$

74. $14y^3 + 94y^2 - 28y$

75. $4yz^3 + 5yz^2 - 6yz$

76. $2yz^3 - 17yz^2 + 8yz$

77. $20b^4 + 41b^3 + 20b^2$

78. $6b^4 - 13b^3 + 6b^2$

79. $9x^3y + 12x^2y + 4xy$

80. $9a^3b - 9a^2b^2 - 10ab^3$

81. $2a^3b - 11a^2b^2 + 5ab^3$

82. $2x^4 + 5x^3y - 12x^2y^2$

83. $12x^2y^2 - 17xy^3 + 6y^4$

84. $6a^3b + 20a^2b^2 - 16ab^3$

2 Factor by grouping.

85. $2t^2 - t - 10$

86. $2t^2 + 5t - 12$

87. $3p^2 - 16p + 5$

88. $6p^2 + 5p + 1$

89. $12y^2 - 7y + 1$

90. $6y^2 - 5y + 1$

91. $5x^2 - 62x - 7$

92. $9x^2 - 13x - 4$

93. $12y^2 + 19y + 5$

94. $5y^2 - 22y + 8$

95. $7a^2 + 47a - 14$

96. $11a^2 - 54a - 5$

97. $4z^2 + 11z + 6$

98. $6z^2 - 25z + 14$

99. $22p^2 + 51p - 10$

100. $14p^2 - 41p + 15$

101. $8y^2 + 17y + 9$

102. $12y^2 - 145y + 12$

103. $6b^2 - 13b + 6$

104. $20b^2 + 37b + 15$

105. $33b^2 + 34b - 35$

106. $15b^2 - 43b + 22$

107. $18y^2 - 39y + 20$

108. $24y^2 + 41y + 12$

109. $15x^2 - 82x + 24$

110. $13z^2 + 49z - 8$

111. $10z^2 - 29z + 10$

112. $15z^2 - 44z + 32$

113. $36z^2 + 72z + 35$

114. $16z^2 + 8z - 35$

115. $14y^2 - 29y + 12$

116. $8y^2 + 30y + 25$

117. $6x^2 + 35x - 6$

118. $4x^2 + 6x + 2$

119. $12x^2 + 33x - 9$

120. $15y^2 - 50y + 35$

121. $30y^2 + 10y - 20$

122. $2x^3 - 11x^2 + 5x$

123. $2x^3 - 3x^2 - 5x$

124. $3a^2 + 5ab - 2b^2$

125. $2a^2 - 9ab + 9b^2$

126. $4y^2 - 11yz + 6z^2$

127. $2y^2 + 7yz + 5z^2$

128. $12 - x - x^2$

129. $2 + x - x^2$

130. $80y^2 - 36y + 4$

131. $24y^2 - 24y - 18$

132. $8z^3 + 14z^2 + 3z$

133. $6z^3 - 23z^2 + 20z$

134. $6x^2y - 11xy - 10y$

135. $8x^2y - 27xy + 9y$

136. $3x^2 - 26xy + 35y^2$

137. $4x^2 + 16xy + 15y^2$

138. $216y^2 - 3y - 3$

139. $360y^2 + 4y - 4$

140. $21 - 20x - x^2$

141. $18 + 17x - x^2$

142. $10t^2 - 5t - 50$

143. $16t^2 + 40t - 96$

144. $3p^3 - 16p^2 + 5p$

145. $6p^3 + 5p^2 + p$

146. $26z^2 + 98z - 24$

147. $30z^2 - 87z + 30$

148. $12a^3 + 14a^2 - 48a$

149. $42a^3 + 45a^2 - 27a$

150. $36p^2 - 9p^3 - p^4$

151. $9x^2y - 30xy^2 + 25y^3$

152. $8x^2y - 38xy^2 + 35y^3$

153. $9x^3y - 24x^2y^2 + 16xy^3$

154. $45a^3b - 78a^2b^2 + 24ab^3$

SUPPLEMENTAL EXERCISES 6.3

Factor.

155. $6y + 8y^3 - 26y^2$

156. $22p^2 - 3p^3 + 16p$

157. $a^3b - 24ab - 2a^2b$

158. $3xy^2 - 14xy + 2xy^3$

159. $25t^2 + 60t - 10t^3$

160. $3xy^3 + 2x^3y - 7x^2y^2$

Simplify. Then factor.

161. $2(y + 2)^2 - (y + 2) - 3$

162. $3(a + 2)^2 - (a + 2) - 4$

163. $10(x + 1)^2 - 11(x + 1) - 6$

164. $4(y - 1)^2 - 7(y - 1) - 2$

Find all integers k such that the trinomial can be factored over the integers.

165. $2x^2 + kx + 3$

166. $2x^2 + kx - 3$

167. $3x^2 + kx + 2$

168. $3x^2 + kx - 2$

169. $2x^2 + kx + 5$

170. $2x^2 + kx - 5$

S E C T I O N **6.4**

Special Factoring

1 Factor the difference of two squares and perfect square trinomials

Recall from Objective 4 in Section 5.3 that the product of the sum and difference of the same two terms equals the square of the first term minus the square of the second term.

$$(a + b)(a - b) = a^2 - b^2$$

The expression $a^2 - b^2$ is the difference of two squares. The pattern above suggests the following rule for factoring the difference of two squares.

Rule for Factoring the Difference of Two Squares

Difference of Two Squares		Sum and Difference of Two Terms
$a^2 - b^2$	$=$	$(a + b)(a - b)$

$a^2 + b^2$ is the *sum* of two squares. It is nonfactorable over the integers.

Example 1 Factor.

        A. $x^2 - 16$ B. $x^2 - 10$ C. $z^6 - 25$

Solution A. $x^2 - 16 = x^2 - 4^2$ ▶ Write $x^2 - 16$ as the difference of two squares.
                 $= (x + 4)(x - 4)$ ▶ The factors are the sum and difference of the terms x and 4.

        B. $x^2 - 10$ is nonfactorable   ▶ Because 10 is not a square, $x^2 - 10$ cannot be over the integers.   written as the difference of two squares.

        C. $z^6 - 25 = (z^3)^2 - 5^2$ ▶ Write $z^6 - 25$ as the difference of two squares.
                 $= (z^3 + 5)(z^3 - 5)$ ▶ The factors are the sum and difference of the terms z^3 and 5.

Problem 1   Factor.

A. $25a^2 - b^2$ B. $6x^2 - 1$ C. $n^8 - 36$

Solution   See page A23.

[handwritten: $5a - b$ $n^4 - 6$
$5a + b$ $n^4 + 6$]

Example 2   Factor: $z^4 - 16$

Solution $z^4 - 16 = (z^2)^2 - (4)^2 =$ ▶ This is the difference of two squares.
$(z^2 + 4)(z^2 - 4) =$ ▶ The factors are the sum and difference of the terms z^2 and 4.

$(z^2 + 4)(z + 2)(z - 2)$ ▶ Factor $z^2 - 4$, which is the difference of two squares. $z^2 + 4$ is nonfactorable over the integers.

Problem 2   Factor: $n^4 - 81$

[handwritten: $n^2 - 9$ $n^2 + 9$ $n - 3$ $n + 3$]

Solution   See page A23.

Recall from Objective 4 in Section 5.3 the pattern for finding the square of a binomial.

$$(a + b)^2 = (a + b)(a + b) = a^2 + ab + ab + b^2$$
$$= a^2 + 2ab + b^2$$

Square of first term————————⌐
Twice the product of the two terms———⌐
Square of last term————————————⌐

The square of a binomial is a perfect square trinomial. The pattern above suggests the following rule for factoring a perfect square trinomial.

Rule for Factoring a Perfect Square Trinomial

Perfect Square Trinomial		Square of a Binomial
$a^2 + 2ab + b^2$	$= (a + b)(a + b) =$	$(a + b)^2$
$a^2 - 2ab + b^2$	$= (a - b)(a - b) =$	$(a - b)^2$

Note in these patterns that the sign in the binomial is the sign of the middle term of the trinomial.

Factor: $4x^2 - 20x + 25$

Check that the first term and the last term are squares.	$4x^2 = (2x)^2,\ 25 = 5^2$

Use the squared terms to factor the trinomial as the square of a binomial. The sign of the binomial is the sign of the middle term of the trinomial.

$(2x - 5)^2$

Check the factorization.

$(2x - 5)^2 =$
$(2x)^2 + 2(2x)(-5) + (-5)^2$
$4x^2 - 20x + 25$

The factorization is correct.

$4x^2 - 20x + 25 = (2x - 5)^2$

Factor: $9x^2 + 30x + 16$

Check that the first term and the last term are squares.

$9x^2 = (3x)^2,\ 16 = 4^2$

Use the squared terms to factor the trinomial as the square of a binomial. The sign of the binomial is the sign of the middle term in the trinomial.

$(3x + 4)^2$

Check the factorization.

$(3x + 4)^2 =$
$(3x)^2 + 2(3x)(4) + 4^2 =$
$9x^2 + 24x + 16$

$9x^2 + 24x + 16 \neq 9x^2 + 30x + 16$
The proposed factorization is not correct.

In this case, the polynomial is not a perfect square trinomial. It may, however, still factor. In fact, $9x^2 + 30x + 16 = (3x + 2)(3x + 8)$. If the trinomial does not check as a perfect square trinomial, try to factor it by another method.

A perfect square trinomial can always be factored using either of the methods presented in Section 6.3. However, noticing that a trinomial is a perfect square trinomial can save you a considerable amount of time.

Example 3   Factor.   A. $9x^2 - 30x + 25$ B. $4x^2 + 37x + 9$

Solution   A. $9x^2 = (3x)^2$, $25 = 5^2$ ▸ Check that the first and last terms are squares.

$(3x - 5)^2$ ▸ Use the squared terms to factor the trinomial as the square of a binomial.

$(3x - 5)^2 =$ ▸ Check the factorization.
$(3x)^2 + 2(3x)(-5) + (-5)^2 =$
$9x^2 - 30x + 25$ ▸ The factorization checks.

$9x^2 - 30x + 25 = (3x - 5)^2$

B. $4x^2 = (2x)^2$, $9 = 3^2$ ▸ Check that the first and last terms are squares.

$(2x + 3)^2$ ▸ Use the squared terms to factor the trinomial as the square of a binomial.

$(2x + 3)^2 =$ ▸ Check the factorization.
$(2x)^2 + 2(2x)(3) + 3^2 =$
$4x^2 + 12x + 9$ ▸ The factorization does not check.

$4x^2 + 37x + 9 =$ ▸ Use another method to factor the
$(4x + 1)(x + 9)$ trinomial.

Problem 3   Factor.   A. $16y^2 + 8y + 1$ B. $x^2 + 14x + 36$

Solution   See page A23.

2   Factor completely

When factoring a polynomial completely, ask yourself the following questions about the polynomial.

1. Is there a common factor? If so, factor out the common factor.

2. Is the polynomial the difference of two squares? If so, factor.

3. Is the polynomial a perfect square trinomial? If so, factor.

4. Is the polynomial a trinomial that is the product of two binomials? If so, factor.

5. Does the polynomial contain four terms? If so, try factoring by grouping.

6. Is each binomial factor nonfactorable over the integers? If not, factor.

Example 4   Factor.   A. $3x^2 - 48$ B. $x^3 - 3x^2 - 4x + 12$ C. $4x^2y^2 + 12xy^2 + 9y^2$

Solution   A. $3x^2 - 48 =$
$\quad\quad 3(x^2 - 16) =$
$\quad\quad 3(x + 4)(x - 4)$

▶ The GCF of the terms is 3.
Factor out the common factor.
▶ Factor the difference of two squares.

B. $x^3 - 3x^2 - 4x + 12 =$
$\quad (x^3 - 3x^2) - (4x - 12) =$
$\quad x^2(x - 3) - 4(x - 3) =$
$\quad (x - 3)(x^2 - 4) =$
$\quad (x - 3)(x + 2)(x - 2)$

▶ The polynomial contains four terms.
Factor by grouping.

▶ Factor the difference of two squares.

C. $4x^2y^2 + 12xy^2 + 9y^2 =$
$\quad y^2(4x^2 + 12x + 9) =$
$\quad y^2(2x + 3)^2$

▶ The GCF of the terms is y^2.
Factor out the common factor.
▶ Factor the perfect square trinomial.

Problem 4   Factor.   A. $12x^3 - 75x$ B. $a^2b - 7a^2 - b + 7$ C. $4x^3 + 28x^2 - 120x$

Solution   See page A23.

EXERCISES 6.4

1   Factor.

1. $x^2 - 4$

2. $x^2 - 9$

3. $a^2 - 81$

4. $a^2 - 49$

5. $4x^2 - 1$

6. $9x^2 - 16$

7. $y^2 + 2y + 1$

8. $y^2 + 14y + 49$

9. $a^2 - 2a + 1$

10. $x^2 + 8x - 16$

11. $z^2 - 18z - 81$

12. $x^2 - 12x + 36$

13. $x^6 - 9$

14. $y^{12} - 64$

15. $25x^2 - 1$

16. $9x^2 - 1$

17. $1 - 49x^2$

18. $1 - 64x^2$

19. $x^2 + 2xy + y^2$

20. $x^2 + 6xy + 9y^2$

21. $4a^2 + 4a + 1$

22. $25x^2 + 10x + 1$

23. $64a^2 - 16a + 1$

24. $9a^2 + 6a + 1$

25. $t^2 + 36$

26. $x^2 + 64$

27. $x^4 - y^2$

28. $b^4 - 16a^2$

29. $9x^2 - 16y^2$

30. $25z^2 - y^2$

31. $16b^2 + 8b + 1$

32. $4a^2 - 20a + 25$

33. $4b^2 + 28b + 49$

34. $9a^2 - 42a + 49$

35. $25a^2 + 30ab + 9b^2$

36. $4a^2 - 12ab + 9b^2$

37. $x^2y^2 - 4$

38. $a^2b^2 - 25$

39. $16 - x^2y^2$

40. $49x^2 + 28xy + 4y^2$

41. $4y^2 - 36yz + 81z^2$

42. $64y^2 - 48yz + 9z^2$

43. $9a^2b^2 - 6ab + 1$

44. $16x^2y^2 - 24xy + 9$

45. $8y^2 - 2$

46. $12a^2 - 48$

47. $3a^3 + 6a^2 + 3a$

48. $4rs^2 - 4rs + r$

49. $m^4 - 256$

50. $81 - t^4$

51. $9x^2 + 13x + 4$

52. $x^2 + 10x + 16$

53. $16y^4 + 48y^3 + 36y^2$

54. $36c^4 - 48c^3 + 16c^2$

55. $y^8 - 81$

56. $32s^8 - 2$

57. $25 - 20p + 4p^2$

58. $9 + 24a + 16a^2$

2   Factor.

59. $2x^2 - 18$

60. $y^3 - 10y^2 + 25y$

61. $x^4 + 2x^3 - 35x^2$

62. $a^4 - 11a^3 + 24a^2$

63. $5b^2 + 75b + 180$

64. $6y^2 - 48y + 72$

65. $3a^2 + 36a + 10$

66. $5a^2 - 30a + 4$

67. $2x^2y + 16xy - 66y$

68. $3a^2b + 21ab - 54b$

69. $x^3 - 6x^2 - 5x$

70. $b^3 - 8b^2 - 7b$

71. $3y^2 - 36$

72. $3y^2 - 147$

73. $20a^2 + 12a + 1$

74. $12a^2 - 36a + 27$

75. $x^2y^2 - 7xy^2 - 8y^2$

76. $a^2b^2 + 3a^2b - 88a^2$

77. $10a^2 - 5ab - 15b^2$

78. $16x^2 - 32xy + 12y^2$

79. $50 - 2x^2$

80. $72 - 2x^2$

81. $12a^3b - a^2b^2 - ab^3$

82. $2x^3y - 7x^2y^2 + 6xy^3$

83. $2ax - 2a + 2bx - 2b$

84. $4ax - 12a - 2bx + 6b$

85. $12a^3 - 12a^2 + 3a$

86. $18a^3 + 24a^2 + 8a$

87. $243 + 3a^2$

88. $75 + 27y^2$

89. $12a^3 - 46a^2 + 40a$

90. $24x^3 - 66x^2 + 15x$

91. $x^3 - 2x^2 - x + 2$

92. $ay^2 - by^2 - a + b$

93. $4a^3 + 20a^2 + 25a$

94. $2a^3 - 8a^2b + 8ab^2$

95. $27a^2b - 18ab + 3b$

96. $a^2b^2 - 6ab^2 + 9b^2$

97. $48 - 12x - 6x^2$

98. $21x^2 - 11x^3 - 2x^4$

99. $ax^2 - 4a + bx^2 - 4b$

100. $a^2x - b^2x - a^2y + b^2y$

101. $x^4 - x^2y^2$

102. $b^4 - a^2b^2$

103. $18a^3 + 24a^2 + 8a$

104. $32xy^2 - 48xy + 18x$

105. $2b + ab - 6a^2b$

106. $20x - 11xy - 3xy^2$

107. $4x - 20 - x^3 + 5x^2$

108. $ay^2 - by^2 - 9a + 9b$

109. $72xy^2 + 48xy + 8x$

110. $4x^2y + 8xy + 4y$

111. $15y^2 - 2xy^2 - x^2y^2$

112. $4x^4 - 38x^3 + 48x^2$

113. $y^3 - 9y$

114. $a^4 - 16$

115. $2x^4y^2 - 2x^2y^2$

116. $6x^5y - 6xy^5$

117. $x^9 - x^5$

118. $8b^5 - 2b^3$

119. $24x^3y + 14x^2y - 20xy$

120. $12x^3y - 60x^2y + 63xy$

121. $4x^4y^2 - 20x^3y^2 - 25x^2y^2$

122. $9x^4y^2 + 24x^3y^2 + 16x^2y^2$

123. $x^3 - 2x^2 - 4x + 8$

124. $24x^2y + 6x^3y - 45x^4y$

125. $8xy^2 - 20x^2y^2 + 12x^3y^2$

126. $45y^2 - 42y^3 - 24y^4$

127. $36a^3b - 62a^2b^2 + 12ab^3$

128. $18a^3b + 57a^2b^2 + 30ab^3$

129. $5x^2y^2 - 11x^3y^2 - 12x^4y^2$

130. $24x^2y^2 - 32x^3y^2 + 10x^4y^2$

131. $2ay^2 - 4a + 2by^2 - 4b$

132. $10x^4y^2 - 15x^3y^3 - 25x^2y^4$

133. $15x^4y^2 - 13x^3y^3 - 20x^2y^4$

SUPPLEMENTAL EXERCISES 6.4

Find all integers k such that the trinomial is a perfect square trinomial.

134. $4x^2 - kx + 9$

135. $25x^2 - kx + 1$

136. $36x^2 + kxy + y^2$

137. $64x^2 + kxy + y^2$

138. $x^2 + 6x + k$

139. $x^2 - 4x + k$

140. $x^2 - 2x + k$

141. $x^2 + 10x + k$

The sum of two cubes or the difference of two cubes can always be factored. Here are the factoring formulas.

Sum of Two Cubes	Difference of Two Cubes
$a^3 + b^3 = (a + b)(a^2 - ab + b^2)$	$a^3 - b^3 = (a - b)(a^2 + ab + b^2)$

Use these formulas to factor each binomial.

142. $x^3 + 8$

143. $a^3 - 64$

144. $27y^3 - 1$

145. $8z^3 + 27$

S E C T I O N **6.5**

Solving Equations

1 Solve equations by factoring

Recall that the Multiplication Property of Zero states that the product of a number and zero is zero.

If a is a real number, than $a \cdot 0 = 0 \cdot a = 0$.

Consider the equation $a \cdot b = 0$. If this is a true equation, then either $a = 0$ or $b = 0$.

Principle of Zero Products

If the product of two factors is zero, then at least one of the factors must be zero.

If $a \cdot b = 0$, then $a = 0$ or $b = 0$.

The Principle of Zero Products is used in solving equations.

Solve: $(x - 2)(x - 3) = 0$

If $(x - 2)(x - 3) = 0$,
then $(x - 2) = 0$ or $(x - 3) = 0$.

$(x - 2)(x - 3) = 0$

Solve each equation for x.

$$x - 2 = 0 \qquad\qquad x - 3 = 0$$
$$x = 2 \qquad\qquad\quad x = 3$$

Check:

$(x - 2)(x - 3) = 0$		$(x - 2)(x - 3) = 0$	
$(2 - 2)(2 - 3)$	0	$(3 - 2)(3 - 3)$	0
$0(-1)$	0	$1(0)$	0
	$0 = 0$		$0 = 0$

A true equation   A true equation

Write the solution.   The solutions are 2 and 3.

An equation of the form $ax^2 + bx + c = 0$, $a \neq 0$, is a **quadratic equation**. A quadratic equation is in **standard form** when the polynomial is in descending order and equal to zero.

$$3x^2 + 2x + 1 = 0$$

$$4x^2 - 3x + 2 = 0$$

A quadratic equation can be solved by using the Principle of Zero Products when the polynomial $ax^2 + bx + c$ is factorable.

Example 1   Solve: $2x^2 + x = 6$

Solution

$$2x^2 + x = 6$$
$$2x^2 + x - 6 = 0 \qquad$$ ▶ Write the equation in standard form.
$$(2x - 3)(x + 2) = 0 \qquad$$ ▶ Factor the trinomial.
$$2x - 3 = 0 \qquad x + 2 = 0 \qquad$$ ▶ Set each factor equal to zero (the Principle of Zero Products).
$$2x = 3 \qquad\qquad x = -2 \qquad$$ ▶ Solve each equation for x.
$$x = \frac{3}{2}$$

Check

$2x^2 + x = 6$		$2x^2 + x = 6$	
$2\left(\frac{3}{2}\right)^2 + \frac{3}{2}$	6	$2(-2)^2 + (-2)$	6
$2\left(\frac{9}{4}\right) + \frac{3}{2}$	6	$2 \cdot 4 - 2$	6
$\frac{9}{2} + \frac{3}{2}$	6	$8 - 2$	6
	$6 = 6$		$6 = 6$

The solutions are $\frac{3}{2}$ and -2.   ▶ Write the solutions.

Problem 1 Solve: $2x^2 - 50 = 0$

Solution See page A24.

Example 2 Solve: $(x - 3)(x - 10) = -10$

Solution $(x - 3)(x - 10) = -10$
$x^2 - 13x + 30 = -10$ ▶ Multiply $(x - 3)(x - 10)$.
$x^2 - 13x + 40 = 0$ ▶ Write the equation in standard form.
$(x - 8)(x - 5) = 0$ ▶ Factor.

$x - 8 = 0 \quad x - 5 = 0$ ▶ Set each factor equal to zero.
$\quad x = 8 \qquad x = 5$ ▶ Solve each equation for x.

The solutions are 8 and 5.   ▶ Write the solutions.

Problem 2 Solve: $(x + 2)(x - 7) = 52$

Solution See page A24.

2 Application problems

Example 3 The sum of the squares of two consecutive positive odd integers is equal to 130. Find the two integers.

Strategy First positive odd integer: n
Second positive odd integer: $n + 2$

The sum of the square of the first positive odd integer and the square of the second positive odd integer is 130.

Solution $n^2 + (n + 2)^2 = 130$
$n^2 + n^2 + 4n + 4 = 130$
$2n^2 + 4n - 126 = 0$
$2(n^2 + 2n - 63) = 0$
$n^2 + 2n - 63 = 0$ ▶ Divide each side of the equation by 2.
$(n - 7)(n + 9) = 0$

$n - 7 = 0 \qquad n + 9 = 0$ ▶ Because -9 is not a positive odd integer,
$\quad n = 7 \qquad\quad n = -9$ it is not a solution.

$n = 7$ ▶ The first positive odd integer is 7.
$n + 2 = 7 + 2 = 9$ ▶ Substitute the value of n into the variable expression for the second positive odd integer and evaluate.

The two integers are 7 and 9.

Problem 3 The sum of the squares of two consecutive positive integers is 85. Find the two integers.

Solution See page A24.

Example 4 A stone is thrown into a well with an initial velocity of 8 ft/s. The well is 440 ft deep. How many seconds later will the stone hit the bottom of the well? Use the equation $d = vt + 16t^2$, where d is the distance in feet, v is the initial velocity, and t is the time in seconds.

Strategy To find the time for the stone to drop to the bottom of the well, replace the variables d and v by their given values and solve for t.

Solution

$$d = vt + 16t^2$$
$$440 = 8t + 16t^2$$
$$-16t^2 - 8t + 440 = 0$$
$$16t^2 + 8t - 440 = 0 \qquad \blacktriangleright \text{Multiply each side of the equation by } -1.$$
$$8(2t^2 + t - 55) = 0$$
$$2t^2 + t - 55 = 0 \qquad \blacktriangleright \text{Divide each side of the equation by 8.}$$
$$(2t + 11)(t - 5) = 0$$

$$2t + 11 = 0 \qquad\qquad t - 5 = 0$$
$$2t = -11 \qquad\qquad\quad t = 5$$
$$t = -\frac{11}{2}$$

$\blacktriangleright$ Because the time cannot be a negative number, $-\frac{11}{2}$ is not a solution.

The time is 5 s.

Problem 4 The length of a rectangle is 3 in. longer than twice the width. The area of the rectangle is 90 in.^{2}. Find the length and width of the rectangle.

Solution See page A24.

EXERCISES 6.5

1 Solve.

1. $(y + 3)(y + 2) = 0$

2. $(y - 3)(y - 5) = 0$

3. $(z - 7)(z - 3) = 0$

4. $(z + 8)(z - 9) = 0$

5. $x(x - 5) = 0$

6. $x(x + 2) = 0$

7. $a(a - 9) = 0$

8. $a(a + 12) = 0$

9. $y(2y + 3) = 0$

10. $t(4t - 7) = 0$

11. $2a(3a - 2) = 0$

12. $4b(2b + 5) = 0$

13. $(b + 2)(b - 5) = 0$ **14.** $(b - 8)(b + 3) = 0$

15. $x^2 - 81 = 0$ **16.** $9x^2 - 1 = 0$

17. $16x^2 - 49 = 0$ **18.** $x^2 + 6x + 8 = 0$

19. $x^2 - 8x + 15 = 0$ **20.** $z^2 + 5z - 14 = 0$

21. $z^2 + z - 72 = 0$ **22.** $x^2 - 5x + 6 = 0$

23. $x^2 - 3x - 10 = 0$ **24.** $y^2 + 4y - 21 = 0$

25. $2y^2 - y - 1 = 0$ **26.** $2a^2 - 9a - 5 = 0$

27. $3a^2 + 14a + 8 = 0$ **28.** $a^2 - 5a = 0$

29. $x^2 - 7x = 0$ **30.** $2a^2 - 8a = 0$

31. $a^2 + 5a = -4$ **32.** $a^2 - 5a = 24$

33. $y^2 - 5y = -6$ **34.** $y^2 - 7y = 8$

35. $2t^2 + 7t = 4$ **36.** $3t^2 + t = 10$

37. $3t^2 - 13t = -4$ **38.** $5t^2 - 16t = -12$

39. $x(x - 12) = -27$ **40.** $x(x - 11) = 12$

41. $y(y - 7) = 18$ **42.** $y(y + 8) = -15$

43. $p(p + 3) = -2$ **44.** $p(p - 1) = 20$

45. $y(y + 4) = 45$ **46.** $y(y - 8) = -15$

47. $x(x + 3) = 28$ **48.** $p(p - 14) = 15$

49. $(x + 8)(x - 3) = -30$ **50.** $(x + 4)(x - 1) = 14$

51. $(y + 3)(y + 10) = -10$ **52.** $(z - 5)(z + 4) = 52$

53. $(z - 8)(z + 4) = -35$ **54.** $(z - 6)(z + 1) = -10$

55. $(a + 3)(a + 4) = 72$ **56.** $(a - 4)(a + 7) = -18$

57. $(z + 3)(z - 10) = -42$ **58.** $(2x + 5)(x + 1) = -1$

59. $(y + 3)(2y + 3) = 5$ **60.** $(y + 5)(3y - 2) = -14$

2   Solve.

61.   The square of a positive number is seven more than six times the positive number. Find the number.

62.   The square of a negative number is fifteen more than twice the negative number. Find the number.

63.   The sum of two numbers is nine. The sum of the squares of the two numbers is forty-one. Find the two numbers.

64.   The sum of two numbers is eight. The sum of the squares of the two numbers is forty. Find the two numbers.

65.   The sum of the squares of two consecutive positive integers is sixty-one. Find the two integers.

66.   The sum of the squares of two consecutive positive even integers is fifty-two. Find the two integers.

67.   The sum of two numbers is twelve. The product of the two numbers is thirty-two. Find the two numbers.

68.   The sum of two numbers is twenty-one. The product of the two numbers is one hundred eight. Find the two numbers.

69.   The product of two consecutive positive integers is two hundred forty. Find the integers.

70.   The product of two consecutive even positive integers is one hundred sixty-eight. Find the integers.

71.   The length of the base of a triangle is three times the height. The area of the triangle is 54 ft^2. Find the base and height of the triangle.

72.   The height of a triangle is 4 m more than twice the length of the base. The area of the triangle is 35 m^2. Find the height of the triangle.

73.   The length of a rectangle is four times the width. The area is 400 in.^{2}. Find the length and width of the rectangle.

74.   The length of a rectangle is two more than twice the width. The area is 144 ft^2. Find the length and width of the rectangle.

75.   The length of each side of a square is extended 2 in. The area of the resulting square is 64 in.^{2}. Find the length of a side of the original square.

76. The length of each side of a square is extended 4 in. The area of the resulting square is 64 in.^{2}. Find the length of a side of the original square.

77. A circle has a radius of 10 in. Find the increase in area when the radius is increased by 2 in. Round to the nearest tenth.

78. The radius of a circle is increased by 3 in., increasing the area by 100 in.^{2}. Find the radius of the original circle. Round to the nearest tenth.

79. The page of a book measures 6 in. by 9 in. A uniform border around the page leaves 28 in.^{2} for type. Find the dimensions of the type area.

80. A small garden measures 8 ft by 10 ft. A uniform border around the garden increases the total area to 168 ft^2. Find the width of the border.

Use the formula $d = vt + 16t^2$, where d is the distance in feet, v is the initial velocity, and t is the time in seconds.

81. An object is released from a plane at an altitude of 1600 ft. The initial velocity is 0 ft/s. How many seconds later will the object hit the ground?

82. An object is released from the top of a building 320 ft high. The initial velocity is 16 ft/s. How many seconds later will the object hit the ground?

Use the formula $S = \frac{n^2 + n}{2}$, where S is the sum of the first n natural numbers.

83. How many consecutive natural numbers beginning with 1 will give a sum of 78?

84. How many consecutive natural numbers beginning with 1 will give a sum of 120?

Use the formula $N = \frac{t^2 - t}{2}$, where N is the number of football games that must be scheduled in a league with t teams if each team is to play every other team once.

85. A league has 28 games scheduled. How many teams are in the league if each team plays every other team once?

86. A league has 45 games scheduled. How many teams are in the league if each team plays every other team once?

SUPPLEMENTAL EXERCISES 6.5

Solve.

87. $2y(y + 4) = -5(y + 3)$

88. $2y(y + 4) = 3(y + 4)$

89. $(a - 3)^2 = 36$

90. $(b + 5)^2 = 16$

91. $p^3 = 9p^2$

92. $p^3 = 7p^2$

93. $(2z - 3)(z + 5) = (z + 1)(z + 3)$

94. $(x + 3)(2x - 1) = (3 - x)(5 - 3x)$

95. Find $3n^2$ if $n(n + 5) = -4$.

96. Find $2n^3$ if $n(n + 3) = 4$.

97. The product of two consecutive negative integers is one hundred fifty-six. Find the sum of the two integers.

98. The length of a rectangle is 7 cm, and the width is 4 cm. If both the length and the width are increased by equal amounts, the area of the rectangle is increased by 42 cm³. Find the length and width of the larger rectangle.

99. A rectangular piece of cardboard is 10 in. longer than it is wide. Squares 2 in. on a side are to be cut from each corner, and then the sides will be folded up to make an open box with a volume of 192 in.³. Find the length and width of the piece of cardboard.

Calculators and Computers

Factoring

Factoring polynomials is an important part of the study of algebra and is a *learned* skill that requires practice. The program FACTORING on the Math ACE Disk will give you additional practice factoring a quadratic polynomial. You may choose to practice polynomials of the form

$$x^2 + bx + c \quad \text{or} \quad ax^2 + bx + c.$$

These choices, along with the option of quitting the program, are given on a menu screen. You may practice for as long as you like with any type of problem. At the end of each problem, you may either return to the menu screen or continue practicing.

The program will present you with a quadratic polynomial to factor. When you have tried to factor the polynomial using paper and pencil, press the RETURN key on the keyboard. The correct factorization will then be displayed.

Something Extra

Prime and Composite Numbers

A **prime number** is a natural number greater than 1 whose only natural-number factors are itself and 1. The number 11 is a prime number because the only natural-number factors of 11 are 11 and 1.

Eratosthenes, a Greek philosopher and astronomer who lived from 270 to 190 B.C., devised a method of identifying prime numbers. It is called the **Sieve of Eratosthenes.** The procedure is illustrated below.

1̶	(2)	(3)	4̶	(5)	6̶	(7)	8̶	9̶	1̶0̶
(11)	1̶2̶	(13)	1̶4̶	1̶5̶	1̶6̶	(17)	1̶8̶	(19)	2̶0̶
2̶1̶	2̶2̶	(23)	2̶4̶	2̶5̶	2̶6̶	2̶7̶	2̶8̶	(29)	3̶0̶
(31)	3̶2̶	3̶3̶	3̶4̶	3̶5̶	3̶6̶	(37)	3̶8̶	3̶9̶	4̶0̶
(41)	4̶2̶	(43)	4̶4̶	4̶5̶	4̶6̶	(47)	4̶8̶	4̶9̶	5̶0̶
5̶1̶	5̶2̶	(53)	5̶4̶	5̶5̶	5̶6̶	5̶7̶	5̶8̶	(59)	6̶0̶
(61)	6̶2̶	6̶3̶	6̶4̶	6̶5̶	6̶6̶	(67)	6̶8̶	6̶9̶	7̶0̶
(71)	7̶2̶	(73)	7̶4̶	7̶5̶	7̶6̶	7̶7̶	7̶8̶	(79)	8̶0̶
8̶1̶	8̶2̶	(83)	8̶4̶	8̶5̶	8̶6̶	8̶7̶	8̶8̶	(89)	9̶0̶
9̶1̶	9̶2̶	9̶3̶	9̶4̶	9̶5̶	9̶6̶	(97)	9̶8̶	9̶9̶	1̶0̶0̶

List all the natural numbers from 1 to 100. Cross out the number 1, because it is not a prime number. The number 2 is prime; circle it. Cross out all the other multiples of 2 (4, 6, 8, . . .), because they are not prime. The number 3 is prime; circle it. Cross out all the other multiples of 3 (6, 9, 12, . . .) that are not already crossed out. The number 4, the next consecutive number in the list, has already been crossed out. The number 5 is prime; circle it. Cross out all the other multiples of 5 that are not already crossed out. Continue in this manner until all the prime numbers less than 100 are circled.

A **composite number** is a natural number greater than 1 that has a natural-number factor other than itself and 1. The number 21 is a composite number because it has factors of 3 and 7. All the numbers crossed out in the table above, except the number 1, are composite numbers.

Solve.

1. Use the Sieve of Eratosthenes to find the prime numbers between 100 and 200.

2. How many prime numbers are even numbers?

3. a. List two prime numbers that are consecutive natural numbers.

    b. Can there be any other pairs of prime numbers that are consecutive natural numbers?

4. Some primes are the sum of a square and 1. For example, $5 = 2^2 + 1$. Find another prime p such that $p = n^2 + 1$, n is a natural number.

5. Find a prime p such that $p = n^2 - 1$, n is a natural number.

6. a. 4! (read "4 factorial") is equal to $4 \cdot 3 \cdot 2 \cdot 1$. Show that $4! + 2$, $4! + 3$, and $4! + 4$ are all composite numbers.

    b. 5! (read "5 factorial") is equal to $5 \cdot 4 \cdot 3 \cdot 2 \cdot 1$. Will $5! + 2$, $5! + 3$, $5! + 4$, $5! + 5$ generate four consecutive composite numbers?

    c. Use the notation 6! to represent a list of five consecutive composite numbers.

Chapter Summary

Key Words

The *greatest common factor* (GCF) of two or more integers is the greatest integer that is a factor of all the integers.

To *factor* a polynomial means to write the polynomial as a product of other polynomials.

To *factor* a trinomial of the form $ax^2 + bx + c$ means to express the trinomial as the product of two binomials.

A polynomial that does not factor using only integers is *nonfactorable over the integers*.

An equation of the form $ax^2 + bx + c = 0$, $a \neq 0$, is a *quadratic equation*.

A quadratic equation is in *standard form* when the polynomial is in descending order and equal to zero. $ax^2 + bx + c = 0$ is in standard form.

Essential Rules

The Principle of Zero Products
If the product of two factors is zero, then at least one of the factors must be zero.

If $a \cdot b = 0$, then $a = 0$ or $b = 0$.

The difference of two squares is the product of the sum and difference of two terms.

Difference of Two Squares	**Sum and Difference of Two Terms**
$a^2 - b^2$	$= \quad (a + b)(a - b)$

A perfect square trinomial is the square of a binomial.

Perfect Square Trinomial	**Square of a Binomial**
$a^2 + 2ab + b^2$	$= \quad (a + b)^2$
$a^2 - 2ab + b^2$	$= \quad (a - b)^2$

General Factoring Strategy

1. Is there a common factor? If so, factor out the common factor.

2. Is the polynomial the difference of two squares? If so, factor.

3. Is the polynomial a perfect square trinomial? If so, factor.

4. Is the polynomial a trinomial that is the product of two binomials? If so, factor.

5. Does the polynomial contain four terms? If so, try factoring by grouping.

6. Is each binomial factor nonfactorable over the integers? If not, factor.

Chapter Review

1. Factor: $14y^9 - 49y^6 + 7y^3$

2. Factor: $3a^2 - 12a + ab - 4b$

3. Factor: $c^2 + 8c + 12$

4. Factor: $a^3 - 5a^2 + 6a$

5. Factor: $6x^2 - 29x + 28$

6. Factor: $3y^2 + 16y - 12$

7. Factor: $18a^2 - 3a - 10$

8. Factor: $a^2b^2 - 1$

9. Factor: $4y^2 - 16y + 16$

10. Solve: $a(5a + 1) = 0$

11. Factor: $12a^2b + 3ab^2$

12. Factor: $b^2 - 13b + 30$

13. Factor: $10x^2 + 25x + 4xy + 10y$

14. Factor: $3a^2 - 15a - 42$

15. Factor: $n^4 - 2n^3 - 3n^2$

16. Factor: $2x^2 - 5x + 6$

17. Factor: $6x^2 - 7x + 2$

18. Factor: $16x^2 + 49$

19. Solve: $(x - 2)(2x - 3) = 0$

20. Factor: $7x^2 - 7$

21. Factor: $3x^5 - 9x^4 - 4x^3$

22. Factor: $4x(x - 3) - 5(3 - x)$

23. Factor: $a^2 + 5a - 14$

24. Factor: $y^2 + 5y - 36$

25. Factor: $5x^2 - 50x - 120$

26. Solve: $(x + 1)(x - 5) = 16$

27. Factor: $7a^2 + 17a + 6$

28. Factor: $4x^2 + 83x + 60$

29. Factor: $9y^4 - 25z^2$

30. Factor: $5x^2 - 5x - 30$

31. Solve: $6 - 6y^2 = 5y$

32. Factor: $12b^3 - 58b^2 + 56b$

33. Factor: $5x^3 + 10x^2 + 35x$

34. Factor: $x^2 - 23x + 42$

35. Factor: $a(3a + 2) - 7(3a + 2)$

36. Factor: $8x^2 - 38x + 45$

37. Factor: $10a^2x - 130ax + 360x$

38. Factor: $2a^2 - 19a - 60$

39. Factor: $21ax - 35bx - 10by + 6ay$

40. Factor: $a^6 - 100$

41. Factor: $16a^2 + 8a + 1$

42. Solve: $4x^2 + 27x = 7$

43. Factor: $20a^2 + 10a - 280$

44. Factor: $6x - 18$

45. Factor: $3x^4y + 2x^3y + 6x^2y$

46. Factor: $d^2 + 3d - 40$

47. Factor: $24x^2 - 12xy + 10y - 20x$

48. Factor: $4x^3 - 20x^2 - 24x$

49. Solve: $x^2 - 8x - 20 = 0$

50. Factor: $3x^2 - 17x + 10$

51. Factor: $16x^2 - 94x + 33$

52. Factor: $9x^2 - 30x + 25$

53. Factor: $12y^2 + 16y - 3$

54. Factor: $3x^2 + 36x + 108$

55. The length of a playing field is twice the width. The area is 5000 yd^2. Find the length and width of the playing field.

56. The length of a hockey field is 20 yd less than twice the width. The area of the field is 6000 yd^2. Find the length and width of the hockey field.

57. The sum of the squares of two consecutive positive integers is forty-one. Find the two integers.

58. The size of a motion picture on the screen is given by the equation $S = d^2$, where d is the distance between the projector and the screen. Find the distance between the projector and the screen when the size of the picture is 400 ft^2.

59. A rectangular garden plot has dimensions 15 ft by 12 ft. A uniform path around the garden increases the total area to 270 ft^2. What is the width of the resulting area?

60. The length of each side of a square is extended 4 ft. The area of the resulting square is 576 ft^2. Find the length of a side of the original square.

Chapter Test

1. Factor: $6x^2y^2 + 9xy^2 + 12y^2$

2. Factor: $6x^3 - 8x^2 + 10x$

3. Factor: $p^2 + 5p + 6$

4. Factor: $a(x - 2) + b(2 - x)$

5. Solve: $(2a - 3)(a + 7) = 0$

6. Factor: $a^2 - 19a + 48$

7. Factor: $x^3 + 2x^2 - 15x$

8. Factor: $8x^2 + 20x - 48$

9. Factor: $ab + 6a - 3b - 18$

10. Solve: $4x^2 - 1 = 0$

11. Factor: $6x^2 + 19x + 8$

12. Factor: $x^2 - 9x - 36$

13. Factor: $2b^2 - 32$

14. Factor: $4a^2 - 12ab + 9b^2$

15. Factor: $px + x - p - 1$

16. Factor: $5x^2 - 45x - 15$

17. Factor: $2x^2 + 4x - 5$

18. Factor: $4x^2 - 49y^2$

19. Solve: $x(x - 8) = -15$

20. Factor: $p^2 + 12p + 36$

21. Factor: $18x^2 - 48xy + 32y^2$

22. Factor: $2y^4 - 14y^3 - 16y^2$

23. The length of a rectangle is 3 cm longer than twice the width. The area of the rectangle is 90 cm^2. Find the length and width of the rectangle.

24. The length of the base of a triangle is three times the height. The area of the triangle is 24 in.^{2}. Find the length of the base of the triangle.

25. The sum of two numbers is 10. The sum of the squares of the two numbers is 58. Find the two numbers.

Chapter 6 / Cumulative Review **257**

Cumulative Review

1. Subtract: $4 - (-5) - 6 - (11)$

2. Simplify: $0.372 \div (-0.046)$. Round to the nearest tenth.

3. Simplify: $(3 - 7)^2 \div (-2) - 3 \cdot (-4)$

4. Evaluate $-2a^2 \div 2b - c$ when $a = -4$, $b = 2$, and $c = -1$.

5. Identify the property that justifies the statement.
$(3 + 8) + 7 = 3 + (8 + 7)$

6. Simplify: $-\frac{3}{4}(-24x^2)$

7. Simplify: $-2[3x - 4(3 - 2x) - 8x]$

8. Solve: $-\frac{5}{7}x = -\frac{10}{21}$

9. Solve: $4 + 3(x - 2) = 13$

10. Solve: $3x - 2 = 12 - 5x$

11. Solve:
$-2 + 4[3x - 2(4 - x) - 3] = 4x + 2$

12. 120% of what number is 42?

13. Simplify:
$(3y^3 - 5y^2 - 6) + (2y^2 - 8y + 1)$

14. Simplify: $(-3a^4b^2)^3$

15. Simplify: $(x + 2)(x^2 - 5x + 4)$

16. Simplify: $(8x^2 + 4x - 3) \div (2x - 3)$

17. Simplify: $(x^{-4}y^2)^3$

18. Factor: $3a - 3b - ax + bx$

19. Factor: $x^2 + 3xy - 10y^2$

20. Factor: $6a^4 + 22a^3 + 12a^2$

21. Factor: $25a^2 - 36b^2$

22. Factor: $12x^2 - 36xy + 27y^2$

23. Solve: $3x^2 + 11x - 20 = 0$

24. The daily high temperatures, in degrees Celsius, during one week were recorded as follows: $-4°$, $-7°$, $2°$, $0°$, $-1°$, $-6°$, $-5°$. Find the average daily high temperature for the week.

25.   The age of a gold coin is 60 years, and the age of a silver coin is 40 years. How many years ago was the gold coin twice the age the silver coin was then?

26.   The width of a rectangle is 40% of the length. The perimeter of the rectangle is 42 cm. Find the length and width of the rectangle.

27.   A board 10 ft long is cut into two pieces. Four times the length of the shorter piece is 2 ft less than three times the length of the longer piece. Find the length of each piece.

28.   An investment of $4000 is made at an annual simple interest rate of 8%. How much additional money must be invested at an annual simple interest rate of 11% so that the total interest earned is $1035?

29.   A stereo that regularly sells for $165 is on sale for $99. Find the discount rate.

30.   Find three consecutive even integers such that five times the middle integer is twelve more than twice the sum of the first and third.

Alexandria

7

Algebraic Fractions

Objectives

- Simplify algebraic fractions
- Multiply algebraic fractions
- Divide algebraic fractions
- Find the least common multiple (LCM) of two or more polynomials
- Express two fractions in terms of the LCM of their denominators
- Add and subtract algebraic fractions with the same denominator
- Add and subtract algebraic fractions with different denominators
- Simplify complex fractions
- Solve equations containing fractions
- Application problems: proportions
- Solve a literal equation for one of the variables
- Work problems
- Uniform motion problems

Measurement of the Circumference of the Earth

Distances on the earth, the circumference of the earth, and the distance to the moon and stars are known to great precision. Eratosthenes, the fifth librarian of Alexandria (230 B.C.), laid the foundation of scientific geography with his determination of the circumference of the earth.

Eratosthenes was familiar with certain astronomical data that enabled him to calculate the circumference of the earth by using a proportion statement.

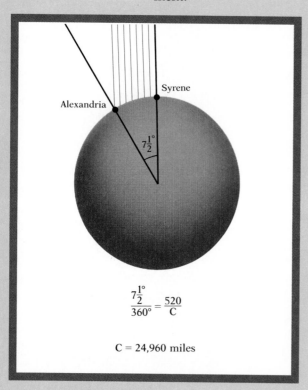

$$\frac{7\frac{1}{2}^{\circ}}{360^{\circ}} = \frac{520}{C}$$

$$C = 24{,}960 \text{ miles}$$

Eratosthenes knew that on a mid-summer day, the sun was directly overhead at Syrene, as shown in the diagram. At the same time, at Alexandria the sun was at a $7\frac{1}{2}^{\circ}$ angle from the zenith. The distance from Syrene to Alexandria was 5000 stadia (about 520 mi).

Knowing that the ratio of the $7\frac{1}{2}^{\circ}$ angle to one revolution (360°) is equal to the ratio of the arc length (520 mi) to the circumference, Eratosthenes was able to write and solve a proportion.

This result, calculated over 2000 years ago, is very close to the accepted value of 24,800 miles.

Multiplication and Division of Algebraic Fractions

1 Simplify algebraic fractions

A fraction in which the numerator or denominator is a variable expression is called an **algebraic fraction.** Examples of algebraic fractions are shown at the right.

$$\frac{5}{z}, \quad \frac{x^2 + 1}{2x - 1}, \quad \frac{y^2 - 3}{3xy + 1}$$

Care must be exercised with algebraic fractions to ensure that when the variables are replaced with numbers, the resulting denominator is not zero.

Consider the algebraic fraction at the right. The value of x cannot be 2 because the denominator would then be zero.

$$\frac{3x + 1}{2x - 4}$$

$$\frac{3 \cdot 2 + 1}{2 \cdot 2 - 4} = \frac{7}{0} \quad \text{Not a real number}$$

An algebraic fraction is in simplest form when the numerator and denominator have no common factors.

To simplify $\frac{x^2 - 4}{x^2 - 2x - 8}$, factor the numerator and denominator. Divide the numerator and denominator by the common factors. Write the answer in simplest form.

$$\frac{x^2 - 4}{x^2 - 2x - 8} = \frac{(x - 2)(x + 2)}{(x - 4)(x + 2)} = \frac{(x - 2)\overset{1}{\cancel{(x + 2)}}}{(x - 4)\underset{1}{\cancel{(x + 2)}}} = \frac{x - 2}{x - 4}$$

To simplify $\frac{10 + 3x - x^2}{x^2 - 4x - 5}$, factor the numerator and denominator.

$$\frac{10 + 3x - x^2}{x^2 - 4x - 5} = \frac{(5 - x)(2 + x)}{(x - 5)(x + 1)}$$

Divide by the common factors. Remember that $5 - x = -(x - 5)$. Therefore,

$$\frac{5 - x}{x - 5} = \frac{-(x - 5)}{x - 5} = \frac{-1}{1} = -1.$$

Write the answer in simplest form.

$$= \frac{\overset{-1}{\cancel{(5 - x)}}(2 + x)}{\underset{1}{\cancel{(x - 5)}}(x + 1)}$$

$$= -\frac{x + 2}{x + 1}$$

261

Example 1   Simplify.   A. $\dfrac{4x^3y^4}{6x^4y}$ B. $\dfrac{9 - x^2}{x^2 + x - 12}$

Solution   A. $\dfrac{4x^3y^4}{6x^4y} = \dfrac{\overset{1}{\cancel{2}} \cdot 2x^3y^4}{\cancel{2} \cdot 3x^4y} = \dfrac{2y^3}{3x}$ ▶ Simplify using the rules of exponents.

B. $\dfrac{9 - x^2}{x^2 + x - 12} = \dfrac{\overset{-1}{\cancel{(3 - x)}}(3 + x)}{\underset{1}{\cancel{(x - 3)}}(x + 4)} = -\dfrac{x + 3}{x + 4}$

Problem 1   Simplify.   A. $\dfrac{6x^5y}{12x^2y^3}$ B. $\dfrac{x^2 + 2x - 24}{16 - x^2}$

Solution   See page A25.

(handwritten notes: -24, $+4$ $x-4$, $x-6$, $\dfrac{x-6}{x-4}$, A, $6x^2y\left(2x^3y^2\right)$)

2   Multiply algebraic fractions

The product of two fractions is a fraction whose numerator is the product of the numerators of the two fractions and whose denominator is the product of the denominators of the two fractions.

$$\frac{a}{b} \cdot \frac{c}{d} = \frac{ac}{bd}$$

$$\frac{2}{3} \cdot \frac{4}{5} = \frac{8}{15}$$

$$\frac{3x}{y} \cdot \frac{2}{z} = \frac{6x}{yz}$$

$$\frac{x + 2}{x} \cdot \frac{3}{x - 2} = \frac{3(x + 2)}{x(x - 2)}$$

Simplify: $\dfrac{x^2 + 3x}{x^2 - 3x - 4} \cdot \dfrac{x^2 - 5x + 4}{x^2 + 2x - 3}$

Factor the numerator and denominator of each fraction.

Multiply.
Divide by the common factors.
Write the answer in simplest form.

$$\frac{x^2 + 3x}{x^2 - 3x - 4} \cdot \frac{x^2 - 5x + 4}{x^2 + 2x - 3} =$$

$$\frac{x(x + 3)}{(x - 4)(x + 1)} \cdot \frac{(x - 4)(x - 1)}{(x + 3)(x - 1)} =$$

$$\frac{x\cancel{(x + 3)}\cancel{(x - 4)}\cancel{(x - 1)}}{\cancel{(x - 4)}(x + 1)\cancel{(x + 3)}\cancel{(x - 1)}} = \frac{x}{x + 1}$$

Example 2   Simplify.   A. $\dfrac{10x^2 - 15x}{12x - 8} \cdot \dfrac{3x - 2}{20x - 25}$ B. $\dfrac{x^2 + x - 6}{x^2 + 7x + 12} \cdot \dfrac{x^2 + 3x - 4}{4 - x^2}$

Solution   A. $\dfrac{10x^2 - 15x}{12x - 8} \cdot \dfrac{3x - 2}{20x - 25} =$

$\dfrac{5x(2x - 3)}{4(3x - 2)} \cdot \dfrac{(3x - 2)}{5(4x - 5)} =$ ▶ Factor the numerator and denominator of each fraction.

$\dfrac{\overset{1}{\cancel{5}}x(2x - 3)\overset{1}{\cancel{(3x - 2)}}}{2 \cdot 2\underset{1}{\cancel{(3x - 2)}}\underset{1}{\cancel{5}}(4x - 5)} =$ ▶ Multiply. Divide by the common factors.

$\dfrac{x(2x - 3)}{4(4x - 5)}$ ▶ Write the answer in simplest form.

B. $\dfrac{x^2 + x - 6}{x^2 + 7x + 12} \cdot \dfrac{x^2 + 3x - 4}{4 - x^2} =$

$\dfrac{(x + 3)(x - 2)}{(x + 3)(x + 4)} \cdot \dfrac{(x + 4)(x - 1)}{(2 - x)(2 + x)} =$

$\dfrac{\overset{1}{\cancel{(x + 3)}}\overset{-1}{\cancel{(x - 2)}}\overset{1}{\cancel{(x + 4)}}(x - 1)}{\underset{1}{\cancel{(x + 3)}}\underset{1}{\cancel{(x + 4)}}\underset{1}{\cancel{(2 - x)}}(2 + x)} =$

$-\dfrac{x - 1}{x + 2}$

Problem 2   Simplify.   A. $\dfrac{12x^2 + 3x}{10x - 15} \cdot \dfrac{8x - 12}{9x + 18}$ B. $\dfrac{x^2 + 2x - 15}{9 - x^2} \cdot \dfrac{x^2 - 3x - 18}{x^2 - 7x + 6}$

Solution   See page A25.

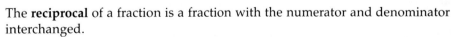

3 ▌ Divide algebraic fractions

The **reciprocal** of a fraction is a fraction with the numerator and denominator interchanged.

$$\text{Fraction} \left\{ \begin{array}{cc} \dfrac{a}{b} & \dfrac{b}{a} \\[2ex] x^2 = \dfrac{x^2}{1} & \dfrac{1}{x^2} \\[2ex] \dfrac{x + 2}{x} & \dfrac{x}{x + 2} \end{array} \right\} \text{Reciprocal}$$

To divide two fractions, multiply by the reciprocal of the divisor.

$$\frac{a}{b} \div \frac{c}{d} = \frac{a}{b} \cdot \frac{d}{c} = \frac{ad}{bc}$$

$$\frac{4}{x} \div \frac{y}{5} = \frac{4}{x} \cdot \frac{5}{y} = \frac{20}{xy}$$

$$\frac{x+4}{x} \div \frac{x-2}{4} = \frac{x+4}{x} \cdot \frac{4}{x-2} = \frac{4(x+4)}{x(x-2)}$$

The basis for the division rule is shown below.

$$\frac{a}{b} \div \frac{c}{d} = \frac{\dfrac{a}{b}}{\dfrac{c}{d}} = \frac{\dfrac{a}{b} \cdot \dfrac{d}{c}}{\dfrac{c}{d} \cdot \dfrac{d}{c}} = \frac{\dfrac{a}{b} \cdot \dfrac{d}{c}}{1} = \frac{a}{b} \cdot \frac{d}{c}$$

Example 3   Simplify.

A. $\dfrac{xy^2 - 3x^2y}{z^2} \div \dfrac{6x^2 - 2xy}{z^3}$ B. $\dfrac{2x^2 + 5x + 2}{2x^2 + 3x - 2} \div \dfrac{3x^2 + 13x + 4}{2x^2 + 7x - 4}$

Solution   A. $\dfrac{xy^2 - 3x^2y}{z^2} \div \dfrac{6x^2 - 2xy}{z^3} = \dfrac{xy^2 - 3x^2y}{z^2} \cdot \dfrac{z^3}{6x^2 - 2xy}$

$$= \frac{xy(\overset{-1}{\cancel{y-3x}}) \cdot z^3}{z^2 \cdot 2x(\underset{1}{\cancel{3x-y}})}$$

$$= -\frac{yz}{2}$$

B. $\dfrac{2x^2 + 5x + 2}{2x^2 + 3x - 2} \div \dfrac{3x^2 + 13x + 4}{2x^2 + 7x - 4} = \dfrac{2x^2 + 5x + 2}{2x^2 + 3x - 2} \cdot \dfrac{2x^2 + 7x - 4}{3x^2 + 13x + 4}$

$$= \frac{(2x+1)(\overset{1}{\cancel{x+2}})}{(\underset{1}{\cancel{2x-1}})(\underset{1}{\cancel{x+2}})} \cdot \frac{(\overset{1}{\cancel{2x-1}})(\overset{1}{\cancel{x+4}})}{(3x+1)(\underset{1}{\cancel{x+4}})}$$

$$= \frac{2x+1}{3x+1}$$

Problem 3   Simplify.

A. $\dfrac{a^2}{4bc^2 - 2b^2c} \div \dfrac{a}{6bc - 3b^2}$ B. $\dfrac{3x^2 + 26x + 16}{3x^2 - 7x - 6} \div \dfrac{2x^2 + 9x - 5}{x^2 + 2x - 15}$

Solution   See page A25.

EXERCISES 7.1

1   Simplify.

1. $\dfrac{9x^3}{12x^4}$

2. $\dfrac{16x^2y}{24xy^3}$

3. $\dfrac{(x + 3)^2}{(x + 3)^3}$

4. $\dfrac{(2x - 1)^5}{(2x - 1)^4}$

5. $\dfrac{3n - 4}{4 - 3n}$

6. $\dfrac{5 - 2x}{2x - 5}$

7. $\dfrac{6y(y + 2)}{9y^2(y + 2)}$

8. $\dfrac{12x^2(3 - x)}{18x(3 - x)}$

9. $\dfrac{6x(x - 5)}{8x^2(5 - x)}$

10. $\dfrac{14x^3(7 - 3x)}{21x(3x - 7)}$

11. $\dfrac{a^2 + 4a}{ab + 4b}$

12. $\dfrac{x^2 - 3x}{2x - 6}$

13. $\dfrac{4 - 6x}{3x^2 - 2x}$

14. $\dfrac{5xy - 3y}{9 - 15x}$

15. $\dfrac{y^2 - 3y + 2}{y^2 - 4y + 3}$

16. $\dfrac{x^2 + 5x + 6}{x^2 + 8x + 15}$

17. $\dfrac{x^2 + 3x - 10}{x^2 + 2x - 8}$

18. $\dfrac{a^2 + 7a - 8}{a^2 + 6a - 7}$

19. $\dfrac{x^2 + x - 12}{x^2 - 6x + 9}$

20. $\dfrac{x^2 + 8x + 16}{x^2 - 2x - 24}$

21. $\dfrac{x^2 - 3x - 10}{25 - x^2}$

22. $\dfrac{4 - y^2}{y^2 - 3y - 10}$

23. $\dfrac{2x^3 + 2x^2 - 4x}{x^3 + 2x^2 - 3x}$

24. $\dfrac{3x^3 - 12x}{6x^3 - 24x^2 + 24x}$

25. $\dfrac{6x^2 - 7x + 2}{6x^2 + 5x - 6}$

26. $\dfrac{2n^2 - 9n + 4}{2n^2 - 5n - 12}$

2   Simplify.

27. $\dfrac{8x^2}{9y^3} \cdot \dfrac{3y^2}{4x^3}$

28. $\dfrac{4a^2b^3}{15x^5y^2} \cdot \dfrac{25x^3y}{16ab}$

29. $\dfrac{12x^3y^4}{7a^2b^3} \cdot \dfrac{14a^3b^4}{9x^2y^2}$

30. $\dfrac{18a^4b^2}{25x^2y^3} \cdot \dfrac{50x^5y^6}{27a^6b^2}$

31. $\dfrac{3x - 6}{5x - 20} \cdot \dfrac{10x - 40}{27x - 54}$

32. $\dfrac{8x - 12}{14x + 7} \cdot \dfrac{42x + 21}{32x - 48}$

33. $\dfrac{3x^2 + 2x}{2xy - 3y} \cdot \dfrac{2xy^3 - 3y^3}{3x^3 + 2x^2}$

34. $\dfrac{4a^2x - 3a^2}{2by + 5b} \cdot \dfrac{2b^3y + 5b^3}{4ax - 3a}$

35. $\dfrac{x^2 + 5x + 4}{x^3 y^2} \cdot \dfrac{x^2 y^3}{x^2 + 2x + 1}$

36. $\dfrac{x^2 + x - 2}{xy^2} \cdot \dfrac{x^3 y}{x^2 + 5x + 6}$

37. $\dfrac{x^4 y^2}{x^2 + 3x - 28} \cdot \dfrac{x^2 - 49}{xy^4}$

38. $\dfrac{x^5 y^3}{x^2 + 13x + 30} \cdot \dfrac{x^2 + 2x - 3}{x^7 y^2}$

39. $\dfrac{2x^2 - 5x}{2xy + y} \cdot \dfrac{2xy^2 + y^2}{5x^2 - 2x^3}$

40. $\dfrac{3a^3 + 4a^2}{5ab - 3b} \cdot \dfrac{3b^3 - 5ab^3}{3a^2 + 4a}$

41. $\dfrac{x^2 - 2x - 24}{x^2 - 5x - 6} \cdot \dfrac{x^2 + 5x + 6}{x^2 + 6x + 8}$

42. $\dfrac{x^2 - 8x + 7}{x^2 + 3x - 4} \cdot \dfrac{x^2 + 3x - 10}{x^2 - 9x + 14}$

43. $\dfrac{x^2 + 2x - 35}{x^2 + 4x - 21} \cdot \dfrac{x^2 + 3x - 18}{x^2 + 9x + 18}$

44. $\dfrac{y^2 + y - 20}{y^2 + 2y - 15} \cdot \dfrac{y^2 + 4y - 21}{y^2 + 3y - 28}$

45. $\dfrac{x^2 - 3x - 4}{x^2 + 6x + 5} \cdot \dfrac{x^2 + 5x + 6}{8 + 2x - x^2}$ $x^2 - 2x - 8$

46. $\dfrac{25 - n^2}{n^2 - 2n - 35} \cdot \dfrac{n^2 - 8n - 20}{n^2 - 3n - 10}$

47. $\dfrac{12x^2 - 6x}{x^2 + 6x + 5} \cdot \dfrac{2x^4 + 10x^3}{4x^2 - 1}$

48. $\dfrac{8x^3 + 4x^2}{x^2 - 3x + 2} \cdot \dfrac{x^2 - 4}{16x^2 + 8x}$

49. $\dfrac{16 + 6x - x^2}{x^2 - 10x - 24} \cdot \dfrac{x^2 - 6x - 27}{x^2 - 17x + 72}$

50. $\dfrac{x^2 - 11x + 28}{x^2 - 13x + 42} \cdot \dfrac{x^2 + 7x + 10}{20 - x - x^2}$

51. $\dfrac{2x^2 + 5x + 2}{2x^2 + 7x + 3} \cdot \dfrac{x^2 - 7x - 30}{x^2 - 6x - 40}$

52. $\dfrac{x^2 - 4x - 32}{x^2 - 8x - 48} \cdot \dfrac{3x^2 + 17x + 10}{3x^2 - 22x - 16}$

53. $\dfrac{2x^2 + x - 3}{2x^2 - x - 6} \cdot \dfrac{2x^2 - 9x + 10}{2x^2 - 3x + 1}$

54. $\dfrac{3y^2 + 14y + 8}{2y^2 + 7y - 4} \cdot \dfrac{2y^2 + 9y - 5}{3y^2 + 16y + 5}$

55. $\dfrac{6x^2 - 11x + 4}{6x^2 + x - 2} \cdot \dfrac{12x^2 + 11x + 2}{8x^2 + 14x + 3}$

56. $\dfrac{6 - x - 2x^2}{4x^2 + 3x - 10} \cdot \dfrac{3x^2 + 7x - 20}{2x^2 + 5x - 12}$

3   Simplify.

57. $\dfrac{4x^2 y^3}{15a^2 b^3} \div \dfrac{6xy}{5a^3 b^5}$

58. $\dfrac{9x^3 y^4}{16a^4 b^2} \div \dfrac{45x^4 y^2}{14a^7 b}$

59. $\dfrac{6x - 12}{8x + 32} \div \dfrac{18x - 36}{10x + 40}$

60. $\dfrac{28x + 14}{45x - 30} \div \dfrac{14x + 7}{30x - 20}$

61. $\dfrac{6x^3 + 7x^2}{12x - 3} \div \dfrac{6x^2 + 7x}{36x - 9}$

62. $\dfrac{5a^2 y + 3a^2}{2x^3 + 5x^2} \div \dfrac{10ay + 6a}{6x^3 + 15x^2}$

63. $\dfrac{x^2 + 4x + 3}{x^2 y} \div \dfrac{x^2 + 2x + 1}{xy^2}$

64. $\dfrac{x^3 y^2}{x^2 - 3x - 10} \div \dfrac{xy^4}{x^2 - x - 20}$

65. $\dfrac{x^2 - 49}{x^4 y^3} \div \dfrac{x^2 - 14x + 49}{x^4 y^3}$

66. $\dfrac{x^2 y^5}{x^2 - 11x + 30} \div \dfrac{xy^6}{x^2 - 7x + 10}$

67. $\dfrac{4ax - 8a}{c^2} \div \dfrac{2y - xy}{c^3}$

68. $\dfrac{3x^2y - 9xy}{a^2b} \div \dfrac{3x^2 - x^3}{ab^2}$

69. $\dfrac{x^2 - 5x + 6}{x^2 - 9x + 18} \div \dfrac{x^2 - 6x + 8}{x^2 - 9x + 20}$

70. $\dfrac{x^2 + 3x - 40}{x^2 + 2x - 35} \div \dfrac{x^2 + 2x - 48}{x^2 + 3x - 18}$

71. $\dfrac{x^2 + 2x - 15}{x^2 - 4x - 45} \div \dfrac{x^2 + x - 12}{x^2 - 5x - 36}$

72. $\dfrac{y^2 - y - 56}{y^2 + 8y + 7} \div \dfrac{y^2 - 13y + 40}{y^2 - 4y - 5}$

73. $\dfrac{8 + 2x - x^2}{x^2 + 7x + 10} \div \dfrac{x^2 - 11x + 28}{x^2 - x - 42}$

74. $\dfrac{x^2 - x - 2}{x^2 - 7x + 10} \div \dfrac{x^2 - 3x - 4}{40 - 3x - x^2}$

75. $\dfrac{2x^2 - 3x - 20}{2x^2 - 7x - 30} \div \dfrac{2x^2 - 5x - 12}{4x^2 + 12x + 9}$

76. $\dfrac{6n^2 + 13n + 6}{4n^2 - 9} \div \dfrac{6n^2 + n - 2}{4n^2 - 1}$

77. $\dfrac{9x^2 - 16}{6x^2 - 11x + 4} \div \dfrac{6x^2 + 11x + 4}{8x^2 + 10x + 3}$

78. $\dfrac{15 - 14x - 8x^2}{4x^2 + 4x - 15} \div \dfrac{4x^2 + 13x - 12}{3x^2 + 13x + 4}$

79. $\dfrac{8x^2 + 18x - 5}{10x^2 - 9x + 2} \div \dfrac{8x^2 + 22x + 15}{10x^2 + 11x - 6}$

80. $\dfrac{10 + 7x - 12x^2}{8x^2 - 2x - 15} \div \dfrac{6x^2 - 13x + 5}{10x^2 - 13x + 4}$

SUPPLEMENTAL EXERCISES 7.1

For what values of x is the algebraic fraction undefined? (*Hint:* Set the denominator equal to zero and solve for x.)

81. $\dfrac{x}{(x + 6)(x - 1)}$

82. $\dfrac{x}{(x - 2)(x + 5)}$

83. $\dfrac{8}{x^2 - 1}$

84. $\dfrac{7}{x^2 - 16}$

85. $\dfrac{x - 4}{x^2 - x - 6}$

86. $\dfrac{x + 5}{x^2 - 4x - 5}$

87. $\dfrac{3x}{x^2 + 6x + 9}$

88. $\dfrac{8x}{x^2 - 12x + 36}$

89. $\dfrac{3}{x(x - 4)}$

90. $\dfrac{5x - 2}{x(x + 3)}$

91. $\dfrac{3x - 8}{3x^2 - 10x - 8}$

92. $\dfrac{4x + 7}{6x^2 - 5x - 4}$

Simplify.

93. $\dfrac{y^2}{x} \cdot \dfrac{x}{2} \div \dfrac{y}{x}$

94. $\dfrac{ab}{3} \cdot \dfrac{a}{b^2} \div \dfrac{a}{4}$

95. $\left(\dfrac{2x}{y}\right)^3 \div \left(\dfrac{x}{3y}\right)^2$

96. $\left(\dfrac{c}{3}\right)^2 \div \left(\dfrac{c}{2} \cdot \dfrac{c}{4}\right)$

97. $\left(\dfrac{a - 3}{b}\right)^2 \left(\dfrac{b}{3 - a}\right)^3$

98. $\left(\dfrac{x - 4}{y^2}\right)^3 \cdot \left(\dfrac{y}{4 - x}\right)^2$

99. $\dfrac{x-2}{x+5} \div \dfrac{x-3}{x+5} \cdot \dfrac{x-3}{x-2}$

100. $\dfrac{b+4}{b-1} \div \dfrac{b+4}{b+2} \cdot \dfrac{b-1}{b-5}$

101. $\dfrac{x^2+3x-40}{x^2+2x-35} \div \dfrac{x^2+2x-48}{x^2+3x-18} \cdot \dfrac{x^2-36}{x^2-9}$

102. $\dfrac{x^2+x-6}{x^2+7x+12} \cdot \dfrac{x^2+3x-4}{x^2+x-2} \div \dfrac{x^2-16}{x^2-4}$

SECTION 7.2

Expressing Fractions in Terms of the Least Common Multiple (LCM)

1 Find the least common multiple (LCM) of two or more polynomials

The **least common multiple (LCM)** of two or more numbers is the smallest number that contains the prime factorization of each number.

The LCM of 12 and 18 is 36. 36 contains the prime factors of 12 and the prime factors of 18.

$12 = 2 \cdot 2 \cdot 3$
$18 = 2 \cdot 3 \cdot 3$

$$\text{LCM} = 36 = \overbrace{2 \cdot \underbrace{2 \cdot 3}_{} \cdot 3}$$

Factors of 12

Factors of 18

The least common multiple of two or more polynomials is the simplest polynomial that contains the factors of each polynomial.

To find the LCM of two or more polynomials, first factor each polynomial completely. The LCM is the product of each factor the greatest number of times it occurs in any one factorization.

The LCM of $4x^2 + 4x$ and $x^2 + 2x + 1$ is the product of the LCM of the numerical coefficients and each variable factor the greatest number of times it occurs in any one factorization.

$4x^2 + 4x = 4x(x+1) = 2 \cdot 2 \cdot x(x+1)$
$x^2 + 2x + 1 = (x+1)(x+1)$

Factors of $4x^2 + 4x$

$$\text{LCM} = 2 \cdot 2 \cdot \overbrace{x(x+1)(x+1)} = 4x(x+1)(x+1)$$

Factors of $x^2 + 2x + 1$

Example 1   Find the LCM of $4x^2y$ and $6xy^2$.

Solution $4x^2y = 2 \cdot 2 \cdot x \cdot x \cdot y$ ▶ Factor each polynomial completely.
$6xy^2 = 2 \cdot 3 \cdot x \cdot y \cdot y$
$\text{LCM} = 2 \cdot 2 \cdot 3 \cdot x \cdot x \cdot y \cdot y$ ▶ Write the product of the LCM of the numerical
$= 12x^2y^2$ coefficients and each variable factor the
greatest number of times it occurs in any one
factorization.

Problem 1   Find the LCM of $8uv^2$ and $12uw$.

Solution   See page A25.

Example 2   Find the LCM of $x^2 - x - 6$ and $9 - x^2$.

Solution $x^2 - x - 6 = (x - 3)(x + 2)$
$9 - x^2 = -(x^2 - 9) = -(x + 3)(x - 3)$
$\text{LCM} = (x - 3)(x + 2)(x + 3)$

Problem 2   Find the LCM of $m^2 - 6m + 9$ and $m^2 - 2m - 3$.

Solution   See page A25.

2   Express two fractions in terms of the
LCM of their denominators

When adding and subtracting fractions, it is often necessary to express two or
more fractions in terms of a common denominator. This common denominator
is the LCM of the denominators of the fractions.

Write the fractions $\dfrac{x + 1}{4x^2}$ and $\dfrac{x - 3}{6x^2 - 12x}$ in terms of the LCM of the denominators.

Find the LCM of the de-   The LCM is $12x^2(x - 2)$.
nominators.

For each fraction, multi- $\dfrac{x + 1}{4x^2} = \dfrac{x + 1}{4x^2} \cdot \dfrac{3(x - 2)}{3(x - 2)} = \dfrac{3x^2 - 3x - 6}{12x^2(x - 2)}$ ◀
ply the numerator and
denominator by the fac-   ⎫
tors   whose   product $\dfrac{x - 3}{6x^2 - 12x} = \dfrac{x - 3}{6x(x - 2)} \cdot \dfrac{2x}{2x} = \dfrac{2x^2 - 6x}{12x^2(x - 2)}$ ◀   ⎬ LCM
with the denominator is   ⎭
the LCM.

Example 3 Write the fractions $\frac{x+2}{3x^2}$ and $\frac{x-1}{8xy}$ in the terms of the LCM of the denominators.

Solution The LCM is $24x^2y$.

$$\frac{x+2}{3x^2} = \frac{x+2}{3x^2} \cdot \frac{8y}{8y} = \frac{8xy+16y}{24x^2y}$$ ▶ The product of $3x^2$ and $8y$ is the LCM.

$$\frac{x-1}{8xy} = \frac{x-1}{8xy} \cdot \frac{3x}{3x} = \frac{3x^2-3x}{24x^2y}$$ ▶ The product of $8xy$ and $3x$ is the LCM.

Problem 3 Write the fractions $\frac{x-3}{4xy^2}$ and $\frac{2x+1}{9y^2z}$ in terms of the LCM of the denominators.

Solution See page A25.

Example 4 Write the fractions $\frac{2x-1}{2x-x^2}$ and $\frac{x}{x^2+x-6}$ in terms of the LCM of the denominators.

Solution $\dfrac{2x-1}{2x-x^2} = \dfrac{2x-1}{-(x^2-2x)} = -\dfrac{2x-1}{x^2-2x}$ ▶ Rewrite $\frac{2x-1}{2x-x^2}$ with a denominator of x^2-2x.

The LCM is $x(x-2)(x+3)$.

$$\frac{2x-1}{2x-x^2} = -\frac{2x-1}{x(x-2)} \cdot \frac{x+3}{x+3} = -\frac{2x^2+5x-3}{x(x-2)(x+3)}$$

$$\frac{x}{x^2+x-6} = \frac{x}{(x-2)(x+3)} \cdot \frac{x}{x} = \frac{x^2}{x(x-2)(x+3)}$$

Problem 4 Write the fractions $\frac{x+4}{x^2-3x-10}$ and $\frac{2x}{25-x^2}$ in terms of the LCM of the denominators.

Solution See page A25.

EXERCISES 7.2

1 Find the LCM of the expressions.

1. $8x^3y$
   $12xy^2$

2. $6ab^2$
   $18ab^3$

3. $10x^4y^2$
   $15x^3y$

4. $12a^2b$
   $18ab^3$

5. $8x^2$
   $4x^2+8x$

6. $6y^2$
$4y + 12$

7. $2x^2y$
$3x^2 + 12x$

8. $4xy^2$
$6xy^2 + 12y^2$

9. $8x^2(x - 1)^2$
$10x^3(x - 1)$

10. $3x + 3$
$2x^2 + 4x + 2$

11. $4x - 12$
$2x^2 - 12x + 18$

12. $(x - 1)(x + 2)$
$(x - 1)(x + 3)$

13. $(2x - 1)(x + 4)$
$(2x + 1)(x + 4)$

14. $(2x + 3)^2$
$(2x + 3)(x - 5)$

15. $(x - 7)(x + 2)$
$(x - 7)^2$

16. $(x - 1)$
$(x - 2)$
$(x - 1)(x - 2)$

17. $(x + 4)(x - 3)$
$x + 4$
$x - 3$

18. $x^2 - x - 6$
$x^2 + x - 12$

19. $x^2 + 3x - 10$
$x^2 + 5x - 14$

20. $x^2 + 5x + 4$
$x^2 - 3x - 28$

21. $x^2 - 10x + 21$
$x^2 - 8x + 15$

22. $x^2 - 2x - 24$
$x^2 - 36$

23. $x^2 + 7x + 10$
$x^2 - 25$

24. $x^2 - 7x - 30$
$x^2 - 5x - 24$

25. $2x^2 - 7x + 3$
$2x^2 + x - 1$

26. $3x^2 - 11x + 6$
$3x^2 + 4x - 4$

27. $2x^2 - 9x + 10$
$2x^2 + x - 15$

28. $6 + x - x^2$
$x + 2$
$x - 3$

29. $15 + 2x - x^2$
$x - 5$
$x + 3$

30. $5 + 4x - x^2$
$x - 5$
$x + 1$

31. $x^2 + 3x - 18$
$3 - x$
$x + 6$

32. $x^2 - 5x + 6$
$1 - x$
$x - 6$

2 Write each fraction in terms of the LCM of the denominators.

33. $\dfrac{4}{x}, \dfrac{3}{x^2}$

34. $\dfrac{5}{ab^2}, \dfrac{6}{ab}$

35. $\dfrac{x}{3y^2}, \dfrac{z}{4y}$

36. $\dfrac{5y}{6x^2}, \dfrac{7}{9xy}$

37. $\dfrac{y}{x(x - 3)}, \dfrac{6}{x^2}$

38. $\dfrac{a}{y^2}, \dfrac{6}{y(y + 5)}$

39. $\dfrac{9}{(x-1)^2}$, $\dfrac{6}{x(x-1)}$

40. $\dfrac{a^2}{y(y+7)}$, $\dfrac{a}{(y+7)^2}$

41. $\dfrac{3}{x-3}$, $-\dfrac{5}{x(3-x)}$

42. $\dfrac{b}{y(y-4)}$, $\dfrac{b^2}{4-y}$

43. $\dfrac{3}{(x-5)^2}$, $\dfrac{2}{5-x}$

44. $\dfrac{3}{7-y}$, $\dfrac{2}{(y-7)^2}$

45. $\dfrac{3}{x^2+2x}$, $\dfrac{4}{x^2}$

46. $\dfrac{2}{y-3}$, $\dfrac{3}{y^3-3y^2}$

47. $\dfrac{x-2}{x+3}$, $\dfrac{x}{x-4}$

48. $\dfrac{x^2}{2x-1}$, $\dfrac{x+1}{x+4}$

49. $\dfrac{3}{x^2+x-2}$, $\dfrac{x}{x+2}$

50. $\dfrac{3x}{x-5}$, $\dfrac{4}{x^2-25}$

51. $\dfrac{5}{2x^2-9x+10}$, $\dfrac{x-1}{2x-5}$

52. $\dfrac{x-3}{3x^2+4x-4}$, $\dfrac{2}{x+2}$

53. $\dfrac{x}{x^2+x-6}$, $\dfrac{2x}{x^2-9}$

54. $\dfrac{x-1}{x^2+2x-15}$, $\dfrac{x}{x^2+6x+5}$

55. $\dfrac{x}{9-x^2}$, $\dfrac{x-1}{x^2-6x+9}$

56. $\dfrac{2x}{10+3x-x^2}$, $\dfrac{x+2}{x^2-8x+15}$

57. $\dfrac{3x}{x-5}$, $\dfrac{x}{x+4}$, $\dfrac{3}{20+x-x^2}$

58. $\dfrac{x+1}{x+5}$, $\dfrac{x+2}{x-7}$, $\dfrac{3}{35+2x-x^2}$

SUPPLEMENTAL EXERCISES 7.2

Write each expression in terms of the LCM of the denominators.

59. $\dfrac{3}{10^2}$, $\dfrac{5}{10^4}$

60. $\dfrac{8}{10^3}$, $\dfrac{9}{10^5}$

61. b, $\dfrac{5}{b}$

62. 3, $\dfrac{2}{n}$

63. $1, \dfrac{y}{y-1}$

64. $x, \dfrac{x}{x^2-1}$

65. $\dfrac{x^2+1}{(x-1)^3}, \dfrac{x+1}{(x-1)^2}, \dfrac{1}{x-1}$

66. $\dfrac{a^2+a}{(a+1)^3}, \dfrac{a+1}{(a+1)^2}, \dfrac{1}{a+1}$

67. $\dfrac{b}{4a^2-4b^2}, \dfrac{a}{8a-8b}$

68. $\dfrac{c}{6c^2+7cd+d^2}, \dfrac{d}{3c^2-3d^2}$

69. $\dfrac{1}{x^2+2x+xy+2y}, \dfrac{1}{x^2+xy-2x-2y}$

70. $\dfrac{1}{ab+3a-3b-b^2}, \dfrac{1}{ab+3a+3b+b^2}$

Solve.

71. When is the LCM of two expressions equal to their product?

SECTION **7.3**

Addition and Subtraction of Algebraic Fractions

1 Add and subtract algebraic fractions with the same denominator

When adding algebraic fractions in which the denominators are the same, add the numerators. The denominator of the sum is the common denominator.

$$\dfrac{a}{b} + \dfrac{c}{b} = \dfrac{a+c}{b}$$

$$\dfrac{5x}{18} + \dfrac{7x}{18} = \dfrac{12x}{18} = \dfrac{2x}{3}$$

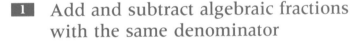

Note that the sum is written in simplest form.

When subtracting algebraic fractions in which the denominators are the same, subtract the numerators. The denominator of the difference is the common denominator. Write the answer in simplest form.

$$\frac{2x}{x-2} - \frac{4}{x-2} = \frac{2x-4}{x-2} = \frac{2(\overset{1}{\cancel{x-2}})}{\underset{1}{\cancel{x-2}}} = 2$$

$$\frac{3x-1}{x^2-5x+4} - \frac{2x+3}{x^2-5x+4} = \frac{(3x-1)-(2x+3)}{x^2-5x+4}$$

$$= \frac{x-4}{x^2-5x+4}$$

$$= \frac{\overset{1}{\cancel{(x-4)}}}{\underset{1}{\cancel{(x-4)}}(x-1)}$$

$$= \frac{1}{x-1}$$

Example 1 Simplify. A. $\dfrac{7}{x^2} + \dfrac{9}{x^2}$ B. $\dfrac{3x^2}{x^2-1} - \dfrac{x+4}{x^2-1}$

Solution A. $\dfrac{7}{x^2} + \dfrac{9}{x^2} = \dfrac{7+9}{x^2}$ ▶ The denominators are the same. Add the numerators.

$$= \frac{16}{x^2}$$

B. $\dfrac{3x^2}{x^2-1} - \dfrac{x+4}{x^2-1} = \dfrac{3x^2-(x+4)}{x^2-1}$ ▶ The denominators are the same. Subtract the numerators.

$$= \frac{3x^2-x-4}{x^2-1}$$

$$= \frac{(3x-4)\overset{1}{\cancel{(x+1)}}}{(x-1)\underset{1}{\cancel{(x+1)}}}$$ ▶ Write the answer in simplest form.

$$= \frac{3x-4}{x-1}$$

Problem 1   Simplify.   A. $\dfrac{3}{xy} + \dfrac{12}{xy}$ B. $\dfrac{2x^2}{x^2 - x - 12} - \dfrac{7x + 4}{x^2 - x - 12}$

Solution   See page A26.

2 **Add and subtract algebraic fractions with different denominators**

Before two fractions with different denominators can be added or subtracted, each fraction must be expressed in terms of a common denominator. This common denominator is the LCM of the denominators of the fractions.

Simplify: $\dfrac{x-3}{x^2 - 2x} + \dfrac{6}{x^2 - 4}$

Find the LCM of the denominators.

$$x^2 - 2x = x(x - 2)$$
$$x^2 - 4 = (x - 2)(x + 2)$$

The LCM is $x(x - 2)(x + 2)$.

$$\frac{x - 3}{x^2 - 2x} + \frac{6}{x^2 - 4} =$$

Write each fraction in terms of the LCM.

$$\frac{x - 3}{x(x - 2)} \cdot \frac{x + 2}{x + 2} + \frac{6}{(x - 2)(x + 2)} \cdot \frac{x}{x} =$$

Multiply the factors in the numerator.

$$\frac{x^2 - x - 6}{x(x - 2)(x + 2)} + \frac{6x}{x(x - 2)(x + 2)}$$

Add the fractions.

$$\frac{x^2 + 5x - 6}{x(x - 2)(x + 2)} =$$

Factor the numerator to determine whether there are common factors in the numerator and denominator.

$$\frac{(x + 6)(x - 1)}{x(x - 2)(x + 2)}$$

Example 2 Simplify.   A. $\dfrac{y}{x} - \dfrac{4y}{3x} + \dfrac{3y}{4x}$ B. $\dfrac{2x}{x-3} - \dfrac{5}{3-x}$

Solution   A. The LCM of the denominators is $12x$.   ▶ Find the LCM of the denomina-
                                                                       tors.

$$\dfrac{y}{x} - \dfrac{4y}{3x} + \dfrac{3y}{4x} =$$

$$\dfrac{y}{x} \cdot \dfrac{12}{12} - \dfrac{4y}{3x} \cdot \dfrac{4}{4} + \dfrac{3y}{4x} \cdot \dfrac{3}{3} =$$ ▶ Write each fraction in terms of
                                                                       the LCM.

$$\dfrac{12y}{12x} - \dfrac{16y}{12x} + \dfrac{9y}{12x} =$$

$$\dfrac{12y - 16y + 9y}{12x} =$$

$$\dfrac{5y}{12x}$$

B. The LCM of $x - 3$ and $3 - x$ is $x - 3$.   ▶ $3 - x = -(x - 3)$

$$\dfrac{2x}{x-3} - \dfrac{5}{3-x} =$$

$$\dfrac{2x}{x-3} - \dfrac{5}{-(x-3)} \cdot \dfrac{-1}{-1} =$$ ▶ Multiply $\dfrac{5}{-(x-3)}$ by $\dfrac{-1}{-1}$ so
                                                                       that the denominator will be
                                                                       $x - 3$.

$$\dfrac{2x}{x-3} - \dfrac{-5}{x-3} =$$

$$\dfrac{2x - (-5)}{x-3} =$$

$$\dfrac{2x + 5}{x-3}$$

Problem 2 Simplify.   A. $\dfrac{z}{8y} - \dfrac{4z}{3y} + \dfrac{5z}{4y}$ B. $\dfrac{5x}{x-2} - \dfrac{3}{2-x}$

Solution   See page A26.

Example 3 Simplify. A. $\dfrac{2x}{2x-3}-\dfrac{1}{x+1}$ B. $\dfrac{x+3}{x^2-2x-8}+\dfrac{3}{4-x}$

Solution A. The LCM is $(2x-3)(x+1)$.

$$\dfrac{2x}{2x-3}-\dfrac{1}{x+1}=\dfrac{2x}{2x-3}\cdot\dfrac{x+1}{x+1}-\dfrac{1}{x+1}\cdot\dfrac{2x-3}{2x-3}$$

$$=\dfrac{2x^2+2x}{(2x-3)(x+1)}-\dfrac{2x-3}{(2x-3)(x+1)}$$

$$=\dfrac{(2x^2+2x)-(2x-3)}{(2x-3)(x+1)}$$

$$=\dfrac{2x^2+3}{(2x-3)(x+1)}$$

B. The LCM is $(x-4)(x+2)$.

$$\dfrac{x+3}{x^2-2x-8}+\dfrac{3}{4-x}=\dfrac{x+3}{(x-4)(x+2)}+\dfrac{3}{-(x-4)}\cdot\dfrac{-1\cdot(x+2)}{-1\cdot(x+2)}$$

$$=\dfrac{x+3}{(x-4)(x+2)}+\dfrac{-3(x+2)}{(x-4)(x+2)}$$

$$=\dfrac{(x+3)+(-3)(x+2)}{(x-4)(x+2)}$$

$$=\dfrac{x+3-3x-6}{(x-4)(x+2)}$$

$$=\dfrac{-2x-3}{(x-4)(x+2)}$$

Problem 3 Simplify. A. $\dfrac{4x}{3x-1}-\dfrac{9}{x+4}$ B. $\dfrac{2x-1}{x^2-25}+\dfrac{2}{5-x}$

Solution See page A26.

EXERCISES 7.3

1 Simplify.

1. $\dfrac{3}{y^2}+\dfrac{8}{y^2}$ **2.** $\dfrac{6}{ab}-\dfrac{2}{ab}$ **3.** $\dfrac{3}{x+4}-\dfrac{10}{x+4}$

4. $\dfrac{x}{x+6} - \dfrac{2}{x+6}$

5. $\dfrac{3x}{2x+3} + \dfrac{5x}{2x+3}$

6. $\dfrac{6y}{4y+1} - \dfrac{11y}{4y+1}$

7. $\dfrac{2x+1}{x-3} + \dfrac{3x+6}{x-3}$

8. $\dfrac{4x+3}{2x-7} + \dfrac{3x-8}{2x-7}$

9. $\dfrac{5x-1}{x+9} - \dfrac{3x+4}{x+9}$

10. $\dfrac{6x-5}{x-10} - \dfrac{3x-4}{x-10}$

11. $\dfrac{x-7}{2x+7} - \dfrac{4x-3}{2x+7}$

12. $\dfrac{2n}{3n+4} - \dfrac{5n-3}{3n+4}$

13. $\dfrac{x}{x^2+2x-15} - \dfrac{3}{x^2+2x-15}$

14. $\dfrac{3x}{x^2+3x-10} - \dfrac{6}{x^2+3x-10}$

15. $\dfrac{2x+3}{x^2-x-30} - \dfrac{x-2}{x^2-x-30}$

16. $\dfrac{3x-1}{x^2+5x-6} - \dfrac{2x-7}{x^2+5x-6}$

17. $\dfrac{4y+7}{2y^2+7y-4} - \dfrac{y-5}{2y^2+7y-4}$

18. $\dfrac{x+1}{2x^2-5x-12} + \dfrac{x+2}{2x^2-5x-12}$

19. $\dfrac{2x^2+3x}{x^2-9x+20} + \dfrac{2x^2-3}{x^2-9x+20} - \dfrac{4x^2+2x+1}{x^2-9x+20}$

2 Simplify.

20. $\dfrac{4}{x} + \dfrac{5}{y}$

21. $\dfrac{7}{a} + \dfrac{5}{b}$

22. $\dfrac{12}{x} - \dfrac{5}{2x}$

23. $\dfrac{5}{3a} - \dfrac{3}{4a}$

24. $\dfrac{1}{2x} - \dfrac{5}{4x} + \dfrac{7}{6x}$

25. $\dfrac{7}{4y} + \dfrac{11}{6y} - \dfrac{8}{3y}$

26. $\dfrac{5}{3x} - \dfrac{2}{x^2} + \dfrac{3}{2x}$

27. $\dfrac{6}{y^2} + \dfrac{3}{4y} - \dfrac{2}{5y}$

28. $\dfrac{2}{x} - \dfrac{3}{2y} + \dfrac{3}{5x} - \dfrac{1}{4y}$

29. $\dfrac{5}{2a} + \dfrac{7}{3b} - \dfrac{2}{b} - \dfrac{3}{4a}$

30. $\dfrac{2x+1}{3x} + \dfrac{x-1}{5x}$

31. $\dfrac{4x-3}{6x} + \dfrac{2x+3}{4x}$

32. $\dfrac{x-3}{6x} + \dfrac{x+4}{8x}$

33. $\dfrac{2x-3}{2x} + \dfrac{x+3}{3x}$

34. $\dfrac{2x+9}{9x} - \dfrac{x-5}{5x}$

35. $\dfrac{3y-2}{12y} - \dfrac{y-3}{18y}$

36. $\dfrac{x+4}{2x} - \dfrac{x-1}{x^2}$

37. $\dfrac{x-2}{3x^2} - \dfrac{x+4}{x}$

38. $\dfrac{x-10}{4x^2} + \dfrac{x+1}{2x}$

39. $\dfrac{x+5}{3x^2} + \dfrac{2x+1}{2x}$

40. $\dfrac{2x+1}{6x^2} - \dfrac{x-4}{4x}$

41. $\dfrac{x+3}{6x} - \dfrac{x-3}{8x^2}$

42. $\dfrac{x+2}{xy} - \dfrac{3x-2}{x^2y}$

43. $\dfrac{3x-1}{xy^2} - \dfrac{2x+3}{xy}$

44. $\dfrac{4x-3}{3x^2y} + \dfrac{2x+1}{4xy^2}$

45. $\dfrac{5x+7}{6xy^2} - \dfrac{4x-3}{8x^2y}$

46. $\dfrac{x-2}{8x^2} - \dfrac{x+7}{12xy}$

47. $\dfrac{3x-1}{6y^2} - \dfrac{x+5}{9xy}$

48. $\dfrac{4}{x-2} + \dfrac{5}{x+3}$

49. $\dfrac{2}{x-3} + \dfrac{5}{x-4}$

50. $\dfrac{6}{x-7} - \dfrac{4}{x+3}$

51. $\dfrac{3}{y+6} - \dfrac{4}{y-3}$

52. $\dfrac{2x}{x+1} + \dfrac{1}{x-3}$

53. $\dfrac{3x}{x-4} + \dfrac{2}{x+6}$

54. $\dfrac{4x}{2x-1} - \dfrac{5}{x-6}$

55. $\dfrac{6x}{x+5} - \dfrac{3}{2x+3}$

56. $\dfrac{2a}{a-7} + \dfrac{5}{7-a}$

57. $\dfrac{4x}{6-x} + \dfrac{5}{x-6}$

58. $\dfrac{x}{x^2-9} + \dfrac{3}{x-3}$

59. $\dfrac{y}{y^2-16} + \dfrac{1}{y-4}$

60. $\dfrac{2x}{x^2-x-6} - \dfrac{3}{x+2}$

61. $\dfrac{5x}{x^2+2x-8} - \dfrac{2}{x+4}$

62. $\dfrac{3x-1}{x^2-10x+25} - \dfrac{3}{x-5}$

63. $\dfrac{2a+3}{a^2-7a+12} - \dfrac{2}{a-3}$

64. $\dfrac{x+4}{x^2-x-42} + \dfrac{3}{7-x}$

65. $\dfrac{x+3}{x^2-3x-10} + \dfrac{2}{5-x}$

66. $\dfrac{x}{2x+4} - \dfrac{2}{x^2+2x}$

67. $\dfrac{x}{2xy-4y^2} - \dfrac{2y}{x^2-2xy}$

68. $\dfrac{2x-5}{6x+9} - \dfrac{4}{2x^2+3x}$

69. $\dfrac{a+5b}{3ab+6b^2} + \dfrac{2b}{a^2+2ab}$

70. $\dfrac{x+2}{4x+16} - \dfrac{2}{x^2+4x}$

71. $\dfrac{2}{3x^2-4x} - \dfrac{2+3x}{12x-16}$

72. $\dfrac{2-x}{x^2+2x-8} + \dfrac{x-4}{x+4}$

73. $\dfrac{x^2-2}{2x^2-x-3} - \dfrac{x-2}{2x-3}$

74. $\dfrac{3x + 2}{3x^2 + 7x + 2} - \dfrac{x - 2}{3x + 1}$

75. $\dfrac{x + 2}{2x^2 + 5x + 2} - \dfrac{3}{2x + 1}$

76. $\dfrac{x - 1}{x^2 - x - 2} + \dfrac{3}{x^2 - 3x + 2}$

77. $\dfrac{x}{x^2 + 5x + 6} - \dfrac{2 - x}{x^2 - x - 6}$

78. $\dfrac{a + 2}{a^2 + a - 2} + \dfrac{3 - a}{a^2 + 2a - 3}$

79. $\dfrac{3 - y}{y^2 - 1} - \dfrac{y + 4}{y^2 - 2y - 3}$

80. $\dfrac{x}{x^2 - 1} - \dfrac{2 + x}{x^2 + 2x + 1}$

81. $\dfrac{2}{3x - 6} - \dfrac{4}{x^2 - 2x} + \dfrac{x - 1}{x - 2}$

82. $\dfrac{3}{x^2 + 4x} - \dfrac{x}{6x + 24} - \dfrac{1}{3x^2 + 12x}$

83. $\dfrac{x}{6x - 9} - \dfrac{3}{4x^2 - 6x} + \dfrac{2 - x}{2x - 3}$

84. $\dfrac{1}{x + 1} + \dfrac{x}{x - 6} - \dfrac{5x - 2}{x^2 - 5x - 6}$

85. $\dfrac{x}{x - 4} + \dfrac{5}{x + 5} - \dfrac{11x - 8}{x^2 + x - 20}$

86. $\dfrac{3x + 1}{x - 1} - \dfrac{x - 1}{x - 3} + \dfrac{x + 1}{x^2 - 4x + 3}$

87. $\dfrac{4x + 1}{x - 8} - \dfrac{3x + 2}{x + 4} - \dfrac{49x + 4}{x^2 - 4x - 32}$

88. $\dfrac{3}{x^2 + x - 12} + \dfrac{2 - x}{2x^2 - x - 15}$

89. $\dfrac{x - 2}{3x - 12} + \dfrac{2}{2x^2 - 8x} - \dfrac{3x + 1}{12x - 48}$

90. $\dfrac{2x + 9}{3 - x} + \dfrac{x + 5}{x + 7} - \dfrac{2x^2 + 3x - 3}{x^2 + 4x - 21}$

91. $\dfrac{3x + 5}{x + 5} - \dfrac{x + 1}{2 - x} - \dfrac{4x^2 - 3x - 1}{x^2 + 3x - 10}$

SUPPLEMENTAL EXERCISES 7.3

Simplify.

92. $y + \dfrac{8}{3y}$

93. $\dfrac{7}{2n} - n$

94. $5 - \dfrac{x - 2}{x + 1}$

95. $\dfrac{a + 4}{a - 1} + 6$

96. $\dfrac{a}{a - b} + \dfrac{b}{b - a} + 1$

97. $\dfrac{y}{x - y} + 2 - \dfrac{x}{y - x}$

98. $b - 3 + \dfrac{5}{b + 4}$

99. $2y - 1 + \dfrac{6}{y + 5}$

100. $\left(\dfrac{n + 1}{n - 1}\right)^2 - 1$

101. $1 - \left(\dfrac{y - 2}{y + 2}\right)^2$

Simplify.

102. $\dfrac{x^2 + x - 6}{x^2 + 2x - 8} \cdot \dfrac{x^2 + 5x + 4}{x^2 + 2x - 3} - \dfrac{2}{x - 1}$

103. $\dfrac{x^2 + 9x + 20}{x^2 + 4x - 5} \div \dfrac{x^2 - 49}{x^2 + 6x - 7} - \dfrac{x}{x - 7}$

104. $\dfrac{x^2 - 9}{x^2 + 6x + 9} \div \dfrac{x^2 + x - 20}{x^2 - x - 12} + \dfrac{1}{x + 1}$

105. $\dfrac{x^2 - 25}{x^2 + 10x + 25} \cdot \dfrac{x^2 - 7x + 10}{x^2 - x - 2} + \dfrac{1}{x + 1}$

SECTION 7.4

Complex Fractions

1   Simplify complex fractions

A **complex fraction** is a fraction whose numerator or denominator contains one or more fractions. Examples of complex fractions are shown at the right.

$$\dfrac{3}{2 - \dfrac{1}{2}}, \quad \dfrac{4 + \dfrac{1}{x}}{3 + \dfrac{2}{x}}, \quad \dfrac{\dfrac{1}{x - 1} + x + 3}{x - 3 + \dfrac{1}{x + 4}}$$

To simplify $\dfrac{1 - \dfrac{4}{x^2}}{1 + \dfrac{2}{x}}$, find the LCM of the denominators of the fractions in the numerator and denominator. (The LCM of x^2 and x is x^2.) Multiply the numerator and denominator of the complex fraction by the LCM. Then simplify.

$$\dfrac{1 - \dfrac{4}{x^2}}{1 + \dfrac{2}{x}} = \dfrac{1 - \dfrac{4}{x^2}}{1 + \dfrac{2}{x}} \cdot \dfrac{x^2}{x^2}$$

$$= \dfrac{1 \cdot x^2 - \dfrac{4}{x^2} \cdot x^2}{1 \cdot x^2 + \dfrac{2}{x} \cdot x^2}$$

$$= \dfrac{x^2 - 4}{x^2 + 2x}$$

$$= \dfrac{(x - 2)\overset{1}{\cancel{(x + 2)}}}{x\,\underset{1}{\cancel{(x + 2)}}}$$

$$= \dfrac{x - 2}{x}$$

Example 1   Simplify.

A. $\dfrac{\dfrac{1}{x}+\dfrac{1}{2}}{\dfrac{1}{x^2}-\dfrac{1}{4}}$ B. $\dfrac{1-\dfrac{2}{x}-\dfrac{15}{x^2}}{1-\dfrac{11}{x}+\dfrac{30}{x^2}}$

A. The LCM of x, 2, x^2, and 4 is $4x^2$.

$$\dfrac{\dfrac{1}{x}+\dfrac{1}{2}}{\dfrac{1}{x^2}-\dfrac{1}{4}}=\dfrac{\dfrac{1}{x}+\dfrac{1}{2}}{\dfrac{1}{x^2}-\dfrac{1}{4}}\cdot\dfrac{4x^2}{4x^2}$$

$$=\dfrac{\dfrac{1}{x}\cdot 4x^2+\dfrac{1}{2}\cdot 4x^2}{\dfrac{1}{x^2}\cdot 4x^2-\dfrac{1}{4}\cdot 4x^2}$$

$$=\dfrac{4x+2x^2}{4-x^2}$$

$$=\dfrac{2x\overset{1}{\cancel{(2+x)}}}{(2-x)\underset{1}{\cancel{(2+x)}}}$$

$$=\dfrac{2x}{2-x}$$

B. The LCM of x and x^2 is x^2.

$$\dfrac{1-\dfrac{2}{x}-\dfrac{15}{x^2}}{1-\dfrac{11}{x}+\dfrac{30}{x^2}}=\dfrac{1-\dfrac{2}{x}-\dfrac{15}{x^2}}{1-\dfrac{11}{x}+\dfrac{30}{x^2}}\cdot\dfrac{x^2}{x^2}$$

$$=\dfrac{1\cdot x^2-\dfrac{2}{x}\cdot x^2-\dfrac{15}{x^2}\cdot x^2}{1\cdot x^2-\dfrac{11}{x}\cdot x^2+\dfrac{30}{x^2}\cdot x^2}$$

$$=\dfrac{x^2-2x-15}{x^2-11x+30}$$

$$=\dfrac{\overset{1}{\cancel{(x-5)}}(x+3)}{\underset{1}{\cancel{(x-5)}}(x-6)}$$

$$=\dfrac{x+3}{x-6}$$

Problem 1   Simplify.

A. $\dfrac{\dfrac{1}{3} - \dfrac{1}{x}}{\dfrac{1}{9} - \dfrac{1}{x^2}}$ B. $\dfrac{1 + \dfrac{4}{x} + \dfrac{3}{x^2}}{1 + \dfrac{10}{x} + \dfrac{21}{x^2}}$

Solution   See page A27.

EXERCISES 7.4

1   Simplify.

1. $\dfrac{1 + \dfrac{3}{x}}{1 - \dfrac{9}{x^2}}$

2. $\dfrac{1 + \dfrac{4}{x}}{1 - \dfrac{16}{x^2}}$

3. $\dfrac{2 - \dfrac{8}{x + 4}}{3 - \dfrac{12}{x + 4}}$

4. $\dfrac{5 - \dfrac{25}{x + 5}}{1 - \dfrac{3}{x + 5}}$

5. $\dfrac{1 + \dfrac{5}{y - 2}}{1 - \dfrac{2}{y - 2}}$

6. $\dfrac{2 - \dfrac{11}{2x - 1}}{3 - \dfrac{17}{2x - 1}}$

7. $\dfrac{4 - \dfrac{2}{x + 7}}{5 + \dfrac{1}{x + 7}}$

8. $\dfrac{5 + \dfrac{3}{x - 8}}{2 - \dfrac{1}{x - 8}}$

9. $\dfrac{\dfrac{3}{x - 2} + 3}{\dfrac{4}{x - 2} + 4}$

10. $\dfrac{\dfrac{3}{2x + 1} - 3}{2 - \dfrac{4x}{2x + 1}}$

11. $\dfrac{2 - \dfrac{3}{x} - \dfrac{2}{x^2}}{2 + \dfrac{5}{x} + \dfrac{2}{x^2}}$

12. $\dfrac{2 + \dfrac{5}{x} - \dfrac{12}{x^2}}{4 - \dfrac{4}{x} - \dfrac{3}{x^2}}$

13. $\dfrac{1 - \dfrac{1}{x} - \dfrac{6}{x^2}}{1 - \dfrac{9}{x^2}}$

14. $\dfrac{1 + \dfrac{4}{x} + \dfrac{4}{x^2}}{1 - \dfrac{2}{x} - \dfrac{8}{x^2}}$

15. $\dfrac{1 - \dfrac{5}{x} - \dfrac{6}{x^2}}{1 + \dfrac{6}{x} + \dfrac{5}{x^2}}$

16. $\dfrac{1 - \dfrac{7}{a} + \dfrac{12}{a^2}}{1 + \dfrac{1}{a} - \dfrac{20}{a^2}}$

17. $\dfrac{1 - \dfrac{6}{x} + \dfrac{8}{x^2}}{\dfrac{4}{x^2} + \dfrac{3}{x} - 1}$

18. $\dfrac{1 + \dfrac{3}{x} - \dfrac{18}{x^2}}{\dfrac{21}{x^2} - \dfrac{4}{x} - 1}$

19. $\dfrac{x - \dfrac{4}{x + 3}}{1 + \dfrac{1}{x + 3}}$

20. $\dfrac{y + \dfrac{1}{y - 2}}{1 + \dfrac{1}{y - 2}}$

21. $\dfrac{1 - \dfrac{x}{2x + 1}}{x - \dfrac{1}{2x + 1}}$

22. $\dfrac{1 - \dfrac{2x - 2}{3x - 1}}{x - \dfrac{4}{3x - 1}}$

23. $\dfrac{x - 5 + \dfrac{14}{x + 4}}{x + 3 - \dfrac{2}{x + 4}}$

24. $\dfrac{a + 4 + \dfrac{5}{a - 2}}{a + 6 + \dfrac{15}{a - 2}}$

25. $\dfrac{x + 3 - \dfrac{10}{x - 6}}{x + 2 - \dfrac{20}{x - 6}}$

26. $\dfrac{x - 7 + \dfrac{5}{x - 1}}{x - 3 + \dfrac{1}{x - 1}}$

27. $\dfrac{y - 6 + \dfrac{22}{2y + 3}}{y - 5 + \dfrac{11}{2y + 3}}$

28. $\dfrac{x + 2 - \dfrac{12}{2x - 1}}{x + 1 - \dfrac{9}{2x - 1}}$

29. $\dfrac{x - \dfrac{2}{2x - 3}}{2x - 1 - \dfrac{8}{2x - 3}}$

30. $\dfrac{x + 3 - \dfrac{18}{2x + 1}}{x - \dfrac{6}{2x + 1}}$

31. $\dfrac{1 - \dfrac{2}{x + 1}}{1 + \dfrac{1}{x - 2}}$

32. $\dfrac{1 - \dfrac{1}{x + 2}}{1 + \dfrac{2}{x - 1}}$

33. $\dfrac{1 - \dfrac{2}{x + 4}}{1 + \dfrac{3}{x - 1}}$

34. $\dfrac{1 + \dfrac{1}{x - 2}}{1 - \dfrac{3}{x + 2}}$

35. $\dfrac{\dfrac{1}{x} - \dfrac{2}{x-1}}{\dfrac{3}{x} + \dfrac{1}{x-1}}$

36. $\dfrac{\dfrac{3}{n+1} + \dfrac{1}{n}}{\dfrac{2}{n+1} + \dfrac{3}{n}}$

37. $\dfrac{\dfrac{3}{2x-1} - \dfrac{1}{x}}{\dfrac{4}{x} + \dfrac{2}{2x-1}}$

38. $\dfrac{\dfrac{4}{3x+1} + \dfrac{3}{x}}{\dfrac{6}{x} - \dfrac{2}{3x+1}}$

39. $\dfrac{\dfrac{3}{b-4} - \dfrac{2}{b+1}}{\dfrac{5}{b+1} - \dfrac{1}{b-4}}$

40. $\dfrac{\dfrac{5}{x-5} - \dfrac{3}{x-1}}{\dfrac{6}{x-1} + \dfrac{2}{x-5}}$

SUPPLEMENTAL EXERCISES 7.4

Simplify.

41. $1 + \dfrac{1}{1 + \dfrac{1}{2}}$

42. $1 + \dfrac{1}{1 + \dfrac{1}{1 + \dfrac{1}{2}}}$

43. $1 - \dfrac{1}{1 - \dfrac{1}{x}}$

44. $1 - \dfrac{1}{1 - \dfrac{1}{y+1}}$

45. $\dfrac{a^{-1} - b^{-1}}{a^{-2} - b^{-2}}$

46. $\dfrac{x^{-2} - y^{-2}}{x^{-2}y^{-2}}$

47. $\left(\dfrac{y}{4} - \dfrac{4}{y}\right) \div \left(\dfrac{4}{y} - 3 + \dfrac{y}{2}\right)$

48. $\left(\dfrac{b}{8} - \dfrac{8}{b}\right) \div \left(\dfrac{8}{b} - 5 + \dfrac{b}{2}\right)$

SECTION 7.5

Equations Containing Fractions

1 Solve equations containing fractions

To solve an equation containing fractions, **clear denominators** by multiplying each side of the equation by the LCM of the denominators. Then solve for the variable.

Solve: $\dfrac{3x-1}{4} + \dfrac{2}{3} = \dfrac{7}{6}$

Find the LCM of the denominators.	The LCM of 4, 3, and 6 is 12.

$$\dfrac{3x-1}{4} + \dfrac{2}{3} = \dfrac{7}{6}$$

Multiply each side of the equation by the LCM of the denominators.

$$12\left(\dfrac{3x-1}{4} + \dfrac{2}{3}\right) = 12 \cdot \dfrac{7}{6}$$

Simplify using the Distributive Property and the Properties of Fractions.

$$12\left(\dfrac{3x-1}{4}\right) + 12 \cdot \dfrac{2}{3} = 12 \cdot \dfrac{7}{6}$$

$$\overset{3}{\cancel{12}}\!\!\Big(\dfrac{3x-1}{\underset{1}{\cancel{4}}}\Big) + \dfrac{\overset{4}{\cancel{12}}}{1} \cdot \dfrac{2}{\underset{1}{\cancel{3}}} = \dfrac{\overset{2}{\cancel{12}}}{1} \cdot \dfrac{7}{\underset{1}{\cancel{6}}}$$

Solve for x.

$$9x - 3 + 8 = 14$$
$$9x + 5 = 14$$
$$9x = 9$$
$$x = 1$$

1 checks as a solution.
The solution is 1.

Example 1   Solve: $\dfrac{4}{x} - \dfrac{x}{2} = \dfrac{7}{2}$

Solution

$$\dfrac{4}{x} - \dfrac{x}{2} = \dfrac{7}{2}$$ ▶ The LCM of x and 2 is $2x$.

$$2x\left(\dfrac{4}{x} - \dfrac{x}{2}\right) = 2x\left(\dfrac{7}{2}\right)$$

$$\dfrac{2x}{1} \cdot \dfrac{4}{x} - \dfrac{2x}{1} \cdot \dfrac{x}{2} = \dfrac{2x}{1} \cdot \dfrac{7}{2}$$

$$8 - x^2 = 7x$$ ▶ This is a quadratic equation.

$$0 = x^2 + 7x - 8$$
$$0 = (x + 8)(x - 1)$$

$$x + 8 = 0 \qquad\qquad x - 1 = 0$$
$$x = -8 \qquad\qquad x = 1$$

Both -8 and 1 check as solutions.
The solutions are -8 and 1.

Problem 1 Solve: $x + \dfrac{1}{3} = \dfrac{4}{3x}$

Solution See page A27.

Occasionally, a value of a variable in a fractional equation makes one of the denominators zero. In this case, that value of the variable is not a solution of the equation.

Solve: $\dfrac{2x}{x-2} = 1 + \dfrac{4}{x-2}$

Find the LCM of the denominators.

The LCM is $x - 2$.

$$\frac{2x}{x-2} = 1 + \frac{4}{x-2}$$

Multiply each side of the equation by the LCM of the denominators.

$$(x-2)\frac{2x}{x-2} = (x-2)\left(1 + \frac{4}{x-2}\right)$$

Simplify using the Distributive Property.

$$(x-2)\left(\frac{2x}{x-2}\right) = (x-2)\cdot 1 + (x-2)\cdot\frac{4}{x-2}$$

Solve for x.

$$2x = x - 2 + 4$$
$$2x = x + 2$$
$$x = 2$$

When x is replaced by 2, the denominators of $\dfrac{2x}{x-2}$ and $\dfrac{4}{x-2}$ are zero.

Therefore, 2 is not a solution of the equation.
The equation has no solution.

Example 2 Solve: $\dfrac{3x}{x-4} = 5 + \dfrac{12}{x-4}$

Solution

$$\frac{3x}{x-4} = 5 + \frac{12}{x-4}$$

▶ The LCM is $x - 4$.

$$\frac{(x-4)}{1}\cdot\frac{3x}{x-4} = \frac{(x-4)}{1}\left(5 + \frac{12}{x-4}\right)$$

$$3x = (x-4)5 + 12$$
$$3x = 5x - 20 + 12$$
$$3x = 5x - 8$$
$$-2x = -8$$
$$x = 4$$

4 does not check as a solution.
The equation has no solution.

Problem 2   Solve: $\dfrac{5x}{x+2} = 3 - \dfrac{10}{x+2}$

Solution   See page A28.

Quantities such as 4 meters, 15 seconds, and 8 gallons are number quantities written with units. In these examples, the units are meters, seconds, and gallons.

A **ratio** is the quotient of two quantities that have the same unit.

The length of a living room is 16 ft, and the width is 12 ft. The ratio of the length to the width is written:

$$\dfrac{16 \text{ ft}}{12 \text{ ft}} = \dfrac{16}{12} = \dfrac{4}{3}$$ A ratio is in simplest form when the two numbers do not have a common factor. Note that the units are not written.

A **rate** is the quotient of two quantities that have different units.

There are 2 lb of salt in 8 gal of water. The salt-to-water rate is:

$$\dfrac{2 \text{ lb}}{8 \text{ gal}} = \dfrac{1 \text{ lb}}{4 \text{ gal}}$$ A rate is in simplest form when the two numbers do not have a common factor. The units are written as part of the rate.

A **proportion** is an equation that states the equality of two ratios or rates.

Examples of proportions are shown at the right.

$$\dfrac{30 \text{ mi}}{4 \text{ h}} = \dfrac{15 \text{ mi}}{2 \text{ h}}$$

$$\dfrac{4}{6} = \dfrac{8}{12}$$

$$\dfrac{3}{4} = \dfrac{x}{8}$$

Because a proportion is an equation containing fractions, the same method used to solve an equation containing fractions is used to solve a proportion. Multiply each side of the equation by the LCM of the denominators. Then solve for the variable.

To solve the proportion $\frac{4}{x} = \frac{2}{3}$, multiply each side of the proportion by the LCM of the denominators.

Solve the equation.

$$\frac{4}{x} = \frac{2}{3}$$

$$3x\left(\frac{4}{x}\right) = 3x\left(\frac{2}{3}\right)$$

$$12 = 2x$$
$$6 = x$$

The solution is 6.

Example 3   Solve.   A. $\dfrac{8}{x+3} = \dfrac{4}{x}$ B. $\dfrac{6}{x+4} = \dfrac{12}{5x-13}$

Solution   A.
$$\frac{8}{x+3} = \frac{4}{x}$$

$$x(x+3)\frac{8}{x+3} = x(x+3)\frac{4}{x}$$ ▶ Multiply each side of the proportion by the LCM of the denominators.

$$8x = (x+3)4$$
$$8x = 4x + 12$$
$$4x = 12$$
$$x = 3$$ ▶ Remember to check the solution because it is a fractional equation.

The solution is 3.

B.
$$\frac{6}{x+4} = \frac{12}{5x-13}$$

$$(5x-13)(x+4)\frac{6}{x+4} = (5x-13)(x+4)\frac{12}{5x-13}$$

$$(5x-13)6 = (x+4)12$$
$$30x - 78 = 12x + 48$$
$$18x - 78 = 48$$
$$18x = 126$$
$$x = 7$$

The solution is 7.

Problem 3   Solve.   A. $\dfrac{2}{x+3} = \dfrac{6}{5x+5}$ B. $\dfrac{5}{2x-3} = \dfrac{10}{x+3}$

Solution   See page A28.

2   Application problems: proportions

Example 4   The monthly loan payment for a car is $29.50 for each $1000 borrowed. At this rate, find the monthly payment for a $9000 car loan.

Strategy   To find the monthly payment, write and solve a proportion using P to represent the monthly car payment.

Solution

$$\frac{\$29.50}{\$1000} = \frac{P}{\$9000}$$

$$9000\left(\frac{29.50}{1000}\right) = 9000\left(\frac{P}{9000}\right)$$

$$265.50 = P$$

The monthly payment is $265.50.

Problem 4   Nine ceramic tiles are required to tile a 4-ft^2 area. At this rate, how many square feet can be tiled with 270 ceramic tiles?

Solution   See page A28.

Example 5   An investment of $1200 earns $96 each year. At the same rate, how much additional money must be invested to earn $128 each year?

Strategy   To find the additional amount of money that must be invested, write and solve a proportion using x to represent the additional money. Then $1200 + x$ is the total amount invested.

Solution

$$\frac{\$1200}{\$96} = \frac{\$1200 + x}{\$128}$$

$$\frac{25}{2} = \frac{1200 + x}{128}$$

$$128\left(\frac{25}{2}\right) = 128\left(\frac{1200 + x}{128}\right)$$

$$1600 = 1200 + x$$

$$400 = x$$

An additional $400 must be invested.

Problem 5   Three ounces of a medication are required for a 120-lb adult. At the same rate, how many additional ounces of the medication are required for a 180-lb adult?

Solution   See page A29.

EXERCISES 7.5

| 1 | Solve.

1. $\dfrac{2x}{3} - \dfrac{5}{2} = -\dfrac{1}{2}$

2. $\dfrac{x}{3} - \dfrac{1}{4} = \dfrac{1}{12}$

3. $\dfrac{x}{3} - \dfrac{1}{4} = \dfrac{x}{4} - \dfrac{1}{6}$

4. $\dfrac{2y}{9} - \dfrac{1}{6} = \dfrac{y}{9} + \dfrac{1}{6}$

5. $\dfrac{2x - 5}{8} + \dfrac{1}{4} = \dfrac{x}{8} + \dfrac{3}{4}$

6. $\dfrac{3x + 4}{12} - \dfrac{1}{3} = \dfrac{5x + 2}{12} - \dfrac{1}{2}$

7. $\dfrac{6}{2a + 1} = 2$

8. $\dfrac{12}{3x - 2} = 3$

9. $\dfrac{9}{2x - 5} = -2$

10. $\dfrac{6}{4 - 3x} = 3$

11. $2 + \dfrac{5}{x} = 7$

12. $3 + \dfrac{8}{n} = 5$

13. $1 - \dfrac{9}{x} = 4$

14. $3 - \dfrac{12}{x} = 7$

15. $\dfrac{2}{y} + 5 = 9$

16. $\dfrac{6}{x} + 3 = 11$

17. $\dfrac{3}{x - 2} = \dfrac{4}{x}$

18. $\dfrac{5}{x + 3} = \dfrac{3}{x - 1}$

19. $\dfrac{2}{3x - 1} = \dfrac{3}{4x + 1}$

20. $\dfrac{5}{3x - 4} = \dfrac{-3}{1 - 2x}$

21. $\dfrac{-3}{2x + 5} = \dfrac{2}{x - 1}$

22. $\dfrac{4}{5y - 1} = \dfrac{2}{2y - 1}$

23. $\dfrac{4x}{x - 4} + 5 = \dfrac{5x}{x - 4}$

24. $\dfrac{2x}{x + 2} - 5 = \dfrac{7x}{x + 2}$

25. $2 + \dfrac{3}{a - 3} = \dfrac{a}{a - 3}$

26. $\dfrac{x}{x + 4} = 3 - \dfrac{4}{x + 4}$

27. $\dfrac{x}{x - 1} = \dfrac{8}{x + 2}$

28. $\dfrac{x}{x + 12} = \dfrac{1}{x + 5}$

29. $\dfrac{2x}{x + 4} = \dfrac{3}{x - 1}$

30. $\dfrac{5}{3n - 8} = \dfrac{n}{n + 2}$

31. $x + \dfrac{6}{x - 2} = \dfrac{3x}{x - 2}$

32. $x - \dfrac{6}{x - 3} = \dfrac{2x}{x - 3}$

33. $\dfrac{x}{x + 2} + \dfrac{2}{x - 2} = \dfrac{x + 6}{x^2 - 4}$

34. $\dfrac{x}{x + 4} = \dfrac{11}{x^2 - 16} + 2$

35. $\dfrac{8}{y} = \dfrac{2}{y-2} + 1$

36. $\dfrac{8}{r} + \dfrac{3}{r-1} = 3$

37. $\dfrac{4}{5} = \dfrac{12}{x}$

38. $\dfrac{x}{12} = \dfrac{3}{4}$

39. $\dfrac{6}{x} = \dfrac{2}{3}$

40. $\dfrac{4}{9} = \dfrac{x}{27}$

41. $\dfrac{16}{9} = \dfrac{64}{x}$

42. $\dfrac{x+3}{12} = \dfrac{5}{6}$

43. $\dfrac{3}{5} = \dfrac{x-4}{10}$

44. $\dfrac{18}{x+4} = \dfrac{9}{5}$

45. $\dfrac{2}{11} = \dfrac{20}{x-3}$

46. $\dfrac{2}{x} = \dfrac{4}{x+1}$

47. $\dfrac{16}{x-2} = \dfrac{8}{x}$

48. $\dfrac{x+3}{4} = \dfrac{x}{8}$

49. $\dfrac{x-6}{3} = \dfrac{x}{5}$

50. $\dfrac{2}{x-1} = \dfrac{6}{2x+1}$

51. $\dfrac{9}{x+2} = \dfrac{3}{x-2}$

52. $\dfrac{2x}{7} = \dfrac{x-2}{14}$

2 Solve.

53. An exit poll showed that 4 out of every 7 voters cast a ballot in favor of an amendment to a city charter. At this rate, how many people voted in favor of the amendment if 35,000 people voted?

54. A quality control inspector found three defective transistors in a shipment of 500 transistors. At this rate, how many transistors would be defective in a shipment of 2000 transistors?

55. An air conditioning specialist recommends two air vents for every 300 ft² of floor space. At this rate, how many air vents are required for a 21,000-ft² office building?

56. In a city of 25,000 homes, a survey was taken to determine the number with cable television. Of the 300 homes surveyed, 210 had cable television. Estimate the number of homes in the city that have cable television.

57. A simple syrup is made by dissolving 2 c of sugar in $\dfrac{2}{3}$ c of boiling water. At this rate, how many cups of sugar are required for 2 c of boiling water?

58. The lighting for a billboard is provided by solar energy. If three energy panels generate 10 watts of power, how many panels are needed to provide 600 watts of power?

59. As part of a conservation effort for a lake, 40 fish were caught, tagged, and then released. Later, 80 fish were caught from the lake. Four of these 80 fish were found to have tags. Estimate the number of fish in the lake.

60. In a wildlife preserve, 10 elk were captured, tagged, and then released. Later, 15 elk were captured and two were found to have tags. Estimate the number of elk in the preserve.

61. A health department estimates that eight vials of a malaria serum will treat 100 people. At this rate, how many vials are required to treat 175 people?

62. A company will accept a shipment of 10,000 computer chips if there are two or fewer defects in a sample of 100 randomly chosen chips. Assume that there are 300 defective chips in the shipment and that the rate of defective chips in the sample is the same as the rate in the shipment. Will the shipment be accepted?

63. A company will accept a shipment of 20,000 precision bearings if there are three or fewer defects in a sample of 100 randomly chosen bearings. Assume that there are 400 defective bearings in the shipment and that the rate of defective bearings in the sample is the same as the rate in the shipment. Will the shipment be accepted?

64. The engine of a small rocket burns 170,000 lb of fuel in 1 min. At this rate, how many pounds of fuel does the rocket burn in 45 s?

65. A caterer estimates that 5 gal of coffee will serve 50 people. How much additional coffee is necessary to serve 70 people?

66. A painter estimates that 5 gal of paint will cover 1200 ft^2 of wall surface. How many additional gallons are required to cover 1680 ft^2?

67. A 50-acre field yields 1100 bushels of wheat annually. How many additional acres must be planted so that the annual yield will be 1320 bushels?

68. The sales tax on a car that sold for $12,000 is $780. At the same rate, how much higher is the sales tax on a car that sells for $13,500?

69. To conserve energy and still allow for as much natural lighting as possible, an architect suggests that the ratio of the area of a window to the area of the total wall surface be 5 to 12. Using this ratio, determine the recommended area of a window to be installed in a wall that measures 8 ft by 12 ft.

70. On a map, two cities are $2\frac{5}{8}$ in. apart. If $\frac{3}{8}$ in. on the map represents 25 mi, find the number of miles in the distance between the two cities.

71. On a map, two cities are $5\frac{5}{8}$ in. apart. If $\frac{3}{4}$ in. on the map represents 100 mi, find the number of miles in the distance between the two cities.

72. A green paint is created by mixing 3 parts of yellow with every 5 parts of blue. How many gallons of yellow paint are needed to make 60 gal of this green paint?

73. A soft drink is made by mixing 4 parts syrup with every 3 parts carbonated water. How many milliliters of syrup are in 280 ml of soft drink?

SUPPLEMENTAL EXERCISES 7.5

Solve.

74. $\frac{2}{3}(x + 2) + \frac{x + 1}{-6} = \frac{1}{2}$

75. $\frac{3}{5}y - \frac{1}{3}(1 - y) = \frac{2y - 5}{15}$

76. $\frac{3}{4}a = \frac{1}{2}(3 - a) + \frac{a - 2}{4}$

77. $\frac{b + 2}{5} = \frac{1}{4}b - \frac{3}{10}(b - 1)$

78. $\frac{x}{2x^2 - x - 1} = \frac{3}{x^2 - 1} + \frac{3}{2x + 1}$

79. $\frac{1}{2y^2 + 3y + 1} = \frac{2}{2y - 1} + \frac{2}{2y + 1}$

80. $\frac{x + 1}{x^2 + x - 2} = \frac{x + 2}{x^2 - 1} + \frac{3}{x + 2}$

81. $\frac{y + 2}{y^2 - y - 2} + \frac{y + 1}{y^2 - 4} = \frac{1}{y + 1}$

82. The sum of a number and its reciprocal is $\frac{26}{5}$. Find the number.

83. The sum of a number and its reciprocal is $\frac{25}{12}$. Find the number.

84. The sum of the multiplicative inverses of two consecutive integers is $\frac{11}{30}$. Find the integers.

85. The sum of the multiplicative inverses of two consecutive integers is $-\frac{7}{12}$. Find the integers.

86. The denominator of a fraction is 2 more than the numerator. If the numerator and denominator of the fraction are increased by 3, the new fraction is $\frac{4}{5}$. Find the original fraction.

87. The numerator of a fraction is 3 less than the denominator. If the numerator and denominator of the fraction are increased by 5, the new fraction is $\frac{4}{5}$. Find the original fraction.

88. Three people put their money together to buy lottery tickets. The first person put in $25, the second person put in $30, and the third person put in $35. One of their tickets was a winning ticket. If they won $4.5 million, what was the first person's share of the winnings?

89. A basketball player has made 5 out of every 6 foul shots attempted. If 42 foul shots are missed in the player's career, how many foul shots were made in the player's career?

90. The "sitting fee" for school pictures is $4. If 10 photos cost $10, including the sitting fee, what would 24 photos cost, including the sitting fee?

91. No one belongs to both the Math Club and the Photography Club, but the two clubs join to hold a car wash. Ten members of the Math Club and 6 members of the Photography Club participate. The profits from the car wash are $120. If each club's profits are proportional to the number of members participating, what share of the profits does the Math Club receive?

SECTION 7.6

Literal Equations

■ **1** Solve a literal equation for one of the variables

A **literal equation** is an equation that contains more than one variable. Examples of literal equations are shown at the right.

$$2x + 3y = 6$$

$$4w - 2x + z = 0$$

Formulas are used to express a relationship among physical quantities. A **formula** is a literal equation that states rules about measurements. Examples of formulas are shown below.

$$\frac{1}{R_1} + \frac{1}{R_2} = \frac{1}{R} \qquad \text{(Physics)}$$

$$s = a + (n - 1)d \qquad \text{(Mathematics)}$$

$$A = P + Prt \qquad \text{(Business)}$$

The Addition and Multiplication Properties can be used to solve a literal equation for one of the variables. The goal is to rewrite the equation so that the letter being solved for is alone on one side of the equation and all other numbers and variables are on the other side.

In solving $A = P(1 + i)$ for i, the goal is to rewrite the equation so that i is on one side of the equation and all other variables are on the other side.

Use the Distributive Property to remove parentheses.	$A = P(1 + i)$ $A = P + Pi$
Subtract P from each side of the equation.	$A - P = P - P + Pi$ $A - P = Pi$
Divide each side of the equation by P.	$\dfrac{A - P}{P} = \dfrac{Pi}{P}$
	$\dfrac{A - P}{P} = i$

Example 1   Solve $3x - 4y = 12$ for y.

Solution

$$3x - 4y = 12$$
$$3x - 3x - 4y = -3x + 12 \qquad \blacktriangleright \text{Subtract } 3x \text{ from each side of the equation.}$$
$$-4y = -3x + 12 \qquad \blacktriangleright \text{Simplify.}$$
$$\frac{-4y}{-4} = \frac{-3x + 12}{-4} \qquad \blacktriangleright \text{Divide each side of the equation by } -4.$$
$$y = \frac{3}{4}x - 3$$

Problem 1   Solve $5x - 2y = 10$ for y.

Solution   See page A29.

Example 2   Solve $I = \dfrac{E}{R + r}$ for R.

Solution

$$I = \frac{E}{R + r}$$
$$(R + r)I = (R + r)\frac{E}{R + r}$$
$$RI + rI = E$$
$$RI + rI - rI = E - rI$$
$$RI = E - rI$$
$$\frac{RI}{I} = \frac{E - rI}{I}$$
$$R = \frac{E - rI}{I}$$

Problem 2   Solve $s = \dfrac{A + L}{2}$ for L.

Solution   See page A29.

Example 3   Solve $L = a(1 + ct)$ for c.

Solution
$$L = a(1 + ct)$$
$$L = a + act$$
$$L - a = a - a + act$$
$$L - a = act$$
$$\frac{L - a}{at} = \frac{act}{at}$$
$$\frac{L - a}{at} = c$$

Problem 3   Solve $S = a + (n - 1)d$ for n.

Solution   See page A29.

Example 4   Solve $S = C - rC$ for C.

Solution
$$S = C - rC$$
$$S = C(1 - r) \qquad \blacktriangleright \text{ Factor } C \text{ from } C - rC.$$
$$\frac{S}{1 - r} = \frac{C(1 - r)}{1 - r} \qquad \blacktriangleright \text{ Divide each side of the equation by the } 1 - r.$$
$$\frac{S}{1 - r} = C$$

Problem 4   Solve $S = C + rC$ for C.

Solution   See page A29.

EXERCISES 7.6   *all odds*

1   Solve for y.

1. $3x + y = 10$ **2.** $2x + y = 5$ **3.** $4x - y = 3$ **4.** $5x - y = 7$

5. $3x + 2y = 6$ **6.** $2x + 3y = 9$ **7.** $2x - 5y = 10$ **8.** $5x - 2y = 4$

9. $2x + 7y = 14$ **10.** $6x - 5y = 10$ **11.** $x + 3y = 6$

12. $x - 4y = 12$ **13.** $x - 3y = 9$ **14.** $7x - 2y - 14 = 0$

15. $2x - 9y - 18 = 0$ **16.** $3x - y + 7 = 0$ **17.** $2x - y + 5 = 0$

Solve for x.

18. $x + 3y = 6$ **19.** $x + 6y = 10$ **20.** $3x - y = 3$

21. $2x - y = 6$ **22.** $2x + 5y = 10$ **23.** $4x + 3y = 12$

24. $x - 2y + 1 = 0$ **25.** $x - 4y - 3 = 0$ **26.** $5x + 4y + 20 = 0$

27. $3x + 5y + 15 = 0$ **28.** $3x - 2y - 15 = 0$ **29.** $5x - 8y + 10 = 0$

Solve the formula for the variable given.

30. $A = \frac{1}{2}bh$; h (Geometry)   **31.** $P = a + b + c$; b (Geometry)

32. $d = rt$; t (Physics)   **33.** $E = IR$; R (Physics)

34. $PV = nRT$; T (Chemistry)   **35.** $A = bh$; h (Geometry)

36. $P = 2l + 2w$; l (Geometry)   **37.** $F = \frac{9}{5}C + 32$; C (Temperature conversion)

38. $A = \frac{1}{2}h(b_1 + b_2)$; b_1 (Geometry)   **39.** $C = \frac{5}{9}(F - 32)$; F (Temperature conversion)

40. $V = \frac{1}{3}Ah$; h (Geometry)   **41.** $P = R - C$; C (Business)

42. $R = \dfrac{C - S}{t}$; S (Business)

43. $P = \dfrac{R - C}{n}$; R (Business)

44. $A = P + Prt$; P (Business)

45. $T = fm - gm$; m (Engineering)

46. $A = Sw + w$; w (Physics)

47. $a = S - Sr$; S (Mathematics)

SUPPLEMENTAL EXERCISES 7.6

The surface area of a right circular cylinder is given by the formula $S = 2\pi rh + 2\pi r^2$, where r is the radius of the base, and h is the height of the cylinder.

48. **a.** Solve the formula $S = 2\pi rh + 2\pi r^2$ for h.

   b. Use your answer to part **a** to find the height of a right circular cylinder when the surface area is 12π in.^{2} and the radius is 1 in.

   c. Use your answer to part **a** to find the height of a right circular cylinder when the surface area is 24π in.^{2} and the radius is 2 in.

Break-even analysis is a method used to determine the sales volume required for a company to break even, or experience neither a profit nor a loss on the sale of its product. The break-even point represents the number of units that must be made and sold for income from sales to equal the cost of the product. The break-even point can be calculated using the formula $B = \dfrac{F}{S - V}$, where F is the fixed costs, S is the selling price per unit, and V is the variable costs per unit.

49. **a.** Solve the formula $B = \dfrac{F}{S - V}$ for S.

   b. Use your answer to part **a** to find the required selling price per unit for a company to break even. The fixed costs are $20,000, the variable costs per unit are $80, and the company plans to make and sell 200 desks.

   c. Use your answer to part **a** to find the required selling price per unit for a company to break even. The fixed costs are $15,000, the variable costs per unit are $50, and the company plans to make and sell 600 cameras.

When markup is based on selling price, the selling price of a product is given by the formula $S = \dfrac{C}{1-r}$, where C is the cost of the product, and r is the markup rate.

50. **a.** Solve the formula $S = \dfrac{C}{1-r}$ for r.

   b. Use your answer to part **a** to find the markup rate on a product when the cost is \$112 and the selling price is \$140.

   c. Use your answer to part **a** to find the markup rate on a product when the cost is \$50.40 and the selling price is \$72.

Resistors are used to control the flow of current. The total resistance of two resistors in a circuit can be given by the formula $R = \dfrac{1}{\dfrac{1}{R_1} + \dfrac{1}{R_2}}$, where R_1 and R_2 are the two resistors in the circuit. Resistance is measured in ohms.

51. **a.** Solve the formula $R = \dfrac{1}{\dfrac{1}{R_1} + \dfrac{1}{R_2}}$ for R_1.

   b. Use your answer to part **a** to find the resistance in R_1 if the resistance in R_2 is 30 ohms and the total resistance is 12 ohms.

   c. Use your answer to part **a** to find the resistance in R_1 if the resistance in R_2 is 15 ohms and the total resistance is 6 ohms.

S E C T I O N 7.7

Application Problems

1 Work problems

If a painter can paint a room in 4 h, then in 1 h the painter can paint $\dfrac{1}{4}$ of the room. The painter's rate of work is $\dfrac{1}{4}$ of the room each hour. The **rate of work** is that part of a task that is completed in one unit of time.

A pipe can fill a tank in 30 min. This pipe can fill $\dfrac{1}{30}$ of the tank in 1 min. The rate of work is $\dfrac{1}{30}$ of the tank each minute. If a second pipe can fill the tank in x min, the rate of work for the second pipe is $\dfrac{1}{x}$ of the tank each minute.

In solving a work problem, the goal is to determine the time it takes to complete a task. The basic equation that is used to solve work problems is:

Rate of work · Time worked = Part of task completed

For example, if a faucet can fill a sink in 6 min, then in 5 min the faucet will fill $\frac{1}{6} \cdot 5 = \frac{5}{6}$ of the sink. In 5 min, the faucet completes $\frac{5}{6}$ of the task.

Solve: A painter can paint a ceiling in 60 min. The painter's apprentice can paint the same ceiling in 90 min. How long will it take to paint the ceiling when they work together?

STRATEGY *for solving a work problem*

■ For each person or machine, write a numerical or variable expression for the rate of work, the time worked, and the part of the task completed. The results can be recorded in a table.

Unknown time to paint the ceiling working together: t

	Rate of work	·	Time worked	=	Part of task completed
Painter	$\dfrac{1}{60}$	·	t	=	$\dfrac{t}{60}$
Apprentice	$\dfrac{1}{90}$	·	t	=	$\dfrac{t}{90}$

■ Determine how the parts of the task completed are related. Use the fact that the sum of the parts of the task completed must equal 1, the complete task.

The sum of the part of the task completed by the painter and the part of the task completed by the apprentice is 1.

$$\frac{t}{60} + \frac{t}{90} = 1$$
$$180\left(\frac{t}{60} + \frac{t}{90}\right) = 180 \cdot 1$$
$$3t + 2t = 180$$
$$5t = 180$$
$$t = 36$$

Working together, they will paint the ceiling in 36 min.

Example 1   A small water pipe takes four times longer to fill a tank than does a large water pipe. With both pipes open, it takes 3 h to fill the tank. Find the time it would take the small pipe, working alone, to fill the tank.

Strategy
- ■ Time for large pipe to fill the tank: t
- ■ Time for small pipe to fill the tank: $4t$

	Rate	Time	Part
Small pipe	$\dfrac{1}{4t}$	3	$\dfrac{3}{4t}$
Large pipe	$\dfrac{1}{t}$	3	$\dfrac{3}{t}$

- ■ The sum of the parts of the task completed by each pipe must equal 1.

Solution
$$\frac{3}{4t} + \frac{3}{t} = 1$$
$$4t\left(\frac{3}{4t} + \frac{3}{t}\right) = 4t \cdot 1$$
$$3 + 12 = 4t$$
$$15 = 4t$$
$$\frac{15}{4} = t \qquad \blacktriangleright \; t \text{ is the time for the large pipe to fill the tank.}$$
$$4t = 4\left(\frac{15}{4}\right) = 15 \qquad \blacktriangleright \; \text{Substitute the value of } t \text{ into the variable expression for the time for the small pipe to fill the tank.}$$

The small pipe working alone takes 15 h to fill the tank.

Problem 1   Two computer printers that work at the same rate are working together to print the payroll checks for a large corporation. After they work together for 3 h, one of the printers quits. The second requires 2 more hours to complete the payroll checks. Find the time it would take one printer, working alone, to print the payroll.

Solution   See pages A29 and A30.

2   Uniform motion problems

A car that travels constantly in a straight line at 30 mph is in uniform motion. **Uniform motion** means the speed of an object does not change.

The basic equation used to solve uniform motion problems is:

Distance = Rate · Time

An alternative form of this equation can be written by solving the equation for time:

$$\frac{\text{Distance}}{\text{Rate}} = \text{Time}$$

This form of the equation is useful when the total time of travel for two objects is known or the times of travel for two objects are equal.

Solve: The speed of a boat in still water is 20 mph. The boat traveled 120 mi down a river in the same amount of time as it took the boat to travel 80 mi up the river. Find the rate of the river's current.

STRATEGY *for solving a uniform motion problem*

■ For each object, write a numerical or variable expression for the distance, rate, and time. The results can be recorded in a table.

The unknown rate of the river's current: r

	Distance	÷	Rate	=	Time
Down river	120	÷	$20 + r$	=	$\dfrac{120}{20 + r}$
Up river	80	÷	$20 - r$	=	$\dfrac{80}{20 - r}$

■ Determine how the times traveled by each object are related. For example, it may be known that the times are equal, or the total time may be known.

The time down the river is equal to the time up the river.

$$\frac{120}{20 + r} = \frac{80}{20 - r}$$

$$(20 + r)(20 - r)\frac{120}{20 + r} = (20 + r)(20 - r)\frac{80}{20 - r}$$

$$(20 - r)120 = (20 + r)80$$

$$2400 - 120r = 1600 + 80r$$

$$-200r = -800$$

$$r = 4$$

The rate of the river's current is 4 mph.

Example 2   A cyclist rode the first 20 mi of a trip at a constant rate. For the next 16 mi, the cyclist reduced the speed by 2 mph. The total time for the 36 mi was 4 h. Find the rate of the cyclist for each leg of the trip.

Strategy ▪ Rate for the first 20 mi: r

	Distance	Rate	Time
First 20 mi	20	r	$\dfrac{20}{r}$
Next 16 mi	16	$r - 2$	$\dfrac{16}{r - 2}$

▪ The total time for the trip was 4 h.

Solution

$$\frac{20}{r} + \frac{16}{r - 2} = 4$$

$$r(r - 2)\left[\frac{20}{r} + \frac{16}{r - 2}\right] = r(r - 2) \cdot 4$$

$$(r - 2)20 + 16r = 4r^2 - 8r$$
$$20r - 40 + 16r = 4r^2 - 8r \qquad \blacktriangleright \text{This is a quadratic equation.}$$
$$36r - 40 = 4r^2 - 8r$$
$$0 = 4r^2 - 44r + 40$$
$$0 = 4(r^2 - 11r + 10)$$
$$0 = r^2 - 11r + 10 \qquad \blacktriangleright \text{Divide both sides by 4.}$$
$$0 = (r - 10)(r - 1) \qquad \blacktriangleright \text{Solve by factoring.}$$

$$r - 10 = 0 \qquad\qquad r - 1 = 0$$
$$r = 10 \qquad\qquad r = 1$$

$$r - 2 = 10 - 2 \qquad\qquad r - 2 = 1 - 2$$
$$= 8 \qquad\qquad\qquad = -1$$

▶ Find the rate for the last 16 mi. The solution $r = 1$ mph is not possible because the rate on the last 16 mi would then be −1 mph.

10 mph was the rate for the first 20 mi.
8 mph was the rate for the next 16 mi.

Problem 2   The total time for a sailboat to sail back and forth across a lake 6 km wide was 3 h. The rate sailing back was twice the rate sailing across the lake. Find the rate of the sailboat going across the lake.

Solution   See page A30.

EXERCISES 7.7

1   Solve.

1. An experienced painter can paint a garage twice as fast as an apprentice. Working together, the painters require 4 h to paint the garage. How long would it take the experienced painter, working alone, to paint the garage?

2. One grocery clerk can stock a shelf in 20 min. A second clerk requires 30 min to stock the same shelf. How long would it take to stock the shelf if the two clerks worked together?

3. One person with a skiploader requires 12 h to remove a large quantity of earth. With a larger skiploader, the same amount of earth can be removed in 4 h. How long would it take to remove the earth if both skiploaders were operated together?

4. One worker can dig the trenches for a sprinkler system in 3 h. A second worker requires 6 h to do the same task. How long would it take to dig the trenches with both people working together?

5. One computer can solve a complex prime factorization problem in 75 h. A second computer can solve the same problem in 50 h. How long would it take both computers, working together, to solve the problem?

6. A new machine makes 10,000 aluminum cans three times faster than an older machine. With both machines operating, it takes 9 h to make 10,000 cans. How long would it take the new machine, working alone, to make 10,000 cans?

7. A small air conditioner will cool a room 2° in 15 min. A larger air conditioner will cool the room 2° in 10 min. How long would it take to cool the room 2° with both air conditioners operating?

8. One printing press can print the first edition of a book in 55 min. A second printing press requires 66 min to print the same number of copies. How long would it take to print the first edition of the book with both presses operating?

9. Two welders working together can complete a job in 6 h. One of the welders, working alone, can complete the task in 10 h. How long would it take the second welder, working alone, to complete the task?

10. Two pipelines can fill a small tank in 30 min. Working alone, the larger pipeline can fill the tank in 45 min. How long would it take the smaller pipeline, working alone, to fill the tank?

11. Working together, two dock workers can load a crate in 6 min. One of the dock workers, working alone, can load the crate in 15 min. How long would it take the other dock worker, working alone, to load the crate?

12. With two harvesters operating, a plot of land can be harvested in 1 h. If only the newer harvester is used, the land can be harvested in 1.5 h. How long would it take to harvest the field using only the older harvester?

13. A cement mason can build a barbecue in 8 h. A second mason requires 12 h to do the same task. After working alone for 4 h, the first mason quits. How long will it take the second mason to complete the job?

14. A mechanic requires 2 h to repair a transmission. An apprentice requires 6 h to make the same repairs. If the mechanic works alone on a transmission for 1 h and then stops, how long will it take the apprentice to complete the repairs?

15. One computer technician can wire a modem in 4 h. A second technician requires 6 h to do the same job. After working alone for 2 h, the first technician quits. How long will it take the second technician to complete the wiring?

16. A wallpaper hanger requires 2 h to hang the wallpaper on one wall of a room. A second wallpaper hanger requires 4 h to hang the same amount of wallpaper. The first wallpaper hanger works alone for 1 h and then quits. How long will it take the second hanger, working alone, to finish papering the wall?

17. Two welders who work at the same rate are riveting the girders of a building. After they work together for 10 h, one of the welders quits. The second welder requires 20 more hours to complete the welds. Find the time it would have taken one of the welders, working alone, to complete the welds.

18. A large heating unit and a small heating unit are being used to heat the water in a pool. The large unit, working alone, requires 8 h to heat the pool. After both units have been operating for 2 h, the large unit is turned off. The small unit requires 9 more hours to heat the pool. How long would it take the small unit, working alone, to heat the pool?

19. Two machines fill cereal boxes at the same rate. After the two machines work together for 7 h, one machine breaks down. The second machine requires 14 more hours to finish filling the boxes. How long would it have taken one of the machines, working alone, to fill the boxes?

20. A large drain and a small drain are opened to drain a pool. The large drain can empty the pool in 6 h. After both drains have been open for 1 h, the large drain becomes clogged and is closed. The smaller drain remains open and requires 9 more hours to empty the pool. How long would it have taken the small drain, working alone, to empty the pool?

2   Solve.

21. A camper drove 90 mi to a recreational area and then hiked 5 mi into the woods. The rate of the camper while driving was nine times the rate while hiking. The time spent hiking and driving was 3 h. Find the rate at which the camper hiked.

22. The president of a company traveled 1800 mi by jet and 300 mi on a prop plane. The rate of the jet was four times the rate of the prop plane. The entire trip took 5 h. Find the rate of the jet plane.

23. A jogger ran 8 mi in the same amount of time as it took a cyclist to ride 20 mi. The rate of the cyclist was 12 mph faster than the rate of the jogger. Find the rate of the jogger and the rate of the cyclist.

24. An express train traveled 600 mi in the same amount of time as it took a freight train to travel 360 mi. The rate of the express train was 20 mph faster than the rate of the freight train. Find the rate of each train.

25. To assess the damage done by a fire, a forest ranger traveled 1080 mi by jet and then an additional 180 mi by helicopter. The rate of the jet was four times the rate of the helicopter. The entire trip took 5 h. Find the rate of the jet.

26. A twin-engined plane flies 800 mi in the same amount of time as it takes a single-engined plane to fly 600 mi. The rate of the twin-engined plane is 50 mph faster than the rate of the single-engined plane. Find the rate of the twin-engined plane.

27. Two planes leave an airport and head for another airport 900 mi away. The rate of the first plane is twice the rate of the second plane. The second plane arrives at the airport 3 h after the first plane. Find the rate of the second plane.

28. A car and a bus leave a town at 1 P.M. and head for a town 300 mi away. The rate of the car is twice the rate of the bus. The car arrives 5 h ahead of the bus. Find the rate of the car.

29. A car is traveling at a rate that is 36 mph faster than the rate of a cyclist. The car travels 384 mi in the same amount of time as it takes the cyclist to travel 96 mi. Find the rate of the car.

30. An engineer traveled 165 mi by car and then an additional 660 mi by plane. The rate of the plane was four times the rate of the car. The total trip took 6 h. Find the rate of the car.

31. A backpacker hiking into a wilderness area walked 9 mi at a constant rate and then reduced this rate by 1 mph. Another 4 mi was hiked at the reduced rate. The time required to hike the 4 mi was 1 h less than the time required to walk the first 9 mi. Find the rate at which the hiker walked the first 9 mi.

32. After sailing 15 mi, a sailor changed direction and increased the boat's speed by 2 mph. An additional 19 mi was sailed at the increased speed. The total sailing time was 4 h. Find the rate of the boat for the first 15 mi.

33. A small motor on a fishing boat can move the boat 6 mph in calm water. Traveling with the current, the boat can travel 24 mi in the same amount of time as it takes to travel 12 mi against the current. Find the rate of the current.

34. A commercial jet can fly 550 mph in calm air. Traveling with the jet stream, the plane can fly 2400 mi in the same amount of time as it can fly 2000 mi against the jet stream. Find the rate of the jet stream.

35. A cruise ship can sail 28 mph in calm water. Sailing with the gulf stream, the ship can sail 170 mi in the same amount of time as it can sail 110 mi against the gulf stream. Find the rate of the gulf stream.

36. Paddling in calm water, a canoeist can paddle at a rate of 8 mph. Traveling with the current, the canoeist went 30 mi in the same amount of time as it took to travel 18 mi against the current. Find the rate of the current.

37. On a recent trip, a trucker traveled 330 mi at a constant rate. Because of road conditions, the trucker then reduced the speed by 25 mph. An additional 30 mi was traveled at the reduced rate. The entire trip took 7 h. Find the rate of the trucker for the first 330 mi.

38. Commuting from work to home, a lab technician traveled 10 mi at a constant rate through congested traffic. Upon reaching the expressway, the technician increased the speed by 20 mph. An additional 20 mi was traveled at the increased speed. The total time for the trip was 1 h. Find the rate of the congested traffic.

39. Rowing with the current of a river, a rowing team can row 25 mi in the same amount of time as it can row 15 mi against the current. The rate of the rowing team in calm water is 20 mph. Find the rate of the current.

40. A plane can fly 180 mph in calm air. Flying with the wind, the plane can fly 600 mi in the same amount of time as it can fly 480 mi against the wind. Find the rate of the wind.

SUPPLEMENTAL EXERCISES 7.7

Solve.

41. One pipe can fill a tank in 2 h, a second pipe can fill the tank in 4 h, and a third pipe can fill the tank in 5 h. How long will it take to fill the tank with all three pipes operating?

42. A mason can construct a retaining wall in 10 h. The mason's more experienced apprentice can do the same job in 15 h. How long would it take the mason's less experienced apprentice to do the job if, working together, all three can complete the wall in 5 h?

43. An Outing Club traveled 32 mi by canoe and then hiked 4 mi. The rate of speed by boat was four times the rate on foot. If the time spent walking was 1 h less than the time spent canoeing, find the amount of time spent traveling by canoe.

44. A motorist drove 120 mi before running out of gas and walking 4 mi to a gas station. The rate of the motorist in the car was ten times the rate walking. The time spent walking was 2 h less than the time spent driving. How long did it take for the motorist to drive the 120 mi?

45. Because of bad weather, a bus driver reduced the usual speed along a 150-mi bus route by 10 mph. The bus arrived only 30 min later than its usual arrival time. How fast does the bus usually travel?

Calculators and Computers

Simplifying an Algebraic Fraction

The program SIMPLIFY AN ALGEBRAIC EXPRESSION on the Math ACE Disk will enable you to practice writing an algebraic fraction in simplest form. The program allows you to choose one of three levels of difficulty. Level one contains the easiest problems, and level three contains the most difficult.

After you have chosen a level of difficulty, the program will display a problem. Using paper and pencil, simplify the expression. Then press the RETURN key. The correct solution will be displayed. After you have completed a problem, you may continue with problems of the same level of difficulty, return to the menu and change the level, or quit the program.

Something Extra

Intensity of Illumination

You are already aware that the standard unit of length in the metric system is the meter (m) and that the standard unit of mass in the metric system is the kilogram (kg). You may not know that the standard unit of light intensity is the candela (s).

The rate at which light falls upon a one-square-unit area of surface is called the **intensity of illumination.** Intensity of illumination is measured in **lumens** (lm). A lumen is defined in the following illustration.

Picture a source of light equal to one candela (1 s) positioned at the center of a hollow sphere that has a radius of 1 m. The rate at which light falls upon 1 m² of the inner surface of the sphere is equal to one lumen (1 lm). If a light source equal to 4 s is positioned at the center of the sphere, each square meter of the inner surface receives four times as much illumination, or 4 lm.

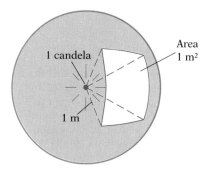

Light rays diverge as they leave a light source. The light that falls upon an area of 1 m² at a distance of 1 m from the source of light spreads out over an area of 4 m² when it is 2 m from the source. The same light spreads out over an area of 9 m² when it is 3 m from the light source and over an area of 16 m² when it is 4 m from the light source. Therefore, as a surface moves farther away from the source of light, the intensity of illumination on the surface decreases from its value at 1 m to $\left(\frac{1}{2}\right)^2$, or $\frac{1}{4}$, that value at 2 m; to $\left(\frac{1}{3}\right)^2$, or $\frac{1}{9}$, that value at 3 m; and to $\left(\frac{1}{4}\right)^2$, or $\frac{1}{16}$, that value at 4 m.

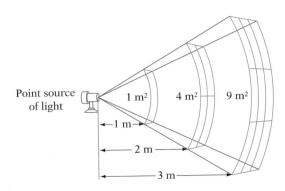

The formula for the intensity of illumination is:

$$I = \frac{s}{r^2},$$

where I is the intensity of illumination in lumens, s is the strength of the light source in candelas, and r is the distance in meters between the light source and the illuminated surface.

A 30-s lamp is 0.5 m above a desk. Find the illumination on the desk.

$$I = \frac{s}{r^2}$$

$$= \frac{30}{(0.5)^2} = 120$$

The illumination on the desk is 120 lm.

Solve.

1. A 100-s light is hanging 5 m above a floor. What is the intensity of illumination on the floor beneath it?

2. A 25-s source of light is 2 m above a desk. Find the intensity of illumination on the desk.

3. How strong a light source is needed to cast 20 lm of light on a surface 4 m from the source?

4. How strong a light source is needed to cast 80 lm of light on a surface 5 m from the source?

5. How far from a desk surface must a 40-s light source be positioned if the desired intensity of illumination is 10 lm?

6. Find the distance between a 36-s light source and a surface if the intensity of illumination on the surface is 0.01 lm.

7. Two lights cast the same intensity of illumination on a wall. One light is 6 m from the wall and has a rating of 36 s. The second light is 8 m from the wall. Find the candela rating of the second light.

8. A 40-s light source and a 10-s light source both throw the same intensity of illumination on a wall. The 10-s light is 6 m from the wall. Find the distance from the 40-s light to the wall.

Chapter Summary

Key Words

An *algebraic fraction* is a fraction in which the numerator or denominator is a variable expression.

An algebraic fraction is in *simplest form* when the numerator and denominator have no common factors.

The *reciprocal* of a fraction is a fraction with the numerator and denominator interchanged.

The *least common multiple (LCM)* of two or more numbers is the smallest number that contains the prime factorization of each number.

A *complex fraction* is a fraction whose numerator or denominator contains one or more fractions.

A *ratio* is the quotient of two quantities that have the same unit.

A *rate* is the quotient of two quantities that have different units.

A *proportion* is an equation that states the equality of two ratios or rates.

A *literal equation* is an equation that contains more than one variable.

A *formula* is a literal equation that states rules about measurements.

Essential Rules

To multiply fractions	$\dfrac{a}{b} \cdot \dfrac{c}{d} = \dfrac{ac}{bd}$
To divide fractions	$\dfrac{a}{b} \div \dfrac{c}{d} = \dfrac{a}{b} \cdot \dfrac{d}{c}$
To add fractions	$\dfrac{a}{b} + \dfrac{c}{b} = \dfrac{a+c}{b}$
To subtract fractions	$\dfrac{a}{b} - \dfrac{c}{b} = \dfrac{a-c}{b}$
Equation for Work Problems	$\begin{array}{c}\text{Rate of} \\ \text{work}\end{array} \cdot \begin{array}{c}\text{Time} \\ \text{worked}\end{array} = \begin{array}{c}\text{Part of task} \\ \text{completed}\end{array}$
Uniform Motion Equation	**Distance = Rate · Time**

Chapter Review

1. Simplify: $\dfrac{8ab^2}{15x^3y} \cdot \dfrac{5xy^4}{16a^2b}$

2. Simplify: $\dfrac{5}{3x - 4} + \dfrac{4}{2x + 3}$

3. Solve $4x + 9y = 18$ for y.

4. Simplify: $\dfrac{16x^5y^3}{-24xy^{10}}$

5. Simplify: $\dfrac{20x^2 - 45x}{6x^3 + 4x^2} \div \dfrac{40x^3 - 90x^2}{12x^2 + 8x}$

6. Simplify: $\dfrac{x - \dfrac{16}{5x - 2}}{3x - 4 - \dfrac{88}{5x - 2}}$

7. Find the LCM of $24a^2b^5$ and $36a^3b$.

8. Simplify: $\dfrac{5x}{3x + 7} - \dfrac{x}{3x + 7}$

9. Write each fraction in terms of the LCM of the denominators.
$$\dfrac{3}{16x}, \dfrac{5}{8x^2}$$

10. Simplify: $\dfrac{2x^2 - 13x - 45}{2x^2 - x - 15}$

11. Simplify: $\dfrac{x^2 - 5x - 14}{x^2 - 3x - 10} \div \dfrac{x^2 - 4x - 21}{x^2 - 9x + 20}$

12. Simplify: $\dfrac{2y}{5y - 7} + \dfrac{3}{7 - 5y}$

13. Simplify: $\dfrac{3x^3 + 10x^2}{10x - 2} \cdot \dfrac{20x - 4}{6x^4 + 20x^3}$

14. Simplify: $\dfrac{5x + 3}{2x^2 + 5x - 3} - \dfrac{3x + 4}{2x^2 + 5x - 3}$

15. Find the LCM of $5x^4(x - 7)^2$ and $15x(x - 7)$.

16. Solve: $\dfrac{6}{x - 7} = \dfrac{8}{x - 6}$

17. Solve: $\dfrac{x + 8}{x + 4} = 1 + \dfrac{5}{x + 4}$

18. Simplify: $\dfrac{12a^2b(4x - 7)}{15ab^2(7 - 4x)}$

19. Simplify:
$$\dfrac{5x - 1}{x^2 - 9} + \dfrac{4x - 3}{x^2 - 9} - \dfrac{8x - 1}{x^2 - 9}$$

20. Write each fraction in terms of the LCM of the denominators.
$$\dfrac{2}{x + 3}, \dfrac{7}{x - 3}$$

21. Solve: $\dfrac{20}{2x + 3} = \dfrac{17x}{2x + 3} - 5$

22. Simplify: $\dfrac{\dfrac{5}{x - 1} - \dfrac{3}{x + 3}}{\dfrac{6}{x + 3} + \dfrac{2}{x - 1}}$

23. Simplify:
$$\dfrac{x - 1}{x + 2} + \dfrac{3x - 2}{5 - x} + \dfrac{5x^2 + 15x - 11}{x^2 - 3x - 10}$$

24. Simplify:
$$\dfrac{18x^2 + 25x - 3}{9x^2 - 28x + 3} \div \dfrac{2x^2 + 9x + 9}{x^2 - 6x + 9}$$

25. Simplify: $\dfrac{x^2 + x - 30}{15 + 2x - x^2}$

26. Solve: $\dfrac{5}{7} + \dfrac{x}{2} = 2 - \dfrac{x}{7}$

27. Simplify: $\dfrac{x + \dfrac{6}{x - 5}}{1 + \dfrac{2}{x - 5}}$

28. Simplify: $\dfrac{3x^2 + 4x - 15}{x^2 - 11x + 28} \cdot \dfrac{x^2 - 5x - 14}{3x^2 + x - 10}$

29. Solve: $\dfrac{x}{5} = \dfrac{x + 12}{9}$

30. Solve: $\dfrac{3}{20} = \dfrac{x}{80}$

31. Simplify: $\dfrac{1 - \dfrac{1}{x}}{1 - \dfrac{8x - 7}{x^2}}$

32. Solve $x - 2y = 15$ for x.

33. Simplify: $\dfrac{6}{a} + \dfrac{9}{b}$

34. Find the LCM of $10x^2 - 11x + 3$ and $20x^2 - 17x + 3$.

35. Solve $i = \dfrac{100m}{c}$ for c.

36. Solve: $\dfrac{15}{x} = \dfrac{3}{8}$

37. Solve: $\dfrac{22}{2x + 5} = 2$

38. Simplify: $\dfrac{x + 7}{15x} + \dfrac{x - 2}{20x}$

39. Simplify:

$\dfrac{16a^2 - 9}{16a^2 - 24a + 9} \cdot \dfrac{8a^2 - 13a - 6}{4a^2 - 5a - 6}$

40. Write each fraction in terms of the LCM of the denominators.

$\dfrac{x}{12x^2 + 16x - 3}, \dfrac{4x^2}{6x^2 + 7x - 3}$

41. Simplify: $\dfrac{3}{4ab} + \dfrac{5}{4ab}$

42. Solve: $\dfrac{20}{x + 2} = \dfrac{5}{16}$

43. Solve: $\dfrac{5x}{3} - \dfrac{2}{5} = \dfrac{8x}{5}$

44. Simplify: $\dfrac{6a^2b^7}{25x^3y} \div \dfrac{12a^3b^4}{5x^2y^2}$

45. A brick mason can construct a patio in 3 h. If the mason works with an apprentice, the patio can be constructed in 2 h. How long would it take the apprentice, working alone, to construct the patio?

46. A weight of 21 lb stretches a spring 14 in. At the same rate, how far would a weight of 12 lb stretch the spring?

47. The rate of a jet is 400 mph in calm air. Traveling with the wind, the jet can fly 2100 mi in the same amount of time as it flies 1900 mi against the wind. Find the rate of the wind.

48. A gardener uses 4 oz of insecticide to make 2 gal of garden spray. At this rate, how much additional insecticide is necessary to make 10 gal of the garden spray?

Chapter 7 / Chapter Test **315**

49. One hose can fill a pool in 15 h. A second hose can fill the pool in 10 h. How long would it take to fill the pool using both hoses?

50. A car travels 315 mi in the same amount of time as it takes a bus to travel 245 mi. The rate of the car is 10 mph greater than that of the bus. Find the rate of the car.

Chapter Test

1. Simplify: $\dfrac{x^2 + 3x + 2}{x^2 + 5x + 4} \div \dfrac{x^2 - x - 6}{x^2 + 2x - 15}$

2. Simplify: $\dfrac{2x}{x^2 + 3x - 10} - \dfrac{4}{x^2 + 3x - 10}$

3. Find the LCM of $6x - 3$ and $2x^2 + x - 1$.

4. Solve: $\dfrac{3}{x + 4} = \dfrac{5}{x + 6}$

5. Simplify: $\dfrac{x^3 y^4}{x^2 - 4x + 4} \cdot \dfrac{x^2 - x - 2}{x^6 y^4}$

6. Simplify: $\dfrac{1 + \dfrac{1}{x} - \dfrac{12}{x^2}}{1 + \dfrac{2}{x} - \dfrac{8}{x^2}}$

7. Write each fraction in terms of the LCM of the denominators.
$\dfrac{3}{x^2 - 2x}, \dfrac{x}{x^2 - 4}$

8. Solve $3x - 8y = 16$ for y.

9. Solve: $\dfrac{6}{x} - 2 = 1$

10. Simplify: $\dfrac{2}{2x - 1} - \dfrac{3}{3x + 1}$

11. Simplify:
$\dfrac{x^2 - x - 56}{x^2 + 8x + 7} \div \dfrac{x^2 - 13x + 40}{x^2 - 4x - 5}$

12. Simplify: $\dfrac{3x}{x^2 + 5x - 24} - \dfrac{9}{x^2 + 5x - 24}$

13. Find the LCM of $3x^2 + 6x$ and $2x^2 + 8x + 8$.

14. Simplify: $\dfrac{x^2 - 7x + 10}{25 - x^2}$

15. Solve: $\dfrac{3x}{x-3} - 2 = \dfrac{10}{x-3}$

16. Solve $f = v + at$ for t.

17. Simplify: $\dfrac{12x^4y^2}{18xy^7}$

18. Simplify: $\dfrac{2}{2x-1} - \dfrac{1}{x+1}$

19. Solve: $\dfrac{2}{x-2} = \dfrac{12}{x+3}$

20. Simplify: $\dfrac{x^5y^3}{x^2-x-6} \cdot \dfrac{x^2-9}{x^2y^4}$

21. Write each fraction in terms of the LCM of the denominators.

$$\dfrac{3y}{x(1-x)}, \ \dfrac{x}{(x+1)(x-1)}$$

22. Simplify: $\dfrac{1 - \dfrac{2}{x} - \dfrac{15}{x^2}}{1 - \dfrac{25}{x^2}}$

23. A salt solution is formed by mixing 4 lb of salt with 10 gal of water. At this rate, how many additional pounds of salt are required for 15 gal of water?

24. A small plane can fly at 110 mph in calm air. Flying with the wind, the plane can fly 260 mi in the same amount of time as it takes to fly 180 mi against the wind. Find the rate of the wind.

25. One pipe can fill a tank in 9 min. A second pipe requires 18 min to fill the tank. How long would it take both pipes, working together, to fill the tank?

Cumulative Review

1. Evaluate $-|-17|$.

2. Evaluate $-\dfrac{3}{4} \cdot (2)^3$.

3. Simplify: $\left(\dfrac{2}{3}\right)^2 \div \left(\dfrac{3}{2} - \dfrac{2}{3}\right) + \dfrac{1}{2}$

4. Evaluate $-a^2 + (a-b)^2$ when $a = -2$ and $b = 3$.

5. Simplify: $-2x - (-3y) + 7x - 5y$

6. Simplify: $2[3x - 7(x - 3) - 8]$

7. Solve: $3 - \frac{1}{4}x = 8$

8. Solve: $3[x - 2(x - 3)] = 2(3 - 2x)$

9. Find $16\frac{2}{3}\%$ of 60.

10. Simplify: $(3xy^4)(-2x^3y)$

11. Simplify: $(a^4b^3)^5$

12. Simplify: $\frac{a^2b^{-5}}{a^{-1}b^{-3}}$

13. Simplify: $(a - 3b)(a + 4b)$

14. Simplify: $\frac{15b^4 - 5b^2 + 10b}{5b}$

15. Simplify: $(x^3 - 8) \div (x - 2)$

16. Factor: $12x^2 - x - 1$

17. Factor: $y^2 - 7y + 6$

18. Factor: $2a^3 + 7a^2 - 15a$

19. Factor: $4b^2 - 100$

20. Solve: $(x + 3)(2x - 5) = 0$

21. Simplify: $\frac{x^2 + 3x - 28}{16 - x^2}$

22. Simplify:
$$\frac{x^2 - 3x - 10}{x^2 - 4x - 12} \div \frac{x^2 - x - 20}{x^2 - 2x - 24}$$

23. Simplify: $\frac{6}{3x - 1} - \frac{2}{x + 1}$

24. Solve: $\frac{4x}{x - 3} - 2 = \frac{8}{x - 3}$

25. Solve $f = v + at$ for a.

26. Translate "the difference between five times a number and eighteen is the opposite of three" into an equation and solve.

27. An investment of $5000 is made at an annual simple interest rate of 7%. How much additional money must be invested at an annual simple interest rate of 11% so that the total interest earned is 9% of the total investment?

28.   A silversmith mixes 60 g of an alloy that is 40% silver with 120 g of another silver alloy. The resulting alloy is 60% silver. Find the percent of silver in the 120-g alloy.

29.   The length of the base of a triangle is 2 in. less than twice the height. The area of the triangle is 30 in.^{2}. Find the base and height of the triangle.

30.   One water pipe can fill a tank in 12 min. A second pipe requires 24 min to fill the tank. How long would it take both pipes, working together, to fill the tank?

8

Graphs and Linear Equations

Objectives

- Graph points on a rectangular coordinate system
- Determine solutions of linear equations in two variables
- Scatter diagrams
- Graph equations of the form $y = mx + b$
- Graph equations of the form $Ax + By = C$
- Find the x- and y-intercepts of a straight line
- Find the slope of a straight line
- Graph a line using the slope and y-intercept
- Find the equation of a line using the equation $y = mx + b$
- Find the equation of a line using the point-slope formula

Magic Squares

A magic square is a square array of distinct integers so arranged that the numbers along any row, column, or main diagonal have the same sum. An example of a magic square is shown at the right.

8	3	4
1	5	9
6	7	2

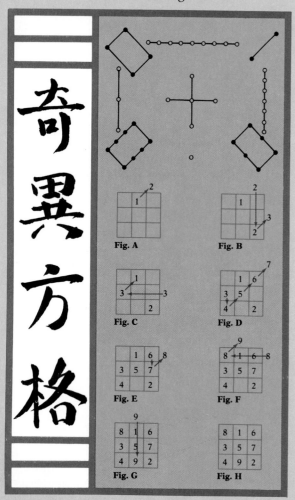

Fig. A

Fig. B

Fig. C

Fig. D

Fig. E

Fig. F

Fig. G

Fig. H

The oldest known example of a magic square comes from China. Estimates are that this magic square is over 4000 years old. It is shown at the left.

There is a simple way to produce a magic square with an odd number of cells. Start by writing a 1 in the top middle cell. The rule then is to proceed diagonally upward to the right with the successive integers.

When the rule takes you outside the square, write the number by shifting either across the square from right to left or down the square from top to bottom, as the case may be. For example, in Fig. B, the second number (2) is outside the square above a column. Since the 2 is above a column, it should be shifted down to the bottom cell in that column. In Fig. C, the 3 is outside the square to the right of a column and should therefore be shifted all the way to the left.

If the rule takes you to a square that is already filled (as shown in Fig. D), then write the number in the cell directly below the last number written. Continue until the entire square is filled.

It is possible to begin a magic square with any integer and proceed by using the above rule and consecutive integers.

For an odd magic square beginning with 1, the sum of a row, column, or diagonal is $\frac{n(n^2 + 1)}{2}$, where n is the number of rows.

The Rectangular Coordinate System

1 Graph points on a rectangular coordinate system

A **rectangular coordinate system** is formed by two number lines, one horizontal and one vertical, that intersect at the zero point of each line. The point of intersection is called the **origin.** The two lines are called the **coordinate axes,** or simply the **axes.**

The axes determine a plane and divide the plane into four regions, called **quadrants.** The quadrants are numbered counterclockwise from I to IV.

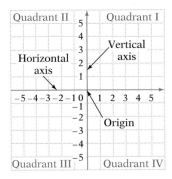

Each point in the plane can be identified by a pair of numbers called an ordered pair. The first number of the pair measures a horizontal distance and is called the **abscissa,** or *x*-**coordinate.** The second number of the pair measures a vertical distance and is called the **ordinate,** or *y*-**coordinate.** The ordered pair (*a, b*) associated with a point is also called the **coordinates** of the point.

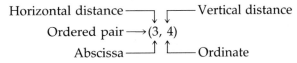

The **graph of an ordered pair** is a point in the plane. The graphs of the points whose coordinates are (2, 3) and (3, 2) are shown at the right. Note that they are different points. The order in which the numbers in an ordered pair appear *is* important.

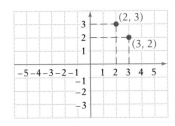

321

Example 1   Graph the ordered pairs $(-2, -3)$, $(3, -2)$, $(1, 3)$, and $(4, 1)$.

Solution

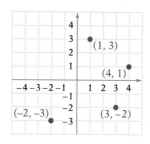

Problem 1   Graph the ordered pairs $(-1, 3)$, $(1, 4)$, $(-4, 0)$, and $(-2, -1)$.

Solution   See page A30.

Example 2   Find the coordinates of each of the points.

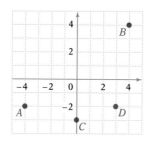

Solution $A(-4, -2)$, $B(4, 4)$, $C(0, -3)$, $D(3, -2)$

Problem 2   Find the coordinates of each of the points.

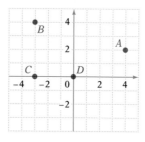

Solution   See page A30.

2 Determine solutions of linear equations in two variables

An equation of the form $y = mx + b$, where m and b are constants, is a **linear equation in two variables.** Examples of linear equations are shown below.

$$y = 3x + 4 \qquad (m = 3, \quad b = 4)$$
$$y = 2x - 3 \qquad (m = 2, \quad b = -3)$$
$$y = -\frac{2}{3}x + 1 \qquad (m = -\frac{2}{3}, b = 1)$$
$$y = -2x \qquad (m = -2, b = 0)$$
$$y = x + 2 \qquad (m = 1, \quad b = 2)$$

In a linear equation, the exponent of each variable is 1. The equations $y = 2x^2 - 1$ and $y = \frac{1}{x}$ are not linear equations.

A **solution of an equation in two variables** is an ordered pair of numbers (x, y) that makes the equation a true statement.

Is $(1, -2)$ a solution of $y = 3x - 5$?

Replace x with 1, the abscissa.
Replace y with -2, the ordinate.

$$
\begin{array}{c|c}
\multicolumn{2}{c}{y = 3x - 5} \\
\hline
-2 & 3(1) - 5 \\
 & 3 - 5 \\
\end{array}
$$
$$-2 = -2$$

Compare the results. If the results are equal, the given ordered pair is a solution. If the results are not equal, the given ordered pair is not a solution.

Yes, $(1, -2)$ is a solution of the equation $y = 3x - 5$.

Besides the ordered pair $(1, -2)$, there are many other ordered pair solutions of the equation $y = 3x - 5$. For example, the method used above can be used to show that $(2, 1)$, $(-1, -8)$, $\left(\frac{2}{3}, -3\right)$, and $(0, -5)$ are also solutions.

Example 3 Is $(-3, 2)$ a solution of $y = 2x + 2$?

Solution
$$
\begin{array}{c|c}
\multicolumn{2}{c}{y = 2x + 2} \\
\hline
2 & 2(-3) + 2 \\
 & -6 + 2 \\
 & -4 \\
\end{array}
$$
$$2 \neq -4$$

▶ Replace x with -3 and y with 2.

No, $(-3, 2)$ is not a solution of $y = 2x + 2$.

Problem 3 Is (2, −4) a solution of $y = -\frac{1}{2}x - 3$?

Solution See page A31.

In general, a linear equation in two variables has an infinite number of solutions. By choosing any value for x and substituting that value into the linear equation, a corresponding value of y can be found.

Find the ordered pair solution of $y = 2x - 5$ corresponding to $x = 1$.

$$y = 2x - 5$$

Substitute 1 for x.
Solve for y.

$$= 2 \cdot 1 - 5$$
$$= 2 - 5$$
$$= -3$$

The ordered pair solution is (1, −3).

Example 4 Find the ordered pair solution of $y = \frac{2}{3}x - 1$ corresponding to $x = 3$.

Solution $y = \frac{2}{3}x - 1$

$$= \frac{2}{3}(3) - 1 \qquad \blacktriangleright \text{Substitute 3 for } x.$$
$$= 2 - 1$$
$$= 1 \qquad \blacktriangleright \text{When } x = 3, y = 1.$$

The ordered pair solution is (3, 1).

Problem 4 Find the ordered pair solution of $y = -\frac{1}{4}x + 1$ corresponding to $x = 4$.

Solution See page A31.

3 Scatter diagrams

There are many situations in which the relationship between two variables may be of interest. For example, a college admissions director wants to know the relationship between SAT scores (one variable) and success in college (the second variable). An employer is interested in the relationship between scores on a pre-employment test and ability to perform a job.

A researcher can investigate the relationship between two variables by means of regression analysis, which is a branch of statistics. The study of the relationship between two variables may begin with a scatter diagram, which is a graph of some of the known data.

The following table gives the record times for events of different lengths at a track meet. Record times are rounded to the nearest second. Lengths are given in meters.

Length, x	100	200	400	800	1000	1500
Time, y	10	20	50	100	130	210

The scatter diagram for these data is shown below. Each ordered pair represents the length of the race and the record time. For example, the ordered pair (400, 50) indicates that the record for the 400-m race is 50 s.

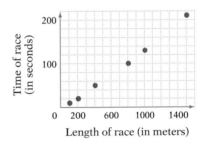

Example 5 To test a heart medicine, a doctor measures the heart rate of five patients before and after they take the medication. The results are recorded in the following table. Graph the scatter diagram for these data.

Before medicine, x	85	80	85	75	90
After medicine, y	75	70	70	80	80

Strategy Graph the ordered pairs on a rectangular coordinate system where the reading on the horizontal axis represents the heart rate before taking the medication and the vertical axis represents the heart rate after taking the medication.

Solution

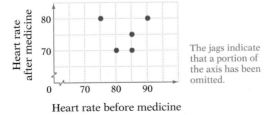

The jags indicate that a portion of the axis has been omitted.

Problem 5   The age of five cars in years and the prices paid for the cars in hundreds of dollars are recorded in the following table. Graph the scatter diagram for these data.

Age, x	4	5	3	5	2
Price, y	27	22	31	18	32

Solution   See page A31.

EXERCISES 8.1

1. Graph the ordered pairs (−2, 1), (3, −5), (−2, 4), and (0, 3).

2. Graph the ordered pairs (5, −1), (−3, −3), (−1, 0), and (1, −1).

3. Graph the ordered pairs (0, 0), (0, −5), (−3, 0), and (0, 2).

4. Graph the ordered pairs (−4, 5), (−3, 1), (3, −4), and (5, 0).

5. Graph the ordered pairs (−1, 4), (−2, −3), (0, 2), and (4, 0).

6. Graph the ordered pairs (5, 2), (−4, −1), (0, 0), and (0, 3).

7. Find the coordinates of each of the points.

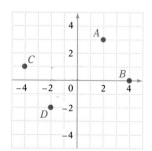

8. Find the coordinates of each of the points.

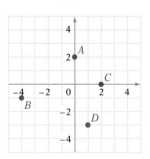

9. Find the coordinates of each of the points.

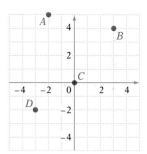

10. Find the coordinates of each of the points.

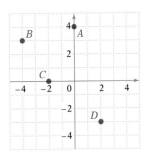

11. a. Name the abscissas of points *A* and *C*.
 b. Name the ordinates of points *B* and *D*.

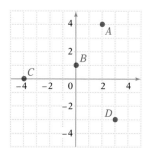

12. a. Name the abscissas of points *A* and *C*.
 b. Name the ordinates of points *B* and *D*.

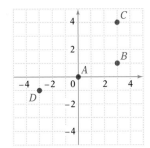

2

13. Is (3, 4) a solution of $y = -x + 7$?

14. Is (2, −3) a solution of $y = x + 5$?

15. Is (−1, 2) a solution of $y = \frac{1}{2}x - 1$?

16. Is (1, −3) a solution of $y = -2x - 1$?

17. Is (4, 1) a solution of $y = \frac{1}{4}x + 1$?

18. Is (−5, 3) a solution of $y = -\frac{2}{5}x + 1$?

19. Is (0, 4) a solution of $y = \frac{3}{4}x + 4$?

20. Is (−2, 0) a solution of $y = -\frac{1}{2}x - 1$?

21. Is (0, 0) a solution of $y = 3x + 2$?

22. Is (0, 0) a solution of $y = -\frac{3}{4}x$?

23. Find the ordered pair solution of $y = 3x - 2$ corresponding to $x = 3$.

24. Find the ordered pair solution of $y = 4x + 1$ corresponding to $x = -1$.

25. Find the ordered pair solution of $y = \frac{2}{3}x - 1$ corresponding to $x = 6$.

26. Find the ordered pair solution of $y = \frac{3}{4}x - 2$ corresponding to $x = 4$.

27.   Find the ordered pair solution of $y = -3x + 1$ corresponding to $x = 0$.

28.   Find the ordered pair solution of $y = \frac{2}{5}x - 5$ corresponding to $x = 0$.

29.   Find the ordered pair solution of $y = \frac{2}{5}x + 2$ corresponding to $x = -5$.

30.   Find the ordered pair solution of $y = -\frac{1}{6}x - 2$ corresponding to $x = 12$.

3   Solve.

31.   The math midterm scores and the final exam scores for six students are given in the following table. Graph the scatter diagram for these data.

Midterm score, x	90	85	75	80	85	70
Final exam score, y	95	75	80	75	90	70

32.   The following table shows, for each of five tennis players, the number of aces served by the player during one set and the number of games that player won in the set. Graph the scatter diagram for these data.

Aces served, x	6	8	4	10	6
Games won, y	6	5	3	6	4

33.   The number of hit records in one year by each of five musical groups and the number of concerts the group performed that year are given in the following table. Graph the scatter diagram for these data.

Hit records, x	11	9	7	12	12
Concerts, y	200	150	200	175	250

34.   The monthly salaries of six employees in a company and the number of years of college completed by each are given in the following table. Graph the scatter diagram for these data.

Years of college, x	4	4	5	4	6	8
Monthly salary (in thousands), y	3	3.5	4	2.5	5	5.5

35.   The number of lift tickets sold during one day and the number of skiing accidents reported that day are recorded in the following table. Graph the scatter diagram for these data.

Ticket sold, x	1500	2500	4000	6000
Accidents reported, y	25	35	50	55

36. A study of six algebra classes measured the number of hours the class met per week and the average score of the class on the final exam. The results are recorded in the following table. Graph the scatter diagram for these data.

Average score, x	72	72	68	75	76	70
Hours in class, y	3	4	3	3	5	4

SUPPLEMENTAL EXERCISES 8.1

In which quadrant is the given point located?

37. $(2, -4)$ **38.** $(-3, 2)$ **39.** $(-1, -6)$

What is the distance from the given point to the horizontal axis?

40. $(-5, 1)$ **41.** $(3, -4)$ **42.** $(-6, 0)$

What is the distance from the given point to the vertical axis?

43. $(-2, 4)$ **44.** $(1, -3)$ **45.** $(5, 0)$

For the given linear equation, find the value of m and the value of b.

46. $y = 4x + 1$ **47.** $y = -3x - 2$

48. $y = \dfrac{5}{6}x + \dfrac{1}{6}$ **49.** $y = -x$

Write the linear equation in the form $y = mx + b$.

50. $\dfrac{x}{4} + \dfrac{y}{3} = 1$ **51.** $\dfrac{x}{2} - \dfrac{y}{5} = 1$

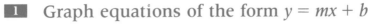

SECTION **8.2**

Graphs of Straight Lines

1 Graph equations of the form $y = mx + b$

The **graph of an equation in two variables** is a drawing of the ordered pair solutions of the equation. For a linear equation in two variables, the graph is a straight line.

To graph a linear equation, find ordered pair solutions of the equation. Do this by choosing any value of x and finding the corresponding value of y. Repeat this procedure, choosing different values for x, until you have found the number of solutions desired.

Because the graph of a linear equation in two variables is a straight line, and a straight line is determined by two points, it is necessary to find only two solutions. However, it is recommended that at least three points be used to ensure accuracy.

To graph $y = 2x + 1$, choose any values of x, and then find the corresponding values of y.
The numbers 0, 2, and -1 were chosen arbitrarily for x.
It is convenient to record these solutions in a table.

x	$y =$	$2x + 1$	y
0		$2 \cdot 0 + 1$	1
2		$2 \cdot 2 + 1$	5
-1		$2(-1) + 1$	-1

The horizontal axis is the x-axis.
The vertical axis is the y-axis.
Graph the ordered pair solutions $(0, 1)$, $(2, 5)$, and $(-1, -1)$.
Draw a line through the ordered pair solutions.

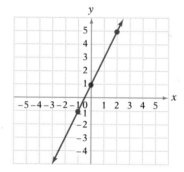

Remember that a graph is a drawing of the ordered pair solutions of the equation. Therefore, every point on the graph is a solution of the equation, and every solution of the equation is a point on the graph.

Note that the points whose coordinates are $(-2, -3)$ and $(1, 3)$ are on the graph, and that these ordered pairs are solutions of the equation $y = 2x + 1$.

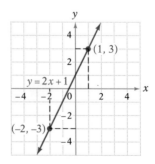

Example 1   Graph $y = 3x - 2$.

Solution

x	y
0	−2
2	4
−1	−5

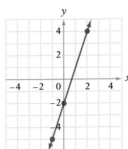

Problem 1   Graph $y = 3x + 1$.

Solution   See page A31.

When m is a fraction in the equation $y = mx + b$, choose values of x that will simplify the evaluation. For example, to graph $y = \frac{1}{3}x - 1$, the numbers 0, 3, and −3 might be chosen for x.

x	y
0	−1
3	0
−3	−2

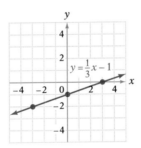

Example 2   Graph $y = \frac{1}{2}x - 1$.

Solution

x	y
0	−1
2	0
−2	−2

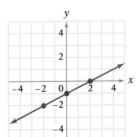

Problem 2   Graph $y = \frac{1}{3}x - 3$.

Solution   See page A31.

See page A31.

2 Graph equations of the form $Ax + By = C$

An equation in the form $Ax + By = C$, where A, B, and C are constants, is also a linear equation. Examples of these equations are shown below.

$$
\begin{array}{ll}
2x + 3y = 6 & (A = 2, B = 3, \quad C = 6) \\
x - 2y = -4 & (A = 1, B = -2, C = -4) \\
2x + y = 0 & (A = 2, B = 1, \quad C = 0) \\
4x - 5y = 2 & (A = 4, B = -5, C = 2)
\end{array}
$$

To graph an equation of the form $Ax + By = C$, first solve the equation for y. Then follow the same procedure used for graphing an equation of the form $y = mx + b$.

Graph $3x + 4y = 12$.

Solve the equation for y.

$$
\begin{aligned}
3x + 4y &= 12 \\
4y &= -3x + 12 \\
y &= -\frac{3}{4}x + 3
\end{aligned}
$$

Find at least three solutions.
Display the ordered pairs in a table.

x	y
0	3
4	0
-4	6

Graph the ordered pairs on a rectangular coordinate system, and draw a straight line through the points.

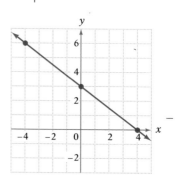

Example 3   Graph.

A. $2x - 5y = 10$ B. $x + 2y = 6$

Solution   A. $2x - 5y = 10$ B. $x + 2y = 6$

$$-5y = -2x + 10$$ $$2y = -x + 6$$

$$y = \frac{2}{5}x - 2$$ $$y = -\frac{1}{2}x + 3$$

x	y
0	-2
5	0
-5	-4

x	y
0	3
-2	4
4	1

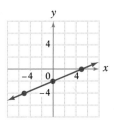
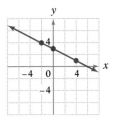

Problem 3   Graph.

A. $5x - 2y = 10$ B. $x - 3y = 9$

Solution   See page A31.

The graph of an equation in which one of the variables is missing is either a horizontal or a vertical line.

The equation $y = 2$ could be written $0 \cdot x + y = 2$. No matter what value of x is chosen, y is always 2. Some solutions to the equation are (3, 2), (−1, 2), (0, 2) and (−4, 2). The graph is shown at the right.

The **graph of** $y = b$ is a horizontal line passing through the point whose coordinates are (0, b).

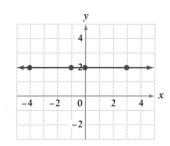

The equation $x = -2$ could be written $x + 0 \cdot y = -2$. No matter what value of y is chosen, x is always -2. Some solutions to the equation are $(-2, 3)$, $(-2, -2)$, $(-2, 0)$, and $(-2, 2)$. The graph is shown at the right.

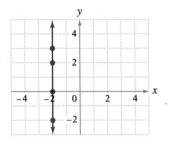

The **graph of $x = a$** is a vertical line passing through the point whose coordinates are $(a, 0)$.

Example 4   Graph.

A. $y = -2$ B. $x = 3$

Solution   A. The graph of an equation of the form $y = b$ is a horizontal line passing through $(0, b)$.

B. The graph of an equation of the form $x = a$ is a vertical line passing through $(a, 0)$.

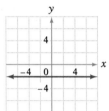

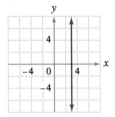

Problem 4   Graph.

A. $y = 3$ B. $x = -4$

Solution   See page A32.

EXERCISES 8.2

1   Graph.

1. $y = 2x - 3$

2. $y = -2x + 2$

3. $y = \dfrac{1}{3}x$

4. $y = -3x$

5. $y = \dfrac{2}{3}x - 1$

6. $y = \dfrac{3}{4}x + 2$

7. $y = -\frac{1}{4}x + 2$

8. $y = -\frac{1}{3}x + 1$

9. $y = -\frac{2}{5}x + 1$

10. $y = -\frac{1}{2}x + 3$

11. $y = 2x - 4$

12. $y = 3x - 4$

13. $y = -x + 2$

14. $y = -x - 1$

15. $y = -\frac{2}{3}x + 1$

16. $y = 5x - 4$

17. $y = -3x + 2$

18. $y = -x + 3$

2   Graph.

19. $3x + y = 3$

20. $2x + y = 4$

21. $2x + 3y = 6$

22. $3x + 2y = 4$

23. $x - 2y = 4$

24. $x - 3y = 6$

25. $y = 4$

$\left(\begin{array}{c} 0, 4 \\ 5, 4 \end{array} \right)$

26. $x = -2$

27. $2x - 3y = 6$

28. $3x - 2y = 8$

29. $2x + 5y = 10$

30. $3x + 4y = 12$

31. $x = 3$ **32.** $y = -4$ **33.** $x - 3y = 6$

34. $3x - y = 6$ **35.** $x - 4y = 8$ **36.** $4x + 3y = 12$

SUPPLEMENTAL EXERCISES 8.2

Solve.

37.   **a.**   Show that the equation $y + 3 = 2(x + 4)$ is a linear equation by writing it in the form $y = mx + b$.

　　　b.   Find the ordered pair solution corresponding to $x = -4$.

38.   **a.**   Show that the equation $y + 4 = -\frac{1}{2}(x + 2)$ is a linear equation by writing it in the form $y = mx + b$.

　　　b.   Find the ordered pair solution corresponding to $x = -2$.

39.   For the linear equation $y = 2x - 3$, what is the increase in y that results when x is increased by 1?

40.   For the linear equation $y = -x - 4$, what is the decrease in y that results when x is increased by 1?

SECTION 8.3

Intercepts and Slopes of Straight Lines

1 Find the x- and y-intercepts of a straight line

The graph of the equation $2x + 3y = 6$ is shown at the right. The graph crosses the x-axis at (3, 0). This point is called the **x-intercept**. The graph also crosses the y-axis at (0, 2). This point is called the **y-intercept**.

Find the x-intercept and the y-intercept of the graph of the equation $2x - 3y = 12$.

To find the x-intercept, let $y = 0$.
(Any point on the x-axis has y-coordinate 0.)

$$2x - 3y = 12$$
$$2x - 3(0) = 12$$
$$2x = 12$$
$$x = 6$$

The x-intercept is $(6, 0)$.

To find the y-intercept, let $x = 0$.
(Any point on the y-axis has x-coordinate 0.)

$$2x - 3y = 12$$
$$2(0) - 3y = 12$$
$$-3y = 12$$
$$y = -4$$

The y-intercept is $(0, -4)$.

Some linear equations can be graphed by finding the x- and y-intercepts and then drawing a line through the two points.

Example 1 Find the x- and y-intercepts for $x - 2y = 4$. Graph the line.

Solution x-intercept: $x - 2y = 4$
                           $x - 2(0) = 4$ ▶ To find the x-intercept, let $y = 0$.
                                   $x = 4$

$(4, 0)$

y-intercept: $x - 2y = 4$
               $0 - 2y = 4$ ▶ To find the y-intercept, let $x = 0$.
                  $-2y = 4$
                    $y = -2$

$(0, -2)$

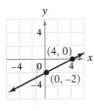

▶ Graph the ordered pairs $(4, 0)$ and $(0, -2)$.
Draw a straight line through the points.

Problem 1 Find the x- and y-intercepts for $4x - y = 4$.
Graph the line.

Solution See page A32.

To find the y-intercept of $y = 3x + 4$, let $x = 0$.

$$y = 3x + 4$$
$$y = 3(0) + 4$$
$$y = 4$$

The y-intercept is $(0, 4)$.

The constant term of $y = 3x + 4$ is the y-coordinate of the y-intercept.

In general, for any equation of the form $y = mx + b$, the y-intercept is $(0, b)$.

Example 2   Find the x- and y-intercepts for $y = 2x - 4$. Graph the line.

Solution x-intercept:
$$y = 2x - 4$$
$$0 = 2x - 4$$
$$-2x = -4$$
$$x = 2$$

▶ To find the x-intercept, let $y = 0$.

$(2, 0)$

y-intercept: $(0, b)$
$$b = -4$$

▶ For any equation of the form $y = mx + b$, the y-intercept is $(0, b)$.

$(0, -4)$

▶ Graph the ordered pairs $(2, 0)$ and $(0, -4)$. Draw a straight line through the points.

Problem 2   Find the x- and y-intercepts for $y = 3x - 6$. Graph the line.

Solution   See page A32.

2   Find the slope of a straight line

The graphs of $y = \frac{2}{3}x + 1$ and $y = 2x + 1$

are shown at the right. Each graph crosses the y-axis at $(0, 1)$, but the graphs have different slants. The **slope** of a line is a measure of the slant of a line. The symbol for slope is m.

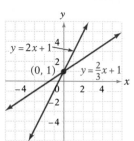

The slope of a line is the ratio of the change in the y-coordinates between any two points to the change in the x-coordinates.

The line containing the points whose coordinates are $(-2, -3)$ and $(6, 1)$ is graphed at the right. The change in y is the difference between the two y-coordinates.

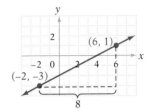

Change in $y = 1 - (-3) = 4$

The change in x is the difference between the two x-coordinates.

Change in $x = 6 - (-2) = 8$

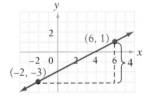

For the line containing the points whose coordinates are $(-2, -3)$ and $(6, 1)$,

$$\text{Slope} = m = \frac{\text{Change in } y}{\text{Change in } x} = \frac{4}{8} = \frac{1}{2}$$

The slope of the line can also be described as the ratio of the vertical change (4 units) to the horizontal change (8 units) from the point whose coordinates are $(-2, -3)$ to the point whose coordinates are $(6, 1)$.

Slope Formula

The slope of a line containing two points, P_1 and P_2, whose coordinates are (x_1, y_1) and (x_2, y_2), is given by

$$\textbf{Slope} = m = \frac{y_2 - y_1}{x_2 - x_1}, \quad x_1 \neq x_2.$$

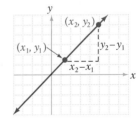

In the slope formula given above, the points P_1 and P_2 are any two points on the line. The slope of a line is constant; therefore, the slope calculated using any two points on the line will be the same.

Find the slope of the line containing the points whose coordinates are $(-1, 1)$ and $(2, 3)$.

Let $P_1 = (-1, 1)$ and $P_2 = (2, 3)$. Then

$x_1 = -1$, $y_1 = 1$, $x_2 = 2$, and $y_2 = 3$.

$$m = \frac{y_2 - y_1}{x_2 - x_1} = \frac{3 - 1}{2 - (-1)} = \frac{2}{3}$$

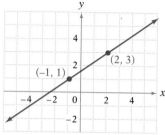

Positive slope

It does not matter which point is named P_1 and which P_2; the slope of the line will be the same.

If $P_1 = (2, 3)$ and $P_2 = (-1, 1)$, then

$$m = \frac{y_2 - y_1}{x_2 - x_1} = \frac{1 - 3}{-1 - 2} = \frac{-2}{-3} = \frac{2}{3}.$$

The slope is a positive number.

A line that slants upward to the right has a **positive slope.**

Find the slope of the line containing the points whose coordinates are $(-3, 4)$ and $(2, -2)$.

Let $P_1 = (-3, 4)$ and $P_2 = (2, -2)$.

$$m = \frac{y_2 - y_1}{x_2 - x_1} = \frac{-2 - 4}{2 - (-3)} = \frac{-6}{5} = -\frac{6}{5}$$

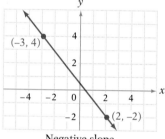

Negative slope

The slope is a negative number.

A line that slants downward to the right has a **negative slope.**

Find the slope of the line containing the points whose coordinates are $(-1, 3)$ and $(4, 3)$.

Let $P_1 = (-1, 3)$ and $P_2 = (4, 3)$.

$$m = \frac{y_2 - y_1}{x_2 - x_1} = \frac{3 - 3}{4 - (-1)} = \frac{0}{5} = 0$$

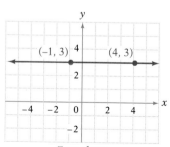

Zero slope

When $y_1 = y_2$, the graph is a horizontal line.

A horizontal line has **zero slope.**

Find the slope of the line containing the points whose coordinates are (2, −2) and (2, 4).

Let $P_1 = (2, -2)$ and $P_2 = (2, 4)$.

$$m = \frac{y_2 - y_1}{x_2 - x_1} = \frac{4 - (-2)}{2 - 2} = \frac{6}{0}$$ Not a real number

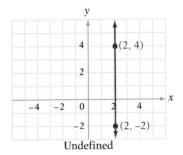

When $x_1 = x_2$, the denominator of $\dfrac{y_2 - y_1}{x_2 - x_1}$ is 0 and the graph is a vertical line. Because division by zero is not defined, the slope of the line is not defined.

The slope of a vertical line is **undefined**.

Remember that zero slope and undefined slope are different. The graph of a line with zero slope is horizontal. The graph of a line with undefined slope is vertical.

Example 3   Find the slope of the line.

A. $P_1(-2, -1)$, $P_2(3, 4)$ B. $P_1(-3, 1)$, $P_2(2, -2)$

C. $P_1(-1, 4)$, $P_2(-1, 0)$ D. $P_1(-1, 2)$, $P_2(4, 2)$

Solution   A. $m = \dfrac{y_2 - y_1}{x_2 - x_1} = \dfrac{4 - (-1)}{3 - (-2)} = \dfrac{5}{5} = 1$ B. $m = \dfrac{y_2 - y_1}{x_2 - x_1} = \dfrac{-2 - 1}{2 - (-3)} = \dfrac{-3}{5}$

The slope is 1.   The slope is $-\dfrac{3}{5}$.

C. $m = \dfrac{y_2 - y_1}{x_2 - x_1} = \dfrac{0 - 4}{-1 - (-1)} = \dfrac{-4}{0}$ D. $m = \dfrac{y_2 - y_1}{x_2 - x_1} = \dfrac{2 - 2}{4 - (-1)} = \dfrac{0}{5} = 0$

The slope is undefined.   The slope is 0.

Problem 3   Find the slope of the line.

A. $P_1(-1, 2)$, $P_2(1, 3)$ B. $P_1(1, 2)$, $P_2(4, -5)$

C. $P_1(2, 3)$, $P_2(2, 7)$ D. $P_1(1, -3)$, $P_2(-5, -3)$

Solution   See page A32.

3 Graph a line using the slope and y-intercept

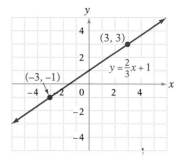

The graph of the equation $y = \frac{2}{3}x + 1$ is shown at the right. The points whose coordinates are $(-3, -1)$ and $(3, 3)$ are on the graph. The slope of the line is:

$$m = \frac{3 - (-1)}{3 - (-3)} = \frac{4}{6} = \frac{2}{3}$$

Note that the slope of the line has the same value as the coefficient of x.

Slope-Intercept Form of a Straight Line

For any equation of the form $y = mx + b$, the slope of the line is m, the coefficient of x. The y-intercept is $(0, b)$. The equation

$$y = mx + b$$

is called the **slope-intercept form of a straight line.**

The slope of the line $y = -\frac{3}{4}x + 1$ is $-\frac{3}{4}$. The y-intercept is $(0, 1)$.

$$y = \boxed{m} \; x + \boxed{b}$$
$$y = \boxed{-\frac{3}{4}} \; x + \boxed{1}$$

Slope $= m = -\frac{3}{4}$ ⟶   ⟵ y-intercept $= (0, b) = (0, 1)$

Find the slope and y-intercept of the line $y = \frac{5}{2}x - 6$.

Slope $= m = \frac{5}{2}$ $\qquad b = -6$

The slope is $\frac{5}{2}$.   The y-intercept is $(0, -6)$.

When the equation of a straight line is in the form $y = mx + b$, the graph can be drawn using the slope and y-intercept. First locate the y-intercept. Use the slope to find a second point on the line. Then draw a line through the two points.

Graph $y = 2x - 3$ by using the slope and y-intercept.

y-intercept $= (0, b) = (0, -3)$

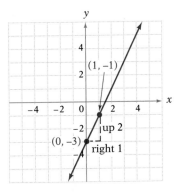

$$m = 2 = \frac{2}{1} = \frac{\text{Change in } y}{\text{Change in } x}$$

Beginning at the y-intercept, move right 1 unit (change in x) and then up 2 units (change in y).

$(1, -1)$ are the coordinates of the second point on the graph.

Draw a line through $(0, -3)$ and $(1, -1)$.

Example 4 Graph $y = -\frac{2}{3}x + 1$ by using the slope and y-intercept.

Solution y-intercept $= (0, b) = (0, 1)$

$$m = -\frac{2}{3} = \frac{-2}{3}$$

▶ A slope of $\frac{-2}{3}$ means to move right 3 units and then down 2 units.

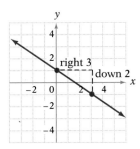

Problem 4 Graph $y = -\frac{1}{4}x - 1$ by using the slope and y-intercept.

Solution See page A32.

Example 5   Graph $2x - 3y = 6$ by using the slope and y-intercept.

Solution $2x - 3y = 6$

$-3y = -2x + 6$ ▶ Solve the equation for y.

$y = \dfrac{2}{3}x - 2$

y-intercept $= (0, -2)$ $m = \dfrac{2}{3}$

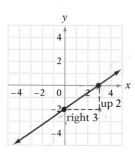

Problem 5   Graph $x - 2y = 4$ by using the slope and y-intercept.

Solution   See page A32.

EXERCISES 8.3

1   Find the x- and y-intercepts.

1. $x - y = 3$

2. $3x + 4y = 12$

3. $y = 2x - 6$

4. $y = 2x + 10$

5. $x - 5y = 10$

6. $3x + 2y = 12$

7. $y = 3x + 12$

8. $y = 5x + 10$

9. $2x - 3y = 0$

10. $3x + 4y = 0$

11. $y = \dfrac{1}{2}x + 3$

12. $y = \dfrac{2}{3}x - 4$

Graph by using x- and y-intercepts.

13. $5x + 2y = 10$

14. $x - 3y = 6$

15. $y = \dfrac{3}{4}x - 3$

16. $y = \dfrac{2}{3}x + 2$

17. $2x + 3y = 6$

18. $x + 2y = 4$

2   Find the slope of the line.

19. $P_1(4, 2),\ P_2(3, 4)$

20. $P_1(2, 1),\ P_2(3, 4)$

21. $P_1(-1, 3),\ P_2(2, 4)$

22. $P_1(-2, 1),\ P_2(2, 2)$

23. $P_1(2, 4),\ P_2(4, -1)$

24. $P_1(1, 3),\ P_2(5, -3)$

25. $P_1(-2, 3),\ P_2(2, 1)$

26. $P_1(5, -2),\ P_2(1, 0)$

27. $P_1(8, -3),\ P_2(4, 1)$

28. $P_1(0, 3),\ P_2(2, -1)$

29. $P_1(3, -4),\ P_2(3, 5)$

30. $P_1(-1, 2),\ P_2(-1, 3)$

31. $P_1(4, -2),\ P_2(3, -2)$

32. $P_1(5, 1),\ P_2(-2, 1)$

33. $P_1(0, -1),\ P_2(3, -2)$

34. $P_1(3, 0),\ P_2(2, -1)$

35. $P_1(-2, 3),\ P_2(1, 3)$

36. $P_1(4, -1),\ P_2(-3, -1)$

37. $P_1(-2, 4),\ P_2(-1, -1)$

38. $P_1(6, -4),\ P_2(4, -2)$

39. $P_1(-2, -3),\ P_2(-2, 1)$

40. $P_1(5, 1),\ P_2(5, -2)$

41. $P_1(-1, 5),\ P_2(5, 1)$

42. $P_1(-1, 5),\ P_2(7, 1)$

3   Graph by using the slope and y-intercept.

43. $y = 3x + 1$

44. $y = -2x - 1$

45. $y = \dfrac{2}{5}x - 2$

46. $y = \dfrac{3}{4}x + 1$

47. $2x + y = 3$

48. $3x - y = 1$

49. $x - 2y = 4$

50. $x + 3y = 6$

51. $y = \dfrac{2}{3}x$

52. $y = \dfrac{1}{2}x$ **53.** $y = -x + 1$ **54.** $y = -x - 3$

55. $3x - 4y = 12$ **56.** $5x - 2y = 10$ **57.** $y = -4x + 2$

58. $y = 5x - 2$ **59.** $4x - 5y = 20$ **60.** $x - 3y = 6$

SUPPLEMENTAL EXERCISES 8.3

Solve.

61. a. Show that the equation $\dfrac{x}{4} + \dfrac{y}{3} = 1$ is a linear equation by writing it in the form $y = mx + b$.

   b. Find the x- and y-intercepts.

62. a. Show that the equation $\dfrac{x}{5} - \dfrac{y}{2} = 1$ is a linear equation by writing it in the form $y = mx + b$.

   b. Find the x- and y-intercepts.

63. What effect does increasing the coefficient of x have on the graph of $y = mx + b$?

64. What effect does decreasing the coefficient of x have on the graph of $y = mx + b$?

65. What effect does increasing the constant term have on the graph of $y = mx + b$?

66. What effect does decreasing the constant term have on the graph of $y = mx + b$?

SECTION 8.4
Equations of Straight Lines

1 Find the equation of a line using the equation $y = mx + b$

When the slope of a line and a point on the line are known, the equation of the line can be written using the slope-intercept form, $y = mx + b$. In the first example shown below, the known point is the y-intercept. In the second example, the known point is a point other than the y-intercept.

Find the equation of the line that has slope 3 and y-intercept (0, 2).

The given slope, 3, is m.	$y = mx + b$
Replace m with 3.	$y = 3x + b$
The given point, (0, 2), is the y-intercept. Replace b with 2.	$y = 3x + 2$

The equation of the line that has slope 3 and y-intercept 2 is $y = 3x + 2$.

Find the equation of the line that has slope $\frac{1}{2}$ and contains the point whose coordinates are (−2, 4).

The given slope, $\frac{1}{2}$, is m.	$y = mx + b$
Replace m with $\frac{1}{2}$.	$y = \frac{1}{2}x + b$
The given point, (−2, 4), is a solution of the equation of the line. Replace x and y in the equation with the coordinates of the point.	$4 = \frac{1}{2}(-2) + b$
Solve for b, the y-intercept.	$4 = -1 + b$ $5 = b$
Write the equation of the line by replacing m and b in the equation by their values.	$y = mx + b$ $y = \frac{1}{2}x + 5$

The equation of the line that has slope $\frac{1}{2}$ and contains the point whose coordinates are (−2, 4) is $y = \frac{1}{2}x + 5$.

Example 1   Find the equation of the line that contains the point whose coordinates are (3, −3) and has slope $\frac{2}{3}$.

Solution $y = mx + b$

$y = \frac{2}{3}x + b$ ▶ Replace m with the given slope.

$-3 = \frac{2}{3}(3) + b$ ▶ Replace x and y in the equation with the coordinates of the given point.

$-3 = 2 + b$ ▶ Solve for b.
$-5 = b$

$y = \frac{2}{3}x - 5$ ▶ Write the equation of the line by replacing m and b in the equation by their values.

Problem 1   Find the equation of the line that contains the point whose coordinates are (4, −2) and has slope $\frac{3}{2}$.

Solution   See page A33.

2   Find the equation of a line using the point-slope formula

An alternative method for finding the equation of a line, given the slope and the coordinates of a point on the line, involves use of the point-slope formula. The point-slope formula is derived from the formula for slope.

Let (x_1, y_1) be the coordinates of the given point on the line, and let (x, y) be the coordinates of any other point on the line.

Formula for slope

$$\frac{y - y_1}{x - x_1} = m$$

Multiply both sides of the equation by $(x - x_1)$.

$$\frac{y - y_1}{x - x_1}(x - x_1) = m(x - x_1)$$

Simplify.

$$y - y_1 = m(x - x_1)$$

Point-Slope Formula

The equation of the line that has slope m and contains the point whose coordinates are (x_1, y_1) can be found by the point-slope formula:

$$y - y_1 = m(x - x_1).$$

Example 2   Use the point-slope formula to find the equation of a line that passes through the point whose coordinates are $(-2, -1)$ and has slope $\frac{3}{2}$.

Solution $(x_1, y_1) = (-2, -1)$ ▶ Let (x_1, y_1) be the given point.

$m = \dfrac{3}{2}$ ▶ m is the given slope.

$y - y_1 = m(x - x_1)$ ▶ This is the point-slope formula.

$y - (-1) = \dfrac{3}{2}[x - (-2)]$ ▶ Substitute -2 for x_1, -1 for y_1, and $\frac{3}{2}$ for m.

$y + 1 = \dfrac{3}{2}(x + 2)$ ▶ Rewrite the equation in the form $y = mx + b$.

$y + 1 = \dfrac{3}{2}x + 3$

$y = \dfrac{3}{2}x + 2$

Problem 2   Use the point-slope formula to find the equation of a line that passes through the point whose coordinates are $(5, 4)$ and has slope $\frac{2}{5}$.

Solution   See page A33.

EXERCISES 8.4

1   Use the slope-intercept form.

1. Find the equation of the line that contains the point whose coordinates are $(0, 2)$ and has slope 2.

2. Find the equation of the line that contains the point whose coordinates are $(0, -1)$ and has slope -2.

3. Find the equation of the line that contains the point whose coordinates are $(-1, 2)$ and has slope -3.

4. Find the equation of the line that contains the point whose coordinates are $(2, -3)$ and has slope 3.

5. Find the equation of the line that contains the point whose coordinates are $(3, 1)$ and has slope $\frac{1}{3}$.

6. Find the equation of the line that contains the point whose coordinates are $(-2, 3)$ and has slope $\frac{1}{2}$.

7. Find the equation of the line that contains the point whose coordinates are (4, −2) and has slope $\frac{3}{4}$.

8. Find the equation of the line that contains the point whose coordinates are (2, 3) and has slope $-\frac{1}{2}$.

9. Find the equation of the line that contains the point whose coordinates are (5, −3) and has slope $-\frac{3}{5}$.

10. Find the equation of the line that contains the point whose coordinates are (5, −1) and has slope $\frac{1}{5}$.

11. Find the equation of the line that contains the point whose coordinates are (2, 3) and has slope $\frac{1}{4}$.

12. Find the equation of the line that contains the point whose coordinates are (−1, 2) and has slope $-\frac{1}{2}$.

13. Find the equation of the line that contains the point whose coordinates are (−3, −5) and has slope $-\frac{2}{3}$.

14. Find the equation of the line that contains the point whose coordinates are (−4, 0) and has slope $\frac{5}{2}$.

2   Use the point-slope formula.

15. Find the equation of the line that passes through the point whose coordinates are (1, −1) and has slope 2.

16. Find the equation of the line that passes through the point whose coordinates are (2, 3) and has slope −1.

17. Find the equation of the line that passes through the point whose coordinates are (−2, 1) and has slope −2.

18. Find the equation of the line that passes through the point whose coordinates are (−1, −3) and has slope −3.

19. Find the equation of the line that passes through the point whose coordinates are (0, 0) and has slope $\frac{2}{3}$.

20. Find the equation of the line that passes through the point whose coordinates are (0, 0) and has slope $-\frac{1}{5}$.

21. Find the equation of the line that passes through the point whose coordinates are (2, 3) and has slope $\frac{1}{2}$.

22. Find the equation of the line that passes through the point whose coordinates are (3, −1) and has slope $\frac{2}{3}$.

23. Find the equation of the line that passes through the point whose coordinates are (−4, 1) and has slope $-\frac{3}{4}$.

24. Find the equation of the line that passes through the point whose coordinates are (−5, 0) and has slope $-\frac{1}{5}$.

25. Find the equation of the line that passes through the point whose coordinates are (−2, 1) and has slope $\frac{3}{4}$.

26. Find the equation of the line that passes through the point whose coordinates are (3, −2) and has slope $\frac{1}{6}$.

27. Find the equation of the line that passes through the point whose coordinates are (−3, −5) and has slope $-\frac{4}{3}$.

28. Find the equation of the line that passes through the point whose coordinates are (3, −1) and has slope $\frac{3}{5}$.

SUPPLEMENTAL EXERCISES 8.4

Is there a linear equation that contains all the given ordered pairs? If there is, find the equation.

29. (5, 1), (4, 2), (0, 6)

30. (−2, −4), (0, −3), (4, −1)

31. (−1, −5), (2, 4), (0, 2)

32. (3, −1), (12, −4), (−6, 2)

The given ordered pairs are solutions to the same linear equation. Find n.

33. (0, 1), (4, 9), (3, n)

34. (2, 2), (−1, 5), (3, n)

35. (2, −2), (−2, −4), (4, n)

36. (1, −2), (−2, 4), (4, n)

The relationship between Celsius and Fahrenheit temperature can be given by a linear equation. Water freezes at 0°C or at 32°F. Water boils at 100°C or at 212°F.

37. Write a linear equation expressing Fahrenheit temperature in terms of Celsius temperature.

38. Graph the equation found in Exercise 37.

alculators and Computers

Graphs of Straight Lines

The program GRAPHS OF STRAIGHT LINES on the Math ACE Disk graphs lines of the form $y = mx + b$, where m is the slope of the line, and b is the y-intercept.

This program will enable you to examine the effect on the graph when changes are made in the slope m while the y-intercept remains constant. You can also examine the effect that changes in the y-intercept have on the graph.

After each line is drawn, you will have the option of erasing the line, drawing another line, or quitting. Here is an example to get you started.

Graph: $y = x + 1$ $(m = 1, \quad b = 1)$
           $y = 2x + 1$ $(m = 2, \quad b = 1)$
           $y = -x + 1$ $(m = -1, b = 1)$
           $y = -2x + 1$ $(m = -2, b = 1)$

For each of these graphs, the y-intercept is 1, and the graph passes through the point whose coordinates are (0, 1).

Now erase the graphs and try these.

Graph: $y = 2x - 1$ $(m = 2, b = -1)$
           $y = 2x + 2$ $(m = 2, b = 2)$
           $y = 2x - 4$ $(m = 2, b = -4)$

The three graphs are parallel because the slope of each line is 2. The y-intercept changes.

Try this program with your own values. Remember that when entering a fractional slope or y-intercept, you must enter it as a decimal. For example, to see the graph of $y = \frac{3}{4}x + 2$, enter the slope $\frac{3}{4}$ as 0.75.

Something Extra

Graphs of Motion

A graph can be useful in analyzing the motion of a body. For example, consider an airplane in uniform motion traveling at 100 m/s. The table at the right shows the distance traveled by the plane at the end of each of five 1-second intervals.

Time (s)	Distance (m)
0	0
1	100
2	200
3	300
4	400
5	500

These data can be graphed on a rectangular coordinate system and a straight line drawn through the points plotted. The travel time is shown along the horizontal axis, and the distance traveled by the plane is shown along the vertical axis. (Note that the units along the two axes are not the same length.)

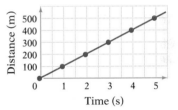

To write the equation for the line graphed above, use the coordinates of any two points on the line to find the slope. The y-intercept is (0, 0).

Let $(x_1, y_1) = (1, 100)$ and $(x_2, y_2) = (2, 200)$.

Then $m = \frac{y_2 - y_1}{x_2 - x_1} = \frac{200 - 100}{2 - 1} = \frac{100}{1} = 100$. $b = 0$.

$y = mx + b$
$y = 100x + 0$
$y = 100x$

Note that the slope of the line, 100, is equal to the speed, 100 m/s.

The slope of a distance-time graph represents the speed of the object. The greater the slope of a distance-time graph, the greater the speed of the object it represents.

The distance-time graphs for two planes are shown at the right. One plane is traveling at 100 m/s, and the other is traveling at 200 m/s. The slope of the line representing the faster plane is greater than the slope of the line representing the slower plane.

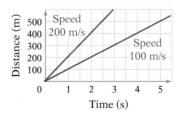

In the graph at the right, the travel time of a plane flying at 100 m/s is shown along the horizontal axis, and the speed of the plane is shown along the vertical axis. Because the speed is constant, the graph is a horizontal line.

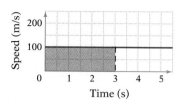

The area between the horizontal line graphed and the horizontal axis is equal to the distance traveled by the plane up to that time. For example, the area of the shaded region on the graph is:

$$\text{Length} \cdot \text{Width} = (3 \text{ s})(100 \text{ m/s}) = \frac{3 \text{ s}}{1} \cdot \frac{100 \text{ m}}{\text{s}} = 300 \text{ m}$$

The distance traveled by the plane in 3 s is equal to 300 m.

Solve.

1. A car in uniform motion is traveling at 20 m/s.
   a. Prepare a distance-time graph for the car for 0 s to 5 s.
   b. Find the slope of the line.
   c. Find the equation of the line.
   d. Prepare a speed-time graph for the car for 0 s to 5 s.
   e. Find the distance traveled by the car after 3 s.

2. One car in uniform motion is traveling at 10 m/s. A second car in uniform motion is traveling at 15 m/s.
   a. Prepare one distance-time graph for both cars for 0 s to 5 s.
   b. Find the slope of each line.
   c. Find the equation of each line graphed.
   d. Assuming that the cars started at the same point at 0 s, find the distance between the cars at the end of 5 s.

3. a. In a distance-time graph, is it possible for the graph to be a horizontal line?
   b. What does a horizontal line reveal about the motion of the object during that time period?

4. a. In a distance-time graph, is it possible for the graph to be a vertical line?
   b. What does a vertical line reveal about the motion of the object?

Chapter Summary

Key Words

A *rectangular coordinate system* is formed by two number lines, one horizontal and one vertical, that intersect at the zero point of each line.

The number lines that make up a rectangular coordinate system are called the *coordinate axes*, or simply the *axes*.

The *origin* is the point of intersection of the two coordinate axes.

A rectangular coordinate system divides the plane into four regions, called *quadrants*.

An *ordered pair* (a, b) is used to locate a point in a plane.

The first number in an ordered pair is called the *abscissa* or x-coordinate.

The second number in an ordered pair is called the *ordinate* or y-coordinate.

The *coordinates* of a point are the numbers in the ordered pair associated with the point.

An equation of the form $y = mx + b$, where m and b are constants, is a *linear equation in two variables*.

The point at which a graph crosses the x-axis is called the *x-intercept*.

The point at which a graph crosses the y-axis is called the *y-intercept*.

The *slope* of a line is a measure of the slant of a line. The symbol for slope is m.

A line that slants upward to the right has a *positive slope*.

A line that slants downward to the right has a *negative slope*.

A horizontal line has *zero slope*.

The slope of a vertical line is *undefined*.

Essential Rules

Slope of a straight line	Slope $= m = \dfrac{y_2 - y_1}{x_2 - x_1}$, $x_2 \neq x_1$
Slope-intercept form of a straight line	$y = mx + b$
Point-slope form of a straight line	$y - y_1 = m(x - x_1)$

Chapter Review

1. Find the ordered pair solution of $y = -\frac{2}{3}x + 2$ corresponding to $x = 3$.

2. Find the equation of the line that contains the point whose coordinates are $(0, -1)$ and has slope 3.

3. Graph $x + 3 = 0$.

4. Graph the ordered pairs $(-3, 1)$ and $(0, 2)$.

5. Find the x- and y-intercepts for $y = \frac{1}{2}x + 1$.

6. Find the slope of the line that contains the points whose coordinates are $(3, -4)$ and $(1, -4)$.

7. Graph $y = 3x + 1$.

8. Graph $3x - 2y = 6$.

9. Find the equation of the line that contains the point whose coordinates are $(-1, 2)$ and has slope $-\frac{2}{3}$.

10. Find the x- and y-intercepts for $6x - 4y = 12$.

11. Graph the line that has slope $-\frac{2}{3}$ and y-intercept $(0, 4)$.

12. Graph $3x + y = 4$.

13. Find the equation of the line that contains the point whose coordinates are $(-3, 1)$ and has slope $\frac{2}{3}$.

14. Find the slope of the line that contains the points whose coordinates are $(2, -3)$ and $(4, 1)$.

15. Graph $y = -\frac{3}{4}x + 3$.

16. Graph the line that has slope 2 and y-intercept -2.

17. Graph the ordered pairs $(-2, -3)$ and $(2, 4)$.

18. Graph $y = x + 1$.

19. Find the x- and y-intercepts for $2x - 3y = 12$.

20. Find the equation of the line that contains the point whose coordinates are $(-1, 0)$ and has slope 2.

21. Graph $y - 3 = 0$.

22. Graph $y = 2x - 1$.

23. Find the slope of the line that contains the points whose coordinates are $(2, -3)$ and $(-3, 4)$.

24. Find the equation of the line that contains the point whose coordinates are $(2, 3)$ and has slope $\frac{1}{2}$.

25. Graph the line that has slope -1 and y-intercept $(0, 2)$.

26. Graph $2x - 3y = 6$.

27. Find the ordered pair solution of $y = 2x - 1$ that corresponds to $x = -2$.

28. Find the equation of the line that contains the point $(0, 2)$ and has slope -3.

29. Find the slope of the line that contains the points whose coordinates are $(3, -2)$ and $(3, 5)$.

30. Find the equation of the line that contains the point whose coordinates are $(2, -1)$ and has slope $\frac{1}{2}$.

31. Is $(-10, 0)$ a solution of $y = \frac{1}{5}x + 2$?

32. Find the ordered pair solution of $y = 4x - 9$ corresponding to $x = 2$.

33. Find the x- and y-intercepts for $4x - 3y = 0$.

34. Find the equation of the line that contains the point whose coordinates are $(-2, 3)$ and has zero slope.

35. Graph $y = \frac{3}{4}x - 2$.

36. Graph the line that has slope -3 and y-intercept $(0, 1)$.

37. Find the equation of the line that contains the point whose coordinates are $(0, -4)$ and has slope 3.

38. Graph $x + 2y = -4$.

39. The following table gives the number of people, in thousands, collecting unemployment benefits in a northeastern state during a four-month period and the number of housing starts, in thousands, in that state during the same period. Graph the scatter diagram of these data.

People collecting unemployment, x	40	45	65	90
Housing starts, y	40	35	25	20

40. A study of six algebra students measured the average number of hours per week a student spent in the math lab and the student's score on the final exam. The results are recorded in the following table. Graph a scatter diagram for these data.

Hours in the math lab, x	3	2	4	5	3	1
Final exam score, y	78	66	75	82	74	63

Chapter Test

1. Find the equation of the line that contains the point whose coordinates are $(9, -3)$ and has slope $-\frac{1}{3}$.

2. Find the slope of the line that contains the points whose coordinates are $(9, 8)$ and $(-2, 1)$.

3. Find the x- and y-intercepts for $3x - 2y = 24$.

4. Find the ordered pair solution of $y = -\frac{4}{3}x - 1$ corresponding to $x = 9$.

5. Graph $5x + 3y = 15$.

6. Graph $y = \frac{1}{4}x + 3$.

7. Find the equation of the line that contains the point whose coordinates are $(2, 1)$ and has slope 4.

8. Is $(6, 3)$ a solution of $y = \frac{2}{3}x + 1$?

9. Graph the line that has slope $\frac{1}{2}$ and y-intercept $(0, -1)$.

10. Graph $3x - 2y = -6$.

11. Graph the ordered pairs $(3, -2)$ and $(0, 4)$.

12. Graph the line that has slope 2 and y-intercept $(0, -4)$.

13. Find the slope of the line that contains the points whose coordinates are $(4, -3)$ and $(-2, -3)$.

14. Find the equation of the line that contains the point whose coordinates are $(0, 7)$ and has slope $-\frac{2}{5}$.

15. Graph $y = -2x - 1$.

16. Graph $y = -\frac{1}{3}x + 3$.

17. Graph $x = -3$.

18. Graph the line that has slope $-\frac{2}{3}$ and y-intercept $(0, 2)$.

19. Find the equation of the line that contains the point whose coordinates are $(5, 0)$ and has slope $\frac{3}{5}$.

20. The distance a house is from a fire station and the amount of fire damage that house sustained in a fire are given in the following table. Graph the scatter diagram for these data.

Distance in miles, x	3.5	4.0	5.5	6.0
Damage in thousands of dollars, y	25	30	40	35

Cumulative Review

1. Simplify: $12 - 18 \div 3 \cdot (-2)^2$

2. Evaluate $\frac{a - b}{a^2 - c}$ when $a = -2$, $b = 3$, and $c = -4$.

3. Simplify: $4(2 - 3x) - 5(x - 4)$

4. Solve: $2x - \frac{2}{3} = \frac{7}{3}$

5. Solve: $3x - 2[x - 3(2 - 3x)] = x - 6$

6. Write $6\frac{2}{3}\%$ as a fraction.

7. Evaluate $\frac{5^{-2}}{5}$.

8. Simplify: $(-2x^2y)^3(2xy^2)^2$

9. Simplify: $\dfrac{-15x^7}{5x^{-2}}$

10. Divide: $(x^2 - 4x - 21) \div (x - 7)$

11. Factor: $5x^2y - 20xy^2$

12. Factor: $5x^2 + 15x + 10$

13. Factor: $x(a + 2) + y(a + 2)$

14. Solve: $x(x - 2) = 8$

15. Simplify: $\dfrac{x^5y^3}{x^2 - x - 6} \cdot \dfrac{x^2 - 9}{x^2y^4}$

16. Simplify: $\dfrac{3x}{x^2 + 5x - 24} - \dfrac{9}{x^2 + 5x - 24}$

17. Solve: $3 - \dfrac{1}{x} = \dfrac{5}{x}$

18. Solve $4x - 5y = 15$ for y.

19. Find the ordered pair solution of $y = 3x - 1$ corresponding to $x = -2$.

20. Find the slope of the line containing the points whose coordinates are $(2, 3)$ and $(-2, 3)$.

21. Find the x- and y-intercepts for $5x + 2y = 20$.

22. Find the equation of the line that contains the point whose coordinates are $(3, 2)$ and has slope -1.

23. Graph $y = \dfrac{1}{2}x + 2$.

24. Graph $3x + y = 2$.

25. The sum of two numbers is 24. Twice the smaller number is three less than the larger number. Find the two numbers.

26. The perimeter of a triangle is 49 ft. The length of the first side is twice the length of the third side, and the length of the second side is 5 ft more than the length of the third side. Find the measure of the first side.

27. A suit that regularly sells for $89 is on sale for 30% off the regular price. Find the sale price.

28. A gold coin is 60 years older than a silver coin. Fifteen years ago, the gold coin was twice as old as the silver coin was then. Find the present ages of the two coins.

29. The real estate tax for a home that costs $50,000 is $625. At this rate, what is the cost of a home for which the real estate tax is $1375?

30. An electrician requires 6 h to wire a garage. An apprentice can do the same job in 10 h. How long would it take to wire the garage when both the electrician and the apprentice are working?

9

Systems of Linear Equations

Objectives

- Solve systems of linear equations by graphing
- Solve systems of linear equations by the substitution method
- Solve systems of linear equations by the addition method
- Rate-of-wind and water-current problems
- Application problems

Input–Output Analysis

The economies of the industrial nations are very complex; they comprise hundreds of different industries, and each industry supplies other industries with goods and services needed in the production process. For example, the steel industry requires coal to produce steel, and the coal industry requires steel (in the form of machinery) to mine and transport coal.

Wassily Leontief, a Russian-born economist, developed a method of describing mathematically the interactions of an economic system. His technique was to examine various sectors of an economy (steel industry, oil, farms, autos, and so on) and determine how each sector interacted with the others. More than five hundred sectors of the economy were studied.

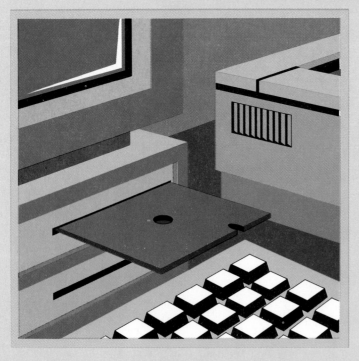

The interaction of each sector with the others was written as a series of equations. This series of equations is called a *system of equations.* Using a computer, economists searched for a solution to the system of equations that would determine the output levels various sectors would have to meet to satisfy the requests from other sectors. The method is called input–output analysis.

Input–output analysis has many applications. For example, it is used today to predict the production needs of large corporations and to determine the effect of price changes on the economy. In recognition of the importance of his ideas, Leontief was awarded the Nobel Prize in Economics in 1973.

This chapter begins the study of systems of equations.

SECTION **9.1**

Solving Systems of Linear Equations by Graphing

1 Solve systems of linear equations by graphing

Equations considered together are called a **system of equations.** A system of equations is shown at the right.

$$2x + y = 3$$
$$x + y = 1$$

A **solution of a system of equations** is an ordered pair that is a solution of each equation of the system.

For example, $(2, -1)$ is a solution of the system of equations given above because it is a solution of each equation in the system.

$2x + y = 3$	
$2(2) + (-1)$	3
$4 + (-1)$	3
	$3 = 3$

$x + y = 1$	
$2 + (-1)$	1
	$1 = 1$

However, $(3, -3)$ is not a solution of this system because it is not a solution of each equation in the system.

$2x + y = 3$	
$2(3) + (-3)$	3
$6 + (-3)$	3
	$3 = 3$

$x + y = 1$	
$3 + (-3)$	1
	$0 \neq 1$

Example 1   Is $(1, -3)$ a solution of the system $3x + 2y = -3$
$$x - 3y = 6?$$

Solution

$3x + 2y = -3$	
$3 \cdot 1 + 2(-3)$	-3
$3 + (-6)$	-3
	$-3 = -3$

$x - 3y = 6$	
$1 - 3(-3)$	6
$1 - (-9)$	6
	$10 \neq 6$

No, $(1, -3)$ is not a solution of the system of equations.

Problem 1   Is $(-1, -2)$ a solution of the system $2x - 5y = 8$
$$-x + 3y = -5?$$

Solution   See page A33.

The solution of a system of linear equations in two variables can be found by graphing the two equations on the same coordinate system. Three possible conditions result.

1. The lines graphed can intersect at one point. The point of intersection of the lines is the ordered pair that is a solution of each equation of the system. It is the solution of the system of equations. The system of equations is **independent.**

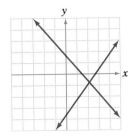

2. The lines graphed can be parallel and not intersect at all. The system of equations is **inconsistent** and has no solution.

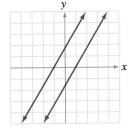

3. The lines graphed can represent the same line. The lines intersect at infinitely many points; therefore, there are infinitely many solutions. The system of equations is **dependent.** The solutions are the ordered pairs that are solutions of either one of the two equations in the system.

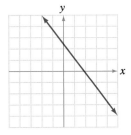

Solve by graphing: $2x + 3y = 6$
$2x + y = -2$

Graph each line.

The equations intersect at one point.
The system of equations is independent.

Find the point of intersection.

The solution is $(-3, 4)$.

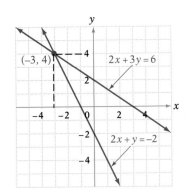

Solve by graphing: $2x - y = 1$
    $6x - 3y = 12$

Graph each line.

The lines are parallel and therefore do not intersect.

The system of equations is inconsistent and has no solution.

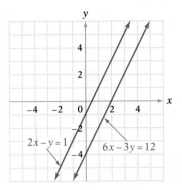

Solve by graphing: $2x + 3y = 6$
    $6x + 9y = 18$

Graph each line.

The two equations represent the same line. The system of equations is dependent and, therefore, has an infinite number of solutions.

The solutions are the ordered pairs that are solutions of the equation $2x + 3y = 6$.

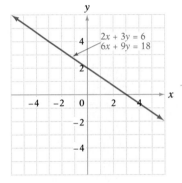

By choosing values for x, substituting these values into the equation $2x + 3y = 6$, and finding the corresponding values for y, some specific ordered pair solutions can be found. For example, $(3, 0)$, $(0, 2)$, and $(6, -2)$ are solutions of this system of equations.

Example 2 Solve by graphing.
A. $x - 2y = 2$ B. $4x - 2y = 6$
       $x + y = 5$ $y = 2x - 3$

Solution A.

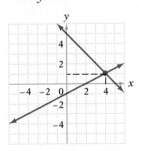

The solution is $(4, 1)$.

B.

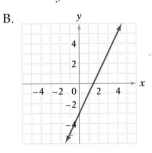

The system of equations is dependent. The solutions are the ordered pairs that satisfy the equation $y = 2x - 3$.

Problem 2   Solve by graphing.

     A. $x + 3y = 3$ B. $y = 3x - 1$

         $-x + y = 5$ $6x - 2y = -6$

Solution   See page A33.

EXERCISES 9.1

1. Is (2, 3) a solution of the system
$$3x + 4y = 18$$
$$2x - y = 1?$$

2. Is (2, −1) a solution of the system
$$x - 2y = 4$$
$$2x + y = 3?$$

3. Is (1, −2) a solution of the system
$$3x - y = 5$$
$$2x + 5y = -8?$$

4. Is (−1, −1) a solution of the system
$$x - 4y = 3$$
$$3x + y = 2?$$

5. Is (4, 3) a solution of the system
$$5x - 2y = 14$$
$$x + y = 8?$$

6. Is (2, 5) a solution of the system
$$3x + 2y = 16$$
$$2x - 3y = 4?$$

7. Is (−1, 3) a solution of the system
$$4x - y = -5$$
$$2x + 5y = 13?$$

8. Is (4, −1) a solution of the system
$$x - 4y = 9$$
$$2x - 3y = 11?$$

9. Is (0, 0) a solution of the system
$$4x + 3y = 0$$
$$2x - y = 1?$$

10. Is (2, 0) a solution of the system
$$3x - y = 6$$
$$x + 3y = 2?$$

11. Is (2, −3) a solution of the system
$$y = 2x - 7$$
$$3x - y = 9?$$

12. Is (−1, −2) a solution of the system
$$3x - 4y = 5$$
$$y = x - 1?$$

13. Is (5, 2) a solution of the system
$$y = 2x - 8$$
$$y = 3x - 13?$$

14. Is (−4, 3) a solution of the system
$$y = 2x + 11$$
$$y = 5x - 19?$$

Solve by graphing.

15. $x - y = 3$
$x + y = 5$

16. $2x - y = 4$
$x + y = 5$

17. $x + 2y = 6$
$x - y = 3$

18. $3x - y = 3$
$2x + y = 2$

19. $3x - 2y = 6$
$y = 3$

20. $x = 2$
$3x + 2y = 4$

21. $x = 3$
      $y = -2$

22. $x + 1 = 0$
       $y - 3 = 0$

23. $y = 2x - 6$
       $x + y = 0$

24. $5x - 2y = 11$
            $y = 2x - 5$

25. $2x + y = -2$
        $6x + 3y = 6$

26. $x + y = 5$
          $3x + 3y = 6$

27. $4x - 2y = 4$
              $y = 2x - 2$

28. $2x + 6y = 6$
             $y = -\dfrac{1}{3}x + 1$

29. $x - y = 5$
          $2x - y = 6$

30. $5x - 2y = 10$
       $3x + 2y = 6$

31. $3x + 4y = 0$
       $2x - 5y = 0$

32. $2x - 3y = 0$
              $y = -\dfrac{1}{3}x$

33. $x - 3y = 3$
        $2x - 6y = 12$

34. $4x + 6y = 12$
       $6x + 9y = 18$

35. $3x + 2y = -4$
             $x = 2y + 4$

36. $5x + 2y = -14$
       $3x - 4y = 2$

37. $4x - y = 5$
        $3x - 2y = 5$

38. $2x - 3y = 9$
       $4x + 3y = -9$

SUPPLEMENTAL EXERCISES 9.1

Write the system of equations given its graph.

39.

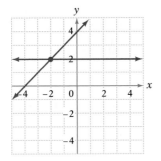

40.

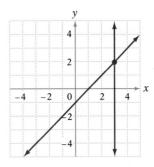

41.

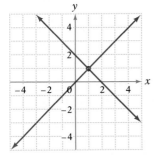

42.

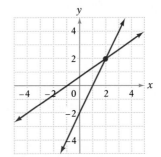

SECTION 9.2

Solving Systems of Linear Equations by the Substitution Method

1   Solve systems of linear equations by the substitution method

A graphical solution of a system of equations may give only an approximate solution of the system. For example, the point whose coordinates are $\left(\frac{1}{4}, \frac{1}{2}\right)$ would be difficult to read from the graph. An algebraic method, called the **substitution method,** can be used to find an exact solution of a system.

In the system of equations at the right, equation (2) states that $y = 3x - 9$. Substitute $3x - 9$ for y in equation (1).

(1) $2x + 5y = -11$
(2) $y = 3x - 9$

$2x + 5(3x - 9) = -11$

Solve for x.

$$2x + 15x - 45 = -11$$
$$17x - 45 = -11$$
$$17x = 34$$
$$x = 2$$

Substitute the value of x into equation (2) and solve for y.

(2)

$$y = 3x - 9$$
$$y = 3 \cdot 2 - 9$$
$$y = 6 - 9$$
$$y = -3$$

The solution is $(2, -3)$.

In the system of equations at the right, equation (1) is the easier equation to solve for one variable in terms of the other.

(1) $5x + y = 4$
(2) $2x - 3y = 5$

Solve equation (1) for y.

$$5x + y = 4$$
$$y = -5x + 4$$

Substitute $-5x + 4$ for y in equation (2).

$2x - 3(-5x + 4) = 5$

Solve for x.

$$2x + 15x - 12 = 5$$
$$17x - 12 = 5$$
$$17x = 17$$
$$x = 1$$

Substitute the value of x in equation (1) and solve for y.

(1)

$$5x + y = 4$$
$$5(1) + y = 4$$
$$5 + y = 4$$
$$y = -1$$

The solution is $(1, -1)$.

Example 1   Solve by substitution: $3x + 4y = -2$ (1)
$-x + 2y = 4$ (2)

Solution $-x + 2y = 4$ ▶ Solve equation (2) for x.
$-x = -2y + 4$
$x = 2y - 4$

$3x + 4y = -2$ ▶ Substitute $2y - 4$ for x in equation (1).
$3(2y - 4) + 4y = -2$
$6y - 12 + 4y = -2$ ▶ Solve for y.
$10y - 12 = -2$
$10y = 10$
$y = 1$

$-x + 2y = 4$ ▶ Substitute the value of y into equation (2).
$-x + 2(1) = 4$
$-x + 2 = 4$ ▶ Solve for x.
$-x = 2$
$x = -2$

The solution is $(-2, 1)$.

Problem 1   Solve by substitution: $7x - y = 4$
$3x + 2y = 9$

Solution   See pages A33 and A34.

Example 2   Solve by substitution: $4x + 2y = 5$ (1)
$y = -2x + 1$ (2)

Solution $4x + 2y = 5$ ▶ Substitute $-2x + 1$ for y in equation (1).
$4x + 2(-2x + 1) = 5$
$4x - 4x + 2 = 5$
$2 = 5$

This is not a true equation. The system of equations is inconsistent. The system does not have a solution.

Note that if equation (1) is solved for y, the resulting equation is $y = -2x + \frac{5}{2}$. Thus the lines $y = -2x + \frac{5}{2}$ and $y = -2x + 1$ have the same slope but different y-intercepts. The lines are parallel.

Problem 2   Solve by substitution: $3x - y = 4$
$y = 3x + 2$

Solution   See page A34.

Example 3   Solve by substitution: $6x - 2y = 4$ (1)
$$y = 3x - 2 \quad (2)$$

Solution
$$6x - 2y = 4 \qquad \blacktriangleright \text{Substitute } 3x - 2 \text{ for } y \text{ in equation (1).}$$
$$6x - 2(3x - 2) = 4$$
$$6x - 6x + 4 = 4$$
$$4 = 4$$

This is a true equation. The system of equations is dependent. The solutions are the ordered pairs that satisfy the equation $y = 3x - 2$.

Note that if equation (1) is solved for y, the resulting equation is $y = 3x - 2$, the same equation as equation (2).

Problem 3   Solve by substitution: $\quad y = -2x + 1$
$$6x + 3y = 3$$

Solution   See page A34.

EXERCISES 9.2

1   Solve by substitution.

1. $2x + 3y = 7$
$\quad x = 2$

2. $\quad y = 3$
$\quad 3x - 2y = 6$

3. $\quad y = x - 3$
$\quad x + y = 5$

4. $\quad y = x + 2$
$\quad x + y = 6$

5. $\quad x = y - 2$
$\quad x + 3y = 2$

6. $\quad x = y + 1$
$\quad x + 2y = 7$

7. $2x + 3y = 9$
$\quad y = x - 2$

8. $3x + 2y = 11$
$\quad y = x + 3$

9. $3x - y = 2$
$\quad y = 2x - 1$

10. $2x - y = -5$
$\quad y = x + 4$

11. $\quad x = 2y - 3$
$\quad 2x - 3y = -5$

12. $\quad x = 3y - 1$
$\quad 3x + 4y = 10$

13. $\quad y = 4 - 3x$
$\quad 3x + y = 5$

14. $\quad y = 2 - 3x$
$\quad 6x + 2y = 7$

15. $\quad x = 3y + 3$
$\quad 2x - 6y = 12$

16. $\quad x = 2 - y$
$\quad 3x + 3y = 6$

17. $3x + 5y = -6$
$\quad x = 5y + 3$

18. $\quad y = 2x + 3$
$\quad 4x - 3y = 1$

19. $x = 4y - 3$
$2x - 3y = 0$

20. $x = 2y$
$-2x + 4y = 6$

21. $y = 2x - 9$
$3x - y = 2$

22. $y = x + 4$
$2x - y = 6$

23. $3x + 4y = 7$
$x = 3 - 2y$

24. $3x - 5y = 1$
$y = 4x - 2$

25. $2x - y = 4$
$3x + 2y = 6$

26. $x + y = 12$
$3x - 2y = 6$

27. $4x - 3y = 5$
$x + 2y = 4$

28. $3x - 5y = 2$
$2x - y = 4$

29. $7x - y = 4$
$5x + 2y = 1$

30. $x - 7y = 4$
$-3x + 2y = 6$

31. $4x - 3y = -1$
$y = 2x - 3$

32. $3x - 7y = 28$
$x = 3 - 4y$

33. $7x + y = 14$
$2x - 5y = -33$

34. $3x + y = 4$
$4x - 3y = 1$

35. $x - 4y = 9$
$2x - 3y = 11$

36. $3x - y = 6$
$x + 3y = 2$

37. $4x - y = -5$
$2x + 5y = 13$

38. $3x - y = 5$
$2x + 5y = -8$

39. $3x + 4y = 18$
$2x - y = 1$

40. $4x + 3y = 0$
$2x - y = 0$

41. $5x + 2y = 0$
$x - 3y = 0$

42. $6x - 3y = 6$
$2x - y = 2$

43. $3x + y = 4$
$9x + 3y = 12$

44. $x - 5y = 6$
$2x - 7y = 9$

45. $x + 7y = -5$
$2x - 3y = 5$

46. $y = 2x + 11$
$y = 5x - 19$

47. $y = 2x - 8$
$y = 3x - 13$

48. $y = -4x + 2$
$y = -3x - 1$

49. $x = 3y + 7$
$x = 2y - 1$

50. $x = 4y - 2$
$x = 6y + 8$

51. $x = 3 - 2y$
$x = 5y - 10$

52. $y = 2x - 7$
     $y = 4x + 5$

53. $3x - y = 11$
     $2x + 5y = -4$

54. $-x + 6y = 8$
     $2x + 5y = 1$

SUPPLEMENTAL EXERCISES 9.2

Rewrite the equation so that the coefficients and constant are the smallest possible integers.

55. $0.5x - 0.8y = 0.7$

56. $1.3x + 2.6y = 9$

57. $1.29x + 3.4y = 8.5$

58. $2x - 4.7y = 3.59$

59. $4.92x - 3.4y = 7.02$

60. $6.2x + 3.09y = 7$

Solve.

61. $0.3x - 0.2y = 0.5$
     $0.1y = 0.2x - 0.4$

62. $0.1x = 0.3y - 0.2$
     $0.4x - 0.1y = 0.3$

63. $0.05x - 0.02y = -0.03$
     $-0.03x + 0.01y = 0.02$

64. $0.1x - 0.6y = -0.4$
     $-0.7x + 0.2y = 0.5$

65. $0.8x - 0.1y = 0.3$
     $0.5x - 0.2y = -0.5$

66. $0.4x + 0.5y = 0.2$
     $0.3x - 0.1y = 1.1$

67. $-0.1x + 0.3y = 1.1$
     $0.4x - 0.1y = -2.2$

68. $1.2x + 0.1y = 1.9$
     $0.1x + 0.3y = 2.2$

69. $1.25x - 0.01y = 1.5$
     $0.24x - 0.02y = -1.52$

For what value of k does the system of equations have no solution?

70. $2x - 3y = 7$
     $kx - 3y = 4$

71. $8x - 4y = 1$
     $2x - ky = 3$

72. $x = 4y + 4$
     $kx - 8y = 4$

S E C T I O N   **9.3**

Solving Systems of Linear Equations by the Addition Method

1 Solve systems of linear equations by the addition method

Another algebraic method for solving a system of equations is called the **addition method.** It is based on the Addition Property of Equations.

In the system of equations at the right, note the effect of adding equation (2) to equation (1). Because $2y$ and $-2y$ are opposites, adding the equations results in an equation with only one variable.

$$(1) \quad 3x + 2y = 4$$
$$(2) \quad 4x - 2y = 10$$
$$7x + 0y = 14$$
$$7x = 14$$

The solution of the resulting equation is the first component of the ordered pair solution of the system.

$$7x = 14$$
$$x = 2$$

The second component is found by substituting the value of x into equation (1) or (2) and then solving for y. Equation (1) is used here.

$$(1) \quad 3x + 2y = 4$$
$$3 \cdot 2 + 2y = 4$$
$$6 + 2y = 4$$
$$2y = -2$$
$$y = -1$$

The solution is $(2, -1)$.

Sometimes adding the two equations does not eliminate one of the variables. In this case, use the Multiplication Property of Equations to rewrite one or both of the equations so that when the equations are added, one of the variables is eliminated.

To do this, first choose which variable to eliminate. The coefficients of that variable must be opposites. Multiply each equation by a constant that will produce coefficients that are opposites.

To eliminate y in the system of equations at the right, multiply equation (1) by 2.

$$(1) \quad 3x + 2y = 7$$
$$(2) \quad 5x - 4y = 19$$

$$2(3x + 2y) = 2 \cdot 7$$
$$5x - 4y = 19$$

Now the coefficients of the y terms are opposites.

$$6x + 4y = 14$$
$$5x - 4y = 19$$

Add the equations.
Solve for x.

$$11x + 0y = 33$$
$$11x = 33$$
$$x = 3$$

Substitute the value of x into one of the equations and solve for y. Equation (2) is used here.

$$(2) \quad 5x - 4y = 19$$
$$5 \cdot 3 - 4y = 19$$
$$15 - 4y = 19$$
$$-4y = 4$$
$$y = -1$$

The solution is $(3, -1)$.

To eliminate x in the system of equations at the right, multiply equation (1) by 2 and equation (2) by -5.

(1) $5x + 6y = 3$
(2) $2x - 5y = 16$

Note how the constants are selected. The negative sign is used so that the coefficients will be opposites.

$$2 \diagdown (5x + 6y) = 2 \cdot 3$$
$$-5 \diagup (2x - 5y) = -5 \cdot 16$$

Now the coefficients of the x terms are opposites.

$$10x + 12y = 6$$
$$-10x + 25y = -80$$

Add the equations.
Solve for y.

$$0x + 37y = -74$$
$$37y = -74$$
$$y = -2$$

Substitute the value of y into one of the equations and solve for x. Equation (1) is used here.

(1) $5x + 6y = 3$
$$5x + 6(-2) = 3$$
$$5x - 12 = 3$$
$$5x = 15$$
$$x = 3$$

The solution is $(3, -2)$.

To solve the system of equations at the right, first write equation (1) in the form $Ax + By = C$.

(1) $5x = 2y - 7$
(2) $3x + 4y = 1$

$$5x - 2y = -7$$
$$3x + 4y = 1$$

Eliminate y. Multiply equation (1) by 2.

$$2(5x - 2y) = 2(-7)$$
$$3x + 4y = 1$$

Now the coefficients of the y terms are opposites.

$$10x - 4y = -14$$
$$3x + 4y = 1$$

Add the equations.
Solve for x.

$$13x + 0y = -13$$
$$13x = -13$$
$$x = -1$$

Substitute the value of x into one of the equations and solve for y.
Equation (1) is used here.

$$5x = 2y - 7$$
$$5(-1) = 2y - 7$$
$$-5 = 2y - 7$$
$$2 = 2y$$
$$1 = y$$

The solution is $(-1, 1)$.

To eliminate y in the system of equations at the right, multiply equation (1) by -2.

(1) $2x + y = 2$
(2) $4x + 2y = 5$

$$-4x - 2y = -4$$
$$4x + 2y = 5$$
$$0x + 0y = 1$$
$$0 = 1$$

Add the equations.
This is not a true equation.

The system of equations is inconsistent.
The system does not have a solution.

Example 1   Solve by the addition method: $2x + 4y = 7$ (1)
$5x - 3y = -2$ (2)

Solution $5(2x + 4y) = 5 \cdot 7$ ▶ Eliminate x.
$-2(5x - 3y) = -2 \cdot (-2)$

$10x + 20y = 35$
$-10x + 6y = 4$
$26y = 39$ ▶ Add the equations.
$y = \dfrac{39}{26} = \dfrac{3}{2}$ ▶ Solve for y.

$2x + 4\left(\dfrac{3}{2}\right) = 7$ ▶ Substitute the value of y in equation (1).

$2x + 6 = 7$ ▶ Solve for x.
$2x = 1$
$x = \dfrac{1}{2}$

The solution is $\left(\dfrac{1}{2}, \dfrac{3}{2}\right)$.

Problem 1   Solve by the addition method: $x - 2y = 1$
$2x + 4y = 0$

Solution   See pages A34 and A35.

Example 2   Solve by the addition method: $6x + 9y = 15$
$4x + 6y = 10$

Solution $4(6x + 9y) = 4 \cdot 15$ ▶ Eliminate x.
$-6(4x + 6y) = -6 \cdot 10$

$24x + 36y = 60$
$-24x - 36y = -60$

▶ Add the equations.

This is a true equation. The system of equations is dependent. The solutions are the ordered pairs that satisfy the equation $6x + 9y = 15$.

Problem 2 Solve by the addition method: $2x - 3y = 4$
$$-4x + 6y = -8$$

Solution See page A35.

EXERCISES 9.3

1 Solve by the addition method.

1. $x + y = 4$
$x - y = 6$

2. $2x + y = 3$
$x - y = 3$

3. $x + y = 4$
$2x + y = 5$

4. $x - 3y = 2$
$x + 2y = -3$

5. $2x - y = 1$
$x + 3y = 4$

6. $x - 2y = 4$
$3x + 4y = 2$

7. $4x - 5y = 22$
$x + 2y = -1$

8. $3x - y = 11$
$2x + 5y = 13$

9. $2x - y = 1$
$4x - 2y = 2$

10. $x + 3y = 2$
$3x + 9y = 6$

11. $4x + 3y = 15$
$2x - 5y = 1$

12. $3x - 7y = 13$
$6x + 5y = 7$

13. $2x - 3y = 1$
$4x - 6y = 2$

14. $2x + 4y = 6$
$3x + 6y = 9$

15. $5x - 2y = -1$
$x + 3y = -5$

16. $4x - 3y = 1$
$8x + 5y = 13$

17. $5x + 7y = 10$
$3x - 14y = 6$

18. $7x + 10y = 13$
$4x + 5y = 6$

19. $3x - 2y = 0$
$6x + 5y = 0$

20. $5x + 2y = 0$
$3x + 5y = 0$

21. $2x - 3y = 16$
$3x + 4y = 7$

22. $3x + 4y = 10$
$4x + 3y = 11$

23. $5x + 3y = 7$
$2x + 5y = 1$

24. $-2x + 7y = 9$
$3x + 2y = -1$

25. $7x - 2y = 13$
$5x + 3y = 27$

26. $12x + 5y = 23$
$2x - 7y = 39$

27. $8x - 3y = 11$
$6x - 5y = 11$

28. $4x - 8y = 36$
$3x - 6y = 27$

29. $5x + 15y = 20$
$2x + 6y = 8$

30. $y = 2x - 3$
$4x + 4y = -1$

31. $3x = 2y + 7$
$5x - 2y = 13$

32. $2y = 4 - 9x$
$9x - y = 25$

33. $2x + 9y = 16$
$5x = 1 - 3y$

34. $3x - 4 = y + 18$
$4x + 5y = -21$

35. $2x + 3y = 7 - 2x$
$7x + 2y = 9$

36. $5x - 3y = 3y + 4$
$4x + 3y = 11$

37. $3x + y = 1$
$5x + y = 2$

38. $2x - y = 1$
$2x - 5y = -1$

39. $4x + 3y = 3$
$x + 3y = 1$

40. $2x - 5y = 4$
$x + 5y = 1$

41. $3x - 4y = 1$
$4x + 3y = 1$

42. $2x - 7y = -17$
$3x + 5y = 17$

43. $2x - 3y = 4$
$-x + 4y = 3$

44. $4x - 2y = 5$
$2x + 3y = 4$

SUPPLEMENTAL EXERCISES 9.3

Solve.

45. $0.2x + 0.3y = 0$
$0.3x + 0.4y = 0.1$

46. $0.04x - 0.03y = -0.02$
$0.05x - 0.02y = 0.01$

47. $0.3x - 0.2y = 1.4$
$0.6x + 0.5y = 1.9$

48. $x - 0.2y = 0.2$
$0.2x + 0.5y = 2.2$

49. $0.5x - 1.2y = 0.3$
$0.2x + y = 1.6$

50. $1.25x - 1.5y = -1.75$
$2.5x - 1.75y = -1$

51. The point of intersection of the graphs of the equations $Ax + 2y = 2$ and $2x + By = 10$ is $(2, -2)$. Find A and B.

52. The point of intersection of the graphs of the equations $Ax - 4y = 9$ and $4x + By = -1$ is $(-1, -3)$. Find A and B.

53. The point of intersection of the graphs of the equations $Ax + 2y = -4$ and $2x + By = 11$ is $(-2, 3)$. Find A and B.

54. The point of intersection of the graphs of the equations $Ax - 3y = 1$ and $4x + By = 11$ is $(2, 3)$. Find A and B.

55. Given that the graphs of the equations $2x - y = 6$, $3x - 4y = 4$, and $Ax - 2y = 0$ all intersect at the same point, find A.

56. Given that the graphs of the equations $3x - 2y = -2$, $2x - y = 0$, and $Ax + y = 8$ all intersect at the same point, find A.

S E C T I O N **9.4**

Application Problems in Two Variables

1 Rate-of-wind and water-current problems

Solving motion problems that involve an object moving with or against a wind or current normally requires two variables.

Solve: Flying with the wind, a small plane can fly 750 mi in 3 h. Against the wind, the plane can fly the same distance in 5 h. Find the rate of the plane in calm air and the rate of the wind.

STRATEGY *for solving rate-of-wind or current problems*

■ Choose one variable to represent the rate of the object in calm conditions and a second variable to represent the rate of the wind or current. Using these variables, express the rate of the object with and against the wind or current. Use the equation $d = rt$ to write expressions for the distance traveled by the object. The results can be recorded in a table.

Rate of plane in calm air: p
Rate of wind: w

	Rate	·	Time	=	Distance
With the wind	$p + w$	·	3	=	$3(p + w)$
Against the wind	$p - w$	·	5	=	$5(p - w)$

■ Determine how the expressions for distance are related.

The distance traveled with the wind is 750 mi.
The distance traveled against the wind is 750 mi.

$3(p + w) = 750$
$5(p - w) = 750$

Solve the system of equations.

$3(p + w) = 750$ $\quad \dfrac{3(p + w)}{3} = \dfrac{750}{3} \quad$ $p + w = 250$

$5(p - w) = 750$ $\quad \dfrac{5(p - w)}{5} = \dfrac{750}{5} \quad$ $p - w = 150$

$$2p = 400$$
$$p = 200$$

Solve for w.
$p + w = 250$
$200 + w = 250$
$w = 50$

The rate of the plane in calm air is 200 mph.
The rate of the wind is 50 mph.

Example 1 A 600-mi trip from one city to another takes 4 h when a plane is flying with the wind. The return trip against the wind takes 5 h. Find the rate of the plane in still air and the rate of the wind.

Strategy ■ Rate of the plane in still air: p
Rate of the wind: w

	Rate	Time	Distance
With wind	$p + w$	4	$4(p + w)$
Against wind	$p - w$	5	$5(p - w)$

■ The distance traveled with the wind is 600 mi.
The distance traveled against the wind is 600 mi.

Solution $4(p + w) = 600$
$5(p - w) = 600$
$$\frac{4(p + w)}{4} = \frac{600}{4}$$
▶ Simplify equation (1) by dividing each side of the equation by 4.
$$\frac{5(p - w)}{5} = \frac{600}{5}$$
▶ Simplify equation (2) by dividing each side of the equation by 5.

$p + w = 150$
$p - w = 120$
$2p = 270$ ▶ Add the two equations.
$p = 135$ ▶ Solve for p, the rate of the plane in still air.

$p + w = 150$
$135 + w = 150$ ▶ Substitute the value of p into one of the equations.
$w = 15$ ▶ Solve for w, the rate of the wind.

The rate of the plane in still air is 135 mph.
The rate of the wind is 15 mph.

Problem 1 A canoeist paddling with the current can travel 24 mi in 3 h. Against the current, it takes 4 h to travel the same distance. Find the rate of the current and the rate of the canoeist in calm water.

Solution See page A35.

2 Application problems

The application problems in this section are varieties of those problems solved earlier in the text. Each of the strategies for the problems in this section will result in a system of equations.

Solve: A jeweler purchased 5 oz of a gold alloy and 20 oz of a silver alloy for a total cost of $700. The next day, at the same prices per ounce, the jeweler purchased 4 oz of the gold alloy and 30 oz of the silver alloy for a total cost of $630. Find the cost per ounce of the silver alloy.

STRATEGY *for solving an application problem in two variables*

■ Choose one variable to represent one of the unknown quantities and a second variable to represent the other unknown quantity. Write numerical or variable expressions for all the remaining quantities. These results can be recorded in two tables, one for each of the conditions.

Cost per ounce of gold: g
Cost per ounce of silver: s

First day

	Amount	·	Unit cost	=	Value
Gold	5	·	g	=	$5g$
Silver	20	·	s	=	$20s$

Second day

	Amount	·	Unit cost	=	Value
Gold	4	·	g	=	$4g$
Silver	30	·	s	=	$30s$

■ Determine a system of equations. The strategies presented in Chapter 4 can be used to determine the relationships between the expressions in the tables. Each table will give one equation of the system.

The total value of the purchase on the first day was $700.
The total value of the purchase on the second day was $630.

$5g + 20s = 700$
$4g + 30s = 630$

Solve the system of equations.

$5g + 20s = 700$	$4(5g + 20s) = 4 \cdot 700$	$20g + 80s = 2800$
$4g + 30s = 630$	$-5(4g + 30s) = -5 \cdot 630$	$-20g - 150s = -3150$
		$-70s = -350$
		$s = 5$

The cost per ounce of the silver alloy was $5.

Example 2   In ten years, an oil painting will be twice as old as a watercolor painting will be then. Ten years ago, the oil painting was three times as old as the watercolor was then. Find the present age of each painting.

Strategy   ■ Present age of the oil painting: x
Present age of the watercolor: y

	Present	Future
Oil	x	$x + 10$
Watercolor	y	$y + 10$

	Present	Past
Oil	x	$x - 10$
Watercolor	y	$y - 10$

■ In ten years, twice the age of the watercolor will be the age of the oil painting. Ten years ago, three times the age of the watercolor was the age of the oil painting.

Solution $2(y + 10) = x + 10$
$3(y - 10) = x - 10$

$2y + 20 = x + 10$ (1)
$3y - 30 = x - 10$ (2)

$2y + 10 = x$ ▶ Solve equation (1) for x in terms of y.
$3y - 30 = (2y + 10) - 10$ ▶ Substitute $2y + 10$ for x in equation (2).
$3y - 30 = 2y$ ▶ Solve equation (2) for y, the present age of the
$y - 30 = 0$ watercolor.
$y = 30$

$2y + 10 = x$ ▶ Substitute the value of y into one of the equa-
$2(30) + 10 = x$ tions.
$60 + 10 = x$ ▶ Solve for x, the present age of the oil painting.
$70 = x$

The present age of the oil painting is 70 years.
The present age of the watercolor is 30 years.

Problem 2   Two coin banks contain only dimes and quarters. In the first bank, the total value of the coins is $3.90. In the second bank, there are twice as many dimes as in the first bank and one half the number of quarters. The total value of the coins in the second bank is $3.30. Find the number of dimes and the number of quarters in the first bank.

Solution   See page A36.

EXERCISES 9.4

1   Solve.

1. A plane flying with the jet stream flew from Los Angeles to Chicago, a distance of 2250 mi, in 5 h. Flying against the jet stream, the plane could fly only 1750 mi in the same amount of time. Find the rate of the plane in calm air and the rate of the wind.

2. A rowing team rowing with the current traveled 40 km in 2 h. Rowing against the current, the team could travel only 16 km in 2 h. Find the team's rowing rate in calm water and the rate of the current.

3. A motorboat traveling with the current went 35 mi in 3.5 h. Traveling against the current, the boat went 12 mi in 3 h. Find the rate of the boat in calm water and the rate of the current.

4. A small plane, flying into a headwind, flew 270 mi in 3 h. Flying with the wind, the plane traveled 260 mi in 2 h. Find the rate of the plane in calm air and the rate of the wind.

5. A plane flying with a tailwind flew 300 mi in 2 h. Against the wind, it took 3 h to travel the same distance. Find the rate of the plane in calm air and the rate of the wind.

6. A rowing team rowing with the current traveled 17 mi in 2 h. Against the current, the team rowed a distance of 7 mi in the same amount of time. Find the rate of the rowing team in calm water and the rate of the current.

7. A seaplane flying with the wind flew from an ocean port to a lake, a distance of 240 mi, in 2 h. Flying against the wind, the trip from the lake to the ocean port took 3 h. Find the rate of the plane in calm air and the rate of the wind.

8. Rowing with the current, a canoeist paddled 14 mi in 2 h. Against the current, the canoeist could paddle only 10 mi in the same amount of time. Find the rate of the canoeist in calm water and the rate of the current.

9. Flying with the wind, a small plane flew 280 mi in 2 h. Flying against the wind, the plane flew 160 mi in 2 h. Find the rate of the plane in calm air and the rate of the wind.

10. With the wind, a quarterback passes a football 140 ft in 2 s. Against the wind, the same pass would have traveled 80 ft in 2 s. Find the rate of the pass and the rate of the wind.

11. Flying with the wind, a plane flew 1000 km in 5 h. Against the wind, the plane could fly only 800 km in the same amount of time. Find the rate of the plane in calm air and the rate of the wind.

12. Traveling with the current, a cruise ship sailed between two islands, a distance of 90 mi, in 3 h. The return trip against the current required 4 h and 30 min. Find the rate of the cruise ship in calm water and the rate of the current.

2   Solve.

13. The manager of a computer software store received two shipments of software. The cost of the first shipment, which contained 12 identical word processing programs and 10 identical spreadsheet programs, was $6190. The second shipment, at the same prices, contained 5 copies of the word processing program and 8 copies of the spreadsheet program. The cost of the second shipment was $3825. Find the cost for one copy of the word processing program.

14. A baker purchased 12 lb of wheat flour and 15 lb of rye flour for a total cost of $18.30. A second purchase, at the same prices, included 15 lb of wheat flour and 10 lb of rye flour. The cost of the second purchase was $16.75. Find the cost per pound of the wheat and rye flours.

15. An investor owned 300 shares of an oil company and 200 shares of a movie company. The quarterly dividend from the two stocks was $165. After the investor sold 100 shares of the oil company and bought an additional 100 shares of the movie company, the quarterly dividend was $185. Find the dividend per share for each stock.

16. For using a computerized financial news network for 25 min during prime time and 35 min during non-prime time, a customer was charged $10.75. A second customer was charged $13.35 for using the network for 30 min of prime time and 45 min of non-prime time. Find the cost per minute for using the financial news network during prime time.

17. A basketball team scored 87 points in two-point baskets and three-point baskets. If the two-point baskets had been three-point baskets and the three-point baskets had been two-point baskets, the team would have scored 93 points. Find the number of two-point baskets and three-point baskets that were scored by the team.

18. A football team scored 30 points in one game with only touchdowns and field goals. If the touchdowns had been field goals and the field goals had been touchdowns, the score would have been 33. Find the number of touchdowns and field goals scored. Use 6 points for a touchdown and 3 points for a field goal.

19. Two coin banks contain only nickels and quarters. The total value of the coins in the first bank is $2.90. In the second bank, there are two more quarters than in the first bank and twice as many nickels. The total value of the coins in the second bank is $3.80. Find the number of nickels and the number of quarters in the first bank.

20. Two coin banks contain only nickels and dimes. The total value of the coins in the first bank is $3. In the second bank, there are 4 more nickels than in the first bank and one half as many dimes. The total value of the coins in the second bank is $2. Find the number of nickels and the number of dimes in the first bank.

21. The total value of the dimes and quarters in a coin bank is $3.30. If the quarters were dimes and the dimes were quarters, the total value of the coins would be $3. Find the number of dimes and the number of quarters in the bank.

22. The total value of the nickels and dimes in a coin bank is $3. If the nickels were dimes and the dimes were nickels, the total value of the coins would be $3.75. Find the number of nickels and the number of dimes in the bank.

23. Two years ago, the sum of the ages of an adult and a child was 40. Two years from now, the age of the adult will be three times the age the child will be then. Find the present ages of the adult and the child.

24. If twice the age of a coin is added to three times the age of a stamp, the result is 300. The difference between four times the age of the coin and twice the age of the stamp is 200. Find the age of each.

SUPPLEMENTAL EXERCISES 9.4

Solve.

25. Two angles are supplementary. The larger angle is 15° more than twice the measure of the smaller angle. Find the measure of the two angles. (Supplementary angles are two angles whose sum is 180°.)

26. Two angles are complementary. The larger angle is four times the measure of the smaller angle. Find the measure of the two angles. (Complementary angles are two angles whose sum is 90°.)

27. The sum of the ages of a gold coin and a silver coin is 45 years. The age of the gold coin ten years from now is equal to twice the age of the silver coin ten years ago. Find the present ages of the two coins.

28. The difference between the ages of an oil painting and a watercolor is 25 years. The age of the oil painting five years from now is equal to twice the age of the watercolor ten years ago. Find the present age of each.

29. An investment club placed a portion of its funds in a 9% annual simple interest account and the remainder in an 8% annual simple interest account. The amount of interest earned for one year was $860. If the amounts placed in each account had been reversed, the interest earned would have been $840. How much was invested in each account?

30. The value of the nickels and dimes in a coin bank is $.25. If the number of nickels and the number of dimes is doubled, the value of the coins would be $.50. How many nickels and dimes are in the bank?

31. An investor has $5000 to invest in two accounts. The first account earns 8% annual simple interest and the second account earns 10% annual simple interest. How much money should be invested in each account so that the annual interest earned is $600?

Calculators and Computers

 ### Systems of Equations

In Chapter 9, three methods of determining the solutions to systems of linear equations are presented: solving by graphing, solving by the substitution method, and solving by the addition method. As stated in the text, solving a system by graphing is not the most efficient method of finding a solution. However, doing so should give you a good visual understanding of the concept of solutions of systems of linear equations.

The substitution method is used most often when one variable is given in terms of the other; for example, in the equation $y = 2x + 5$, y is given in terms of x. Therefore, the value $2x + 5$ is easily substituted for y in another equation.

The addition method is used when neither of the equations is easily solved for one of the variables. Students sometimes find this method difficult at first because of the number of steps involved in finding a solution. Remember that the goal here is the same as it was when finding the solution to one equation: to rewrite the equation in the form *variable = constant*.

By using the addition method, the system of equations $\begin{aligned} ax + by &= c \\ dx + ey &= f \end{aligned}$ can be solved.

The solution is $x = \dfrac{ce - bf}{ae - bd}$ and $y = \dfrac{af - cd}{ae - bd}$, $ae - bd \neq 0$.

Using this solution, a system of equations can be solved by using a calculator. It is helpful to observe that the denominators for each expression are identical. The calculation for the denominator is done first and then stored in the calculator's memory. If the value of the denominator is zero, the system is dependent or inconsistent, and this calculator method cannot be used.

Solve: $2x - 5y = 9$
$\quad\quad 4x + 3y = 2$

Make a list of the values of a, b, c, d, e, and f.

$$a = 2 \quad\quad b = -5 \quad\quad c = 9$$
$$d = 4 \quad\quad e = 3 \quad\quad f = 2$$

Calculate the denominator $D = ae - bd$. $D = 2 \cdot 3 - (-5) \cdot 4 = 6 + 20 = 26$

Store the result in memory.   Press $\boxed{\text{M+}}$.

Find x. Replace the letters by the given values. $x = \dfrac{ce - bf}{D} = \dfrac{9 \cdot 3 - (-5) \cdot 2}{26}$

Calculate x. $9\boxed{\times}3\boxed{-}\boxed{(}5\boxed{+/-}\boxed{\times}2\boxed{)}\boxed{=}\boxed{\div}\boxed{\text{MR}}\boxed{=}$

The result in the display should be 1.423077.

Find y. Replace the letters by the given values. $y = \dfrac{af - cd}{D} = \dfrac{2 \cdot 2 - 9 \cdot 4}{26}$

Calculate y. $2\boxed{\times}2\boxed{-}\boxed{(}9\boxed{\times}4\boxed{)}\boxed{=}\boxed{\div}\boxed{\text{MR}}\boxed{=}$

The result in the display should be −1.230769.

The solution of the system is (1.423077, −1.230769).

The keys $\boxed{\text{M+}}$ (store in memory) and $\boxed{\text{MR}}$ (recall from memory) were used for this illustration. Some calculators use the keys $\boxed{\text{STO}}$ (store in memory) and $\boxed{\text{RCL}}$ (recall from memory). If your calculator uses these keys, use them in place of the keys shown in the illustration.

Something Extra

Break-even Analysis

Break-even analysis is a method used to determine the sales volume required for a company to break even, or experience neither a profit nor a loss on the sale of its product. The break-even point represents the number of units that must be made and sold in order for income from sales to equal the cost of the product.

The break-even point can be determined by graphing two equations on the same coordinate grid. The first equation is $R = SN$, where R is the revenue earned, S is the selling price per unit, and N is the number of units sold. The second equation is $T = VN + F$, where T is the total cost, F is the fixed costs, V is the variable costs per unit, and N is the number of units sold. The break-even point is the point where the graphs of the two equations intersect.

Solve.

1. a. A company manufactures and sells digital watches. The fixed costs are $20,000, the variable costs per unit are $25, and the selling price per watch is $125. Write two equations for this information.

    b. Graph the two equations on the same coordinate grid, using only Quadrant I. The horizontal axis is the number of units sold. Use the model shown at the right.

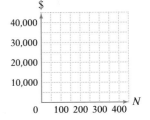

    c. How many watches must the company sell in order to break even?

2. a. A company manufactures and sells basketballs. The fixed costs are $12,000, the variable costs per unit are $10, and the selling price per basketball is $50. Write two equations for this information.

    b. Graph the two equations on the same coordinate grid, using only Quadrant I. The horizontal axis is the number of units sold. Use the model graph shown in Exercise 1.

    c. How many basketballs must the company sell in order to break even?

3. a. A company manufactures and sells calculators. The fixed costs are $10,000, the variable costs per unit are $5, and the selling price per calculator is $25. How many calculators must the company sell in order to experience neither a profit nor a loss?

    b. If the company changes the selling price per calculator to $30, what is the break-even point?

Chapter Summary

Key Words

Equations considered together are called a *system of equations.*

A *solution of a system of equations* in two variables is an ordered pair that is a solution of each equation of the system.

An *independent system of equations* has one solution. An *inconsistent system of equations* has no solution. A *dependent system of equations* has an infinite number of solutions.

The *substitution method* and the *addition method* are two algebraic methods of solving systems of equations.

Essential Rules

A system of equations can be solved by the graphing method, the substitution method, or the addition method.

Chapter Review

1. Solve by substitution:
$4x + 7y = 3$
$x = y - 2$

2. Solve by graphing:
$3x + y = 3$
$x = 2$

3. Solve by the addition method:
$3x + 8y = -1$
$x - 2y = -5$

4. Solve by substitution:
$8x - y = 2$
$y = 5x + 1$

5. Solve by graphing:
$x + y = 2$
$x - y = 0$

6. Solve by the addition method:
$4x - y = 9$
$2x + 3y = -13$

7. Solve by substitution:
$-2x + y = -4$
$x = y + 1$

8. Solve by the addition method:
$8x - y = 25$
$32x - 4y = 100$

9. Solve by the addition method:
$5x - 15y = 30$
$2x + 6y = 0$

10. Solve by graphing:
$3x - y = 6$
$y = -3$

11. Solve by the addition method:
$7x - 2y = 0$
$2x + y = -11$

12. Solve by substitution:
$x - 5y = 4$
$y = x - 4$

13. Is $(-1, -3)$ a solution of the system
$5x + 4y = -17$
$2x - y = 1$?

14. Solve by the addition method:
$6x + 4y = -3$
$12x - 10y = -15$

15. Solve by the addition method:
$5x + 2y = -9$
$12x - 7y = 2$

16. Solve by graphing:
$x - 3y = 12$
$y = x - 6$

17. Solve by the addition method:
$5x + 7y = 21$
$20x + 28y = 63$

18. Solve by substitution:
$9x + 12y = -1$
$x - 4y = -1$

19. Solve by graphing:
$4x - 2y = 8$
$y = 2x - 4$

20. Solve by the addition method:
$3x + y = -2$
$-9x - 3y = 6$

21. Solve by the addition method:
$11x - 2y = 4$
$25x - 4y = 2$

22. Solve by substitution:
$4x + 3y = 12$
$y = -\frac{4}{3}x + 4$

23. Solve by graphing:
$y = -\frac{1}{4}x + 3$
$2x - y = 6$

24. Solve by the addition method:
$2x - y = 5$
$10x - 5y = 20$

25. Solve by substitution:
$6x + 5y = -2$
$y = 2x - 2$

26. Is $(-2, 0)$ a solution of the system
$-x + 9y = 2$
$6x - 4y = 12$?

27. Solve by the addition method:
$6x - 18y = 7$
$9x + 24y = 2$

28. Solve by substitution:
$12x - 9y = 18$
$y = \frac{4}{3}x - 3$

29. Solve by substitution:
$9x - y = -3$
$18x - y = 0$

30. Solve by graphing:
$x + 2y = 3$
$y = -\frac{1}{2}x + 1$

31. Solve by the addition method:
$7x - 9y = 9$
$3x - y = 1$

32. Solve by substitution:
$7x + 3y = -16$
$x - 2y = 5$

33. Solve by substitution:
$5x - 3y = 6$
$x - y = 2$

34. Solve by the addition method:
$6x + y = 12$
$9x + 2y = 18$

35. Solve by the addition method:
$$5x + 12y = 4$$
$$x + 6y = 8$$

36. Solve by substitution:
$$6x - y = 0$$
$$7x - y = 1$$

37. Flying with the wind, a plane can travel 800 mi in 4 h. Against the wind, the plane requires 5 h to fly the same distance. Find the rate of the plane in calm air and the rate of the wind.

38. Admission to a movie theater is $7 for adults and $5 for children. If the receipts from 200 tickets were $1120, how many adult's tickets and how many children's tickets were sold?

39. A canoeist traveling with the current traveled the 30 mi between two riverside campsites in 3 h. The return trip took 5 h. Find the rate of the canoeist in still water and the rate of the current.

40. A small wood carving company mailed 190 advertisements, some requiring 25¢ in postage and others requiring 45¢ in postage. If the total cost for the mailings was $59.50, how many mailings that required 25¢ were sent?

41. A boat traveling with the current went 48 km in 3 h. Against the current, the boat traveled 24 km in 2 h. Find the rate of the boat in calm water and the rate of the current.

42. A music shop sells some compact discs for $15 and some for $10. A customer spent $120 on 10 compact discs. How many at each price did the customer purchase?

43. With a tailwind, a flight crew flew 420 km in 3 h. Flying against the tailwind, the crew flew 440 km in 4 h. Find the rate of the plane in calm air and the rate of the wind.

44. A paddle boat can travel 4 mi downstream in 1 h. Paddling upstream, against the current, the paddle boat traveled 2 mi in 1 h. Find the rate of the boat in calm water and the rate of the current.

45. A silo contains a mixture of lentils and corn. If 50 bushels of lentils were added to this mixture, there would be twice as many bushels of lentils as corn. If 150 bushels of corn were added to the original mixture, there would be the same amount of corn as lentils. How many bushels of each are in the silo?

46. Flying with the wind, a small plane flew 360 mi in 3 h. Against the wind, the plane took 4 h to fly the same distance. Find the rate of the plane in calm air and the rate of the wind.

47. A coin purse contains 80¢ in nickels and dimes. If there are 12 coins in all, how many dimes are in the coin purse?

48. A pilot flying with the wind flew 2100 mi from one city to another in 6 h. The return trip against the wind took 7 h. Find the rate of the plane in calm air and the rate of the wind.

49. An investor buys 1500 shares of stock, some costing $6 per share and the rest costing $25 per share. If the total cost of the stock is $12,800, how many shares of each did the investor buy?

50. Rowing with the wind, a sculling team went 24 mi in 2 h. Rowing against the current, the team went 18 mi in 3 h. Find the rate of the sculling team in calm water and the rate of the current.

Chapter Test

1. Solve by substitution:
$4x - y = 11$
$y = 2x - 5$

2. Solve by the addition method:
$4x + 3y = 11$
$5x - 3y = 7$

3. Is $(-2, 3)$ a solution of the system
$2x + 5y = 11$
$x + 3y = 7$?

4. Solve by substitution:
$x = 2y + 3$
$3x - 2y = 5$

5. Solve by the addition method:
$2x - 5y = 6$
$4x + 3y = -1$

6. Solve by graphing:
$3x + 2y = 6$
$5x + 2y = 2$

7. Solve by substitution:
$4x + 2y = 3$
$y = -2x + 1$

8. Solve by substitution:
$3x + 5y = 1$
$2x - y = 5$

9. Solve by the addition method:
$7x + 3y = 11$
$2x - 5y = 9$

10. Solve by substitution:
$3x - 5y = 13$
$x + 3y = 1$

11. Solve by the addition method:
$5x + 6y = -7$
$3x + 4y = -5$

12. Is $(2, 1)$ a solution of the system
$3x - 2y = 8$
$4x + 5y = 3$?

13. Solve by substitution:
$3x - y = 5$
$y = 2x - 3$

14. Solve by the addition method:
$3x + 2y = 2$
$5x - 2y = 14$

15. Solve by graphing:
$3x + 2y = 6$
$3x - 2y = 6$

16. Solve by substitution:
$x = 3y + 1$
$2x + 5y = 13$

17. Solve by the addition method:
$$5x + 4y = 7$$
$$3x - 2y = 13$$

18. Solve by graphing:
$$3x + 6y = 2$$
$$y = -\frac{1}{2}x + \frac{1}{3}$$

19. Solve by substitution:
$$4x - 3y = 1$$
$$2x + y = 3$$

20. Solve by the addition method:
$$5x - 3y = 29$$
$$4x + 7y = -5$$

21. Solve by substitution:
$$3x - 5y = -23$$
$$x + 2y = -4$$

22. Solve by the addition method:
$$9x - 2y = 17$$
$$5x + 3y = -7$$

23. With the wind, a plane flies 240 mi in 2 h. Against the wind, the plane requires 3 h to fly the same distance. Find the rate of the plane in calm air and the rate of the wind.

24. With the current, a motorboat can travel 48 mi in 3 h. Against the current, the boat requires 4 h to travel the same distance. Find the rate of the boat in calm water and the rate of the current.

25. Two coin banks contain only dimes and nickels. In the first bank, the total value of the coins is $5.50. In the second bank, there are one half as many dimes as in the first bank and 10 fewer nickels. The total value of the coins in the second bank is $3. Find the number of dimes and the number of nickels in the first bank.

Cumulative Review

1. Evaluate $\dfrac{a^2 - b^2}{2a}$ when $a = 4$ and $b = -2$.

2. Solve: $-\dfrac{3}{4}x = \dfrac{9}{8}$

3. Solve: $4 - 3(2 - 3x) = 7x - 9$

4. Solve: $3[2 - 4(x + 1)] = 6x - 2$

5. Simplify: $(2a^2 - 3a + 1)(2 - 3a)$

6. Simplify: $\dfrac{(-2x^2y)^4}{-8x^3y^2}$

7. Simplify: $(4b^2 - 8b + 4) \div (2b - 3)$

8. Simplify: $\dfrac{8x^{-2}y^5}{-2xy^{-4}}$

9. Factor: $4x^2y^4 - 64y^2$

10. Solve: $(x - 5)(x + 2) = -6$

11. Simplify: $\dfrac{x^2 - 6x + 8}{2x^3 + 6x^2} \div \dfrac{2x - 8}{4x^3 + 12x^2}$

12. Simplify: $\dfrac{x - 1}{x + 2} + \dfrac{2x + 1}{x^2 + x - 2}$

13. Simplify: $\dfrac{x + 4 - \dfrac{7}{x - 2}}{x + 8 + \dfrac{21}{x - 2}}$

14. Solve: $\dfrac{x}{2x - 3} + 2 = \dfrac{-7}{2x - 3}$

15. Solve $A = P + Prt$ for r.

16. Find the x- and y-intercepts of $3x - 6y = 12$.

17. Graph $3x - 2y = 6$.

18. Graph $x = 3$.

19. Find the slope of the line containing the points whose coordinates are $(2, -3)$ and $(-3, 4)$.

20. Find the equation of the line that contains the point whose coordinates are $(-2, 3)$ and has slope $-\dfrac{3}{2}$.

21. Is $(2, 0)$ a solution of the system
$5x - 3y = 10$
$4x + 7y = 8$?

22. Solve by substitution:
$2x - 3y = -7$
$x + 4y = 2$

23. Solve by graphing:
$2x + 3y = 6$
$3x + y = 2$

24. Solve by the addition method:
$5x - 2y = 8$
$4x + 3y = 11$

25. A total of $8750 is invested in two simple interest accounts. On one account, the annual simple interest rate is 9.6%. On the second account, the annual simple interest rate is 7.2%. How much should be invested in each account so that the total interest earned by each account is the same?

26. A passenger train leaves a train depot one-half hour after a freight train leaves the same depot. The freight train is traveling 8 mph slower than the passenger train. Find the rate of each train if the passenger train overtakes the freight train in 3 h.

27. The length of each side of a square is extended 4 in. The area of the resulting square is 144 in.^{2}. Find the length of a side of the original square.

28. A plane can travel 160 mph in calm air. Flying with the wind, the plane can fly 570 mi in the same amount of time as it takes to fly 390 mi against the wind. Find the rate of the wind.

29. With the current, a motorboat can travel 24 mi in 2 h. Against the current, the boat requires 3 h to travel the same distance. Find the rate of the boat in calm water.

30. Two coin banks contain only dimes and nickels. In the first bank, the total value of the coins is $5.25. In the second bank, there are one half as many dimes as in the first bank and 15 fewer nickels. The total value of the coins in the second bank is $2.50. Find the number of dimes in the first bank.

10

Inequalities

Objectives

- Write sets using the roster method
- Write sets using set builder notation
- Graph the solution set of an inequality on the number line
- Solve inequalities using the Addition Property of Inequalities
- Solve inequalities using the Multiplication Property of Inequalities
- Application problems
- Solve general inequalities
- Application problems
- Graph inequalities in two variables

Calculations of Pi

There are many early references of estimated values for pi. One of the earliest is from the Rhind Papyrus, which was found in Egypt in the 1800s. Scientists have estimated that these tablets were written around 1600 B.C. The Rhind Papyrus contains the estimate 3.1604 for pi.

One of the most famous calculations of pi was made around 240 B.C. by Archimedes. The calculation was based on finding the perimeter of inscribed and circumscribed six-sided polygons (or hexagons). Once the perimeter for a hexagon figure was calculated, known formulas could be used to calculate the perimeter of polygons with twice that number of sides. Continuing in this way, Archimedes calculated the perimeters for the polygons with 12, 24, 48, and 96 sides. His calculations resulted in a value of pi between $3\frac{10}{71}$ and $3\frac{1}{7}$. You might recognize $3\frac{1}{7}$, or $\frac{22}{7}$, as an approximation for pi still used today.

After Archimedes' work, calculations to improve the accuracy of pi were continued. One French mathematician, using Archimedes' method, estimated pi by using a polygon of 393,216 sides. A mathematician from the Netherlands estimated pi by using a polygon with over one million sides.

Around the 1650s, new mathematical methods were developed to estimate the value of pi. These methods started yielding estimates of pi that were accurate to over 70 places. By the 1850s an estimate for pi was accurate to 200 places.

Today, using more refined mathematical methods and computers, estimates of the value of pi now exceed one million places.

In 1914, an issue of *Scientific American* contained the following short note:

> "See, I have a rhyme assisting my feeble brain,
> its tasks oftimes resisting."

Can you see what this note has to do with the estimates for the value of pi?

(Each word length represents a digit in the approximation 3.141592653579.)

Sets

1 Write sets using the roster method

A **set** is a collection of objects. The objects in a set are called the **elements** of the set.

The **roster method** of writing sets encloses a list of the elements in braces.

The set of the last three letters of the alphabet is written {x, y, z}.

The set of the positive integers less than 5 is written {1, 2, 3, 4}.

Example 1 Use the roster method to write the set.
A. the set of integers between 0 and 10
B. the set of natural numbers
C. the set of odd positive integers

Solution A. $A = \{1, 2, 3, 4, 5, 6, 7, 8, 9\}$ ▶ A set is designated by a capital letter. Note that 0 and 10 are not elements of the set.

B. $A = \{1, 2, 3, 4,...\}$ ▶ The three dots mean that the pattern of numbers continues without end.

C. $A = \{1, 3, 5,...\}$

Problem 1 Use the roster method to write the set.
A. the set of odd positive integers less than 12
B. the set of even positive integers
C. the set of odd negative integers greater than −10

Solution See page A36.

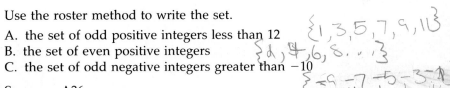

The symbol ε means "is an element of."

2 ε B is read "2 is an element of set B."

Given $A = \{3, 5, 9\}$, then 3 ε A, 5 ε A, and 9 ε A.

The **empty set,** or **null set,** is the set that contains no elements. The symbol Ø or { } is used to represent the empty set.

The set of people who have run a two-minute mile is the empty set.

The **union** of two sets, written $A \cup B$, is the set that contains the elements of A and the elements of B.

Given $A = \{1, 2, 3, 4\}$ and $B = \{3, 4, 5, 6\}$, the union of A and B contains all the elements of A and all the elements of B. The elements that are in both A and B are listed only once.

$A \cup B = \{1, 2, 3, 4, 5, 6\}$

The **intersection** of two sets, written $A \cap B$, is the set that contains the elements that are common to both A and B.

Given $A = \{1, 2, 3, 4\}$ and $B = \{3, 4, 5, 6\}$, the intersection of A and B contains the elements common to A and B.

$A \cap B = \{3, 4\}$

Example 2   Find $D \cup E$, given $D = \{6, 8, 10, 12\}$ and $E = \{-8, -6, 10, 12\}$.

Solution $D \cup E = \{-8, -6, 6, 8, 10, 12\}$ ▶ The union of D and E contains all the elements of D and all the elements of E.

Problem 2   Find $A \cup B$, given $A = \{-2, -1, 0, 1, 2\}$ and $B = \{0, 1, 2, 3, 4\}$.

Solution   See page A36.

$\{-2, -1, 0, 1, 2, 3, 4\}$

Example 3   Find $A \cap B$, given $A = \{5, 6, 9, 11\}$ and $B = \{5, 9, 13, 15\}$.

Solution $A \cap B = \{5, 9\}$ ▶ The intersection of A and B contains the elements common to A and B.

Problem 3   Find $A \cap B$, given $A = \{10, 12, 14, 16\}$ and $B = \{10, 16, 20, 26\}$.

Solution   See page A37.

Example 4   Find $A \cap B$, given $A = \{1, 2, 3, 4\}$ and $B = \{8, 9, 10, 11\}$.

Solution $A \cap B = \varnothing$

Problem 4   Find $A \cap B$, given $A = \{-5, -4, -3, -2\}$ and $B = \{2, 3, 4, 5\}$.

Solution   See page A37.

2 Write sets using set builder notation

Another method of representing sets is called **set builder notation.** Using this notation, the set of all positive integers less than 10 would be written

$\{x|x < 10,\ x$ is a positive integer$\}$,   read "the set of all x such that x is less than 10, x is a positive integer."

Using set builder notation, the set of real numbers greater than 4 would be written

$\{x|x > 4,\ x$ is a real number$\}$,   read "the set of all x such that x is greater than 4, x is a real number."

Example 5   Use set builder notation to write the set.
A. the set of negative integers greater than -100
B. the set of real numbers less than 60

Solution   A. $\{x|x > -100,\ x$ is a negative integer$\}$
B. $\{x|x < 60,\ x$ is a real number$\}$

Problem 5   Use set builder notation to write the set.
A. the set of positive even integers less than 59
B. the set of real numbers greater than -3

Solution   See page A37.

3 Graph the solution set of an inequality on the number line

An expression that contains the symbol $>$, $<$, $\geq$ (is greater than or equal to) or $\leq$ (is less than or equal to) is called an **inequality.** An inequality expresses the relative order of two mathematical expressions. The expressions can be either numerical or variable expressions.

$$\left.\begin{array}{l} 4 > 2 \\ 3x \leq 7 \\ x^2 - 2x > y + 4 \end{array}\right\} \text{Inequalities}$$

The **solution set of an inequality** is a set of real numbers and can be graphed on the number line.

The graph of the solution set of $x > 1$ is shown at the right. The solution set is the real numbers greater than 1. The circle on the graph indicates that 1 is not included in the solution set.

The graph of the solution set of $x \geq 1$ is shown at the right. The dot at 1 indicates that 1 is included in the solution set.

The graph of the solution set of $x < -1$ is shown at the right. The numbers less than -1 are to the left of -1 on the number line.

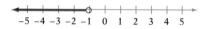

Example 6 Graph the solution set of $x < 3$.

Solution

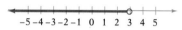

▶ The solution set is the numbers less than 3.

Problem 6 Graph the solution set of $x > -2$.

Solution See page A37.

The union of two sets is the set that contains all the elements of each set.

The graph of the set $\{x|x > 4\} \cup \{x|x < 1\}$ is shown at the right. The solution set is the numbers greater than 4 and the numbers less than 1.

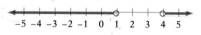

The intersection of two sets is the set that contains the elements common to both sets.

The graph of the set $\{x|x > -1\} \cap \{x|x < 2\}$ is shown at the right. The solution set is the numbers between -1 and 2.

Example 7 Graph.

A. $\{x|x > -2\} \cap \{x|x < 1\}$ B. $\{x|x \leq 5\} \cup \{x|x \geq -3\}$ C. $\{x|x > 3\} \cup \{x|x < 1\}$

Solution A.

▶ The set is the numbers between -2 and 1.

B.

▶ The set is the real numbers.

C.

▶ The set is the numbers greater than 3 and the numbers less than 1.

Problem 7   Graph.

A. $\{x|x > -1\} \cup \{x|x < -3\}$ B. $\{x|x < 5\} \cup \{x|x \geq -2\}$ C. $\{x|x \leq 4\} \cap \{x|x \geq -4\}$

Solution   See page A37.

EXERCISES 10.1

1   Use the roster method to write the set.

1. the integers between 15 and 22

2. the integers between -10 and -4

3. the odd integers between 8 and 18

4. the even integers between -11 and -1

5. the letters of the alphabet between a and d

6. the letters of the alphabet between p and v

7. all perfect squares less than 50

8. positive integers less than 20 that are divisible by 4

Find $A \cup B$.

9. $A = \{3, 4, 5\}$ $B = \{4, 5, 6\}$

10. $A = \{-3, -2, -1\}$ $B = \{-2, -1, 0\}$

11. $A = \{-10, -9, -8\}$ $B = \{8, 9, 10\}$

12. $A = \{a, b, c\}$ $B = \{x, y, z\}$

13. $A = \{a, b, d, e\}$ $B = \{c, d, e, f\}$

14. $A = \{m, n, p, q\}$ $B = \{m, n, o\}$

15. $A = \{1, 3, 7, 9\}$ $B = \{7, 9, 11, 13\}$

16. $A = \{-3, -2, -1\}$ $B = \{-1, 1, 2\}$

Find $A \cap B$.

17. $A = \{3, 4, 5\}$ $B = \{4, 5, 6\}$

18. $A = \{-4, -3, -2\}$ $B = \{-6, -5, -4\}$

19. $A = \{-4, -3, -2\}$ $B = \{2, 3, 4\}$

20. $A = \{1, 2, 3, 4\}$ $B = \{1, 2, 3, 4\}$

21. $A = \{a, b, c, d, e\}$ $B = \{c, d, e, f, g\}$

22. $A = \{m, n, o, p\}$ $B = \{k, l, m, n\}$

23. $A = \{1, 7, 9, 11\}$ $B = \{7, 11, 17\}$

24. $A = \{3, 6, 9, 12\}$ $B = \{6, 12, 18\}$

2   Use set builder notation to write the set.

25. the negative integers greater than −5

26. the positive integers less than 5

27. the integers greater than 30

28. the integers less than −70

29. the even integers greater than 5

30. the odd integers less than −2

31. the real numbers greater than 8

32. the real numbers less than 57

33. the real numbers greater than −5

34. the real numbers less than −63

3   Graph.

35. $x > 2$

36. $x \geq -1$

37. $x \leq 0$

38. $x < 4$

39. $\{x|x > -2\} \cup \{x|x < -4\}$

40. $\{x|x > 4\} \cup \{x|x < -2\}$

41. $\{x|x > -2\} \cap \{x|x < 4\}$

42. $\{x|x > -3\} \cap \{x|x < 3\}$

43. $\{x|x \geq -2\} \cup \{x|x < 4\}$

44. $\{x|x > 0\} \cup \{x|x \leq 4\}$

SUPPLEMENTAL EXERCISES 10.1

Use the roster method to write $A \cap B$.

45. $A = \{x|x < 15, x \text{ is a positive integer}\}$
$B = \{x|x > 10, x \text{ is a positive integer}\}$

46. $A = \{x|x > -9, x \text{ is a negative integer}\}$
$B = \{x|x < 0, x \text{ is a negative integer}\}$

Use set builder notation to write $A \cup B$.

47. $A = \{x|x > 8, x \text{ is a positive integer}\}$
$B = \{x|x > 20, x \text{ is a positive integer}\}$

48. $A = \{x|x > -6, x \text{ is a real number}\}$
$B = \{x|x > 1, x \text{ is a positive integer}\}$

Graph the solution set.

49. $\left\{x|x < -\frac{1}{2}\right\} \cap \left\{x|x \geq -\frac{5}{2}\right\}$

50. $\left\{x|x < \frac{9}{2}\right\} \cap \left\{x|x \geq -\frac{3}{2}\right\}$

51. $\left\{x|x > \frac{7}{2}\right\} \cup \left\{x|x < -\frac{3}{2}\right\}$

52. $\left\{x|x \leq \frac{5}{2}\right\} \cup \left\{x|x \geq -\frac{7}{2}\right\}$

53. $|x| < 3$

54. $|x| < 4$

55. $|x| > 2$ **56.** $|x| > 1$

SECTION **10.2**

The Addition and Multiplication Properties of Inequalities

1 Solve inequalities using the Addition Property of Inequalities

The **solution set of an inequality** is a set of numbers, each element of which, when substituted for the variable, results in a true inequality.

The inequality at the right is true if the variable is replaced by 7, 9.3, or $\frac{15}{2}$.

$$x + 3 > 8$$

$$\left. \begin{array}{l} 7 + 3 > 8 \\ 9.3 + 3 > 8 \\ \dfrac{15}{2} + 3 > 8 \end{array} \right\} \quad \text{True inequalities}$$

The inequality $x + 3 > 8$ is false if the variable is replaced by 4, 1.5, or $-\frac{1}{2}$.

$$\left. \begin{array}{l} 4 + 3 > 8 \\ 1.5 + 3 > 8 \\ -\dfrac{1}{2} + 3 > 8 \end{array} \right\} \quad \text{False inequalities}$$

There are many values of the variable x that make the inequality $x + 3 > 8$ true. The solution set of $x + 3 > 8$ is any number greater than 5.

The graph of the solution set of $x + 3 > 8$

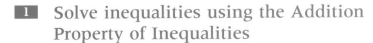

In solving an inequality, the goal is to rewrite the given inequality in the form *variable > constant* or *variable < constant*. The Addition Property of Inequalities is used to rewrite an inequality in this form.

Addition Property of Inequalities

The same number can be added to each side of an inequality without changing the solution set of the inequality.

If $a > b$, then $a + c > b + c$.
If $a < b$, then $a + c < b + c$.

The Addition Property of Inequalities also holds true for an inequality containing the symbol $\geq$ or $\leq$.

The Addition Property of Inequalities is used when, in order to rewrite an inequality in the form *variable > constant* or *variable < constant*, a term must be removed from one side of the inequality. Add the opposite of the term to each side of the inequality.

Because subtraction is defined in terms of addition, the Addition Property of Inequalities allows the same number to be subtracted from each side of an inequality without changing the solution set of the inequality.

To rewrite the inequality at the right, subtract 4 from each side of the inequality.
Simplify.

$$x + 4 < 5$$
$$x + 4 - 4 < 5 - 4$$
$$x < 1$$

The graph of the solution set of $x + 4 < 5$

$$-6\ -5\ -4\ -3\ -2\ -1\ \ 0\ \ 1\ \ 2\ \ 3\ \ 4\ \ 5\ \ 6$$

To solve $5x - 6 \le 4x - 4$, subtract $4x$ from each side of the inequality.
Simplify.

$$5x - 6 \le 4x - 4$$
$$5x - 4x - 6 \le 4x - 4x - 4$$
$$x - 6 \le -4$$

Add 6 to each side of the inequality.
Simplify.

$$x - 6 + 6 \le -4 + 6$$
$$x \le 2$$

Example 1 Solve and graph the solution set of $x + 5 > 3$.

Solution
$$x + 5 > 3$$
$$x + 5 - 5 > 3 - 5 \qquad \blacktriangleright \text{ Subtract 5 from each side of the inequality.}$$
$$x > -2$$

$$-5\ -4\ -3\ -2\ -1\ \ 0\ \ 1\ \ 2\ \ 3\ \ 4\ \ 5$$

Problem 1 Solve and graph the solution set of $x + 2 < -2$.

Solution See page A37.

Example 2 Solve: $7x - 14 \le 6x - 16$

Solution
$$7x - 14 \le 6x - 16$$
$$7x - 6x - 14 \le 6x - 6x - 16 \qquad \blacktriangleright \text{ Subtract } 6x \text{ from each side of the inequality.}$$
$$x - 14 \le -16$$
$$x - 14 + 14 \le -16 + 14 \qquad \blacktriangleright \text{ Add 14 to each side of the inequality.}$$
$$x \le -2$$

Problem 2   Solve: $5x + 3 > 4x + 5$

Solution   See page A37.

2 Solve inequalities using the
Multiplication Property of Inequalities

In solving an inequality, the goal is to rewrite the given inequality in the form *variable* > *constant* or *variable* < *constant*. The Multiplication Property of Inequalities is used when, in order to rewrite an inequality in this form, a coefficient must be removed from one side of the inequality.

Multiplication Property of Inequalities

Rule 1

Each side of an inequality can be multiplied by the same positive number without changing the solution set of the inequality.

If $a > b$ and $c > 0$, then $ac > bc$.   If $a < b$ and $c > 0$, then $ac < bc$.

$5 > 4$ $6 < 9$
$5(2) > 4(2)$ $6(3) < 9(3)$
$10 > 8$ A true inequality $18 < 27$ A true inequality

Multiplication Property of Inequalities

Rule 2

If each side of an inequality is multiplied by the same negative number and the inequality symbol is reversed, then the solution set of the inequality is not changed.

If $a > b$ and $c < 0$, then $ac < bc$.   If $a < b$ and $c < 0$, then $ac > bc$.

$5 > 4$ $6 < 9$
$5(-2) < 4(-2)$ $6(-3) > 9(-3)$
$-10 < -8$ A true inequality $-18 > -27$ A true inequality

The Multiplication Property of Inequalities also holds true for an inequality containing the symbol $\geq$ or $\leq$.

To rewrite the inequality at the right, multiply each side of the inequality by the reciprocal of the coefficient $-\frac{3}{2}$. Because $-\frac{2}{3}$ is a negative number, the inequality symbol must be reversed.

$$-\frac{3}{2}x \le 6$$

$$-\frac{2}{3}\left(-\frac{3}{2}x\right) \ge -\frac{2}{3}(6)$$

Simplify.

$$x \ge -4$$

The graph of the solution set of $-\frac{3}{2}x \le 6$

Recall that division is defined in terms of multiplication. Therefore, the Multiplication Property of Inequalities allows each side of an inequality to be divided by the same number. When each side of an inequality is divided by a positive number, the inequality symbol remains the same. When each side of an inequality is divided by a negative number, the inequality symbol must be reversed.

Solve: $-5x > 8$

$$-5x > 8$$

Divide each side of the inequality by -5. Because   is a negative number, the inequality symbol must be reversed.

$$\frac{-5x}{-5} < \frac{8}{-5}$$

Simplify.

$$x < -\frac{8}{5}$$

Example 3   Solve and graph the solution set of $7x < -14$.

Solution $7x < -14$

$\dfrac{7x}{7} < \dfrac{-14}{7}$ ▶ Divide each side of the inequality by 7.

$x < -2$

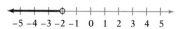

Problem 3   Solve and graph the solution set of $3x < 9$.

Solution   See page A37.

Example 4   Solve: $-\frac{5}{8}x \le \frac{5}{12}$

Solution $-\frac{5}{8}x \le \frac{5}{12}$

$-\frac{8}{5}\left(-\frac{5}{8}x\right) \ge -\frac{8}{5}\left(\frac{5}{12}\right)$ ▶ Multiply each side of the inequality by the reciprocal of $-\frac{5}{8}$. $-\frac{8}{5}$ is a negative number. Reverse the inequality symbol.

$x \ge -\frac{2}{3}$

Problem 4   Solve: $-\frac{3}{4}x \ge 18$

Solution   See page A37.

3 Application problems

Example 5   A student must have at least 450 points out of 500 points on five tests to receive an A in a course. One student's results on the first four tests were 93, 79, 87, and 94. What scores on the last test will enable this student to receive an A in the course?

Strategy   To find the scores, write and solve an inequality using N to represent the score on the last test.

Solution

Total number of points on the five tests	is greater than or equal to	450

$93 + 79 + 87 + 94 + N \ge 450$
$353 + N \ge 450$
$353 - 353 + N \ge 450 - 353$
$N \ge 97$

The student's score on the last test must be equal to or greater than 97.

Problem 5   An appliance dealer will make a profit on the sale of a television set if the cost of the new set is less than 70% of the selling price. What minimum selling price will enable the dealer to make a profit on a television set that costs the dealer $340?

Solution   See page A37.

EXERCISES 10.2

1 Solve and graph the solution set.

1. $x + 1 < 3$

2. $y + 2 < 2$

3. $x - 5 > -2$

4. $x - 3 > -2$

5. $n + 4 \geq 7$

6. $x + 5 \geq 3$

7. $x - 6 \leq -10$

8. $y - 8 \leq -11$

9. $5 + x \geq 4$

10. $-2 + n \geq 0$

Solve.

11. $y - 3 \geq -12$

12. $x + 8 \geq -14$

13. $3x - 5 < 2x + 7$

14. $5x + 4 < 4x - 10$

15. $8x - 7 \geq 7x - 2$

16. $3n - 9 \geq 2n - 8$

17. $2x + 4 < x - 7$

18. $9x + 7 < 8x - 7$

19. $4x - 8 \leq 2 + 3x$

20. $5b - 9 < 3 + 4b$

21. $6x + 4 \geq 5x - 2$

22. $7x - 3 \geq 6x - 2$

23. $2x - 12 > x - 10$

24. $3x + 9 > 2x + 7$

25. $d + \dfrac{1}{2} < \dfrac{1}{3}$

26. $x - \dfrac{3}{8} < \dfrac{5}{6}$

27. $x + \dfrac{5}{8} \geq -\dfrac{2}{3}$

28. $y + \dfrac{5}{12} \geq -\dfrac{3}{4}$

29. $x - \dfrac{3}{8} < \dfrac{1}{4}$

30. $y + \dfrac{5}{9} \leq \dfrac{5}{6}$

31. $2x - \dfrac{1}{2} < x + \dfrac{3}{4}$

32. $6x - \dfrac{1}{3} \leq 5x - \dfrac{1}{2}$

33. $3x + \dfrac{5}{8} > 2x + \dfrac{5}{6}$

34. $4b - \dfrac{7}{12} \geq 3b - \dfrac{9}{16}$

35. $3.8x < 2.8x - 3.8$

36. $1.2x < 0.2x - 7.3$

37. $x + 5.8 \leq 4.6$

38. $n - 3.82 \leq 3.95$

39. $x - 3.5 < 2.1$

40. $x - 0.23 \leq 0.47$

41. $1.33x - 1.62 > 0.33x - 3.1$

42. $2.49x + 1.35 \geq 1.49x - 3.45$

2   Solve and graph the solution set.

43. $3x < 12$

44. $8x \leq -24$

45. $5y \geq 15$

46. $24x > -48$

47. $16x \leq 16$

48. $3x > 0$

49. $-8x > 8$

50. $-2n \leq -8$

51. $-6b > 24$

52. $-4x < 8$

Solve.

53. $-5y \geq 20$

54. $3x < 5$

55. $7x > 2$

56. $-8x \leq -40$

57. $-6x \leq -40$

58. $10x > -25$

59. $-3x \geq \dfrac{6}{7}$

60. $-5x \geq \dfrac{10}{3}$

61. $-2x < -\dfrac{6}{7}$

62. $6x \leq -1$

63. $2x \leq -5$

64. $\dfrac{5}{6}n < 15$

65. $\dfrac{2}{3}x < -12$

66. $\dfrac{5}{6}x < -20$

67. $-\dfrac{3}{8}x < 6$

68. $\dfrac{3}{4}x < 12$

69. $\dfrac{2}{3}y \geq 4$

70. $\dfrac{5}{8}x \geq 10$

71. $-\dfrac{2}{3}x \leq 4$

72. $-\dfrac{3}{7}x \leq 6$

73. $-\dfrac{2}{11}b \geq -6$

74. $-\dfrac{4}{7}x \geq -12$

75. $\dfrac{2}{3}n < \dfrac{1}{2}$

76. $\dfrac{3}{5}x > \dfrac{7}{10}$

77. $-\dfrac{2}{3}x \geq \dfrac{4}{7}$

78. $-\dfrac{3}{8}x \geq \dfrac{9}{14}$

79. $-\dfrac{3}{5}x < -\dfrac{6}{7}$

80. $-\dfrac{4}{5}x < -\dfrac{8}{15}$

81. $-\dfrac{3}{4}y \geq -\dfrac{5}{8}$

82. $-\dfrac{8}{9}x \geq -\dfrac{16}{27}$

83. $\dfrac{2}{3}x \leq \dfrac{9}{14}$

84. $\dfrac{7}{12}x \leq \dfrac{9}{14}$

85. $-\dfrac{3}{5}y < \dfrac{9}{10}$

86. $-\frac{5}{12}y < \frac{1}{6}$

87. $\frac{2}{3}n \geq -\frac{8}{15}$

88. $\frac{5}{3}n \geq -\frac{9}{10}$

89. $\frac{21}{16}y \leq -\frac{3}{8}$

90. $\frac{9}{10}y \leq -\frac{12}{25}$

91. $-\frac{5}{18}n \geq \frac{16}{27}$

92. $-\frac{5}{7}x \geq -\frac{3}{8}$

93. $-\frac{7}{2}x \leq -\frac{6}{5}$

94. $\frac{7}{10}y > -\frac{4}{15}$

95. $\frac{12}{7}y < -\frac{12}{29}$

96. $-\frac{3}{35}t > -\frac{40}{49}$

97. $-0.27x < 0.135$

98. $-0.63x < 4.41$

99. $8.4y \geq -6.72$

100. $3.7y \geq -1.48$

101. $1.5x \leq 6.30$

102. $2.3x \leq 5.29$

103. $-3.5d > 7.35$

104. $-0.24x > 0.768$

105. $4.25m > -34$

106. $-3.9x \geq -19.5$

107. $0.035x < -0.0735$

108. $0.07x < -0.378$

109. $-11.7x \leq 4.68$

110. $0.685y \geq -2.1235$

111. $1.38n > -0.966$

112. $-5.24x < 41.92$

113. $-0.066b < 0.1914$

114. $13.58x \leq 95.06$

115. $18.92x < 264.88$

---

3   Solve.

116. Three-fifths of a number is greater than two-thirds. Find the smallest integer that satisfies the inequality.

117. The circumference of an official major league baseball is between 9.00 in. and 9.25 in. Find the possible diameters of a major league baseball to the nearest hundredth of an inch. Recall that $C = \pi d$. Use 3.14 for π.

118. To be eligible for a basketball tournament, a basketball team must win at least 60% of its remaining games. If the team has 17 games remaining, how many games must the team win to qualify for the tournament?

119.   To avoid a tax penalty, at least 90% of a self-employed person's total annual income tax liability must be paid by April 15. What amount of income tax must be paid by April 15 by a person with an annual income tax liability of $3500?

120.   Computer software engineers are fond of saying that software takes at least twice as long to develop as they think it will. Applying this saying, how many hours will it take to develop a software product that an engineer thinks can be finished in 50 h?

121.   A health official recommends a maximum cholesterol level of 220 units. A patient has a cholesterol level of 275. By how many units must this patient's cholesterol level be reduced to satisfy the recommended maximum level?

122.   A government agency recommends a minimum daily intake of 60 mg of vitamin C. For breakfast, you drink a glass of orange juice containing 10 mg of vitamin C. How many additional milligrams of vitamin C do you need that day to satisfy the recommended daily intake?

123.   A service organization will receive a bonus of $200 for collecting more than 1850 lb of aluminum cans during its four collection drives. On the first three drives, the organization collected 505 lb, 493 lb, and 412 lb. How many pounds of cans must the organization collect on the fourth drive to receive the bonus?

124.   A professor scores all tests with a maximum of 100 points. To earn an A in this course, a student must have an average of 92 on four tests. One student's grades on the first three tests were 89, 86, and 90. Can this student earn an A grade?

125.   A student must have an average of 80 points on five tests to receive a B in a course. The student's grades on the first four tests were 75, 83, 86, and 78. What scores on the last test will enable this student to receive a B in the course?

126.   A car sales representative receives a commission that is the greater of $250 or 8% of the selling price of a car. What dollar amounts in the sale price of a car will make the commission offer more attractive than the $250 fee?

127.   A sales representative for a stereo store has the option of a monthly salary of $2000 or a 35% commission on the selling price of each item sold by the representative. What dollar amounts in sales will make the commission more attractive than the monthly salary?

SUPPLEMENTAL EXERCISES 10.2

Use the roster method to list the set of positive integers that are solutions of the inequality.

128. $3x < 14$

129. $y - 3 < 4$

130. $-\dfrac{2}{3}x > -4$

131. $7y - 5 \le 6y - 3$

132. $-\dfrac{3}{5}n \ge -\dfrac{15}{9}$

133. $b + 3.5 \le 8.2$

Given $a > b$, and a and b are real numbers, for which real numbers c is the statement true? Use set builder notation to write the answer.

134. $ac > bc$

135. $ac < bc$

136. $\dfrac{a}{c} > \dfrac{b}{c}$

137. $\dfrac{a}{c} < \dfrac{b}{c}$

SECTION 10.3
General Inequalities

1 Solve general inequalities

In solving an inequality, it is often necessary to apply both the Addition and Multiplication Properties of Inequalities.

Solve: $3x - 2 < 5x + 4$

$$3x - 2 < 5x + 4$$

Subtract $5x$ from each side of the inequality. Simplify.

$$3x - 5x - 2 < 5x - 5x + 4$$
$$-2x - 2 < 4$$

Add 2 to each side of the inequality. Simplify.

$$-2x - 2 + 2 < 4 + 2$$
$$-2x < 6$$

Divide each side of the inequality by -2. Because   is a negative number, the inequality symbol must be reversed. Simplify.

$$\dfrac{-2x}{-2} > \dfrac{6}{-2}$$
$$x > -3$$

Example 1   Solve: $7x - 3 \le 3x + 17$

Solution
$$7x - 3 \le 3x + 17$$
$$7x - 3x - 3 \le 3x - 3x + 17$$ ▶ Subtract $3x$ from each side of the inequality.
$$4x - 3 \le 17$$
$$4x - 3 + 3 \le 17 + 3$$ ▶ Add 3 to each side of the inequality.
$$4x \le 20$$
$$\frac{4x}{4} \le \frac{20}{4}$$ ▶ Divide each side of the inequality by 4.
$$x \le 5$$

Problem 1   Solve: $5 - 4x > 9 - 8x$

Solution   See page A38.

When an inequality contains parentheses, one of the steps in solving the inequality requires the use of the Distributive Property.

To solve $-2(x - 7) > 3 - 4(2x - 3)$, use the Distributive Property to remove parentheses. Simplify.	$-2(x - 7) > 3 - 4(2x - 3)$ $-2x + 14 > 3 - 8x + 12$ $-2x + 14 > 15 - 8x$
Add $8x$ to each side of the inequality. Simplify.	$-2x + 8x + 14 > 15 - 8x + 8x$ $6x + 14 > 15$
Subtract 14 from each side of the inequality. Simplify.	$6x + 14 - 14 > 15 - 14$ $6x > 1$
Divide each side of the inequality by 6.	$\dfrac{6x}{6} > \dfrac{1}{6}$
Simplify.	$x > \dfrac{1}{6}$

Example 2   Solve: $3(3 - 2x) \ge -5x - 2(3 - x)$

Solution
$$3(3 - 2x) \ge -5x - 2(3 - x)$$
$$9 - 6x \ge -5x - 6 + 2x$$ ▶ Use the Distributive Property.
$$9 - 6x \ge -3x - 6$$
$$9 - 6x + 3x \ge -3x + 3x - 6$$ ▶ Add $3x$ to each side of the inequality.
$$9 - 3x \ge -6$$
$$9 - 9 - 3x \ge -6 - 9$$ ▶ Subtract 9 from each side of the inequality.
$$-3x \ge -15$$
$$\frac{-3x}{-3} \le \frac{-15}{-3}$$ ▶ Divide each side of the inequality by -3. Reverse the inequality symbol.
$$x \le 5$$

Problem 2   Solve: $8 - 4(3x + 5) \le 6(x - 8)$

Solution   See page A38.

---

2   Application problems

Example 3   The base of a triangle is 8 in., and the height is $(3x - 5)$ in. Express as an integer the maximum height of the triangle when the area is less than 112 in.^{2}. (The area of a triangle is equal to one-half the base times the height.)

Strategy   To find the maximum height:
- Replace the variables in the area formula by the given values and solve for x.
- Replace the variable in the expression $3x - 5$ with the value found for x.

Solution

One half the base times the height	is less than	112 in.^{2}

$$\frac{1}{2}(8)(3x - 5) < 112$$
$$4(3x - 5) < 112$$
$$12x - 20 < 112$$
$$12x - 20 + 20 < 112 + 20$$
$$12x < 132$$
$$\frac{12x}{12} < \frac{132}{12}$$
$$x < 11$$

$3x - 5 < 28$ ▶ Substitute the value of x into the variable expression for the height. $3x - 5 = 3(11) - 5 = 28$. Note that the height <u>is less than</u> 28 because $x < 11$.

The maximum height of the triangle is 27 in.

Problem 3   Company A rents cars for $9 a day and 10¢ for every mile driven. Company B rents cars for $12 a day and 8¢ per mile driven. You want to rent a car for one week. What is the maximum number of miles you can drive a Company A car if it is to cost you less than a Company B car?

Solution   See page A38.

EXERCISES 10.3

1   Solve.

1. $4x - 8 < 2x$

2. $7x - 4 < 3x$

3. $2x - 8 > 4x$

4. $3y + 2 > 7y$

5. $8 - 3x \le 5x$

6. $10 - 3x \le 7x$

7. $3x + 2 \ge 5x - 8$

8. $2n - 9 \ge 5n + 4$

9. $5x - 2 < 3x - 2$

10. $8x - 9 > 3x - 9$

11. $0.1(180 + x) > x$

12. $x > 0.2(50 + x)$

13. $0.15x + 55 > 0.10x + 80$

14. $-3.6b + 16 < 2.8b + 25.6$

15. $2(3x - 1) > 3x + 4$

16. $5(2x + 7) > -4x - 7$

17. $3(2x - 5) \ge 8x - 5$

18. $5x - 8 \ge 7x - 9$

19. $2(2y - 5) \le 3(5 - 2y)$

20. $2(5x - 8) \le 7(x - 3)$

21. $5(2 - x) > 3(2x - 5)$

22. $4(3d - 1) > 3(2 - 5d)$

23. $5(x - 2) > 9x - 3(2x - 4)$

24. $3x - 2(3x - 5) > 4(2x - 1)$

25. $4 - 3(3 - n) \le 3(2 - 5n)$

26. $15 - 5(3 - 2x) \le 4(x - 3)$

27. $2x - 3(x - 4) \ge 4 - 2(x - 7)$

28. $4 + 2(3 - 2y) \le 4(3y - 5) - 6y$

29. $\frac{1}{2}(9x - 10) \le -\frac{1}{3}(12 - 6x)$

30. $\frac{1}{4}(8 - 12d) < \frac{2}{5}(10d + 15)$

31. $\frac{2}{3}(9t - 15) + 4 < 6 + \frac{3}{4}(4 - 12t)$

32. $\frac{3}{8}(16 - 8c) - 9 \ge \frac{3}{5}(10c - 15) + 7$

33. $3[4(n - 2) - (1 - n)] > 5(n - 4)$

34. $2(m + 7) \le 4[3(m - 2) - 5(1 + m)]$

2   Solve.

35. Four times the sum of a number and five is less than six times the number. Find the smallest integer that satisfies the inequality.

36. The sales agent for a jewelry company is offered a flat monthly salary of $3200 or a salary of $1000 plus an 11% commission on the selling price of each item sold by the agent. If the agent chooses the $3200 salary, what dollar amount does the agent expect to sell in one month?

37. A baseball player is offered an annual salary of $200,000 or a base salary of $100,000 plus a bonus of $1000 for each hit over 100 hits. How many hits must the baseball player make to earn more than $200,000?

38. A computer bulletin board service charges a flat fee of $10 per month or $4 per month plus $.10 for each minute the service is used. For how many minutes must a person use this service for the cost to exceed $10?

39. A site licensing fee for a computer program is $1500. The fee allows a company to use the program at any computer terminal within the company. Alternatively, a company can choose to pay $200 for each computer terminal it has. How many computer terminals must a company have for the site licensing fee to be the more economical plan?

40. For a product to be labeled orange juice, a state agency requires that at least 80% of the drink be real orange juice. How many ounces of artificial flavors can be added to 32 oz of real orange juice if the drink is to be labeled orange juice?

41. Grade A hamburger cannot contain more than 20% fat. How much fat can a butcher mix with 300 lb of lean meat to meet the 20% requirement?

42. A shuttle service taking skiers to a ski area charges $8 per person each way. Four skiers are debating whether to take the shuttle bus or rent a car for $45 plus $.25 per mile. Assuming that the skiers will share the cost of the car and that they want the least expensive method of transportation, how far away is the ski area if they choose to take the shuttle service?

43. A residential water bill is based on a flat fee of $10 plus a charge of $.75 for each 1000 gal of water used. Find the number of gallons of water a family can use and have a monthly water bill that is less than $55.

44. Company A rents cars for $25 per day and 8¢ for every mile driven. Company B rents cars for $15 per day and 14¢ for every mile driven. You want to rent a car for one day. Find the maximum number of miles you can drive a Company B car if it is to cost you less than a Company A car.

45. A maintenance crew requires between 30 min and 45 min to prepare an aircraft for its next flight. How many aircraft can this crew prepare for flight in a 6-h period of time?

46. A rectangle is 8 ft wide and $(2x + 7)$ ft long. Express as an integer the maximum length of the rectangle when the area is less than 152 ft^2.

47. Find three positive consecutive odd integers such that three times the sum of the first two is less than four times the third integer. (*Hint:* There is more than one solution.)

48. Find three positive consecutive even integers such that four times the sum of the first two is less than or equal to five times the third integer. (*Hint:* There is more than one solution.)

SUPPLEMENTAL EXERCISES 10.3

Place the correct symbol, $<$ or $>$, between the given two numbers and between the opposites of the two numbers. (Example: Given 6   9, $6 < 9$ and $-6 > -9$.)

49. 3   5

50. -2 7

51. -8 -4

52. -1 0

53. 19   36

54. -77 -25

Based on the answers to Exercises 49–54, use an inequality symbol to complete the statement.

55. If a and b are real numbers, and $a < b$ is true, then $-a$ _____ $-b$ is true.

Use the roster method to list the set of positive integers that are solutions of the inequality.

56. $3x + 4 \le 13$

57. $4y < 6 + 2y$

58. $7 - 2b \le 15 - 5b$

59. $13 - 8a \ge 2 - 6a$

60. $2(2c - 3) < 5(6 - c)$

61. $-6(2 - d) \ge 4(4d - 9)$

Use the roster method to list the set of integers that are common to the solution sets of the two inequalities.

62. $5x - 12 \le x + 8$
     $3x - 4 \ge 2 + x$

63. $6x - 5 > 9x - 2$
     $5x - 6 < 8x + 9$

64. $4(x - 2) \le 3x + 5$
     $7(x - 3) \ge 5x - 1$

65. $3(x + 2) < 2(x + 4)$
     $4(x + 5) > 3(x + 6)$

Solve.

66. The charges for a long-distance telephone call are $1.21 for the first three minutes and $.42 for each additional minute or fraction of a minute. What is the largest whole number of minutes a call could last if it is to cost you less than $6?

67. Your class decides to publish a calendar to raise money. The initial cost, regardless of the number of calendars printed, is $900. After the initial cost, each calendar costs $1.50 to produce. What is the minimum number of calendars your class must sell at $6 per calendar to make a profit of at least $1200?

SECTION 10.4
Graphing Linear Inequalities

1 Graph inequalities in two variables

The graph of the linear equation $y = x - 2$ separates a plane into three sets:

the set of points on the line,

the set of points above the line,

the set of points below the line.

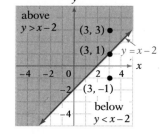

The point whose coordinates are (3, 1) is a solution of $y = x - 2$.

$$y = x - 2$$
$$1 \mid 3 - 2$$
$$1 = 1$$

The point whose coordinates are (3, 3) is a solution of $y > x - 2$.

$$y > x - 2$$
$$3 \mid 3 - 2$$
$$3 > 1$$

Any point above the line is a solution of $y > x - 2$.

The point whose coordinates are (3, −1) is a solution of $y < x - 2$.

$$y < x - 2$$
$$-1 \mid 3 - 2$$
$$-1 < 1$$

Any point below the line is a solution of $y < x - 2$.

The solution set of $y = x - 2$ is all points on the line. The solution set of $y > x - 2$ is all points above the line. The solution set of $y < x - 2$ is all points below the line. The solution set of an inequality in two variables is a **half plane**.

The following illustrates the procedure for graphing a linear inequality.

Graph the solution set of $2x + 3y \leq 6$.

Solve the inequality for y.

$$2x + 3y \leq 6$$
$$2x - 2x + 3y \leq -2x + 6$$
$$3y \leq -2x + 6$$
$$\frac{3y}{3} \leq \frac{-2x + 6}{3}$$
$$y \leq -\frac{2}{3}x + 2$$

Change the inequality to an equality and graph the line. If the inequality is $\geq$ or $\leq$, the line is in the solution set and is shown by a **solid line.** If the inequality is $>$ or $<$, the line is not a part of the solution set and is shown by a **dotted line.**

$$y = -\frac{2}{3}x + 2$$

If the inequality is $>$ or $\geq$, shade the **upper half plane.** If the inequality is $<$ or $\leq$, shade the **lower half plane.**

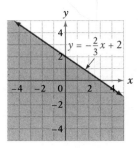

Example 1   Graph the solution set of $3x + y > -2$.

Solution
$$3x + y > -2$$
$$3x - 3x + y > -3x - 2$$
$$y > -3x - 2$$

▶ Solve the inequality for y.

▶ Graph $y = -3x - 2$ as a dotted line. Shade the upper half plane.

Problem 1   Graph the solution set of $x - 3y < 2$.

Solution   See page A38.

Example 2   Graph the solution set of $y > 3$.

Solution   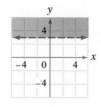   ▶ The inequality is solved for y.
Graph $y = 3$ as a dashed line.
Shade the upper half plane.

Problem 2   Graph the solution set of $x < 3$.

Solution   See page A38.

EXERCISES 10.4

1   Graph the solution set.

1. $x + y > 4$ **2.** $x - y > -3$ **3.** $2x - y < -3$ **4.** $3x - y < 9$

5. $2x + y \geq 4$ **6.** $3x + y \geq 6$ **7.** $y \leq -2$ **8.** $y > 3$

9. $3x - 2y < 8$ **10.** $5x + 4y > 4$ **11.** $-3x - 4y \geq 4$ **12.** $-5x - 2y \geq 8$

13. $6x + 5y \leq -10$ **14.** $2x + 2y \leq -4$ **15.** $-4x + 3y < -12$ **16.** $-4x + 5y < 15$

SUPPLEMENTAL EXERCISES 10.4

Write the inequality given its graph.

17.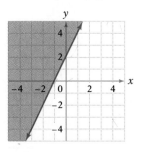

18.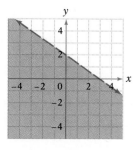

19.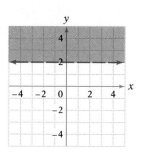

20.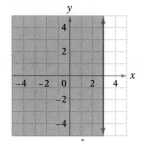

Graph the solution set.

21. $y - 5 < 4(x - 2)$

22. $y + 3 < 6(x + 1)$

23. $\dfrac{x}{4} + \dfrac{y}{2} > 1$

24. $\dfrac{x}{6} - \dfrac{y}{3} > 1$

25. $3x - 2(y + 1) \le y - (5 - x)$

26. $2x - 3(y + 1) \ge y - (4 - x)$

Calculators and Computers

First-Degree Inequalities in One Variable

The program FIRST-DEGREE INEQUALITIES on the Math ACE Disk provides inequalities for you to solve. There are three levels of difficulty. The first level is the easiest, and the third level is the most difficult.

Once you choose the level of difficulty, the program will display a problem. Using paper and pencil, solve the inequality. When you are ready, press the RETURN key, and compare your solution with the one on the screen.

After you complete a problem, you may continue practicing at the same level, return to the menu and select a different level, or quit the program.

Something Extra

Measurements as Approximations

From arithmetic, you know the rules for rounding decimals.

If the digit to the right of the given place value is less than 5, that digit and all digits to the right are dropped.

> 6.31 rounded to the nearest tenth is 6.3.

If the digit to the right of the given place value is greater than or equal to 5, increase the given place value by one and drop all digits to its right.

> 6.28 rounded to the nearest tenth is 6.3.

Given the rules for rounding numbers, what range of values can the number 6.3 represent? The smallest possible value of 6.3 is 6.25; any number smaller than that would not have been rounded up to 6.3. What about the largest possible value of 6.3? The number 6.34 would be rounded down to 6.3. So would the numbers 6.349, 6.3499, 6.349999, and so on. Therefore, we cannot name the largest possible value of 6.3. We can say that the number must be less than 6.35. Any number less than 6.35 would be rounded down to 6.3. The exact value of 6.3 is greater than or equal to 6.25 and less than 6.35.

The dimensions of a rectangle are given as 4.3 cm by 3.2 cm. Using the smallest and the largest possible values of the length and of the width, we can represent the possible values of the area, A, of the rectangle as follows:

$$4.25(3.15) \le A < 4.35(3.25)$$
$$13.3875 \quad \le A < 14.1375$$

The area is greater than or equal to 13.3875 cm^2 and less than 14.1375 cm^2.

Solve.

1. The measurements of the three sides of a triangle are given as 8.37 m, 5.42 m, and 9.61 m. Find the possible lengths of the perimeter of the triangle.

2. The length of a side of a square is given as 4.7 cm. Find the possible values for the area of the square.

3. What are the possible values for the area of a rectangle whose dimensions are 6.5 cm by 7.8 cm?

4. The length of a box is 40 cm, the width is 25 cm, and the height is 8 cm. What are the possible values of the volume of the box?

Chapter Summary

Key Words

A *set* is a collection of objects. The objects of a set are called the *elements* of the set.

The *empty set*, or *null set*, written Ø or { }, is the set that contains no elements.

The *union* of two sets, written $A \cup B$, is the set that contains all the elements of A and all the elements of B. (The elements that are in both set A and set B are listed only once.)

The *intersection* of two sets, written $A \cap B$, is the set that contains the elements common to both A and B.

An *inequality* is an expression that contains the symbol $<$, $>$, $\leq$, or $\geq$.

The *solution set of an inequality* is a set of numbers, each element of which, when substituted for the variable, results in a true inequality.

Essential Rules

Addition Property of Inequalities
The same number can be added to each side of an inequality without changing the solution set of the inequality.

If $a > b$, then $a + c > b + c$.
If $a < b$, then $a + c < b + c$.

Multiplication Property of Inequalities

Rule 1 Each side of an inequality can be multiplied by the same **positive number** without changing the solution set of the inequality.

If $a > b$ and $c > 0$, then $ac > bc$.
If $a < b$ and $c > 0$, then $ac < bc$.

Rule 2 If each side of an inequality is multiplied by the same **negative number** and the inequality symbol is reversed, then the solution set of the inequality is not changed.

If $a > b$ and $c < 0$, then $ac < bc$.
If $a < b$ and $c < 0$, then $ac > bc$.

Chapter Review

1. Use set builder notation to write the set of odd integers greater than -8.

2. Use the roster method to write the set of odd positive integers less than 9.

3. Solve: $3x + 17 < 2x - 1$

4. Find $A \cup B$, given $A = \{6, 8, 10\}$ and $B = \{2, 4, 6\}$.

5. Graph the solution set of $x \leq -2$.

6. Solve and graph the solution set of $x - 3 > -1$.

7. Graph the set $\{x | x > -1\} \cap \{x | x \leq 2\}$.

8. Solve and graph the solution set of $2 + x < -2$.

9. Solve: $7x - 2(x + 3) \geq x + 10$

10. Solve: $\frac{5}{9}x < 1$

11. Use set builder notation to write the set of even integers less than 5.

12. Solve: $12 - 4(x - 1) \leq 5(x - 4)$

13. Graph the solution set of $3x + 2y \leq 12$.

14. Graph the solution set of $6x - y > 6$.

15. Find $A \cap B$, given $A = \{0, 2, 4, 6, 8\}$ and $B = \{-2, -4\}$.

16. Use set builder notation to write the set of real numbers less than 14.

17. Solve and graph the solution set of $5x \leq -10$.

18. Graph the set $\{x|x < 5\} \cap \{x|x \geq -2\}$.

19. Solve: $6x - 9 < 4x + 3(x + 3)$

20. Solve: $8x \geq -3$

21. Graph the solution set of $x > 3$.

22. Solve and graph the solution set of $3x > -12$.

23. Solve: $5x - 4 \geq 4x + 8$

24. Use the roster method to write the set of negative integers greater than or equal to -4.

25. Solve: $4x - 12 < x + 24$

26. Find $C \cap D$, given $C = \{1, 5, 9, 13\}$ and $D = \{1, 3, 5, 7, 9\}$.

27. Graph the set $\{x|x > 0\} \cup \{x|x > -3\}$.

28. Solve: $-\dfrac{3}{4}x > \dfrac{2}{3}$

29. Graph the set $\{x|x < 2\} \cup \{x|x > 5\}$.

30. Solve and graph the solution set of $x - 4 \le -3$.

31. Solve: $2x - 3 > x + 15$

32. Solve: $5x - 6 > 19$

33. Use set builder notation to write the set of positive integers less than 20.

34. Solve: $-15x \le 45$

35. Graph the solution set of $5x + 2y < 6$.

36. Graph the solution set of $y > 2$.

37. Solve: $5 - 4(x + 9) > 11(12x - 9)$

38. Solve and graph the solution set of $4x \ge 16$.

39. Solve: $3x + 4 \ge -8$

40. Find $E \cup F$, given $E = \{1, 2, 3\}$ and $F = \{1, 2, 3, 4, 5\}$.

41. Use the roster method to write the set of even integers between 5 and 13.

42. Use set builder notation to write the set of negative integers greater than -5.

43. Graph the set $\{x|x \le 3\} \cup \{x|x > -1\}$.

44. Graph the solution set of $x - y \le 5$.

45. The product of three and the difference between a number and four is less than one more than the product of five and the number. Find the smallest integer that satisfies the inequality.

46. Florist A charges a $3 delivery fee plus $21 per bouquet delivered. Florist B charges a $15 delivery fee plus $18 per bouquet delivered. An organization wants to send each resident of a small nursing home a bouquet for Valentine's Day. How many residents are in the nursing home if it is more economical for the organization to use Florist B?

47. A student's grades on five sociology exams were 68, 82, 90, 73, and 95. What is the lowest score this student can receive on the sixth exam and still have earned a total of at least 480 points?

48. A booster club had a checking account balance of $148.95. The club treasurer wrote checks for $43.50 and $127.69. How much money must be deposited in the account so that the balance is not below $5?

49. Three-fourths of a number is greater than or equal to negative sixteen. Find the smallest number that satisfies the inequality.

50. The width of a rectangle is 12 ft. The length is $(3x + 5)$ ft. Express as an integer the minimum length of the rectangle if the area is greater than 276 ft^2.

▰Chapter Test

1. Use set builder notation to write the set of real numbers greater than -23.

2. Solve: $x + \dfrac{1}{3} > \dfrac{5}{6}$

3. Graph the solution set of $x > -2$.

4. Solve and graph the solution set of $\dfrac{2}{3}x \geq 2$.

5. Solve: $6x - 3(2 - 3x) \leq 4(2x - 7)$

6. Solve: $3(x - 7) \geq 5x - 12$

7. Use the roster method to write the set of even positive integers between 3 and 9.

8. Solve: $-\dfrac{3}{8}x \leq 6$

9. Graph the solution set of $3x + y > 4$.

10. Graph the solution set of $4x - 5y \geq 15$.

11. Find $A \cap B$ given $A = \{6, 8, 10, 12\}$ and $B = \{12, 14, 16\}$.

12. Solve: $3(2x - 5) \geq 8x - 9$

13. Use set builder notation to write the set of positive integers less than 50.

14. Find $A \cup B$ given $A = \{0, 1, 2\}$ and $B = \{-2, -1, 0\}$.

15. Use the roster method to write the set of odd positive integers less than 10.

16. Solve: $15 - 3(5x - 7) < 2(7 - 2x)$

17. Graph the solution set of $\frac{3}{8}x > -\frac{3}{4}$.

18. Graph the set $\{x | x < 5\} \cap \{x | x > 0\}$.

19. Graph the set $\{x | x \geq 1\} \cup \{x | x < -2\}$.

20. Graph the solution set of $-2 + x \leq -3$.

21. Graph the solution set of $2x - y \geq 2$.

22. Five more than a number is less than -3. Find the largest integer that satisfies the inequality.

23. A rectangle is 15 ft long and $(2x - 4)$ ft wide. Express as an integer the maximum width of the rectangle when the area is less than 180 ft^2. (The area of a rectangle is equal to its length times its width.)

24. Three-fifths of a number is less than negative fifteen. Find the largest integer that satisfies the inequality.

25. Company A rents cars for $6 a day and 25¢ for every mile driven. Company B rents cars for $15 a day and 10¢ per mile. You want to rent a car for 6 days. What is the maximum number of miles you can drive a Company A car if it is to cost you less than a Company B car?

Cumulative Review

1. Simplify: $12 - 2(7 - 5)^2 \div 4$

2. Evaluate $a^2 - 4b$ when $a = -2$ and $b = 3$.

3. Simplify: $2[5a - 3(2 - 5a) - 8]$

4. Solve: $\frac{5}{8} - 4x = \frac{1}{8}$

5. Solve: $2x - 3[x - 2(x - 3)] = 2$

6. Simplify: $(-3a)(-2a^3b^2)^2$

7. Simplify: $-3y^2(-2y^2 - 4y + 5)$

8. Simplify: $\frac{(27a^3b^2)^{-1}}{(-3ab^2)^{-3}}$

9. Simplify: $(16x^2 - 12x - 2) \div (4x - 1)$

10. Factor: $9x^2 + 48xy + 64y^2$

11. Factor: $4x^2 - 21x + 5$

12. Factor: $27a^2x^2 - 3a^2$

13. Simplify: $\frac{x^2 - 2x}{x^2 - 2x - 8} \div \frac{x^3 - 5x^2 + 6x}{x^2 - 7x + 12}$

14. Simplify: $\frac{4a}{2a - 3} - \frac{2a}{a + 3}$

15. Solve: $\frac{5y}{6} - \frac{5}{9} = \frac{y}{3} - \frac{5}{6}$

16. Solve $R = \frac{C - S}{t}$ for C.

17. Graph $y = 2x - 3$.

18. Graph $y = -2$.

19. Find the slope of the line containing the points whose coordinates are $(3, -4)$ and $(-2, 5)$.

20. Find the equation of the line that contains the point whose coordinates are $(1, -3)$ and has slope $-\frac{3}{2}$.

21. Solve by substitution:
$$x = 3y + 1$$
$$2x + 5y = 13$$

22. Solve by the addition method:
$$9x - 2y = 17$$
$$5x + 3y = -7$$

23. Graph the set $\{x|x > 2\} \cup \{x|x < 0\}$.

24. Graph the solution set of $\frac{4}{5}x \geq -4$.

25. Solve: $9 - 2(4x - 5) < 3(7 - 6x)$

26. Graph the solution set $x + 3y > 2$.

27. A file cabinet that normally sells for $99 is on sale for 20% off. Find the sale price.

28. A drawer contains 13¢ stamps and 18¢ stamps. The number of 13¢ stamps is two less than the number of 18¢ stamps. The total value of all the stamps is $2.53. How many 13¢ stamps are in the drawer?

29. In a lake, 100 fish are caught, tagged, and then released. Later, 150 fish are caught. Three of the 150 fish are found to have tags. Estimate the number of fish in the lake.

30. A rectangle is 6 ft wide and $(2x + 3)$ ft long. Express as an integer the maximum length of the rectangle when the area is less than 150 ft². (The area of a rectangle is equal to its length times its width.)

11

Radical Expressions

Objectives

- Simplify numerical radical expressions
- Simplify variable radical expressions
- Add and subtract radical expressions
- Multiply radical expressions
- Divide radical expressions
- Solve equations containing one or more radical expressions
- Application problems

A Table of Square Roots

The practice of finding the square root of a number has existed for at least two thousand years. Because the process of finding a square root is tedious and time-consuming, it is convenient to have tables of square roots. There is one such table in the back of this book.

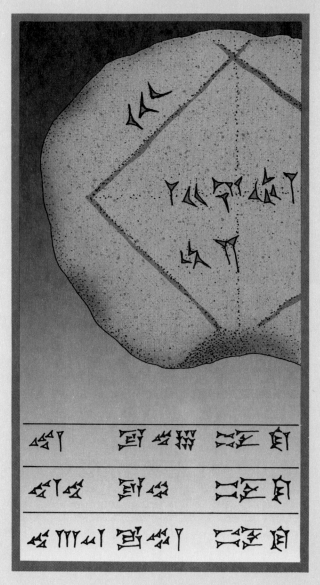

But this table is not the first square root table ever written (nor is it likely to be the last). The table shown here is part of an old Babylonian clay tablet that was written around 350 B.C. It is an incomplete table of square roots written in a style called *cuneiform*.

The number base of the Babylonians was 60 instead of 10, as we use today. The symbol **Y** was used for 1, and 10 was written as **◄**. Some examples of numbers using this system are given below.

$$\text{(symbol)} = 9 \qquad \text{(symbol)} = 40$$

A translation of the first couple of lines of the table is given next to that line. The number in parentheses is the equivalent base-10 number that would be used today. You might try to translate the third line. The answer appears at the bottom of this page.

$40 \times 60 + 1$ (= 2401), which is the square of 49.

$41 \times 60 + 40$ (= 2500), which is the square of 50.

Answer: $43 \times 60 + 21$ (= 2601), which is the square of 51.

Introduction to Radical Expressions

1 Simplify numerical radical expressions

A **square root** of a positive number x is a number whose square is x.

A square root of 16 is 4 because $4^2 = 16$.
A square root of 16 is -4 because $(-4)^2 = 16$.

Every positive number has two square roots, one a positive number and one a negative number. The symbol $\sqrt{}$, called a **radical,** is used to indicate the positive or **principal square root** of a number. For example, $\sqrt{16} = 4$ and $\sqrt{25} = 5$. The number under the radical sign is called the **radicand.**

When the negative square root of a number is to be found, a negative sign is placed in front of the radical. For example, $-\sqrt{16} = -4$ and $-\sqrt{25} = -5$.

The square of an integer is a **perfect square.** 49, 81, and 144 are examples of perfect squares.

$$7^2 = 49$$
$$9^2 = 81$$
$$12^2 = 144$$

An integer that is a perfect square can be written as the product of prime factors, each of which has an even exponent when expressed in exponential form.

$$49 = 7 \cdot 7 = 7^2$$
$$81 = 3 \cdot 3 \cdot 3 \cdot 3 = 3^4$$
$$144 = 2 \cdot 2 \cdot 2 \cdot 2 \cdot 3 \cdot 3 = 2^4 3^2$$

To find the square root of a perfect square written in exponential form, remove the radical sign, and divide each exponent by 2.

To simplify $\sqrt{625}$, write the prime factorization of the radicand in exponential form.

$$\sqrt{625} = \sqrt{5^4}$$

Remove the radical sign, and divide the exponent by 2.

$$= 5^2$$

Simplify.

$$= 25$$

The square root of a negative number is not a real number because the square of a real number is always positive.

$\sqrt{-4}$ is not a real number.

$\sqrt{-25}$ is not a real number.

If a number is not a perfect square, its square root can only be approximated. For example, 2 and 7 are not perfect squares. Thus their square roots can only be approximated. These numbers are **irrational numbers.** Their decimal representations never terminate or repeat.

$$\sqrt{2} \approx 1.4142135...$$

$$\sqrt{7} \approx 2.6457513...$$

The approximate square roots of the positive integers up to 200 can be found in the Appendix on page A2. The square roots have been rounded to the nearest thousandth.

A radical expression is in simplest form when the radicand contains no factor that is a perfect square. The Product Property of Square Roots is used to simplify radical expressions.

The Product Property of Square Roots

If a and b are positive real numbers, then $\sqrt{ab} = \sqrt{a} \cdot \sqrt{b}$.

To simplify $\sqrt{96}$, write the prime factorization of the radicand in exponential form.	$\sqrt{96} = \sqrt{2^5 \cdot 3}$
Write the radicand as a product of a perfect square and factors that do not contain a perfect square. Remember that a perfect square has an even exponent when expressed in exponential form.	$= \sqrt{2^4(2 \cdot 3)}$
Use the Product Property of Square Roots.	$= \sqrt{2^4} \sqrt{2 \cdot 3}$
Simplify.	$= 2^2 \sqrt{2 \cdot 3}$
	$= 4\sqrt{6}$

$\sqrt{96}$ and $4\sqrt{6}$ represent the same number.
$4\sqrt{6}$ is the simplest form of $\sqrt{96}$.

Example 1   Simplify: $3\sqrt{90}$

Solution $3\sqrt{90} = 3\sqrt{2 \cdot 3^2 \cdot 5}$ ▶ Write the prime factorization of the radicand in exponential form.

$= 3\sqrt{3^2(2 \cdot 5)}$ ▶ Write the radicand as a product of a perfect square and factors that do not contain a perfect square.

$= 3\sqrt{3^2}\sqrt{2 \cdot 5}$ ▶ Use the Product Property of Square Roots.

$= 3 \cdot 3\sqrt{10}$ ▶ Simplify.

$= 9\sqrt{10}$

Problem 1   Simplify: $-5\sqrt{32}$

  Solution   See page A39.

Example 2   Simplify: $\sqrt{252}$

  Solution $\sqrt{252} = \sqrt{2^2 \cdot 3^2 \cdot 7}$
    $= \sqrt{2^2 \cdot 3^2}\,\sqrt{7}$
    $= 2 \cdot 3\sqrt{7}$
    $= 6\sqrt{7}$

Problem 2   Simplify: $\sqrt{216}$

  Solution   See page A39.

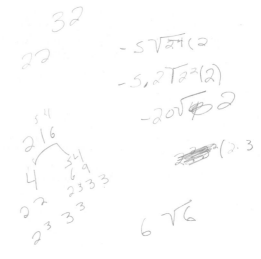

A radical expression can be simplified and then a decimal approximation found by using the table on page A2 or a calculator.

For example, to find the decimal approximation of $\sqrt{125}$ to the nearest thousandth, write the prime factorization of the radicand in exponential form.

$$\sqrt{125} = \sqrt{5^3}$$

Write the radicand as a product of a perfect square and factors that do not contain a perfect square.

$$= \sqrt{5^2 \cdot 5}$$

Use the Product Property of Square Roots.

$$= \sqrt{5^2}\,\sqrt{5}$$

Simplify.

$$= 5\sqrt{5}$$

Replace the radical expression by a decimal approximation.

$$\approx 5(2.236)$$

Simplify.

$$\approx 11.180$$

2   Simplify variable radical expressions

Variable expressions that contain radicals do not always represent real numbers.

The variable expression at the right does not represent a real number when x is a negative number, for example, -4.

$\sqrt{x^3}$

$\sqrt{(-4)^3} = \sqrt{-64}$ Not a real number

Now consider the expression $\sqrt{x^2}$ and evaluate this expression for $x = -2$ and $x = 2$.

$$\sqrt{x^2}$$
$$\sqrt{(-2)^2} = \sqrt{4} = 2 = |-2|$$
$$\sqrt{2^2} \quad = \sqrt{4} = 2 = |2|$$

This suggests the following.

The Square Root of a^2

For any real number a, $\sqrt{a^2} = |a|$.
If $a \geq 0$, then $\sqrt{a^2} = a$.

In order to avoid variable expressions that do not represent real numbers, and so that absolute value signs are not needed for certain expressions, the variables in this chapter will represent *positive* numbers unless otherwise stated.

A variable or a product of variables written in exponential form is a **perfect square** when each exponent is an even number.

To find the square root of a perfect square, remove the radical sign, and divide each exponent by 2.

For example, to simplify $\sqrt{a^6}$, remove the radical sign, and divide the exponent by 2.

$$\sqrt{a^6} = a^3$$

A variable expression is in simplest form when the radicand contains no factor that is a perfect square.

To simplify $\sqrt{x^7}$, write x^7 as the product of x and a perfect square.

$$\sqrt{x^7} = \sqrt{x^6 \cdot x}$$

Use the Product Property of Square Roots.

$$= \sqrt{x^6}\,\sqrt{x}$$

Simplify the perfect square.

$$= x^3\sqrt{x}$$

Example 3   Simplify: $\sqrt{b^{15}}$

Solution $\sqrt{b^{15}} = \sqrt{b^{14} \cdot b} = \sqrt{b^{14}} \cdot \sqrt{b} = b^7\sqrt{b}$

Problem 3   Simplify: $\sqrt{y^{19}}$

Solution   See page A39.

$\sqrt{y^{19}} = \sqrt{y^{18} \cdot b} = \sqrt{y^{18}} \cdot \sqrt{b} = y^9\sqrt{y}$

y^{19}

$y^{18} \cdot y$

$y^9 \sqrt{y}$

Simplify: $3x\sqrt{8x^3y^{13}}$

Write the prime factorization of the coefficient of the radicand in exponential form.

$$3x\sqrt{8x^3y^{13}} = 3x\sqrt{2^3x^3y^{13}}$$

Write the radicand as a product of a perfect square and factors that do not contain a perfect square.

$$= 3x\sqrt{2^2x^2y^{12}(2xy)}$$

Use the Product Property of Square Roots.

$$= 3x\sqrt{2^2x^2y^{12}}\sqrt{2xy}$$

Simplify.

$$= 3x \cdot 2xy^6\sqrt{2xy}$$

$$= 6x^2y^6\sqrt{2xy}$$

Example 4 Simplify.   A. $\sqrt{24x^5}$ B. $2a\sqrt{18a^3b^{10}}$

Solution   A. $\sqrt{24x^5} = \sqrt{2^3 \cdot 3 \cdot x^5}$ B. $2a\sqrt{18a^3b^{10}} = 2a\sqrt{2 \cdot 3^2 \cdot a^3 \cdot b^{10}}$
$$= \sqrt{2^2x^4(2 \cdot 3x)} \qquad\qquad\qquad = 2a\sqrt{3^2a^2b^{10}(2a)}$$
$$= \sqrt{2^2x^4}\sqrt{2 \cdot 3x} \qquad\qquad\qquad = 2a\sqrt{3^2a^2b^{10}}\sqrt{2a}$$
$$= 2x^2\sqrt{6x} \qquad\qquad\qquad\qquad = 2a \cdot 3ab^5\sqrt{2a}$$
$$\qquad\qquad\qquad\qquad\qquad\qquad = 6a^2b^5\sqrt{2a}$$

Problem 4 Simplify.   A. $\sqrt{45b^7}$ B. $3a\sqrt{28a^9b^{18}}$

Solution   See page A39. $3^2 b^{76}(5b)$

$3.\ b^3 \qquad 3b^3\sqrt{\cdot 5b}$

$3b^3\sqrt{5b}$

Simplify $\sqrt{25(x+2)^2}$.

Write the prime factorization of the coefficient in exponential form.

$$\sqrt{25(x+2)^2} = \sqrt{5^2(x+2)^2}$$

Simplify.

$$= 5(x+2)$$
$$= 5x + 10$$

Example 5 Simplify: $\sqrt{16(x+5)^2}$

Solution $\sqrt{16(x+5)^2} = \sqrt{2^4(x+5)^2} = 2^2(x+5) = 4(x+5) = 4x + 20$

Problem 5 Simplify: $\sqrt{25(a+3)^2}$

Solution See page A39. $25(a+3)^2$

$2\ 5^2(a+3)^2$

$5^2\ \overline{(a+3)}$

$\sqrt{a+3}$

$\sqrt{a+3}$

$\sqrt{5a+15}$

EXERCISES 11.1

1   Simplify.

1. $\sqrt{16}$ 4

2. $\sqrt{64}$ 8

3. $\sqrt{49}$ 7

4. $\sqrt{144}$ 12

5. $\sqrt{32}$ 2^5 $2^2\sqrt{2}$

6. $\sqrt{50}$

7. $\sqrt{8}$ $2^0\sqrt{2}$

8. $\sqrt{12}$ $2^2, 3$ $2\sqrt{3}$

9. $6\sqrt{18}$

10. $-3\sqrt{48}$

11. $5\sqrt{40}$

12. $2\sqrt{28}$

13. $\sqrt{15}$

14. $\sqrt{21}$

15. $\sqrt{29}$

16. $\sqrt{13}$

17. $-9\sqrt{72}$

18. $11\sqrt{80}$

19. $\sqrt{45}$

20. $\sqrt{225}$

21. $\sqrt{0}$

22. $\sqrt{210}$

23. $6\sqrt{128}$

24. $9\sqrt{288}$

25. $\sqrt{105}$

26. $\sqrt{55}$

27. $\sqrt{900}$

28. $\sqrt{300}$

29. $5\sqrt{180}$

30. $7\sqrt{98}$

31. $\sqrt{250}$

32. $\sqrt{120}$

33. $\sqrt{96}$

34. $\sqrt{160}$

35. $\sqrt{324}$

36. $\sqrt{444}$

Find the decimal approximation to the nearest thousandth.

37. $\sqrt{240}$

38. $\sqrt{300}$

39. $\sqrt{288}$

40. $\sqrt{600}$

41. $\sqrt{256}$

42. $\sqrt{324}$

43. $\sqrt{275}$

44. $\sqrt{450}$

45. $\sqrt{245}$

46. $\sqrt{525}$

47. $\sqrt{352}$

48. $\sqrt{363}$

2   Simplify.

49. $\sqrt{x^6}$ x^3

50. $\sqrt{x^{12}}$ x^6

51. $\sqrt{y^{15}}$ $y^7\sqrt{y}$

52. $\sqrt{y^{11}}$ $y^5\sqrt{y}$

53. $\sqrt{a^{20}}$

54. $\sqrt{a^{16}}$

55. $\sqrt{x^4y^4}$

56. $\sqrt{x^{12}y^8}$

57. $\sqrt{4x^4}$
58. $\sqrt{25y^8}$
59. $\sqrt{24x^2}$
60. $\sqrt{x^3y^{15}}$

61. $\sqrt{x^3y^7}$
62. $\sqrt{a^{15}b^5}$
63. $\sqrt{a^3b^{11}}$
64. $\sqrt{24y^7}$

65. $\sqrt{60x^5}$
66. $\sqrt{72y^7}$
67. $\sqrt{49a^4b^8}$
68. $\sqrt{144x^2y^8}$

69. $\sqrt{18x^5y^7}$
70. $\sqrt{32a^5b^{15}}$
71. $\sqrt{40x^{11}y^7}$
72. $\sqrt{72x^9y^3}$

73. $\sqrt{80a^9b^{10}}$
74. $\sqrt{96a^5b^7}$
75. $2\sqrt{16a^2b^3}$
76. $5\sqrt{25a^4b^7}$

77. $x\sqrt{x^4y^2}$
78. $y\sqrt{x^3y^6}$
79. $4\sqrt{20a^4b^7}$
80. $5\sqrt{12a^3b^4}$

81. $3x\sqrt{12x^2y^7}$
82. $4y\sqrt{18x^5y^4}$
83. $2x^2\sqrt{8x^2y^3}$
84. $3y^2\sqrt{27x^4y^3}$

85. $\sqrt{25(a+4)^2}$
86. $\sqrt{81(x+y)^4}$
87. $\sqrt{4(x+2)^4}$
88. $\sqrt{9(x+2)^2}$

89. $\sqrt{x^2+4x+4}$
90. $\sqrt{b^2+8b+16}$
91. $\sqrt{y^2+2y+1}$
92. $\sqrt{a^2+6a+9}$

SUPPLEMENTAL EXERCISES 11.1

Simplify.

93. $\sqrt{0.16a^4}$
94. $\sqrt{0.0025a^3b^5}$
95. $\dfrac{1}{5x}\sqrt{25x^3}$
96. $-\dfrac{1}{3}\sqrt{36x^2y^4}$

97. $-\dfrac{3y}{4}\sqrt{64x^4y^2}$
98. $\sqrt{a^2b+a^3}$
99. $\sqrt{x^2y^3+x^3y^2}$
100. $\sqrt{4a^5b^4-4a^4b^5}$

Solve.

101. Use the roster method to list the whole numbers between $\sqrt{8}$ and $\sqrt{90}$.

102. The Table of Square Roots on page A2 lists $\sqrt{3}$ as 1.732. Show that this value is not exact.

103. The area of a square is 76 cm². Find the length of a side of the square. Round to the nearest tenth.

Assuming x can be any real number, for what values of x is the radical expression a real number? Write the answer as an inequality or write "all real numbers."

104. $\sqrt{x}$ **105.** $\sqrt{4x}$ **106.** $\sqrt{x-2}$ **107.** $\sqrt{x+5}$

108. $\sqrt{6-4x}$ **109.** $\sqrt{5-2x}$ **110.** $\sqrt{x^2+7}$ **111.** $\sqrt{x^2+1}$

SECTION 11.2
Addition and Subtraction of Radical Expressions

1   Add and subtract radical expressions

The Distributive Property is used to simplify the sum or difference of radical expressions with the same radicand.

$$5\sqrt{2} + 3\sqrt{2} = (5+3)\sqrt{2} = 8\sqrt{2}$$
$$6\sqrt{2x} - 4\sqrt{2x} = (6-4)\sqrt{2x} = 2\sqrt{2x}$$

Radical expressions that are in simplest form and have different radicands cannot be simplified by the Distributive Property.

$2\sqrt{3} + 4\sqrt{2}$ cannot be simplified by the Distributive Property.

To simplify the sum or difference of radical expressions, simplify each term. Then use the Distributive Property.

Simplify $4\sqrt{8} - 10\sqrt{2}$.

Simplify each term.

$$\begin{aligned} 4\sqrt{8} - 10\sqrt{2} &= 4\sqrt{2^3} - 10\sqrt{2} \\ &= 4\sqrt{2^2 \cdot 2} - 10\sqrt{2} \\ &= 4\sqrt{2^2}\sqrt{2} - 10\sqrt{2} \\ &= 4 \cdot 2\sqrt{2} - 10\sqrt{2} \\ &= 8\sqrt{2} - 10\sqrt{2} \end{aligned}$$

Simplify the expression by using the Distributive Property.

$$\begin{aligned} &= (8-10)\sqrt{2} \\ &= -2\sqrt{2} \end{aligned}$$

Example 1 Simplify.   A. $5\sqrt{2} - 3\sqrt{2} + 12\sqrt{2}$ B. $3\sqrt{12} - 5\sqrt{27}$

Solution A. $5\sqrt{2} - 3\sqrt{2} + 12\sqrt{2} = (5 - 3 + 12)\sqrt{2}$ ▶ Use the Distributive Property.
$$= 14\sqrt{2}$$

B. $3\sqrt{12} - 5\sqrt{27} = 3\sqrt{2^2 \cdot 3} - 5\sqrt{3^3}$
$$= 3\sqrt{2^2 \cdot 3} - 5\sqrt{3^2 \cdot 3}$$
$$= 3\sqrt{2^2}\sqrt{3} - 5\sqrt{3^2}\sqrt{3}$$
$$= 3 \cdot 2\sqrt{3} - 5 \cdot 3\sqrt{3}$$
$$= 6\sqrt{3} - 15\sqrt{3}$$
$$= -9\sqrt{3}$$

Problem 1 Simplify.   A. $9\sqrt{3} + 3\sqrt{3} - 18\sqrt{3}$ B. $2\sqrt{50} - 5\sqrt{32}$

Solution See page A39.

To simplify $8\sqrt{18x} - 2\sqrt{32x}$, simplify each term. Then subtract the radical expressions.

$$8\sqrt{18x} - 2\sqrt{32x} = 8\sqrt{2 \cdot 3^2 x} - 2\sqrt{2^5 x}$$
$$= 8\sqrt{3^2}\sqrt{2x} - 2\sqrt{2^4}\sqrt{2x}$$
$$= 8 \cdot 3\sqrt{2x} - 2 \cdot 2^2\sqrt{2x}$$
$$= 24\sqrt{2x} - 8\sqrt{2x}$$
$$= 16\sqrt{2x}$$

Example 2 Simplify.   A. $3\sqrt{12x^3} - 2x\sqrt{3x}$ B. $2x\sqrt{8y} - 3\sqrt{2x^2 y} + 2\sqrt{32x^2 y}$

Solution A. $3\sqrt{12x^3} - 2x\sqrt{3x} = 3\sqrt{2^2 \cdot 3 \cdot x^3} - 2x\sqrt{3x}$
$$= 3\sqrt{2^2 \cdot x^2}\sqrt{3x} - 2x\sqrt{3x}$$
$$= 3 \cdot 2x\sqrt{3x} - 2x\sqrt{3x}$$
$$= 6x\sqrt{3x} - 2x\sqrt{3x}$$
$$= 4x\sqrt{3x}$$

B. $2x\sqrt{8y} - 3\sqrt{2x^2 y} + 2\sqrt{32x^2 y} = 2x\sqrt{2^3 y} - 3\sqrt{2x^2 y} + 2\sqrt{2^5 x^2 y}$
$$= 2x\sqrt{2^2}\sqrt{2y} - 3\sqrt{x^2}\sqrt{2y} + 2\sqrt{2^4 x^2}\sqrt{2y}$$
$$= 2x \cdot 2\sqrt{2y} - 3 \cdot x\sqrt{2y} + 2 \cdot 2^2 x\sqrt{2y}$$
$$= 4x\sqrt{2y} - 3x\sqrt{2y} + 8x\sqrt{2y}$$
$$= 9x\sqrt{2y}$$

Problem 2 Simplify.

A. $y\sqrt{28y} + 7\sqrt{63y^3}$ B. $2\sqrt{27a^5} - 4a\sqrt{12a^3} + a^2\sqrt{75a}$

Solution See page A39.

EXERCISES 11.2

1 Simplify.

1. $2\sqrt{2} + \sqrt{2}$

2. $3\sqrt{5} + 8\sqrt{5}$

3. $-3\sqrt{7} + 2\sqrt{7}$

4. $4\sqrt{5} - 10\sqrt{5}$

5. $-3\sqrt{11} - 8\sqrt{11}$

6. $-3\sqrt{3} - 5\sqrt{3}$

7. $2\sqrt{x} + 8\sqrt{x}$

8. $3\sqrt{y} + 2\sqrt{y}$

9. $8\sqrt{y} - 10\sqrt{y}$

10. $-5\sqrt{2a} + 2\sqrt{2a}$

11. $-2\sqrt{3b} - 9\sqrt{3b}$

12. $-7\sqrt{5a} - 5\sqrt{5a}$

13. $3x\sqrt{2} - x\sqrt{2}$

14. $2y\sqrt{3} - 9y\sqrt{3}$

15. $2a\sqrt{3a} - 5a\sqrt{3a}$

16. $-5b\sqrt{3x} - 2b\sqrt{3x}$

17. $3\sqrt{xy} - 8\sqrt{xy}$

18. $-4\sqrt{xy} + 6\sqrt{xy}$

19. $\sqrt{45} + \sqrt{125}$

20. $\sqrt{32} - \sqrt{98}$

21. $2\sqrt{2} + 3\sqrt{8}$

22. $4\sqrt{128} - 3\sqrt{32}$

23. $5\sqrt{18} - 2\sqrt{75}$

24. $5\sqrt{75} - 2\sqrt{18}$

25. $5\sqrt{4x} - 3\sqrt{9x}$

26. $-3\sqrt{25y} + 8\sqrt{49y}$

27. $3\sqrt{3x^2} - 5\sqrt{27x^2}$

28. $-2\sqrt{8y^2} + 5\sqrt{32y^2}$

29. $2x\sqrt{xy^2} - 3y\sqrt{x^2y}$

30. $4a\sqrt{b^2a} - 3b\sqrt{a^2b}$

31. $3x\sqrt{12x} - 5\sqrt{27x^3}$

32. $2a\sqrt{50a} + 7\sqrt{32a^3}$

33. $4y\sqrt{8y^3} - 7\sqrt{18y^5}$

34. $2a\sqrt{8ab^2} - 2b\sqrt{2a^3}$

35. $b^2\sqrt{a^5b} + 3a^2\sqrt{ab^5}$

36. $y^2\sqrt{x^5y} + x\sqrt{x^3y^5}$

37. $4\sqrt{2} - 5\sqrt{2} + 8\sqrt{2}$

38. $3\sqrt{3} + 8\sqrt{3} - 16\sqrt{3}$

39. $5\sqrt{x} - 8\sqrt{x} + 9\sqrt{x}$

40. $\sqrt{x} - 7\sqrt{x} + 6\sqrt{x}$

41. $8\sqrt{2} - 3\sqrt{y} - 8\sqrt{2}$

42. $8\sqrt{3} - 5\sqrt{2} - 5\sqrt{3}$

43. $8\sqrt{8} - 4\sqrt{32} - 9\sqrt{50}$

44. $2\sqrt{12} - 4\sqrt{27} + \sqrt{75}$

45. $-2\sqrt{3} + 5\sqrt{27} - 4\sqrt{45}$

46. $-2\sqrt{8} - 3\sqrt{27} + 3\sqrt{50}$

47. $4\sqrt{75} + 3\sqrt{48} - \sqrt{99}$

48. $2\sqrt{75} - 5\sqrt{20} + 2\sqrt{45}$

49. $\sqrt{25x} - \sqrt{9x} + \sqrt{16x}$

50. $\sqrt{4x} - \sqrt{100x} - \sqrt{49x}$

51. $3\sqrt{3x} + \sqrt{27x} - 8\sqrt{75x}$

52. $5\sqrt{5x} + 2\sqrt{45x} - 3\sqrt{80x}$

53. $2a\sqrt{75b} - a\sqrt{20b} + 4a\sqrt{45b}$

54. $2b\sqrt{75a} - 5b\sqrt{27a} + 2b\sqrt{20a}$

55. $x\sqrt{3y^2} - 2y\sqrt{12x^2} + xy\sqrt{3}$

56. $a\sqrt{27b^2} + 3b\sqrt{147a^2} - ab\sqrt{3}$

57. $3\sqrt{ab^3} + 4a\sqrt{a^2b} - 5b\sqrt{4ab}$

58. $5\sqrt{a^3b} + a\sqrt{4ab} - 3\sqrt{49a^3b}$

59. $3a\sqrt{2ab^2} - \sqrt{a^2b^2} + 4b\sqrt{3a^2b}$

60. $2\sqrt{4a^2b^2} - 3a\sqrt{9ab^2} + 4b\sqrt{a^2b}$

SUPPLEMENTAL EXERCISES 11.2

Simplify.

61. $5\sqrt{x + 2} + 3\sqrt{x + 2}$

62. $8\sqrt{a + 5} - 4\sqrt{a + 5}$

63. $\frac{1}{2}\sqrt{8x^2y} + \frac{1}{3}\sqrt{18x^2y}$

64. $\frac{1}{4}\sqrt{48ab^2} + \frac{1}{5}\sqrt{75ab^2}$

65. $\frac{a}{3}\sqrt{54ab^3} + \frac{b}{4}\sqrt{96a^3b}$

66. $\frac{x}{6}\sqrt{72xy^5} + \frac{y}{7}\sqrt{98x^3y^3}$

Solve.

67. The lengths of the sides of a triangle are $4\sqrt{3}$ cm, $2\sqrt{3}$ cm, and $2\sqrt{15}$ cm. Find the perimeter of the triangle.

68. The length of a rectangle is $3\sqrt{2}$ cm. The width is $\sqrt{2}$ cm. Find the perimeter of the rectangle.

69. The length of a rectangle is $4\sqrt{5}$ cm. The width is $\sqrt{5}$ cm. Find the decimal approximation of the perimeter. Round to the nearest tenth.

70. Use the expressions $\sqrt{9 + 16}$ and $\sqrt{9} + \sqrt{16}$ to show that $\sqrt{a + b} \neq \sqrt{a} + \sqrt{b}$.

71. Is the equation $\sqrt{a^2 + b^2} = \sqrt{a} + \sqrt{b}$ true for all real numbers a and b?

S E C T I O N   **11.3**

Multiplication and Division of Radical Expressions

1   Multiply radical expressions

The Product Property of Square Roots is used to multiply variable radical expressions.

$$\sqrt{2x}\sqrt{3y} = \sqrt{2x \cdot 3y}$$
$$= \sqrt{6xy}$$

Simplify $\sqrt{2x^2}\sqrt{32x^5}$.

Use the Product Property of Square Roots.

$$\sqrt{2x^2}\sqrt{32x^5} = \sqrt{2x^2 \cdot 32x^5}$$

Multiply the radicands.

$$= \sqrt{64x^7}$$

Simplify.

$$= \sqrt{2^6x^7}$$
$$= \sqrt{2^6x^6}\sqrt{x}$$
$$= 2^3x^3\sqrt{x}$$
$$= 8x^3\sqrt{x}$$

Example 1   Simplify: $\sqrt{3x^4}\sqrt{2x^2y}\sqrt{6xy^2}$

Solution $\sqrt{3x^4}\sqrt{2x^2y}\sqrt{6xy^2} = \sqrt{3x^4 \cdot 2x^2y \cdot 6xy^2}$
$$= \sqrt{36x^7y^3}$$
$$= \sqrt{2^23^2x^7y^3}$$
$$= \sqrt{2^23^2x^6y^2}\sqrt{xy}$$
$$= 2 \cdot 3x^3y\sqrt{xy}$$
$$= 6x^3y\sqrt{xy}$$

Problem 1   Simplify: $\sqrt{5a}\sqrt{15a^3b^4}\sqrt{3b^5}$

Solution   See page A39.

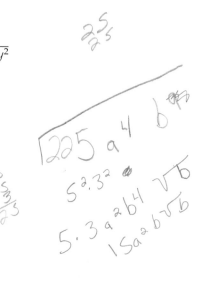

When the expression $(\sqrt{x})^2$ is simplified by using the Product Property of Square Roots, the result is x.

$$(\sqrt{x})^2 = \sqrt{x}\sqrt{x}$$
$$= \sqrt{x \cdot x}$$
$$= \sqrt{x^2}$$
$$= x$$

For $a > 0$, $(\sqrt{a})^2 = \sqrt{a^2} = a$.

Simplify $\sqrt{2x}(x + \sqrt{2x})$.

Use the Distributive Property
to remove parentheses.

$$\sqrt{2x}(x + \sqrt{2x}) = \sqrt{2x}(x) + \sqrt{2x}\sqrt{2x}$$
$$= x\sqrt{2x} + (\sqrt{2x})^2$$
$$= x\sqrt{2x} + 2x$$

Example 2   Simplify: $\sqrt{3ab}(\sqrt{3a} + \sqrt{9b})$

Solution $\sqrt{3ab}(\sqrt{3a} + \sqrt{9b}) = \sqrt{3ab}(\sqrt{3a}) + \sqrt{3ab}(\sqrt{9b})$
$$= \sqrt{3^2a^2b} + \sqrt{3^3ab^2}$$
$$= \sqrt{3^2a^2}\sqrt{b} + \sqrt{3^2b^2}\sqrt{3a}$$
$$= 3a\sqrt{b} + 3b\sqrt{3a}$$

Problem 2   Simplify: $\sqrt{5x}(\sqrt{5x} - \sqrt{25y})$

Solution   See page A39.

To simplify $(\sqrt{2} - 3x)(\sqrt{2} + x)$, use the FOIL method to remove parentheses.

$$(\sqrt{2} - 3x)(\sqrt{2} + x) = \sqrt{2}\sqrt{2} + x\sqrt{2} - 3x\sqrt{2} - 3x^2$$
$$= (\sqrt{2})^2 + (x - 3x)\sqrt{2} - 3x^2$$
$$= 2 - 2x\sqrt{2} - 3x^2$$

Example 3   Simplify: $(2\sqrt{x} - \sqrt{y})(5\sqrt{x} - 2\sqrt{y})$

Solution $(2\sqrt{x} - \sqrt{y})(5\sqrt{x} - 2\sqrt{y}) = 10(\sqrt{x})^2 - 4\sqrt{xy} - 5\sqrt{xy} + 2(\sqrt{y})^2$
$$= 10x - 9\sqrt{xy} + 2y$$

Problem 3   Simplify: $(3\sqrt{x} - \sqrt{y})(5\sqrt{x} - 2\sqrt{y})$

Solution   See page A39.

The expressions $a + b$ and $a - b$, which are the sum and difference of two terms, are called **conjugates** of each other.

The product of conjugates is the difference of two squares.

$$(a + b)(a - b) = a^2 - b^2$$
$$(2 + \sqrt{7})(2 - \sqrt{7}) = 2^2 - (\sqrt{7})^2 = 4 - 7 = -3$$
$$(3 + \sqrt{y})(3 - \sqrt{y}) = 3^2 - (\sqrt{y})^2 = 9 - y$$

Example 4   Simplify: $(\sqrt{a} - \sqrt{b})(\sqrt{a} + \sqrt{b})$

Solution $(\sqrt{a} - \sqrt{b})(\sqrt{a} + \sqrt{b}) = (\sqrt{a})^2 - (\sqrt{b})^2 = a - b$

Problem 4   Simplify: $(2\sqrt{x} + 7)(2\sqrt{x} - 7)$

Solution   See page A39.

2   Divide radical expressions

The square root of a quotient is equal to the quotient of the square roots.

The Quotient Property of Square Roots

If a and b are positive real numbers, then $\sqrt{\dfrac{a}{b}} = \dfrac{\sqrt{a}}{\sqrt{b}}$.

To simplify $\sqrt{\dfrac{4x^2}{z^6}}$, rewrite the radical expression as the quotient of the square roots.

$$\sqrt{\frac{4x^2}{z^6}} = \frac{\sqrt{4x^2}}{\sqrt{z^6}} = \frac{\sqrt{2^2x^2}}{\sqrt{z^6}}$$

Simplify.

$$= \frac{2x}{z^3}$$

To simplify $\sqrt{\dfrac{24x^3y^7}{3x^7y^2}}$, simplify the radicand.

$$\sqrt{\frac{24x^3y^7}{3x^7y^2}} = \sqrt{\frac{8y^5}{x^4}}$$

Rewrite the radical expression as the quotient of the square roots.

$$= \frac{\sqrt{8y^5}}{\sqrt{x^4}}$$

Simplify.

$$= \frac{\sqrt{2^3y^5}}{\sqrt{x^4}}$$

$$= \frac{\sqrt{2^2y^4}\sqrt{2y}}{\sqrt{x^4}}$$

$$= \frac{2y^2\sqrt{2y}}{x^2}$$

To simplify $\dfrac{\sqrt{4x^2y}}{\sqrt{xy}}$, use the Quotient

Property of Square Roots.

$$\frac{\sqrt{4x^2y}}{\sqrt{xy}} = \sqrt{\frac{4x^2y}{xy}}$$

Simplify the radicand.

$$= \sqrt{4x}$$

Simplify the radical expression.

$$= \sqrt{2^2}\sqrt{x}$$
$$= 2\sqrt{x}$$

A radical expression is not in simplest form if a radical remains in the denominator. The procedure used to remove a radical from the denominator is called **rationalizing the denominator.**

The expression $\dfrac{2}{\sqrt{3}}$ has a radical expression in the

denominator. Multiply the expression by $\dfrac{\sqrt{3}}{\sqrt{3}}$,

which equals 1.

$$\frac{2}{\sqrt{3}} = \frac{2}{\sqrt{3}} \cdot \frac{\sqrt{3}}{\sqrt{3}}$$

$$= \frac{2\sqrt{3}}{(\sqrt{3})^2}$$

Simplify.

$$= \frac{2\sqrt{3}}{3}$$

Thus $\dfrac{2}{\sqrt{3}} = \dfrac{2\sqrt{3}}{3}$, but $\dfrac{2}{\sqrt{3}}$ is not in simplest form. $\dfrac{2\sqrt{3}}{3}$ is in simplest form, because no radical remains in the denominator and the radical in the numerator contains no perfect square factors other than 1.

Simplify: $\dfrac{\sqrt{2} + \sqrt{18y^2}}{\sqrt{2}}$

Multiply the numerator and de-nominator by $\sqrt{2}$.

$$\frac{\sqrt{2} + \sqrt{18y^2}}{\sqrt{2}} = \frac{\sqrt{2} + \sqrt{18y^2}}{\sqrt{2}} \cdot \frac{\sqrt{2}}{\sqrt{2}}$$

Simplify.

$$= \frac{(\sqrt{2})^2 + \sqrt{36y^2}}{(\sqrt{2})^2}$$

$$= \frac{2 + 6y}{2}$$

$$= 1 + 3y$$

Simplify: $\dfrac{\sqrt{2y}}{\sqrt{y}+3}$

Multiply the numerator and denominator by $\sqrt{y}-3$, the conjugate of $\sqrt{y}+3$.

$$\dfrac{\sqrt{2y}}{\sqrt{y}+3}=\dfrac{\sqrt{2y}}{\sqrt{y}+3}\cdot\dfrac{\sqrt{y}-3}{\sqrt{y}-3}$$

Simplify.

$$=\dfrac{\sqrt{2y^2}-3\sqrt{2y}}{(\sqrt{y})^2-3^2}$$

$$=\dfrac{y\sqrt{2}-3\sqrt{2y}}{y-9}$$

Example 5   Simplify.   A. $\dfrac{\sqrt{4x^2y^5}}{\sqrt{3x^4y}}$ B. $\dfrac{\sqrt{2}}{\sqrt{2}-\sqrt{x}}$ C. $\dfrac{2+3\sqrt{5}}{3-\sqrt{5}}$

Solution   A. $\dfrac{\sqrt{4x^2y^5}}{\sqrt{3x^4y}}=\sqrt{\dfrac{2^2x^2y^5}{3x^4y}}=\sqrt{\dfrac{2^2y^4}{3x^2}}=\dfrac{2y^2}{x\sqrt{3}}=\dfrac{2y^2}{x\sqrt{3}}\cdot\dfrac{\sqrt{3}}{\sqrt{3}}=\dfrac{2y^2\sqrt{3}}{3x}$

B. $\dfrac{\sqrt{2}}{\sqrt{2}-\sqrt{x}}=\dfrac{\sqrt{2}}{\sqrt{2}-\sqrt{x}}\cdot\dfrac{\sqrt{2}+\sqrt{x}}{\sqrt{2}+\sqrt{x}}=\dfrac{2+\sqrt{2x}}{2-x}$

C. $\dfrac{3-\sqrt{5}}{2+3\sqrt{5}}=\dfrac{3-\sqrt{5}}{2+3\sqrt{5}}\cdot\dfrac{2-3\sqrt{5}}{2-3\sqrt{5}}$

$$=\dfrac{6-9\sqrt{5}-2\sqrt{5}+3\cdot5}{4-9\cdot5}$$

$$=\dfrac{21-11\sqrt{5}}{-41}$$

$$=-\dfrac{21-11\sqrt{5}}{41}$$

Problem 5   Simplify.   A. $\sqrt{\dfrac{15x^6y^7}{3x^7y^9}}$ B. $\dfrac{\sqrt{y}}{\sqrt{y}+3}$ C. $\dfrac{5+\sqrt{y}}{1-2\sqrt{y}}$

Solution   See page A40.

EXERCISES 11.3

1 Simplify.

1. $\sqrt{5} \cdot \sqrt{5}$

2. $\sqrt{11} \cdot \sqrt{11}$

3. $\sqrt{3} \cdot \sqrt{12}$

4. $\sqrt{2} \cdot \sqrt{8}$

5. $\sqrt{x} \cdot \sqrt{x}$

6. $\sqrt{y} \cdot \sqrt{y}$

7. $\sqrt{xy^3} \cdot \sqrt{x^5y}$

8. $\sqrt{a^3b^5} \cdot \sqrt{ab^5}$

9. $\sqrt{3a^2b^5} \cdot \sqrt{6ab^7}$

10. $\sqrt{5x^3y} \cdot \sqrt{10x^2y}$

11. $\sqrt{6a^3b^2} \cdot \sqrt{24a^5b}$

12. $\sqrt{8ab^5} \cdot \sqrt{12a^7b}$

13. $\sqrt{2}(\sqrt{2} - \sqrt{3})$

14. $3(\sqrt{12} - \sqrt{3})$

15. $\sqrt{x}(\sqrt{x} - \sqrt{y})$

16. $\sqrt{b}(\sqrt{a} - \sqrt{b})$

17. $\sqrt{5}(\sqrt{10} - \sqrt{x})$

18. $\sqrt{6}(\sqrt{y} - \sqrt{18})$

19. $\sqrt{8}(\sqrt{2} - \sqrt{5})$

20. $\sqrt{10}(\sqrt{20} - \sqrt{a})$

21. $(\sqrt{x} - 3)^2$

22. $(2\sqrt{a} - y)^2$

23. $\sqrt{3a}(\sqrt{3a} - \sqrt{3b})$

24. $\sqrt{5x}(\sqrt{10x} - \sqrt{x})$

25. $\sqrt{2ac} \cdot \sqrt{5ab} \cdot \sqrt{10cb}$

26. $\sqrt{3xy} \cdot \sqrt{6x^3y} \cdot \sqrt{2y^2}$

27. $(3\sqrt{x} - 2y)(5\sqrt{x} - 4y)$

28. $(5\sqrt{x} + 2\sqrt{y})(3\sqrt{x} - \sqrt{y})$

29. $(\sqrt{x} - \sqrt{y})(\sqrt{x} + \sqrt{y})$

30. $(\sqrt{3x} + y)(\sqrt{3x} - y)$

31. $(2\sqrt{x} + \sqrt{y})(5\sqrt{x} + 4\sqrt{y})$

32. $(5\sqrt{x} - 2\sqrt{y})(3\sqrt{x} - 4\sqrt{y})$

2 Simplify.

33. $\dfrac{\sqrt{32}}{\sqrt{2}}$

34. $\dfrac{\sqrt{45}}{\sqrt{5}}$

35. $\dfrac{\sqrt{98}}{\sqrt{2}}$

36. $\dfrac{\sqrt{48}}{\sqrt{3}}$

37. $\dfrac{\sqrt{27a}}{\sqrt{3a}}$

38. $\dfrac{\sqrt{72x^5}}{\sqrt{2x}}$

39. $\dfrac{\sqrt{15x^3y}}{\sqrt{3xy}}$

40. $\dfrac{\sqrt{40x^5y^2}}{\sqrt{5xy}}$

41. $\dfrac{\sqrt{2a^5b^4}}{\sqrt{98ab^4}}$

42. $\dfrac{\sqrt{48x^5y^2}}{\sqrt{3x^3y}}$

43. $\dfrac{1}{\sqrt{3}}$

44. $\dfrac{1}{\sqrt{8}}$

45. $\dfrac{15}{\sqrt{75}}$

46. $\dfrac{6}{\sqrt{72}}$

47. $\dfrac{6}{\sqrt{12x}}$

48. $\dfrac{14}{\sqrt{7y}}$

49. $\dfrac{8}{\sqrt{32x}}$

50. $\dfrac{15}{\sqrt{50x}}$

51. $\dfrac{3}{\sqrt{x}}$

52. $\dfrac{4}{\sqrt{2x}}$

53. $\dfrac{\sqrt{8x^2y}}{\sqrt{2x^4y^2}}$

54. $\dfrac{\sqrt{4x^2}}{\sqrt{9y}}$

55. $\dfrac{\sqrt{16a}}{\sqrt{49ab}}$

56. $\dfrac{5\sqrt{8}}{4\sqrt{50}}$

57. $\dfrac{5\sqrt{18}}{9\sqrt{27}}$

58. $\dfrac{\sqrt{12a^3b}}{\sqrt{24a^2b^2}}$

59. $\dfrac{\sqrt{3xy}}{\sqrt{27x^3y^2}}$

60. $\dfrac{\sqrt{9xy^2}}{\sqrt{27x}}$

61. $\dfrac{\sqrt{4x^2y}}{\sqrt{3xy^3}}$

62. $\dfrac{\sqrt{16x^3y^2}}{\sqrt{8x^3y}}$

63. $\dfrac{1}{\sqrt{2}-3}$

64. $\dfrac{5}{\sqrt{7}-3}$

65. $\dfrac{3}{5+\sqrt{5}}$

66. $\dfrac{7}{\sqrt{2}-7}$

67. $\dfrac{\sqrt{xy}}{\sqrt{x}-\sqrt{y}}$

68. $\dfrac{\sqrt{x}}{\sqrt{x}-\sqrt{y}}$

69. $\dfrac{5\sqrt{x^2y}}{\sqrt{75xy^2}}$

70. $\dfrac{3\sqrt{ab}}{a\sqrt{6b}}$

71. $\dfrac{\sqrt{2}}{\sqrt{2}-\sqrt{3}}$

72. $\dfrac{1+\sqrt{2}}{1-\sqrt{2}}$

73. $\dfrac{\sqrt{5}}{\sqrt{2}-\sqrt{5}}$

74. $\dfrac{\sqrt{6}}{\sqrt{3}-\sqrt{2}}$

75. $\dfrac{\sqrt{x}}{\sqrt{x}+3}$

76. $\dfrac{\sqrt{y}}{2-\sqrt{y}}$

77. $\dfrac{5\sqrt{3}-7\sqrt{3}}{4\sqrt{3}}$

78. $\dfrac{10\sqrt{7}-2\sqrt{7}}{2\sqrt{7}}$

79. $\dfrac{5\sqrt{8}-3\sqrt{2}}{\sqrt{2}}$

80. $\dfrac{5\sqrt{12}-\sqrt{3}}{\sqrt{27}}$

81. $\dfrac{3\sqrt{2}-8\sqrt{2}}{\sqrt{2}}$

82. $\dfrac{5\sqrt{3}-2\sqrt{3}}{2\sqrt{3}}$

83. $\dfrac{2\sqrt{8}+3\sqrt{2}}{\sqrt{32}}$

84. $\dfrac{3-\sqrt{6}}{5-2\sqrt{6}}$

85. $\dfrac{6-2\sqrt{3}}{4+3\sqrt{3}}$

86. $\dfrac{\sqrt{2}+2\sqrt{6}}{2\sqrt{2}-3\sqrt{6}}$

87. $\dfrac{2\sqrt{3}-\sqrt{6}}{5\sqrt{3}+2\sqrt{6}}$

88. $\dfrac{3+\sqrt{x}}{2-\sqrt{x}}$

89. $\dfrac{\sqrt{a}-4}{2\sqrt{a}+2}$

90. $\dfrac{3+2\sqrt{y}}{2-\sqrt{y}}$

91. $\dfrac{2+\sqrt{y}}{\sqrt{y}-3}$

92. $\dfrac{\sqrt{x}+\sqrt{y}}{\sqrt{x}-\sqrt{y}}$

SUPPLEMENTAL EXERCISES 11.3

Simplify.

93. $\sqrt{0.6} \cdot \sqrt{0.6}$

94. $-\sqrt{1.3} \cdot \sqrt{1.3}$

95. $\sqrt{\dfrac{5}{8}} \cdot \sqrt{\dfrac{5}{8}}$

96. $\sqrt{\dfrac{25}{49}}$

97. $-\sqrt{\dfrac{16}{81}}$

98. $-\sqrt{\dfrac{1}{144}}$

99. $\sqrt{1\dfrac{7}{9}}$

100. $\sqrt{1\dfrac{9}{16}}$

101. $\sqrt{2\dfrac{1}{4}}$

102. $-\sqrt{2\dfrac{7}{9}}$

103. $-\sqrt{6\dfrac{1}{4}}$

104. $\sqrt{3\dfrac{1}{16}}$

Solve.

105. Show that 2 is a solution of the equation $\sqrt{x+2} + \sqrt{x-1} = 3$.

106. Is 16 a solution of the equation $\sqrt{x} - \sqrt{x+9} = 1$?

107. Show that $(1 + \sqrt{6})$ and $(1 - \sqrt{6})$ are solutions of the equation $x^2 - 2x - 5 = 0$.

S E C T I O N 11.4
Solving Equations Containing Radical Expressions

1 Solve equations containing one or more radical expressions

An equation that contains a variable expression in a radicand is a **radical equation**.

$$\left.\begin{array}{l} \sqrt{x} = 4 \\ \sqrt{x+2} = \sqrt{x-7} \end{array}\right\} \quad \begin{array}{l} \text{Radical} \\ \text{equations} \end{array}$$

The following property of equality states that if two numbers are equal, then the squares of the numbers are equal. This property is used to solve radical equations.

Property of Squaring Both Sides of an Equation

If a and b are real numbers and $a = b$, then $a^2 = b^2$.

Solve: $\sqrt{x-2} - 7 = 0$

Rewrite the equation with the radical on one side of the equation and the constant on the other side.

$$\sqrt{x-2} - 7 = 0$$
$$\sqrt{x-2} = 7$$

Square both sides of the equation.

$$(\sqrt{x-2})^2 = 7^2$$

Solve the resulting equation.

$$x - 2 = 49$$
$$x = 51$$

Check the solution.
When both sides of an equation are squared, the resulting equation may have a solution that is not a solution of the original equation.

Check: $\sqrt{x-2} - 7 = 0$
$$\sqrt{51-2} - 7 = 0$$
$$\sqrt{49} - 7 = 0$$
$$\sqrt{7^2} - 7 = 0$$
$$7 - 7 = 0$$
$$0 = 0$$

The solution is 51.

Example 1   Solve: $\sqrt{3x} + 2 = 5$

Solution
$$\sqrt{3x} + 2 = 5$$
$$\sqrt{3x} = 3$$

▶ Rewrite the equation with the radical on one side of the equation and the constant on the other side.

$$(\sqrt{3x})^2 = 3^2$$
▶ Square both sides of the equation.
$$3x = 9$$ ▶ Solve for x.
$$x = 3$$

Check
$$\sqrt{3x} + 2 = 5$$
$$\sqrt{3 \cdot 3} + 2 = 5$$
$$\sqrt{3^2} + 2 = 5$$
$$3 + 2 = 5$$
$$5 = 5$$

▶ Both sides of the equations were squared. The solution must be checked.

▶ This is a true equation. The solution checks.

The solution is 3.

Problem 1   Solve: $\sqrt{4x} + 3 = 7$

Solution   See page A40.

Example 2   Solve: $0 = 3 - \sqrt{2x - 3}$

Solution
$$0 = 3 - \sqrt{2x - 3}$$
$$\sqrt{2x - 3} = 3$$
▶ Rewrite the equation so that the radical is alone on one side of the equation.

$$(\sqrt{2x - 3})^2 = 3^2$$
▶ Square both sides of the equation.
$$2x - 3 = 9$$
$$2x = 12$$
$$x = 6$$

Check
$$0 = 3 - \sqrt{2x - 3}$$
$$0 = 3 - \sqrt{2 \cdot 6 - 3}$$
$$0 = 3 - \sqrt{12 - 3}$$
$$0 = 3 - \sqrt{9}$$
$$0 = 3 - \sqrt{3^2}$$
$$0 = 3 - 3$$
$$0 = 0$$
▶ This is a true equation. The solution checks.

The solution is 6.

Problem 2   Solve: $\sqrt{3x - 2} - 5 = 0$

Solution   See page A40.

Solve: $\sqrt{5 + x} + \sqrt{x} = 5$

Solve for one of the radical expressions.	$\sqrt{5 + x} + \sqrt{x} = 5$ $\sqrt{5 + x} = 5 - \sqrt{x}$
Square each side.	$(\sqrt{5 + x})^2 = (5 - \sqrt{x})^2$
Recall that $(a - b)^2 = a^2 - 2ab + b^2$. Simplify.	$5 + x = 25 - 10\sqrt{x} + x$ $-20 = -10\sqrt{x}$
This is still a radical equation.	$2 = \sqrt{x}$
Square each side.	$2^2 = (\sqrt{x})^2$ $4 = x$
4 checks as the solution.	The solution is 4.

Example 3   Solve: $\sqrt{x} - \sqrt{x-5} = 1$

Solution $\sqrt{x} - \sqrt{x-5} = 1$

$\sqrt{x} = 1 + \sqrt{x-5}$

▶ Solve for one of the radical expressions.

$(\sqrt{x})^2 = (1 + \sqrt{x-5})^2$

▶ Square each side.

$x = 1 + 2\sqrt{x-5} + (x-5)$

▶ Simplify.

$4 = 2\sqrt{x-5}$

$2 = \sqrt{x-5}$

▶ This is a radical equation.

$2^2 = (\sqrt{x-5})^2$

▶ Square each side.

$4 = x - 5$

▶ Simplify.

$9 = x$

Check $\sqrt{x} - \sqrt{x-5} = 1$

$\sqrt{9} - \sqrt{9-5} = 1$

$3 - \sqrt{4} = 1$

$3 - 2 = 1$

$1 = 1$

The solution is 9.

Problem 3   Solve: $\sqrt{x} + \sqrt{x+9} = 9$

Solution   See page A40.

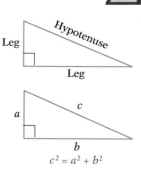

2   ## Application problems

A right triangle contains one 90° angle. The side opposite the 90° angle is called the **hypotenuse.** The other two sides are called **legs.**

Pythagoras, a Greek mathematician, discovered that the square of the hypotenuse of a right triangle is equal to the sum of the squares of the two legs. This is called the **Pythagorean Theorem.**

Leg

Hypotenuse

Leg

a

c

b

$c^2 = a^2 + b^2$

The Pythagorean Theorem

If a and b are the legs of a right triangle and c is the hypotenuse, then

$$a^2 + b^2 = c^2.$$

Using this theorem, the hypotenuse of the right triangle shown at the right can be found. The two legs are known. Use the formula

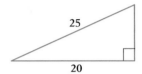

$$\text{Hypotenuse} = \sqrt{(\text{leg})^2 + (\text{leg})^2}$$
$$c = \sqrt{a^2 + b^2}$$
$$= \sqrt{(5)^2 + (12)^2}$$
$$= \sqrt{25 + 144}$$
$$= \sqrt{169}$$
$$= 13$$

The unknown leg of the right triangle shown at the right can be found. One leg and the hypotenuse are known. Use the formula

$$\text{Leg} = \sqrt{(\text{hypotenuse})^2 - (\text{leg})^2}$$
$$a = \sqrt{c^2 - b^2}$$
$$= \sqrt{(25)^2 - (20)^2}$$
$$= \sqrt{625 - 400}$$
$$= \sqrt{225}$$
$$= 15$$

Example 4 and Problem 4 illustrate the use of the Pythagorean Theorem. Example 5 and Problem 5 illustrate other applications of radical equations.

Example 4   A guy wire is attached to a point 22 m above the ground on a telephone pole. The wire is anchored to the ground at a point 9 m from the base of the pole. Find the length of the guy wire. Round to the nearest hundredth.

Strategy   To find the length of the guy wire, use the Pythagorean Theorem. One leg is the distance from the bottom of the wire to the base of the telephone pole. The other leg is the distance from the top of the wire to the base of the telephone pole. The guy wire is the hypotenuse. Solve the Pythagorean Theorem for the hypotenuse.

Solution $c = \sqrt{a^2 + b^2}$
$c = \sqrt{(22)^2 + (9)^2}$
$c = \sqrt{484 + 81}$
$c = \sqrt{565}$
$c \approx 23.77$

The guy wire has a length of 23.77 m.

Problem 4   A ladder 12 ft long is resting against a building. How high on the building will the ladder reach when the bottom of the ladder is 5 ft from the building? Round to the nearest hundredth.

Solution   See pages A40 and A41.

Example 5   How far would a submarine periscope have to be above the water to locate a ship 5 mi away? The equation for the distance in miles that the lookout can see is $d = 1.4\sqrt{h}$, where h is the height in feet above the surface of the water. Round to the nearest hundredth.

Strategy   To find the height above water, replace d in the equation with the given value, and solve for h.

Solution $1.4\sqrt{h} = d$

$1.4\sqrt{h} = 5$

$$\sqrt{h} = \frac{5}{1.4}$$

$$(\sqrt{h})^2 = \left(\frac{5}{1.4}\right)^2$$

$$h = \frac{25}{1.96}$$

$$h \approx 12.76$$

The periscope must be 12.76 ft above the water.

Problem 5   Find the length of a pendulum that makes one swing in 1.5 s. The equation for the time for one swing is $T = 2\pi \sqrt{\dfrac{L}{32}}$, where T is the time in seconds, and L is the length in feet. Use 3.14 for π. Round to the nearest hundredth.

Solution   See page A41.

EXERCISES 11.4

1   Solve and check.

1. $\sqrt{x} = 5$

2. $\sqrt{y} = 7$

3. $\sqrt{a} = 12$

4. $\sqrt{a} = 9$

5. $\sqrt{5x} = 5$

6. $\sqrt{3x} = 4$

7. $\sqrt{4x} = 8$

8. $\sqrt{6x} = 3$

9. $\sqrt{2x} - 4 = 0$

10. $3 - \sqrt{5x} = 0$

11. $\sqrt{4x} + 5 = 2$

12. $\sqrt{3x} + 9 =$

13. $\sqrt{3x-2}=4$

14. $\sqrt{5x+6}=1$

15. $\sqrt{2x+1}=7$

16. $\sqrt{5x+4}=3$

17. $0=2-\sqrt{3-x}$

18. $0=5-\sqrt{10+x}$

19. $\sqrt{5x+2}=0$

20. $\sqrt{3x-7}=0$

21. $\sqrt{3x}-6=-4$

22. $\sqrt{5x}+8=23$

23. $0=\sqrt{3x-9}-6$

24. $0=\sqrt{2x+7}-3$

25. $\sqrt{5x-1}=\sqrt{3x+9}$

26. $\sqrt{3x+4}=\sqrt{12x-14}$

27. $\sqrt{5x-3}=\sqrt{4x-2}$

28. $\sqrt{5x-9}=\sqrt{2x-3}$

29. $\sqrt{x^2-5x+6}=\sqrt{x^2-8x+9}$

30. $\sqrt{x^2-2x+4}=\sqrt{x^2+5x-12}$

31. $\sqrt{x}=\sqrt{x+3}-1$

32. $\sqrt{x+5}=\sqrt{x}+1$

33. $\sqrt{2x+5}=5-\sqrt{2x}$

34. $\sqrt{2x}+\sqrt{2x+9}=9$

35. $\sqrt{3x}-\sqrt{3x+7}=1$

36. $\sqrt{x}-\sqrt{x+9}=1$

2   Solve. Round to the nearest hundredth.

37. The two legs of a right triangle measure 5 cm and 9 cm. Find the length of the hypotenuse.

38. The two legs of a right triangle measure 8 in. and 4 in. Find the length of the hypotenuse.

39. The hypotenuse of a right triangle measures 12 ft. One leg of the triangle measures 7 ft. Find the length of the other leg of the triangle.

40. The hypotenuse of a right triangle measures 20 cm. One leg of the triangle measures 16 cm. Find the length of the other leg of the triangle.

41. The infield of a baseball diamond is a square. The distance between successive bases is 90 ft. The pitcher's mound is on the diagonal between home plate and second base at a distance of 60.5 ft from home plate. Is the pitcher's mound more or less than halfway between home plate and second base?

42. The infield of a softball diamond is a square. The distance between successive bases is 60 ft. The pitcher's mound is on the diagonal between home plate and second base at a distance of 46 ft from home plate. Is the pitcher's mound more or less than halfway between home plate and second base?

43. The measure of a television screen is given by the length of a diagonal across the screen. A 17-in. television has a width of 14.6 in. Find the height of the screen.

44. The measure of a television screen is given by the length of a diagonal across the screen. A 33-in. big-screen television has a width of 26.4 in. Find the height of the screen.

45. Five added to the square root of the product of four and a number is equal to seven. Find the number.

46. The product of a number and the square root of three is equal to the square root of twenty-seven. Find the number.

47. Two added to the square root of the sum of a number and five is equal to six. Find the number.

48. The product of a number and the square root of seven is equal to the square root of twenty-eight. Find the number.

49. How far would a submarine periscope have to be above the water to locate a ship 4 mi away? The equation for the distance in miles that the lookout can see is $d = 1.4\sqrt{h}$, where h is the height in feet above the surface of the water.

50. How far would a submarine periscope have to be above the water to locate a ship 7 mi away? The equation for the distance in miles that the lookout can see is $d = 1.4\sqrt{h}$, where h is the height in feet above the surface of the water.

51. A stone is dropped from a bridge and hits the water 2 s later. How high is the bridge? The equation for the distance an object falls in T seconds is $T = \sqrt{\dfrac{d}{16}}$, where d is the distance in feet.

52. A stone is dropped into a mine shaft and hits the bottom 3.5 s later. How deep is the mine shaft? The equation for the distance an object falls in T seconds is $T = \sqrt{\dfrac{d}{16}}$, where d is the distance in feet.

53. Find the length of a pendulum that makes one swing in 3 s. The equation for the time of one swing of a pendulum is $T = 2\pi\sqrt{\dfrac{L}{32}}$, where T is the time in seconds, and L is the length in feet.

54. Find the length of a pendulum that makes one swing in 2 s. The equation for the time of one swing of a pendulum is $T = 2\pi\sqrt{\dfrac{L}{32}}$, where T is the time in seconds, and L is the length in feet.

SUPPLEMENTAL EXERCISES 11.4

Solve.

55. $\sqrt{\dfrac{3x-2}{4}} = 2$

56. $\sqrt{\dfrac{5y+2}{3}} = 3$

57. $\sqrt{\dfrac{4n-4}{3}} = 2$

58. $\sqrt{\dfrac{8n}{3}} - 1 = 3$

59. $\sqrt{\dfrac{4x}{3}} - 1 = 1$

60. $\sqrt{\dfrac{3y}{5}} - 1 = 2$

61. In the coordinate plane, a triangle is formed by drawing lines between the points (0, 0) and (5, 0), (5, 0) and (5, 12), and (5, 12) and (0, 0). Find the number of units in the perimeter of the triangle.

62. The hypotenuse of a right triangle is $5\sqrt{2}$ cm, and one leg is $4\sqrt{2}$ cm.
a. Find the perimeter of the triangle.
b. Find the area of the triangle.

Calculators and Computers

Simplifying Radical Expressions

Chapter 11 presents simplification of numerical and variable radical expressions and operations with radical expressions (addition, subtraction, multiplication, and division). These concepts are followed by solving equations containing one or more radical expressions and solving application problems that involve radical expressions.

Just as expressions with the same variable part can be added or subtracted, expressions with the same radicand can be added or subtracted.

$$2x + 3x = 5x \qquad\qquad 2\sqrt{x} + 3\sqrt{x} = 5\sqrt{x}$$

If the variable parts of the radicands are different, the expressions cannot be added or subtracted.

$$2x + 3y = 2x + 3y \qquad\qquad 2\sqrt{x} + 3\sqrt{y} = 2\sqrt{x} + 3\sqrt{y}$$

Expressions with the same or different variable parts or radicands can be multiplied.

$$(2x)(3x) = 6x^2 \qquad\qquad (2\sqrt{x})(3\sqrt{x}) = 6\sqrt{x^2} = 6x$$
$$(2x)(3y) = 6xy \qquad\qquad (2\sqrt{x})(3\sqrt{y}) = 6\sqrt{xy}$$

A computer program can be written to simplify radical expressions. The program RADICAL EXPRESSIONS on the Math ACE Disk will enable you to practice simplifying radical expressions. The program will display a radical expression. Then, using pencil and paper, simplify the expression. When you are finished, press the RETURN key. The correct solution will be displayed on the screen.

After you complete a problem, you may either continue to practice or quit the program.

Something Extra

The Square Root Algorithm

An algorithm is a method or procedure that repeats the same sequence of steps over and over. For example, to divide two whole numbers, the division algorithm is used.

$$
\begin{array}{r}
2 \\
4\overline{)936} \\
-8 \\
\hline
13
\end{array}
$$

1. Estimate the quotient. $9 \div 4 = 2$
2. Multiply the quotient times the divisor. $2 \cdot 4 = 8$
3. Subtract. $9 - 8 = 1$
4. Bring down the next digit. 3

$$
\begin{array}{r}
234 \\
4\overline{)936} \\
-8 \\
\hline
13 \\
-12 \\
\hline
16 \\
-16 \\
\hline
0
\end{array}
$$

The division problem is completed by repeating these four steps until there are no numbers to bring down. The algorithm consists of the four steps: estimate, multiply, subtract, and bring down.

Newton's method of approximating a square root is an algorithm for finding the square root of a number. The two steps in the algorithm are:

1. Divide
2. Average

In the following example, the square root algorithm is used to approximate $\sqrt{41}$. A calculator was used to perform the operations.

Approximate $\sqrt{41}$ to the nearest thousandth.

Because $6^2 = 36$ and $7^2 = 49$, and $36 < 41 < 49$, $\sqrt{41}$ is between 6 and 7. We will start with 6 as an approximation for $\sqrt{41}$.

Step 1.	Divide 41 by 6.	$41 \div 6 = 6.8333333$
Step 2.	Find the average of 6 and 6.8333333.	$\dfrac{6 + 6.8333333}{2} = 6.4166667$
Repeat Step 1.	Divide 41 by 6.4166667.	$41 \div 6.4166667 = 6.3896104$
Repeat Step 2.	Find the average of 6.4166667 and 6.3896104.	$\dfrac{6.4166667 + 6.3896104}{2} = 6.4031386$
Repeat Step 1.	Divide 41 by 6.4031386.	$41 \div 6.4031386 = 6.4031099$

The answers obtained in the last two steps are the same to three decimal places.

$\sqrt{41} = 6.403$, to the nearest thousandth.

Use the square root algorithm to approximate each square root to the nearest thousandth.

1. $\sqrt{52}$ 2. $\sqrt{68}$ 3. $\sqrt{87}$ 4. $\sqrt{119}$

Chapter Summary

Key Words

A *square root* of a positive number x is a number whose square is x.

The *principal square root* of a number is the positive square root.

The symbol $\sqrt{}$ is called a *radical* and is used to indicate the principal square root of a number.

The *radicand* is the expression under the radical sign.

The square of an integer is a *perfect square*.

If a number is not a perfect square, its square root can only be approximated. Such numbers are *irrational numbers*. Their decimal representations never terminate or repeat.

Conjugates are binomial expressions that differ only in the sign of the second term. (The expressions $a + b$ and $a - b$ are conjugates.)

Rationalizing the denominator is the procedure used to remove a radical from the denominator of a fraction.

A *radical equation* is an equation that contains a variable expression in a radicand.

Essential Rules

The Product Property of Square Roots — If a and b are positive real numbers, then $\sqrt{ab} = \sqrt{a}\sqrt{b}.$

The Quotient Property of Square Roots — If a and b are positive real numbers, then $\sqrt{\dfrac{a}{b}} = \dfrac{\sqrt{a}}{\sqrt{b}}.$

Property of Squaring Both Sides of an Equation — If a and b are real numbers and $a = b$, then $a^2 = b^2.$

Pythagorean Theorem — $c^2 = a^2 + b^2$

Chapter Review

1. Simplify: $5\sqrt{3} - 16\sqrt{3}$

2. Simplify: $\dfrac{2x}{\sqrt{3} - \sqrt{5}}$

3. Simplify: $\sqrt{x^2 + 16x + 64}$

4. Solve: $3 - \sqrt{7x} = 5$

5. Simplify: $\dfrac{5\sqrt{y} - 2\sqrt{y}}{3\sqrt{y}}$

6. Simplify: $6\sqrt{7} + \sqrt{7}$

7. Simplify: $(6\sqrt{a} + 5\sqrt{b})(2\sqrt{a} + 3\sqrt{b})$

8. Simplify: $\sqrt{49(x + 3)^4}$

9. Simplify: $2\sqrt{36}$

10. Solve: $\sqrt{b} = 4$

11. Simplify: $9x\sqrt{5} - 5x\sqrt{5}$

12. Simplify: $(\sqrt{5ab} - \sqrt{7})(\sqrt{5ab} + \sqrt{7})$

13. Solve: $\sqrt{2x + 9} = \sqrt{8x - 9}$

14. Simplify: $\sqrt{35}$

15. Simplify: $2x\sqrt{60x^3y^3} + 3x^2y\sqrt{15xy}$

16. Simplify: $\dfrac{\sqrt{3x^3y}}{\sqrt{27xy^5}}$

17. Simplify: $(3\sqrt{x} - \sqrt{y})^2$

18. Simplify: $\sqrt{(a + 4)^2}$

19. Simplify: $5\sqrt{48}$

20. Simplify: $3\sqrt{12x} + 5\sqrt{48x}$

21. Simplify: $\dfrac{8}{\sqrt{x} - 3}$

22. Simplify: $\sqrt{6a}(\sqrt{3a} + \sqrt{2a})$

23. Simplify: $3\sqrt{18a^5b}$

24. Simplify: $-3\sqrt{120}$

25. Simplify: $\sqrt{20a^5b^9} - 2ab^2\sqrt{45a^3b^5}$

26. Simplify: $\dfrac{\sqrt{98x^7y^9}}{\sqrt{2x^3y}}$

27. Solve: $\sqrt{5x + 1} = \sqrt{20x - 8}$

28. Simplify: $\sqrt{c^{18}}$

29. Simplify: $\sqrt{450}$

30. Simplify: $6a\sqrt{80b} - \sqrt{180a^2b} + 5a\sqrt{b}$

31. Simplify: $\dfrac{16}{\sqrt{a}}$

32. Solve: $6 - \sqrt{2y} = 2$

33. Simplify: $\sqrt{a^3b^4c}\sqrt{a^7b^2c^3}$

34. Simplify: $7\sqrt{630}$

35. Simplify: $y\sqrt{24y^6}$

36. Simplify: $\dfrac{\sqrt{250}}{\sqrt{10}}$

37. Solve: $\sqrt{x^2 + 5x + 4} = \sqrt{x^2 + 7x - 6}$

38. Simplify: $(4\sqrt{y} - \sqrt{5})(2\sqrt{y} + 3\sqrt{5})$

39. Find the decimal approximation of $\sqrt{9900}$ to the nearest thousandth.

40. Simplify: $\sqrt{7} \cdot \sqrt{7}$

41. Simplify: $5x\sqrt{150x^7}$

42. Simplify:
$$2x^2\sqrt{18x^2y^5} + 6y\sqrt{2x^6y^3} - 9xy^2\sqrt{8x^4y}$$

43. Simplify: $\dfrac{\sqrt{54a^3}}{\sqrt{6a}}$

44. Simplify: $4\sqrt{250}$

45. Solve: $\sqrt{5x} = 10$

46. Simplify: $\dfrac{3a\sqrt{3} + 2\sqrt{12a^2}}{\sqrt{27}}$

47. Simplify: $\sqrt{2} \cdot \sqrt{50}$

48. Simplify: $\sqrt{36x^4y^5}$

49. Simplify: $4y\sqrt{243x^{17}y^9}$

50. Solve: $\sqrt{x+1} - \sqrt{x-2} = 1$

51. Simplify: $\sqrt{400}$

52. Simplify: $-4\sqrt{8x} + 7\sqrt{18x} - 3\sqrt{50x}$

53. Solve: $0 = \sqrt{10x+4} - 8$

54. Simplify: $\sqrt{3}(\sqrt{12} - \sqrt{3})$

55. The square root of the sum of two consecutive odd integers is equal to 10. Find the larger integer.

56. The weight of an object is related to the object's distance from the surface of the earth. An equation for this relationship is $d = 4000\sqrt{\dfrac{W_0}{W_a}}$, where W_0 is the object's weight on the surface of the earth and W_a is the object's weight at a distance of d miles above the earth's surface. A space explorer weighs 36 lb when 8000 mi above the earth's surface. Find the explorer's weight on the surface of the earth.

57. The hypotenuse of a right triangle measures 18 cm. One leg of the triangle measures 11 cm. Find the length of the other leg of the triangle. Round to the nearest hundredth.

58. A bicycle will overturn if it rounds a corner too sharply or too quickly. The equation for the maximum velocity at which a cyclist can turn a corner without tipping over is given by the equation $v = 4\sqrt{r}$, where v is the velocity of the bicycle in miles per hour, and r is the radius of the corner in feet. Find the radius of the sharpest corner that a cyclist can safely turn when riding at a speed of 20 mph.

59. A tsunami is a great sea wave produced by underwater earthquakes or volcanic eruption. The velocity of a tsunami as it approaches land is approximated by the equation $v = 3\sqrt{d}$, where v is the velocity in feet per second, and d is the depth of the water in feet. Find the depth of the water when the velocity of a tsunami is 30 ft/s.

60. A guy wire is attached to a point 25 ft above the ground on a telephone pole. The wire is anchored to the ground at a point 8 ft from the base of the pole. Find the length of the guy wire. Round to the nearest hundredth.

Chapter Test

1. Simplify: $\sqrt{121x^8y^2}$

2. Simplify: $5\sqrt{8} - 3\sqrt{50}$

3. Simplify: $\sqrt{3x^2y}\sqrt{6x^2}\sqrt{2x}$

4. Simplify: $\sqrt{45}$

5. Simplify: $\sqrt{72x^7y^2}$

6. Simplify: $3\sqrt{8y} - 2\sqrt{72x} + 5\sqrt{18x}$

7. Simplify: $(\sqrt{y} + 3)(\sqrt{y} + 5)$

8. Simplify: $\dfrac{3\sqrt{x^3} - 4\sqrt{9x}}{3\sqrt{x}}$

9. Simplify: $\dfrac{\sqrt{162}}{\sqrt{2}}$

10. Solve: $\sqrt{5x - 6} = 7$

11. Find the decimal approximation of $\sqrt{500}$ to the nearest thousandth.

12. Simplify: $\sqrt{32a^5b^{11}}$

13. Simplify: $\sqrt{a}(\sqrt{a} - \sqrt{b})$

14. Simplify: $\sqrt{8x^3y}\sqrt{10xy^4}$

15. Simplify: $\dfrac{\sqrt{98a^6b^4}}{\sqrt{2a^3b^2}}$

16. Solve: $\sqrt{9x} + 3 = 18$

17. Simplify: $\sqrt{192x^{13}y^5}$

18. Simplify: $2a\sqrt{2ab^3} + b\sqrt{8a^3b} - 5ab\sqrt{ab}$

19. Simplify: $(\sqrt{a} - 2)(\sqrt{a} + 2)$

20. Simplify: $\dfrac{3}{2 - \sqrt{5}}$

21. Solve: $3 = 8 - \sqrt{5x}$

22. Simplify: $\sqrt{3}(\sqrt{6} - \sqrt{x^2})$

23. Simplify: $3\sqrt{a} - 9\sqrt{a}$

24. Simplify: $\sqrt{108}$

25. Find the decimal approximation of $\sqrt{63}$ to the nearest thousandth.

26. Simplify: $\dfrac{\sqrt{108a^7b^3}}{\sqrt{3a^4b}}$

27. Solve: $\sqrt{x} - \sqrt{x+3} = 1$

28. The square root of the sum of two consecutive integers is equal to 9. Find the smaller integer.

29. Find the length of a pendulum that makes one swing in 1.5 s. The equation for the time of one swing of a pendulum is $T = 2\pi\sqrt{\dfrac{L}{32}}$, where T is the time in seconds, and L is the length in feet. Round to the nearest hundredth.

30. A ladder 10 ft long is resting against a building. How high on the building will the ladder reach when the bottom of the ladder is 2 ft from the building? Round to the nearest hundredth.

Cumulative Review

1. Simplify: $\left(\dfrac{2}{3}\right)^2 \cdot \left(\dfrac{3}{4} - \dfrac{3}{2}\right) + \left(\dfrac{1}{2}\right)^2$

2. Simplify: $-3[x - 2(3 - 2x) - 5x] + 2x$

3. Solve:
$2x - 4[3x - 2(1 - 3x)] = 2(3 - 4x)$

4. Simplify: $(-3x^2y)(-2x^3y^{-4})$

5. Simplify: $\dfrac{12b^4 - 6b^2 + 2}{-6b^2}$

6. Factor: $12x^3y^2 - 9x^2y^3$

7. Factor: $9b^2 + 3b - 20$

8. Factor: $2a^3 - 16a^2 + 30a$

9. Simplify: $\dfrac{3x^3 - 6x^2}{4x^2 + 4x} \cdot \dfrac{3x - 9}{9x^3 - 45x^2 + 54x}$

10. Simplify: $\dfrac{1 - \dfrac{2}{x} - \dfrac{15}{x^2}}{1 - \dfrac{9}{x^2}}$

11. Simplify: $\dfrac{x+2}{x-4} - \dfrac{6}{(x-4)(x-3)}$

12. Solve: $\dfrac{x}{2x-5} - 2 = \dfrac{3x}{2x-5}$

13. Find the slope of the line containing the points whose coordinates are $(2, -5)$ and $(-4, 3)$.

14. Find the equation of the line that contains the point whose coordinates are $(-2, -3)$ and has slope $\dfrac{1}{2}$.

15. Solve by substitution:
$$4x - 3y = 1$$
$$2x + y = 3$$

16. Solve by the addition method:
$$5x + 4y = 7$$
$$3x - 2y = 13$$

17. Solve by graphing:
$$3x - 2y = 8$$
$$4x + 5y = 3$$

18. Find $A \cup B$ given $A = \{-4, -2, 0, 2, 4\}$ and $B = \{-8, -4, 0, 4, 8\}$.

19. Solve: $3(x - 7) \geq 5x - 12$

20. Graph the solution set of $2x + y < -2$.

21. Use set builder notation to write the set of real numbers greater than 30.

22. Simplify: $2\sqrt{27a} - 5\sqrt{49a} + 8\sqrt{48a}$

23. Simplify: $\dfrac{\sqrt{320}}{\sqrt{5}}$

24. Solve: $\sqrt{2x - 3} - 5 = 0$

25. The selling price of a book is $29.40. The markup rate used by the bookstore is 20% of the cost. Find the cost of the book.

26. How many ounces of pure water must be added to 40 oz of a 12% salt solution to make a salt solution that is 5% salt?

27. The sum of two numbers is twenty-one. The product of the two numbers is one hundred four. Find the two numbers.

28. A small water pipe takes twice as long to fill a tank as does a larger water pipe. With both pipes open, it takes 16 h to fill the tank. Find the time it would take the small pipe, working alone, to fill the tank.

29. Three-eighths of a number is less than negative twelve. Find the largest integer that satisfies the inequality.

30. The square root of the sum of two consecutive integers is equal to 7. Find the smaller integer.

12

Quadratic Equations

Objectives

- Solve quadratic equations by factoring
- Solve quadratic equations by taking square roots
- Solve quadratic equations by completing the square
- Solve quadratic equations by using the quadratic formula
- Graph a quadratic equation of the form $y = ax^2 + bx + c$
- Application problems

Algebraic Symbolism

The way in which an algebraic expression or equation is written has gone through several stages of development. First there was the *rhetoric*, which was in vogue until the late 13th century. In this method, an expression would be written out in sentences. The word *res* was used to represent an unknown.

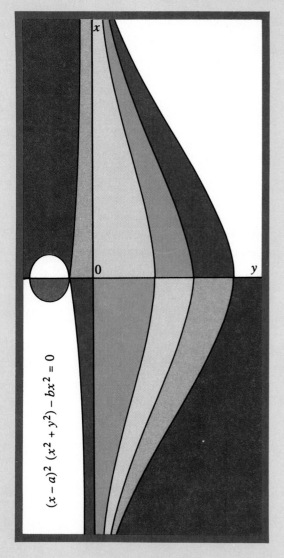

$(x - a)^2 (x^2 + y^2) - bx^2 = 0$

Rhetoric: From the additive *res* in the additive *res* results in a square *res*. From the three in an additive *res* comes three additive *res*, and from the subtractive four in the additive *res* comes subtractive four *res*. From three in subtractive four comes subtractive twelve.

Modern: $(x + 3)(x - 4) = x^2 - x - 12$

The second stage was the *syncoptic*, which was a shorthand in which abbreviations were used for words.

Syncoptic: *a* 6 in *b* quad $-$ *c* plano 4 in *b* + *b* cub

Modern: $6ab^2 - 4cb + b^3$

The current modern stage, called the *symbolic* stage, began with the use of exponents rather than words to symbolize exponential expressions. This occurred near the beginning of the 17th century with the publication of the book *La Geometrie* by René Descartes. Modern notation is still evolving as mathematicians continue to search for convenient methods to symbolize concepts.

Solving Quadratic Equations by Factoring or by Taking Square Roots

1 Solve quadratic equations by factoring

In Section 5 of Chapter 6, we solved quadratic equations by factoring. In this section, we will review that material and then solve quadratic equations by taking square roots.

An equation of the form $ax^2 + bx + c = 0$, where a, b, and c are constants and $a \neq 0$, is a **quadratic equation.**

$$4x^2 - 3x + 1 = 0, \quad a = 4, \ b = -3, \ c = 1$$

$$3x^2 - 4 = 0, \quad a = 3, \ b = 0, \quad c = -4$$

A quadratic equation is also called a **second-degree equation.**

A quadratic equation is in **standard form** when the polynomial is in descending order and equal to zero.

Recall that the Principle of Zero Products states that if the product of two factors is zero, then at least one of the factors must be zero.

If $a \cdot b = 0$, then $a = 0$ or $b = 0$.

The Principle of Zero Products can be used in solving quadratic equations.

Example 1   Solve by factoring: $2x^2 - x = 1$

Solution
$$2x^2 - x = 1$$
$$2x^2 - x - 1 = 0 \qquad \blacktriangleright \text{Write the equation in standard form.}$$
$$(2x + 1)(x - 1) = 0 \qquad \blacktriangleright \text{Factor.}$$

$$2x + 1 = 0 \qquad x - 1 = 0 \qquad \blacktriangleright \text{Let each factor equal zero.}$$
$$2x = -1 \qquad\quad x = 1 \qquad \blacktriangleright \text{Solve each equation for } x.$$
$$x = -\frac{1}{2}$$

The solutions are $-\frac{1}{2}$ and 1.

473

Problem 1   Solve by factoring: $3t = t^2 - 10$

Solution   See page A41.

Example 2   Solve by factoring: $x^2 = 10x - 25$

Solution
$$x^2 = 10x - 25$$
$$x^2 - 10x + 25 = 0 \qquad \blacktriangleright \text{Write the equation in standard form.}$$
$$(x - 5)(x - 5) = 0 \qquad \blacktriangleright \text{Factor.}$$
$$x - 5 = 0 \qquad x - 5 = 0 \qquad \blacktriangleright \text{Let each factor equal zero.}$$
$$x = 5 \qquad\qquad x = 5$$

The solution is 5.

The factorization in Example 2 produced two identical factors. Because $x - 5$ occurs twice in the factored form of the equation, 5 is a **double root** of the equation.

Problem 2   Solve by factoring: $(x + 5)(x - 1) = 3x + 1$

Solution   See page A41.

2   Solve quadratic equations by taking square roots

Consider a quadratic equation of the form $x^2 = a$. This equation can be solved by factoring.

$$x^2 = 25$$
$$x^2 - 25 = 0$$
$$(x + 5)(x - 5) = 0$$
$$x + 5 = 0 \qquad x - 5 = 0$$
$$x = -5 \qquad x = 5$$

The solutions are -5 and 5.

Solutions that are plus or minus the same number are frequently written using $\pm$. For the example above, this would be written: "The solutions are ± 5." Because the solutions ± 5 can be written as $\pm\sqrt{25}$, an alternative method of solving this equation is suggested.

Principle of Taking the Square Root of Each Side of an Equation

If $x^2 = a$, then $x = \pm\sqrt{a}$.

Solve by taking square roots: $x^2 = 25$

Take the square root of each side the equation.	$x^2 = 25$ $\sqrt{x^2} = \sqrt{25}$
Simplify.	$x = \pm\sqrt{25} = \pm 5$
Write the solutions.	The solutions are 5 and -5.

Solve by taking square roots: $3x^2 = 36$

Solve for x^2.	$3x^2 = 36$ $x^2 = 12$
Take the square root of each side of the equation.	$\sqrt{x^2} = \sqrt{12}$
Simplify.	$x = \pm\sqrt{12} = \pm 2\sqrt{3}$

Check:

$$
\begin{array}{c|c}
\multicolumn{2}{c}{3x^2 = 36} \\
\hline
3(2\sqrt{3})^2 & 36 \\
3(12) & 36 \\
36 = 36 &
\end{array}
\qquad
\begin{array}{c|c}
\multicolumn{2}{c}{3x^2 = 36} \\
\hline
3(-2\sqrt{3})^2 & 36 \\
3(12) & 36 \\
36 = 36 &
\end{array}
$$

These are true equations.
The solutions check.
Write the solutions.

The solutions are $2\sqrt{3}$ and $-2\sqrt{3}$.

Example 3   Solve by taking square roots.   A. $2x^2 - 72 = 0$ B. $x^2 + 16 = 0$

Solution   A. $2x^2 - 72 = 0$
$$2x^2 = 72 \qquad \blacktriangleright \text{Solve for } x^2.$$
$$x^2 = 36$$
$$\sqrt{x^2} = \sqrt{36} \qquad \blacktriangleright \text{Take the square root of each side of the equation.}$$
$$x = \pm 6$$

The solutions are 6 and -6.

B. $x^2 + 16 = 0$
$$x^2 = -16$$
$$\sqrt{x^2} = \sqrt{-16} \qquad \blacktriangleright \sqrt{-16} \text{ is not a real number.}$$

The equation has no real number solution.

Problem 3   Solve by taking square roots.   A. $4x^2 - 96 = 0$ B. $x^2 + 81 = 0$

  Solution   See page A42.

An equation containing the square of a binomial can be solved by taking square roots.

Solve by taking square roots: $2(x - 1)^2 - 36 = 0$

Solve for $(x - 1)^2$.
$$2(x - 1)^2 - 36 = 0$$
$$2(x - 1)^2 = 36$$
$$(x - 1)^2 = 18$$

Take the square root of each side of the equation.
$$\sqrt{(x - 1)^2} = \sqrt{18}$$

Simplify.
$$x - 1 = \pm\sqrt{18}$$
$$x - 1 = \pm 3\sqrt{2}$$

Solve for x.
$$x - 1 = 3\sqrt{2} \qquad x - 1 = -3\sqrt{2}$$
$$x = 1 + 3\sqrt{2} \qquad x = 1 - 3\sqrt{2}$$

$1 + 3\sqrt{2}$ and $1 - 3\sqrt{2}$ check as solutions.

Write the solutions.   The solutions are $1 + 3\sqrt{2}$ and $1 - 3\sqrt{2}$.

Example 4   Solve by taking square roots: $(x - 6)^2 = 12$

  Solution
$$(x - 6)^2 = 12$$
$$\sqrt{(x - 6)^2} = \sqrt{12}$$
  ▶ Take the square root of each side of the equation.

$$x - 6 = \pm\sqrt{12}$$
$$x - 6 = \pm 2\sqrt{3}$$

$$x - 6 = 2\sqrt{3} \qquad x - 6 = -2\sqrt{3}$$
$$x = 6 + 2\sqrt{3} \qquad x = 6 - 2\sqrt{3}$$
  ▶ Solve for x.

The solutions are $6 + 2\sqrt{3}$ and $6 - 2\sqrt{3}$.

Problem 4   Solve by taking square roots: $(x + 5)^2 = 20$

  Solution   See page A42.

EXERCISES 12.1

1 Solve by factoring.

1. $x^2 + 2x - 15 = 0$ **2.** $t^2 + 3t - 10 = 0$ **3.** $z^2 - 4z + 3 = 0$ **4.** $s^2 - 5s + 4 = 0$

5. $p^2 + 3p + 2 = 0$ **6.** $v^2 + 6v + 5 = 0$ **7.** $x^2 - 6x + 9 = 0$ **8.** $y^2 - 8y + 16 = 0$

9. $6x^2 - 9x = 0$ **10.** $r^2 - 10 = 3r$ **11.** $t^2 - 12 = 4t$ **12.** $3v^2 - 5v + 2 = 0$

13. $2p^2 - 3p - 2 = 0$ **14.** $3s^2 + 8s = 3$ **15.** $3x^2 + 5x = 12$ **16.** $9z^2 = 12z - 4$

17. $6r^2 = 12 - r$ **18.** $4t^2 = 4t + 3$ **19.** $5y^2 + 11y = 12$ **20.** $4v^2 - 4v + 1 = 0$

21. $9s^2 - 6s + 1 = 0$ **22.** $x^2 - 9 = 0$ **23.** $t^2 - 16 = 0$ **24.** $4y^2 - 1 = 0$

25. $9z^2 - 4 = 0$ **26.** $x + 15 = x(x - 1)$

27. $p + 18 = p(p - 2)$ **28.** $r^2 - r - 2 = (2r - 1)(r - 3)$

29. $s^2 + 5s - 4 = (2s + 1)(s - 4)$ **30.** $x^2 + x + 5 = (3x + 2)(x - 4)$

2 Solve by taking square roots.

31. $x^2 = 36$ **32.** $y^2 = 49$ **33.** $v^2 - 1 = 0$ **34.** $z^2 - 64 = 0$

35. $4x^2 - 49 = 0$ **36.** $9w^2 - 64 = 0$ **37.** $9y^2 = 4$ **38.** $4z^2 = 25$

39. $16v^2 - 9 = 0$ **40.** $25x^2 - 64 = 0$ **41.** $y^2 + 81 = 0$ **42.** $z^2 + 49 = 0$

43. $w^2 - 24 = 0$ **44.** $v^2 - 48 = 0$ **45.** $(x - 1)^2 = 36$ **46.** $(y + 2)^2 = 49$

47. $2(x + 5)^2 = 8$

48. $4(z - 3)^2 = 100$

49. $2(x + 1)^2 = 50$

50. $3(x - 4)^2 = 27$

51. $4(x + 5)^2 = 64$

52. $9(x - 3)^2 = 81$

53. $2(x - 9)^2 = 98$

54. $5(x + 5)^2 = 125$

55. $12(x + 3)^2 = 27$

56. $8(x - 4)^2 = 50$

57. $9(x - 1)^2 - 16 = 0$

58. $4(y + 3)^2 - 81 = 0$

59. $49(v + 1)^2 - 25 = 0$

60. $81(y - 2)^2 - 64 = 0$

61. $(x - 4)^2 - 20 = 0$

62. $(y + 5)^2 - 50 = 0$

63. $(x + 1)^2 + 36 = 0$

64. $(y - 7)^2 + 49 = 0$

65. $2\left(z - \dfrac{1}{2}\right)^2 = 12$

66. $3\left(v + \dfrac{3}{4}\right)^2 = 36$

67. $4\left(x - \dfrac{2}{3}\right)^2 = 16$

SUPPLEMENTAL EXERCISES 12.1

For the given quadratic equation, find the values of a, b, and c.

68. $3x^2 - 4x + 1 = 0$

69. $2x^2 - 5 = 0$

70. $6x^2 - 3x = 0$

Write the quadratic equation in standard form.

71. $x + 5 = x(x - 3)$

72. $2(x + 3)^2 = 5$

73. $4\left(x - \dfrac{1}{2}\right)^2 = 6$

74. $3\left(\dfrac{x - 3}{3}\right)^2 = 27$

75. $\dfrac{x^2}{2} - \dfrac{x}{4} = 5$

76. $\dfrac{2x^2}{3} - 4 = \dfrac{x}{2}$

Solve for x.

77. $\dfrac{3x^2}{2} = 4x - 2$

78. $\dfrac{2x^2}{5} = 3x - 5$

79. $\dfrac{2x^2}{9} + x = 2$

80. $ax^2 - bx = 0$, $a > 0$ and $b > 0$

81. $ax^2 - b = 0$, $a > 0$ and $b > 0$

82. $x^2 = x$

SECTION **12.2**

Solving Quadratic Equations by Completing the Square

1 Solve quadratic equations by completing the square

Recall that a perfect square trinomial is the square of a binomial.

Perfect square trinomial		Square of a binomial
$x^2 + 6x + 9$	$=$	$(x + 3)^2$
$x^2 - 10x + 25$	$=$	$(x - 5)^2$
$x^2 + 8x + 16$	$=$	$(x + 4)^2$

For each perfect square trinomial, the square of $\frac{1}{2}$ of the coefficient of x equals the constant term.

$$x^2 + 6x + 9, \quad \left(\frac{1}{2} \cdot 6\right)^2 = 9$$

$$x^2 - 10x + 25, \quad \left[\frac{1}{2}(-10)\right]^2 = 25 \qquad \left(\frac{1}{2} \text{ coefficient of } x\right)^2 = \text{Constant term}$$

$$x^2 + 8x + 16, \quad \left(\frac{1}{2} \cdot 8\right)^2 = 16$$

This relationship can be used to write the constant term for a perfect square trinomial. Adding to a binomial the constant term that makes it a perfect square trinomial is called **completing the square.**

Complete the square of $x^2 - 8x$. Write the resulting perfect square trinomial as the square of a binomial.

$$x^2 - 8x$$

Find the constant term.

$$\left[\frac{1}{2}(-8)\right]^2 = 16$$

Complete the square of $x^2 - 8x$ by adding the constant term.

$$x^2 - 8x + 16$$

Write the resulting perfect square trinomial as the square of a binomial.

$$x^2 - 8x + 16 = (x - 4)^2$$

Complete the square of $y^2 + 5y$. Write the resulting perfect square trinomial as the square of a binomial.

$$y^2 + 5y$$

Find the constant term.

$$\left(\frac{1}{2} \cdot 5\right)^2 = \left(\frac{5}{2}\right)^2 = \frac{25}{4}$$

Complete the square of $y^2 + 5y$ by adding the constant term.

$$y + 5y + \frac{25}{4}$$

Write the resulting perfect square trinomial as the square of a binomial.

$$y^2 + 5y + \frac{25}{4} = \left(y + \frac{5}{2}\right)^2$$

A quadratic equation that cannot be solved by factoring can be solved by completing the square. Add to each side of the equation the term that completes the square. Rewrite the quadratic equation in the form $(x + a)^2 = b$. Take the square root of each side of the equation, and then solve for x.

Solve by completing the square: $x^2 - 6x - 3 = 0$

Add the opposite of the constant term to each side of the equation.

$$x^2 - 6x - 3 = 0$$
$$x^2 - 6x = 3$$

Find the constant term that completes the square of $x^2 - 6x$.

$$\left[\frac{1}{2}(-6)\right]^2 = 9$$

Add this term to each side of the equation.

$$x^2 - 6x + 9 = 3 + 9$$

Factor the perfect square trinomial.

$$(x - 3)^2 = 12$$

Take the square root of each side of the equation.

$$\sqrt{(x - 3)^2} = \sqrt{12}$$

Simplify.

$$x - 3 = \pm\sqrt{12}$$
$$x - 3 = \pm 2\sqrt{3}$$

Solve for x.

$$x - 3 = 2\sqrt{3} \qquad x - 3 = -2\sqrt{3}$$
$$x = 3 + 2\sqrt{3} \qquad x = 3 - 2\sqrt{3}$$

Check:

$$\begin{array}{c|c}
x^2 - 6x - 3 = 0 \\
\hline
(3 + 2\sqrt{3})^2 - 6(3 + 2\sqrt{3}) - 3 & 0 \\
9 + 12\sqrt{3} + 12 - 18 - 12\sqrt{3} - 3 & \\
& 0 = 0
\end{array}$$

$$\begin{array}{c|c}
x^2 - 6x - 3 = 0 \\
\hline
(3 - 2\sqrt{3})^2 - 6(3 - 2\sqrt{3}) - 3 & 0 \\
9 - 12\sqrt{3} + 12 - 18 + 12\sqrt{3} - 3 & \\
& 0 = 0
\end{array}$$

Write the solution.

The solutions are $3 + 2\sqrt{3}$ and $3 - 2\sqrt{3}$.

Solve by completing the square: $2x^2 - x - 1 = 0$

Add the opposite of the constant term to each side of the equation.

$$2x^2 - x - 1 = 0$$
$$2x^2 - x = 1$$

To complete the square, the coefficient of the x^2 term must be 1. Multiply each term by the reciprocal of the coefficient of x^2.

$$\frac{1}{2}(2x^2 - x) = \frac{1}{2} \cdot 1$$
$$x^2 - \frac{1}{2}x = \frac{1}{2}$$

Find the constant term that completes the square of $x^2 - \frac{1}{2}x$.

$$\left[\frac{1}{2}\left(-\frac{1}{2}\right)\right]^2 = \left(-\frac{1}{4}\right)^2 = \frac{1}{16}$$

Add this term to each side of the equation.

$$x^2 - \frac{1}{2}x + \frac{1}{16} = \frac{1}{2} + \frac{1}{16}$$

Factor the perfect square trinomial.

$$\left(x - \frac{1}{4}\right)^2 = \frac{9}{16}$$

Take the square root of each side of the equation.

$$\sqrt{\left(x - \frac{1}{4}\right)^2} = \sqrt{\frac{9}{16}}$$

Simplify.

$$x - \frac{1}{4} = \pm\sqrt{\frac{9}{16}}$$

$$x - \frac{1}{4} = \pm\frac{3}{4}$$

Solve for x.

$$x - \frac{1}{4} = \frac{3}{4} \qquad x - \frac{1}{4} = -\frac{3}{4}$$

$$x = 1 \qquad\qquad x = -\frac{1}{2}$$

Check:

$$
\begin{array}{c|c}
2x^2 - x - 1 = 0 & \\
\hline
2(1)^2 - 1 - 1 & 0 \\
2(1) - 1 - 1 & \\
2 - 1 - 1 & \\
0 = 0 &
\end{array}
$$

$$
\begin{array}{c|c}
2x^2 - x - 1 = 0 & \\
\hline
2\left(-\frac{1}{2}\right)^2 - \left(-\frac{1}{2}\right) - 1 & 0 \\
2\left(\frac{1}{4}\right) - \left(-\frac{1}{2}\right) - 1 & \\
\frac{1}{2} + \frac{1}{2} - 1 & \\
0 = 0 &
\end{array}
$$

Write the solution.

The solutions are 1 and $-\frac{1}{2}$.

Example 1   Solve by completing the square:
$2x^2 - 4x - 1 = 0$

Solution $2x^2 - 4x - 1 = 0$

$\qquad\qquad 2x^2 - 4x = 1$ ▶ Add the opposite of -1 to each side of the equation.

$\qquad \dfrac{1}{2}\left(2x^2 - 4x\right) = \dfrac{1}{2} \cdot 1$ ▶ The coefficient of the x^2 term must be 1.

$\qquad\qquad x^2 - 2x = \dfrac{1}{2}$

$\qquad x^2 - 2x + 1 = \dfrac{1}{2} + 1$ ▶ Complete the square of $x^2 - 2x$. Add 1 to each side of the equation.

$\qquad\qquad (x - 1)^2 = \dfrac{3}{2}$ ▶ Factor the perfect square trinomial.

$\qquad\qquad \sqrt{(x - 1)^2} = \sqrt{\dfrac{3}{2}}$ ▶ Take the square root of each side of the equation.

$\qquad\qquad x - 1 = \pm\sqrt{\dfrac{3}{2}}$

$\qquad\qquad x - 1 = \pm\dfrac{\sqrt{6}}{2}$

$x - 1 = \dfrac{\sqrt{6}}{2} \qquad\qquad x - 1 = -\dfrac{\sqrt{6}}{2}$

$\quad x = 1 + \dfrac{\sqrt{6}}{2} \qquad\qquad x = 1 - \dfrac{\sqrt{6}}{2}$

$\qquad = \dfrac{2}{2} + \dfrac{\sqrt{6}}{2} \qquad\qquad = \dfrac{2}{2} - \dfrac{\sqrt{6}}{2}$

$\qquad = \dfrac{2 + \sqrt{6}}{2} \qquad\qquad = \dfrac{2 - \sqrt{6}}{2}$

The solutions are $\dfrac{2 + \sqrt{6}}{2}$ and $\dfrac{2 - \sqrt{6}}{2}$.

Problem 1   Solve by completing the square:
$3x^2 - 6x - 2 = 0$

Solution   See page A42.

Example 2   Solve by completing the square: $x^2 + 6x + 4 = 0$
Approximate the solutions to the nearest thousandth.

Solution $x^2 + 6x + 4 = 0$

$x^2 + 6x = -4$

$x^2 + 6x + 9 = -4 + 9$ ▶ Complete the square of $x^2 + 6x$. Add 9 to each
side of the equation.

$(x + 3)^2 = 5$

$\sqrt{(x + 3)^2} = \sqrt{5}$

$x + 3 = \pm \sqrt{5}$

$x + 3 = \sqrt{5}$ $x + 3 = -\sqrt{5}$

$x = -3 + \sqrt{5}$ $x = -3 - \sqrt{5}$

$\approx -3 + 2.236$ $\approx -3 - 2.236$

≈ -0.764 ≈ -5.236

The solutions are approximately -0.764 and -5.236.

Problem 2   Solve by completing the square: $x^2 + 8x + 8 = 0$
Approximate the solutions to the nearest thousandth.

Solution   See page A43

EXERCISES 12.2

1   Solve by completing the square.

1. $x^2 + 2x - 3 = 0$

2. $y^2 + 4y - 5 = 0$

3. $z^2 - 6z - 16 = 0$

4. $w^2 + 8w - 9 = 0$

5. $x^2 = 4x - 4$

6. $z^2 = 8z - 16$

7. $v^2 - 6v + 13 = 0$

8. $x^2 + 4x + 13 = 0$

9. $y^2 + 5y + 4 = 0$

10. $v^2 - 5v - 6 = 0$

11. $w^2 + 7w = 8$

12. $y^2 + 5y = -4$

13. $v^2 + 4v + 1 = 0$

14. $y^2 - 2y - 5 = 0$

15. $x^2 + 6x = 5$

16. $w^2 - 8w = 3$

17. $z^2 = 2z + 1$

18. $y^2 = 10y - 20$

19. $p^2 + 3p = 1$

20. $r^2 + 5r = 2$

21. $t^2 - 3t = -2$

22. $x^2 + 6x + 4 = 0$

23. $y^2 - 8y - 1 = 0$

24. $y^2 + y - 2 = 0$

25. $r^2 - 8r = -2$

26. $s^2 + 6s = 5$

27. $x^2 = 4 - 2x$

28. $y^2 = 4y + 12$

29. $x^2 = 1 - 3x$

30. $w^2 = 3w + 5$

31. $x^2 - x - 1 = 0$

32. $x^2 - 7x = -3$

33. $y^2 - 5y + 3 = 0$

34. $z^2 - 5z = -2$

35. $v^2 + v - 3 = 0$

36. $x^2 - x = 1$

37. $y^2 = 7 - 10y$

38. $v^2 = 14 + 16v$

39. $r^2 - 3r = 5$

40. $s^2 + 3s = -1$

41. $t^2 - t = 4$

42. $y^2 + y - 4 = 0$

43. $x^2 - 3x + 5 = 0$

44. $z^2 + 5z + 7 = 0$

45. $2t^2 - 3t + 1 = 0$

46. $2x^2 - 7x + 3 = 0$

47. $2r^2 + 5r = 3$

48. $2y^2 - 3y = 9$

49. $2s^2 = 7s - 6$

50. $2x^2 = 3x + 20$

51. $2v^2 = v + 1$

52. $2z^2 = z + 3$

53. $3r^2 + 5r = 2$

54. $3t^2 - 8t = 3$

55. $3y^2 + 8y + 4 = 0$

56. $3z^2 - 10z - 8 = 0$

57. $4x^2 + 4x - 3 = 0$

58. $4v^2 + 4v - 15 = 0$

59. $6s^2 + 7s = 3$

60. $6z^2 = z + 2$

61. $6p^2 = 5p + 4$

62. $6t^2 = t - 2$

63. $4v^2 - 4v - 1 = 0$

64. $2s^2 - 4s - 1 = 0$ **65.** $4z^2 - 8z = 1$ **66.** $3r^2 - 2r = 2$

67. $3y - 6 = (y - 1)(y - 2)$ **68.** $7s + 55 = (s + 5)(s + 4)$ **69.** $4p + 2 = (p - 1)(p + 3)$

Solve by completing the square. Approximate the solutions to the nearest thousandth.

70. $y^2 + 3y = 5$ **71.** $w^2 + 5w = 2$ **72.** $2z^2 - 3z = 7$

73. $2x^2 + 3x = 11$ **74.** $4x^2 + 6x - 1 = 0$ **75.** $4x^2 + 2x - 3 = 0$

SUPPLEMENTAL EXERCISES 12.2

Solve.

76. $\dfrac{x^2}{4} - \dfrac{x}{2} = 3$ **77.** $\dfrac{x^2}{6} - \dfrac{x}{3} = 1$ **78.** $\dfrac{2x^2}{3} = 2x + 3$

79. $\dfrac{3x^2}{2} = 3x + 2$ **80.** $\sqrt{x + 2} = x - 4$ **81.** $\sqrt{x + 7} = 5 - x$

82. $\sqrt{2x + 5} - 3 = x$ **83.** $\sqrt{3x + 4} - x = 2$ **84.** $\dfrac{x}{3} + \dfrac{3}{x} = \dfrac{8}{3}$

85. $\dfrac{x}{4} + \dfrac{2}{x} = \dfrac{3}{2}$ **86.** $\dfrac{x + 1}{2} + \dfrac{3}{x - 1} = 4$ **87.** $\dfrac{x - 2}{3} + \dfrac{2}{x + 2} = 4$

SECTION 12.3

Solving Quadratic Equations by Using the Quadratic Formula

1 Solve quadratic equations by using the quadratic formula

Any quadratic equation can be solved by completing the square. Applying this method to the standard form of a quadratic equation produces a formula that can be used to solve any quadratic equation. This is shown on the next page.

To solve $ax^2 + bx + c = 0$, $a \neq 0$, by completing the square, subtract the constant term from each side of the equation.

$$ax^2 + bx + c = 0$$
$$ax^2 + bx + c - c = 0 - c$$
$$ax^2 + bx = -c$$

Multiply each side of the equation by the reciprocal of a, the coefficient of x^2.

$$\frac{1}{a}(ax^2 + bx) = \frac{1}{a}(-c)$$
$$x^2 + \frac{b}{a}x = -\frac{c}{a}$$

Complete the square by adding $\left(\frac{1}{2} \cdot \frac{b}{a}\right)^2$ to each side of the equation.

$$x^2 + \frac{b}{a}x \quad \left(\frac{1}{2} \cdot \frac{b}{a}\right)^2 = \left(\frac{1}{2} \cdot \frac{b}{a}\right)^2 - \frac{c}{a}$$
$$x^2 + \frac{b}{a}x + \frac{b^2}{4a^2} = \frac{b^2}{4a^2} - \frac{c}{a}$$

Simplify the right side of the equation.

$$x^2 + \frac{b}{a}x + \frac{b^2}{4a^2} = \frac{b^2}{4a^2} - \left(\frac{c}{a} \cdot \frac{4a}{4a}\right)$$
$$x^2 + \frac{b}{a}x + \frac{b^2}{4a^2} = \frac{b^2}{4a^2} - \frac{4ac}{4a^2}$$
$$x^2 + \frac{b}{a}x + \frac{b^2}{4a^2} = \frac{b^2 - 4ac}{4a^2}$$

Factor the perfect square trinomial on the left side of the equation.

$$\left(x + \frac{b}{2a}\right)^2 = \frac{b^2 - 4ac}{4a^2}$$

Take the square root of each side of the equation.

$$\sqrt{\left(x + \frac{b}{2a}\right)^2} = \sqrt{\frac{b^2 - 4ac}{4a^2}}$$
$$\left(x + \frac{b}{2a}\right) = \pm\frac{\sqrt{b^2 - 4ac}}{2a}$$

Solve for x.

$$x + \frac{b}{2a} = \frac{\sqrt{b^2 - 4ac}}{2a} \qquad\qquad x + \frac{b}{2a} = -\frac{\sqrt{b^2 - 4ac}}{2a}$$
$$x = -\frac{b}{2a} + \frac{\sqrt{b^2 - 4ac}}{2a} \qquad\qquad x = -\frac{b}{2a} - \frac{\sqrt{b^2 - 4ac}}{2a}$$
$$x = \frac{-b + \sqrt{b^2 - 4ac}}{2a} \qquad\qquad x = \frac{-b - \sqrt{b^2 - 4ac}}{2a}$$

The Quadratic Formula

The solution of $ax^2 + bx + c = 0$, $a \neq 0$, is

$$x = \frac{-b + \sqrt{b^2 - 4ac}}{2a} \quad \text{or} \quad x = \frac{-b - \sqrt{b^2 - 4ac}}{2a}.$$

The quadratic formula is frequently written in the form

$$x = \frac{-b \pm \sqrt{b^2 - 4ac}}{2a}.$$

Solve by using the quadratic formula: $2x^2 = 4x - 1$

$$2x^2 = 4x - 1$$

Write the equation in standard form.
$a = 2$, $b = -4$, and $c = 1$.

$$2x^2 - 4x + 1 = 0$$

Replace a, b, and c in the quadratic formula by their values.

$$x = \frac{-b \pm \sqrt{b^2 - 4ac}}{2a}$$

Simplify.

$$= \frac{-(-4) \pm \sqrt{(-4)^2 - 4 \cdot 2 \cdot 1}}{2 \cdot 2} = \frac{4 \pm \sqrt{16 - 8}}{4}$$

$$= \frac{4 \pm \sqrt{8}}{4} = \frac{4 \pm 2\sqrt{2}}{4} = \frac{2 \pm \sqrt{2}}{2}$$

$\dfrac{2 + \sqrt{2}}{2}$ and $\dfrac{2 - \sqrt{2}}{2}$ check as solutions.

Write the solutions.

The solutions are $\dfrac{2 + \sqrt{2}}{2}$ and $\dfrac{2 - \sqrt{2}}{2}$.

Example 1   Solve by using the quadratic formula: $2x^2 - 3x + 1 = 0$

Solution $2x^2 - 3x + 1 = 0$ ▶ $a = 2$, $b = -3$, $c = 1$

$$x = \frac{-(-3) \pm \sqrt{(-3)^2 - 4(2)(1)}}{2 \cdot 2}$$

$$= \frac{3 \pm \sqrt{9 - 8}}{4} = \frac{3 \pm \sqrt{1}}{4} = \frac{3 \pm 1}{4}$$

$$x = \frac{3 + 1}{4} \qquad\qquad x = \frac{3 - 1}{4}$$

$$= \frac{4}{4} = 1 \qquad\qquad = \frac{2}{4} = \frac{1}{2}$$

The solutions are 1 and $\dfrac{1}{2}$.

Problem 1   Solve by using the quadratic formula.
A. $3x^2 + 4x - 4 = 0$ B. $2x^2 - 3x + 4 = 0$

Solution   See page A43.

Example 2 Solve by using the quadratic formula: $2x^2 = 8x - 5$

Solution

$$2x^2 = 8x - 5$$

$$2x^2 - 8x + 5 = 0$$ ▶ Write the equation in standard form.

$a = 2, b = -8, c = 5$

$$x = \frac{-(-8) \pm \sqrt{(-8)^2 - 4(2)(5)}}{2 \cdot 2}$$

$$= \frac{8 \pm \sqrt{64 - 40}}{4} = \frac{8 \pm \sqrt{24}}{4}$$

$$= \frac{8 \pm 2\sqrt{6}}{4} = \frac{2(4 \pm \sqrt{6})}{2 \cdot 2}$$

$$= \frac{4 \pm \sqrt{6}}{2}$$

The solutions are $\frac{4 + \sqrt{6}}{2}$ and $\frac{4 - \sqrt{6}}{2}$.

Problem 2 Solve by using the quadratic formula: $x^2 + 2x = 1$

Solution See page A43.

EXERCISES 12.3

1 Solve by using the quadratic formula.

1. $x^2 - 4x - 5 = 0$

2. $y^2 + 3y + 2 = 0$

3. $z^2 - 2z - 15 = 0$

4. $v^2 + 5v + 4 = 0$

5. $z^2 + 6z - 7 = 0$

6. $s^2 + 3s - 10 = 0$

7. $t^2 + t - 6 = 0$

8. $x^2 - x - 2 = 0$

9. $y^2 = 2y + 3$

10. $w^2 = 3w + 18$

11. $r^2 = 5 - 4r$

12. $z^2 = 3 - 2z$

13. $2y^2 - y - 1 = 0$

14. $2t^2 - 5t + 3 = 0$

15. $w^2 + 3w + 5 = 0$

16. $x^2 - 2x + 6 = 0$

17. $p^2 - p = 0$

18. $2v^2 + v = 0$

19. $4t^2 - 9 = 0$

20. $4s^2 - 25 = 0$

21. $4y^2 + 4y = 15$

22. $4r^2 + 4r = 3$

23. $3t^2 = 7t + 6$

24. $3x^2 = 10x + 8$

25. $5z^2 + 11z = 12$

26. $4v^2 = v + 3$

27. $6s^2 - s - 2 = 0$

28. $6y^2 + 5y - 4 = 0$

29. $2x^2 + x + 1 = 0$

30. $3r^2 - r + 2 = 0$

31. $t^2 - 2t = 5$

32. $y^2 - 4y = 6$

33. $t^2 + 6t - 1 = 0$

34. $z^2 + 4z + 1 = 0$

35. $w^2 = 4w + 9$

36. $y^2 = 8y + 3$

37. $4t^2 - 4t - 1 = 0$

38. $4x^2 - 8x - 1 = 0$

39. $v^2 + 6v + 1 = 0$

40. $2x^2 - x - 1 = 0$

41. $4x^2 + 3x - 1 = 0$

42. $3x^2 - 6x + 2 = 0$

43. $3y^2 - 5y + 2 = 0$

44. $2x^2 - x = 3$

45. $5x^2 - 6x = 3$

46. $3t^2 = 2t + 3$

47. $4n^2 = 7n - 2$

48. $4p^2 = -12p - 9$

49. $3y^2 + 6y = -3$

50. $2y^2 + 3 = 8y$

51. $5x^2 - 1 = x$

52. $3y^2 - 4 = 5y$

53. $6x^2 - 5 = 3x$

54. $3x^2 = x + 3$

55. $2n^2 = 7 - 3n$

56. $5d^2 - 2d - 8 = 0$

57. $x^2 - 7x - 10 = 0$

58. $s^2 + 4s - 8 = 0$

59. $4t^2 - 12t - 15 = 0$

60. $4w^2 - 20w + 5 = 0$

61. $9y^2 + 6y - 1 = 0$ **62.** $9s^2 - 6s - 2 = 0$ **63.** $4p^2 + 4p + 1 = 0$

64. $9z^2 + 12z + 4 = 0$ **65.** $2x^2 = 4x - 5$ **66.** $3r^2 = 5r - 6$

67. $4p^2 + 16p = -11$ **68.** $4y^2 - 12y = -1$ **69.** $4x^2 = 4x + 11$

70. $4s^2 + 12s = 3$ **71.** $9v^2 = -30v - 23$ **72.** $9t^2 = 30t + 17$

Solve by using the quadratic formula. Approximate the solutions to the nearest thousandth.

73. $x^2 - 2x - 21 = 0$ **74.** $y^2 + 4y - 11 = 0$ **75.** $s^2 - 6s - 13 = 0$

76. $w^2 + 8w - 15 = 0$ **77.** $2p^2 - 7p - 10 = 0$ **78.** $3t^2 - 8t - 1 = 0$

79. $4z^2 + 8z - 1 = 0$ **80.** $4x^2 + 7x + 1 = 0$ **81.** $5v^2 - v - 5 = 0$

SUPPLEMENTAL EXERCISES 12.3

Solve. Remember to check solutions to radical equations.

82. $\dfrac{x^2}{4} - \dfrac{x}{2} = 5$ **83.** $\dfrac{3x^2}{2} + 2x = 1$ **84.** $\sqrt{x + 3} = x - 3$

85. $x = 6 + \sqrt{2x + 3}$ **86.** $\dfrac{x}{4} + \dfrac{3}{x} = \dfrac{5}{2}$ **87.** $\dfrac{x + 1}{5} - \dfrac{4}{x - 1} = 2$

In the quadratic formula, the quantity $b^2 - 4ac$ is called the **discriminant.** The discriminant determines whether or not a quadratic equation will have real number solutions. If $b^2 - 4ac$ is nonnegative, the equation has real number solutions. If $b^2 - 4ac$ is negative, the equation has no real number solutions. Use the discriminant to determine whether or not the equation has real number solutions.

88. $3x^2 - 4x + 7 = 0$ **89.** $2x^2 + 5x - 6 = 0$ **90.** $4x^2 - 2x - 1 = 0$

91. $2x^2 - 2x + 1 = 0$ **92.** $3x^2 - 3x + 4 = 0$ **93.** $x^2 + 5x + 7 = 0$

SECTION 12.4

Graphing Quadratic Equations in Two Variables

1 Graph a quadratic equation of the form $y = ax^2 + bx + c$

An equation of the form $y = ax^2 + bx + c$, $a \neq 0$, is a **quadratic equation in two variables.** Examples of quadratic equations in two variables are shown at the right.

$$y = 3x^2 - x + 1$$
$$y = -x^2 - 3$$
$$y = 2x^2 - 5x$$

The graph of a quadratic equation in two variables is a **parabola.** The graph is "cup-shaped" and opens either up or down. The graphs of two parabolas are shown below.

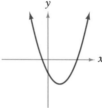

Parabola that opens up

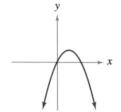

Parabola that opens down

Graph $y = x^2 - 2x - 3$.

Find several solutions of the equation. Because the graph is not a straight line, several solutions must be found in order to determine the cup shape.

Display the ordered pair solutions in a table.

x	y
0	−3
1	−4
−1	0
2	−3
3	0

Graph the ordered pair solutions on a rectangular coordinate system.

Draw a parabola through the points.

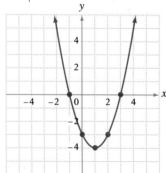

The graph of $y = -2x^2 + 1$ is shown below.

x	y
0	1
1	−1
−1	−1
2	−7
−2	−7

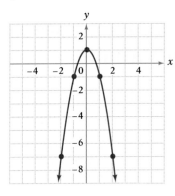

Note that the graph of $y = x^2 - 2x - 3$, shown on the previous page, **opens up** and that the coefficient of x^2 is **positive**. The graph of $y = -2x^2 + 1$ **opens down**, and the coefficient of x^2 is **negative**.

Example 1   Graph.   A. $y = x^2 - 2x$ B. $y = -x^2 + 4x - 4$

Solution   A. $a = 1$. a is positive.   B. $a = -1$. a is negative.
The parabola opens up.   The parabola opens down.

x	y
0	0
1	−1
−1	3
2	0
3	3

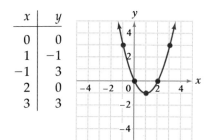

x	y
0	−4
1	−1
2	0
3	−1
4	−4

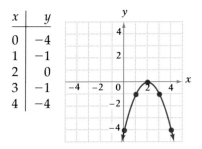

Problem 1   Graph.   A. $y = x^2 + 2$ B. $y = -x^2 - 2x + 3$

Solution   See page A44.

EXERCISES 12.4

1   Graph.

1. $y = x^2$

2. $y = -x^2$

3. $y = -x^2 + 1$

4. $y = x^2 - 1$

5. $y = 2x^2$

6. $y = \frac{1}{2}x^2$

7. $y = -\frac{1}{2}x^2 + 1$

8. $y = 2x^2 - 1$

9. $y = x^2 - 4x$

10. $y = x^2 + 4x$

11. $y = x^2 - 2x + 3$

12. $y = x^2 - 4x + 2$

13. $y = -x^2 + 2x + 3$

14. $y = x^2 + 2x + 1$

15. $y = -x^2 + 4x - 4$

16. $y = -x^2 + 6x - 9$

17. $y = 2x^2 + x - 3$

18. $y = -2x^2 - 3x + 3$

SUPPLEMENTAL EXERCISES 12.4

Determine whether the graph of the equation opens up or down.

19. $y = -\dfrac{1}{3}x^2 + 5$ **20.** $y = x^2 - 2x + 3$ **21.** $y = -x^2 + 4x - 1$

Show that the equation is a quadratic equation in two variables by writing it in the form $y = ax^2 + bx + c$.

22. $y + 1 = (x - 4)^2$ **23.** $y - 2 = 3(x + 1)^2$ **24.** $y - 4 = 2(x - 3)^2$

The solutions of the equation $ax^2 + bx + c = 0$ can be found by graphing the equation $y = ax^2 + bx + c$. The solutions are the x-intercepts of the parabola (the points at which the graph crosses the x-axis). For example, the equation $y = x^2 - 1$ crosses the x-axis at $(1, 0)$ and $(-1, 0)$. The solutions of the equation $x^2 - 1 = 0$ are 1 and -1. Solve the equations by graphing.

25. $x^2 - 4 = 0$ **26.** $-x^2 + 4 = 0$ **27.** $x^2 - x = 0$

28. $-x^2 + x = 0$ **29.** $x^2 - 4x + 3 = 0$ **30.** $-x^2 - 5x + 6 = 0$

Solve.

31. The point whose coordinates are (x_1, y_1) lies in Quadrant I and is a point on the graph of the equation $y = 2x^2 - 2x + 1$. Given $y_1 = 13$, find x_1.

32. The point whose coordinates are (x_1, y_1) lies in Quadrant II and is a point on the graph of the equation $y = 2x^2 - 3x - 2$. Given $y_1 = 12$, find x_1.

33. Graph $y = x^2 + 4x + 5$. By examining the graph, determine whether the solutions of the equation $x^2 + 4x + 5 = 0$ are real numbers.

34. Graph $y = -2x^2 - 4x + 5$. By examining the graph, determine whether the solutions of the equation $-2x^2 - 4x + 5 = 0$ are real numbers.

S E C T I O N **12.5**

Application Problems

 Application problems

The application problems in this section are varieties of those problems solved earlier in the text. Each of the strategies for the problems in this section results in a quadratic equation.

Solve: It took a motorboat a total of 7 h to travel 48 mi down a river and then 48 mi back again. The rate of the current was 2 mph. Find the rate of the motorboat in calm water.

STRATEGY for solving an application problem

<blockquote>
■ Determine the type of problem. For example, is it a distance-rate problem, a geometry problem, a work problem, or an age problem?
</blockquote>

The problem is a distance-rate problem.

<blockquote>
■ Choose a variable to represent the unknown quantity. Write numerical or variable expressions for all the remaining quantities. These results can be recorded in a table.
</blockquote>

The unknown rate of the motorboat: r

	Distance	÷	Rate	=	Time
Down river	48	÷	$r + 2$	=	$\dfrac{48}{r + 2}$
Up river	48	÷	$r - 2$	=	$\dfrac{48}{r - 2}$

<blockquote>
■ Determine how the quantities are related. If necessary, review the strategies presented in Chapter 4.
</blockquote>

The total time of the trip was 7 h.

$$\frac{48}{r + 2} + \frac{48}{r - 2} = 7$$

$$(r + 2)(r - 2)\left(\frac{48}{r + 2} + \frac{48}{r - 2}\right) = (r + 2)(r - 2)7$$

$$(r - 2)48 + (r + 2)48 = 7r^2 - 28$$

$$48r - 96 + 48r + 96 = 7r^2 - 28$$

$$96r = 7r^2 - 28$$

$$0 = 7r^2 - 96r - 28$$

$$0 = (7r + 2)(r - 14)$$

$$7r + 2 = 0 \qquad\qquad r - 14 = 0$$
$$7r = -2 \qquad\qquad\quad r = 14$$
$$r = -\frac{2}{7}$$

The solution $r = -\frac{2}{7}$ is not possible because the rate cannot be negative.

The rate of the motorboat in calm water is 14 mph.

Example 1   Working together, a painter and the painter's apprentice can paint a room in 4 h. The apprentice working alone requires 6 more hours to paint the room than the painter requires working alone. How long does it take the painter working alone to paint the room?

Strategy   ▪ This is a work problem.
▪ Time for the painter to paint the room: t
Time for the apprentice to paint the room: $t + 6$

	Rate	Time	Part
Painter	$\dfrac{1}{t}$	4	$\dfrac{4}{t}$
Apprentice	$\dfrac{1}{t+6}$	4	$\dfrac{4}{t+6}$

▪ The sum of the parts of the task completed must equal 1.

Solution
$$\frac{4}{t} + \frac{4}{t+6} = 1$$
$$t(t+6)\left(\frac{4}{t} + \frac{4}{t+6}\right) = t(t+6) \cdot 1$$
$$(t+6)4 + t(4) = t(t+6)$$
$$4t + 24 + 4t = t^2 + 6t$$
$$0 = t^2 - 2t - 24$$
$$0 = (t-6)(t+4)$$

$t - 6 = 0 \qquad t + 4 = 0$
$t = 6 \qquad\quad t = -4$ ▶ The solution $t = -4$ is not possible.

Working alone, the painter requires 6 h to paint the room.

Problem 1   The length of a rectangle is 3 m more than the width. The area is 40 m². Find the width.

Solution   See page A44.

EXERCISES 12.5

1   Solve.

1. The area of the batter's box on a major league baseball field is 24 ft². The length of the batter's box is 2 ft more than the width. Find the length and width of the batter's box. ($A = lw$)

2. The length of the batter's box on a softball field is 1 ft more than twice the width. The area of the batter's box is 21 ft². Find the length and width of the batter's box. ($A = lw$)

3. The length of a children's playground is twice the width. The area is 5000 ft². Find the length and width of the playground. $(A = lw)$

4. The length of the singles tennis court is 24 ft more than twice the width. The area is 2106 ft². Find the length and width of the singles tennis court. $(A = lw)$

5. The sum of the squares of two positive odd integers is 130. Find the two integers.

6. The sum of the squares of two consecutive positive even integers is 164. Find the two integers.

7. The sum of two integers is 12. The product of the two integers is 35. Find the two integers.

8. The difference between two integers is 4. The product of the two integers is 60. Find the integers.

9. Twice an integer equals the square of the integer. Find the integer.

10. The square of an integer equals the integer. Find the integer.

11. A silver coin is twice the age of a gold coin. Three years ago, the product of the sum of their ages and the difference between their ages was 45. Find the present ages of the coins.

12. An oil painting is twice the age of a watercolor. One year ago, the product of their ages was 10. Find the present ages of the oil painting and the watercolor.

13. One coin is two years older than a second coin. Two years ago, the product of their ages was 24. Find the present ages of the coins.

14. One car is three times the age of a second car. Eight years ago, the product of their ages was 19. Find the present ages of the cars.

15. One computer takes 21 min longer to calculate the value of a complex equation than a second computer. Working together, these computers can complete the calculation in 10 min. How long would it take each computer, working alone, to calculate the value?

16. A tank has two drains. One drain takes 16 min longer to empty the tank than does a second drain. With both drains open, the tank is emptied in 6 min. How long would it take each drain, working alone, to empty the tank?

17. It takes 6 h longer to cross a channel in a ferryboat when one engine of the boat is used than when a second engine is used. Using both engines, the ferryboat can make the crossing in 4 h. How long would it take each engine, working alone, to power the ferryboat across the channel?

18. An apprentice mason takes 8 h longer to build a small fireplace than an experienced mason. Working together, they can build the fireplace in 3 h. How long would it take the experienced mason, working alone, to build the fireplace?

19. It took a small plane 2 h more to fly 375 mi against the wind than it took the plane to fly the same distance with the wind. The rate of the wind was 25 mph. Find the rate of the plane in calm air.

20. It took a motorboat 1 h longer to travel 36 mi against the current than it took the boat to travel 36 mi with the current. The rate of the current was 3 mph. Find the rate of the boat in calm water.

21. A cruise ship sailed through a 20-mi inland passageway at a constant rate before increasing its speed by 15 mph. Another 75 mi was traveled at the increased rate. The total time for the 95-mi trip was 5 h. Find the rate of the ship during the last 75 mi.

22. A motorist traveled 150 mi at a constant rate before decreasing the speed by 15 mph. Another 35 mi was driven at the decreased speed. The total time for the 185-mi trip was 4 h. Find the motorist's rate during the first 150 mi.

SUPPLEMENTAL EXERCISES 12.5

Solve.

23. The sum of the squares of four consecutive integers is 86. Find the four integers.

24. The hypotenuse of a right triangle is $\sqrt{13}$ cm. One leg is 1 cm shorter than twice the length of the other leg. Find the lengths of the legs of the right triangle.

25. The radius of a large pizza is 1 in. less than twice the radius of a small pizza. The difference between the areas of the two pizzas is 33π in.^{2}. Find the radius of the large pizza.

26. The perimeter of a rectangular garden is 54 ft. The area of the garden is 180 ft^2. Find the length and width of the garden.

Calculators and Computers

 Checking Solutions to Quadratic Equations

A calculator can be used to check solutions to quadratic equations. Here are some examples.

Solve and check: $x^2 - 6x - 16 = 0$

Solve by factoring. The solutions are 8 and -2.

To check the solutions, replace x by 8. $8^2 - 6 \cdot 8 - 16 \overset{?}{=} 0$

Enter the following keystrokes: $8 \;\boxed{x^2}\; \boxed{-}\; \boxed{(}\; 6 \;\boxed{\times}\; 8 \;\boxed{)}\; \boxed{-}\; 16 \;\boxed{=}$

The result in the display should be zero. The solution is correct.

The solution -2 can be checked in a similar manner.

One note about the calculation—the parentheses keys are used to ensure that multiplication is completed before subtraction.

Solve and check: $2x^2 - x - 9 = 0$

Use the quadratic formula. The solutions are $\dfrac{1 + \sqrt{73}}{4}$ and $\dfrac{1 - \sqrt{73}}{4}$.

To check the solutions, first evaluate the expression $\dfrac{1 + \sqrt{73}}{4}$, and store the result in the calculator's memory.

$$\boxed{(}\; 1 \;\boxed{+}\; 73 \;\boxed{\sqrt{}}\; \boxed{)}\; \boxed{\div}\; 4 \;\boxed{=}\; \boxed{M+}$$

Now replace x in the equation by the solution, and determine whether the left and right sides of the equation are equal. The $\boxed{MR}$ key recalls the solution from memory.

$$2 \;\boxed{\times}\; \boxed{MR}\; \boxed{x^2}\; \boxed{-}\; \boxed{MR}\; \boxed{-}\; 9 \;\boxed{=}$$

Is the result in the display zero? Probably not! Nonetheless, the answer is very close to zero. The result in our display was 9.000000 − 10. The −10 at the end of the display means that the decimal point should be moved ten places to the left. That makes the number 0.0000000009, which is indeed close to zero.

The reason the answer was not exactly zero is that $\sqrt{73}$ is an irrational number and therefore has an infinitely long decimal representation. The calculator, on the other hand, can store only 8 or 9 significant digits. Thus the calculator is using only an approximation of $\sqrt{73}$, so when the solution is checked, the result is not exactly zero.

The solution $\dfrac{1 - \sqrt{73}}{4}$ can be checked in a similar manner.

Something Extra

Profit

A company's revenue, R, is the total amount of money earned by the company by selling its products. The cost, C, is the total amount of money spent by the company to manufacture and sell its products. A company's profit, P, is the difference between the revenue and cost: $P = R - C$. A company's revenue and cost may be represented by equations.

A company manufactures and sells wood stoves. The total monthly cost, in dollars, to produce n wood stoves is $C = 30n + 2000$. Write a variable expression for the company's monthly profit if the revenue obtained from selling all n wood stoves is $R = 150n - 0.4n^2$.

Substitute the variable expressions for revenue and cost into the equation $P = R - C$.

$$P = R - C$$
$$P = (150n - 0.4n^2) - (30n + 2000)$$

Simplify the right-hand side of the equation.

$$P = 150n - 0.4n^2 - 30n - 2000$$
$$P = -0.4n^2 + 120n - 2000$$

The company's monthly profit is $P = 0.4n^2 + 120n - 2000$.

How many wood stoves must the company manufacture and sell in order to make a profit of $6000 a month?

Substitute 6000 for P in the profit equation.

$$P = -0.4n^2 + 120n - 2000$$
$$6000 = -0.4n^2 + 120n - 2000$$

Write the equation in standard form.

$$0 = -0.4n^2 + 120n - 8000$$

$-0.4 = -\dfrac{4}{10}$. Multiply each side of the equation by $-\dfrac{10}{4}$.

$$0 = n^2 - 300n + 20{,}000$$

Factor.

$$0 = (n - 100)(n - 200)$$
$$n = 100 \quad n = 200$$

The company will make a monthly profit of $6000 if either 100 wood stoves are manufactured and sold or 200 wood stoves are manufactured and sold.

The graph of the profit equation $P = -0.4n^2 + 120n - 2000$ is shown at the right. The highest point on the graph is the point at which the maximum profit can be made. For this example, a maximum profit of $7000 occurs when the company manufactures and sells 150 wood stoves.

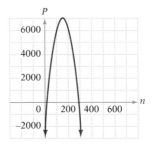

Solve.

1. The total cost, in dollars, for a company to produce and sell n guitars per month is $C = 240n + 1200$. The company's revenue from selling all n guitars is $R = 400n - 2n^2$.

   a. How many guitars must the company produce and sell each month in order to make a monthly profit of $1200?

   b. Graph the profit equation. What is the maximum monthly profit the company can make?

2. A company's total monthly cost, in dollars, for manufacturing and selling n videotapes per month is $C = 35n + 2000$. The company's revenue from selling all n videotapes is $R = 175n - 0.2n^2$.

   a. How many videotapes must be produced and sold each month in order for the company to make a monthly proft of $18,000?

   b. Graph the profit equation. How many videotapes must be produced and sold in order for the company to make the maximum monthly profit?

Chapter Summary

Key Words

A *quadratic equation* is an equation that can be written in the form $ax^2 + bx + c = 0$, where a, b, and c are constants and $a \neq 0$. A quadratic equation is also called a *second-degree equation.*

A quadratic equation is in *standard form* when the polynomial is in descending order and equal to zero.

Adding to a binomial the constant term that makes it a perfect square trinomial is called *completing the square.*

The *graph* of an equation of the form $y = ax^2 + bx + c$ is a parabola.

Essential Rules

The Quadratic Formula $\qquad\qquad x = \dfrac{-b \pm \sqrt{b^2 - 4ac}}{2a}$

Chapter Review

1. Solve: $b^2 - 16 = 0$

2. Solve: $x^2 - x - 3 = 0$

3. Solve: $x^2 - 3x - 5 = 0$

4. Solve: $49x^2 = 25$

5. Graph $y = -\frac{1}{4}x^2$.

6. Graph $y = -3x^2$.

7. Solve: $x^2 = 10x - 2$

8. Solve: $6x(x + 1) = x - 1$

9. Solve: $4y^2 + 9 = 0$

10. Solve: $5x^2 + 20x + 12 = 0$

11. Solve: $x^2 - 4x + 1 = 0$

12. Solve: $x^2 - x = 30$

13. Solve: $6x^2 + 13x - 28 = 0$

14. Solve: $x^2 = 40$

15. Solve: $3x^2 - 4x = 1$

16. Solve: $x^2 - 2x - 10 = 0$

17. Solve: $x^2 - 12x + 27 = 0$

18. Solve: $(x - 7)^2 = 81$

19. Graph $y = 2x^2 + 1$.

20. Graph $y = \frac{1}{2}x^2 - 1$.

21. Solve: $(y + 4)^2 - 25 = 0$

22. Solve: $4x^2 + 16x = 7$

23. Solve: $24x^2 + 34x + 5 = 0$

24. Solve: $x^2 = 4x - 8$

25. Solve: $x^2 - 5 = 8x$

26. Solve: $25(2x^2 - 2x + 1) = (x + 3)^2$

27. Solve: $\left(x - \frac{1}{2}\right)^2 = \frac{9}{4}$

28. Solve: $16x^2 = 30x - 9$

29. Solve: $x^2 + 7x = 3$

30. Solve: $12x^2 + 10 = 29x$

31. Solve: $4(x - 3)^2 = 20$

32. Solve: $x^2 + 8x - 3 = 0$

33. Graph $y = x^2 - 3x$.

34. Graph $y = x^2 - 4x + 3$.

35. Solve: $(x + 9)^2 = x + 11$

36. Solve: $(x - 2)^2 - 24 = 0$

37. Solve: $x^2 + 6x + 12 = 0$

38. Solve: $x^2 + 6x - 2 = 0$

39. Solve: $18x^2 - 52x = 6$

40. Solve: $2x^2 + 5x = 1$

41. Graph $y = -x^2 + 4x - 5$.

42. Solve: $x^2 + 3 = 9x$

43. Solve: $2x^2 + 5 = 7x$

44. Graph $y = 4 - x^2$.

45. It took an air balloon half an hour more to fly 70 mi against the wind than it took to fly 40 mi with the wind. The rate of the wind was 5 mph. Find the rate of the air balloon in calm air.

46. The height of a triangle is 2 m more than twice the length of the base. The area of the triangle is 20 m^2. Find the height of the triangle and the length of the base.

47. The sum of the squares of two consecutive positive odd integers is thirty-four. Find the two integers.

48. In 5 h, two campers rowed 12 mi down a stream and then rowed back to their campsite. The rate of the stream's current was 1 mph. Find the rate at which the campers rowed.

49. One car is two years older than a second car. Two years ago, the product of their ages was 120. Find the present ages of the two cars.

50. A smaller drain takes 8 h longer to empty a tank than does a larger drain. Working together, the drains can empty the tank in 3 h. How long would it take each drain, working alone, to empty the tank?

Chapter Test

1. Solve: $3(x + 4)^2 - 60 = 0$

2. Solve: $2x^2 + 8x = 3$

3. Solve: $3x^2 + 7x = 20$

4. Solve: $x^2 - 3x = 6$

5. Solve: $x^2 + 4x - 16 = 0$

6. Graph $y = x^2 + 2x - 4$.

7. Solve: $x^2 + 4x + 2 = 0$

8. Solve: $x^2 + 3x = 8$

9. Solve: $2x^2 - 5x - 3 = 0$

10. Solve: $2x^2 - 6x + 1 = 0$

11. Solve: $3x^2 - x = 1$

12. Solve: $2(x - 5)^2 = 36$

13. Solve: $x^2 - 6x - 5 = 0$

14. Solve: $x^2 - 5x = 1$

15. Solve: $x^2 - 5x = 2$

16. Solve: $6x^2 - 17x = -5$

17. Solve: $x^2 + 3x - 7 = 0$

18. Solve: $2x^2 - 4x - 5 = 0$

19. Solve: $2x^2 - 3x - 2 = 0$

20. Graph $y = x^2 - 2x - 3$.

21. Solve: $3x^2 - 2x = 3$

22. The length of a rectangle is 2 ft less than twice the width. The area of the rectangle is 40 ft². Find the length and width of the rectangle.

23. It took a motorboat one hour more to travel 60 mi against the current than it took to go 60 mi with the current. The rate of the current was 1 mph. Find the rate of the boat in calm water.

24. The sum of the squares of three consecutive odd integers is 83. Find the middle odd integer.

25. A jogger ran 7 mi at a constant rate and then reduced the rate by 3 mph. An additional 8 mi was run at the reduced rate. The total time spent jogging the 15 mi was 3 h. Find the rate for the last 8 mi.

Cumulative Review

1. Simplify:
$2x - 3[2x - 4(3 - 2x) + 2] - 3$

2. Solve: $-\frac{3}{5}x = -\frac{9}{10}$

3. Solve:
$2x - 3(4x - 5) = -3x - 6$

4. Simplify: $\dfrac{(2a^{-2}b)^2}{-3a^{-5}b^4}$

5. Simplify: $(x^2 - 8) \div (x - 2)$

6. Factor: $4y(x - 4) - 3(x - 4)$

7. Factor: $3x^3 + 2x^2 - 8x$

8. Simplify: $\dfrac{3x^2 - 6x}{4x - 6} \div \dfrac{2x^2 + x - 6}{6x^2 - 24x}$

9. Simplify: $\dfrac{x}{2(x - 1)} - \dfrac{1}{(x - 1)(x + 1)}$

10. Simplify: $\dfrac{1 - \frac{7}{x} + \frac{12}{x^2}}{2 - \frac{1}{x} - \frac{15}{x^2}}$

11. Solve: $\dfrac{x}{x + 6} = \dfrac{3}{x}$

12. Find the x- and y-intercepts of the line $4x - 3y = 12$.

13. Find the equation of the line that contains the point whose coordinates are $(-3, 2)$ and has slope $-\frac{4}{3}$.

14. Solve by substitution:
$3x - y = 5$
$y = 2x - 3$

15. Solve by the addition method:
$3x + 2y = 2$
$5x - 2y = 14$

16. Solve: $2x - 3(2 - 3x) > 2x - 5$

17. Simplify: $(\sqrt{a} - \sqrt{2})(\sqrt{a} + \sqrt{2})$

18. Simplify: $\dfrac{\sqrt{12x^2} - \sqrt{27}}{\sqrt{3}}$

19. Solve: $3 = 8 - \sqrt{5x}$

20. Solve: $2x^2 - 7x = -3$

21. Solve: $3(x - 2)^2 = 36$

22. Solve: $3x^2 - 4x - 5 = 0$

23. Graph $2x - 3y > 6$.

24. Graph $y = x^2 - 3x - 2$.

25. Find the cost per pound of a mixture made from 20 lb of cashews that cost $3.50 per pound and 50 lb of peanuts that cost $1.75 per pound.

26. A stock investment of 100 shares paid a dividend of $215. At this rate, how many additional shares are required to earn a dividend of $752.50?

27. A 720-mi trip from one city to another takes 3 h when a plane is flying with the wind. The return trip, against the wind, takes 4.5 h. Find the rate of the plane in calm air and the rate of the wind.

28. A student received a 70, a 91, an 85, and a 77 on four tests in a math class. What scores on the fifth test will enable the student to receive a minimum of 400 points?

29. A guy wire is attached to a point 30 m above the ground on a telephone pole. The wire is anchored to the ground at a point 10 m from the base of the pole. Find the length of the guy wire. Round to the nearest hundredth.

30. The sum of the squares of three consecutive odd integers is 155. Find the middle odd integer.

Final Exam

1. Evaluate $-|-3|$.

2. Subtract: $-15 - (-12) - 3$

3. Simplify: $-2^4 \cdot (-2)^4$

4. Simplify: $-7 - \dfrac{12 - 15}{2 - (-1)} \cdot (-4)$

5. Evaluate $\dfrac{a^2 - 3b}{2a - 2b^2}$ when $a = 3$ and $b = -2$.

6. Simplify: $6x - (-4y) - (-3x) + 2y$

7. Simplify: $(-15z)(-\frac{2}{5})$

8. Simplify: $-2[5 - 3(2x - 7) - 2x]$

9. Solve: $20 = -\frac{2}{5}x$

10. Solve: $4 - 2(3x + 1) = 3(2 - x) + 5$

11. Write $\frac{1}{8}$ as a percent.

12. Find 19% of 80.

13. Simplify: $(2x^2 - 5x + 1) - (5x^2 - 2x - 7)$

14. Simplify: $(-3xy^3)^4$

15. Simplify: $(3x^2 - x - 2)(2x + 3)$

16. Simplify: $\dfrac{(-2x^2y^3)^3}{(-4x^{-1}y^4)^2}$

17. Simplify: $\dfrac{12x^3y^2 - 16x^2y^2 - 20y^2}{4xy^2}$

18. Simplify: $(5x^2 - 2x - 1) \div (x + 2)$

19. Simplify: $(4x^{-2}y)^3(2xy^{-2})^{-2}$

20. Factor: $2x^2 - x - 3$

21. Factor: $x^2 - 5x - 6$

22. Factor: $6x^2 - 5x - 6$

23. Factor: $8x^3 - 28x^2 + 12x$

24. Factor: $25x^2 - 16$

25. Factor: $2a(4 - x) - 6(x - 4)$

26. Factor: $75y - 12x^2y$

27. Solve: $2x^2 = 7x - 3$

28. Simplify: $\dfrac{2x^2 - 3x + 1}{4x^2 - 2x} \cdot \dfrac{4x^2 + 4x}{x^2 - 2x + 1}$

29. Simplify: $\dfrac{5}{x + 3} - \dfrac{3x}{2x - 5}$

30. Simplify: $\dfrac{x - \dfrac{3}{2x - 1}}{1 - \dfrac{2}{2x - 1}}$

31. Solve: $\dfrac{5x}{3x - 5} - 3 = \dfrac{7}{3x - 5}$

32. Solve $a = 3a - 2b$ for a.

33. Find the slope of the line that contains the points whose coordinates are $(-1, -3)$ and $(2, -1)$.

34. Find the equation of the line that contains the point whose coordinates are $(3, -4)$ and has slope $-\dfrac{2}{3}$.

35. Solve by substitution:
$y = 4x - 7$
$y = 2x + 5$

36. Solve by the addition method:
$4x - 3y = 11$
$2x + 5y = -1$

37. Solve: $4 - x \geq 7$

38. Solve: $2 - 2(y - 1) \leq 2y - 6$

39. Simplify: $\sqrt{49x^6}$

40. Simplify: $2\sqrt{27a} + 8\sqrt{48a}$

41. Simplify: $\dfrac{\sqrt{3}}{\sqrt{5} - 2}$

42. Solve: $\sqrt{x + 4} - \sqrt{x - 1} = 1$

43. Solve: $3x^2 - x = 4$

44. Solve: $4x^2 - 2x - 1 = 0$

45. Graph the line with slope $-\dfrac{1}{2}$ and y-intercept $(0, -3)$.

46. Graph $y = x^2 - 4x + 3$.

47. Translate and simplify "the sum of twice a number and three times the difference between the number and two."

48. Because of depreciation, the value of an office machine is now $2400. This is 80% of its original value. Find the original value of the machine.

49. The manufacturer's cost for a laser printer is $900. The manufacturer sells the printer for $1485. Find the markup rate.

50. An investment of $3000 is made at an annual simple interest rate of 8%. How much additional money must be invested at 11% so that the total interest earned is 10% of the total investment?

51. A grocer mixes 4 lb of peanuts that cost $2 per pound with 2 lb of walnuts that cost $5 per pound. What is the cost per pound of the resulting mixture?

52. A pharmacist mixes 20 L of a solution that is 60% acid with 30 L of a solution that is 20% acid. What is the percent concentration of acid in the resulting mixture?

53. A small plane flew at a constant rate for 1 h. The pilot then doubled the plane's speed. An additional 1.5 h was flown at the increased speed. If the entire flight was 860 km, how far did the plane travel during the first hour?

54. One angle of a triangle is 10° more than the measure of the second angle. The third angle is 10° more than the measure of the first angle. Find the measure of each angle of the triangle.

55. A coin bank contains quarters and dimes. There are three times as many dimes as quarters. The total value of the coins in the bank is $11. Find the number of dimes in the bank.

56. The length of a rectangle is 5 m more than the width. The area of the rectangle is 50 m². Find the dimensions of the rectangle.

57. A paint formula requires 2 oz of dye for every 15 oz of base paint. How many ounces of dye are required for 120 oz of base paint?

58. It takes a chef 1 h to prepare a dinner. The chef's apprentice can prepare the same dinner in 1.5 h. How long would it take the chef and the apprentice, working together, to prepare the dinner?

59. With the current, a motorboat travels 50 mi in 2.5 h. Against the current, it takes twice as long to travel 50 mi. Find the rate of the boat in calm water and the rate of the current.

60. It took a plane $\frac{1}{2}$ h more to fly 500 mi against the wind than it took to fly the same distance with the wind. The rate of the plane in calm air is 225 mph. Find the rate of the wind.

APPENDIX

Table of Square Roots

Decimal approximations have been rounded to the nearest thousandth.

N	√N	N	√N	N	√N	N	√N	N	√N
1	1	41	6.403	81	9	121	11	161	12.689
2	1.414	42	6.481	82	9.055	122	11.045	162	12.728
3	1.732	43	6.557	83	9.110	123	11.091	163	12.767
4	2	44	6.633	84	9.165	124	11.136	164	12.806
5	2.236	45	6.708	85	9.220	125	11.180	165	12.845
6	2.449	46	6.782	86	9.274	126	11.225	166	12.884
7	2.646	47	6.856	87	9.327	127	11.269	167	12.923
8	2.828	48	6.928	88	9.381	128	11.314	168	12.961
9	3	49	7	89	9.434	129	11.358	169	13
10	3.162	50	7.071	90	9.487	130	11.402	170	13.038
11	3.317	51	7.141	91	9.539	131	11.446	171	13.077
12	3.464	52	7.211	92	9.592	132	11.489	172	13.115
13	3.606	53	7.280	93	9.644	133	11.533	173	13.153
14	3.742	54	7.348	94	9.695	134	11.576	174	13.191
15	3.873	55	7.416	95	9.747	135	11.619	175	13.229
16	4	56	7.483	96	9.798	136	11.662	176	13.266
17	4.123	57	7.550	97	9.849	137	11.705	177	13.304
18	4.243	58	7.616	98	9.899	138	11.747	178	13.342
19	4.359	59	7.681	99	9.950	139	11.790	179	13.379
20	4.472	60	7.746	100	10	140	11.832	180	13.416
21	4.583	61	7.810	101	10.050	141	11.874	181	13.454
22	4.690	62	7.874	102	10.100	142	11.916	182	13.491
23	4.796	63	7.937	103	10.149	143	11.958	183	13.528
24	4.899	64	8	104	10.198	144	12	184	13.565
25	5	65	8.062	105	10.247	145	12.042	185	13.601
26	5.099	66	8.124	106	10.296	146	12.083	186	13.638
27	5.196	67	8.185	107	10.344	147	12.124	187	13.675
28	5.292	68	8.246	108	10.392	148	12.166	188	13.711
29	5.385	69	8.307	109	10.440	149	12.207	189	13.748
30	5.477	70	8.367	110	10.488	150	12.247	190	13.784
31	5.568	71	8.426	111	10.536	151	12.288	191	13.820
32	5.657	72	8.485	112	10.583	152	12.329	192	13.856
33	5.745	73	8.544	113	10.630	153	12.369	193	13.892
34	5.831	74	8.602	114	10.677	154	12.410	194	13.928
35	5.916	75	8.660	115	10.724	155	12.450	195	13.964
36	6	76	8.718	116	10.770	156	12.490	196	14
37	6.083	77	8.775	117	10.817	157	12.530	197	14.036
38	6.164	78	8.832	118	10.863	158	12.570	198	14.071
39	6.245	79	8.888	119	10.909	159	12.610	199	14.107
40	6.325	80	8.944	120	10.954	160	12.649	200	14.142

SOLUTIONS to Chapter 1 Problems

SECTION 1.1 *pages 3–5*

Problem 1   **A.** $5 > -13$ **B.** $-8 > -22$ **C.** $17 > -6$ **D.** $-13 > -15$

Problem 2   **A.** 9 **B.** -62

Problem 3   **A.** $|-5| = 5$ **B.** $-|-9| = -9$

SECTION 1.2 *pages 8–11*

Problem 1   **A.** $-162 + 98$ **B.** $-154 + (-37)$ **C.** $-36 + 17 + (-21)$
$\qquad\qquad\qquad\;\; -64 \qquad\qquad\qquad\; -191 \qquad\qquad\qquad -19 + (-21)$
$\qquad\qquad\qquad\qquad\qquad\qquad\qquad\qquad\qquad\qquad\qquad\qquad -40$

Problem 2 $-8 - 14$ $\qquad\qquad\qquad\qquad$ **Problem 3** $3 - (-4) - 15$
$\qquad\qquad\; -8 + (-14) \qquad\qquad\qquad\qquad\qquad\qquad\qquad\; 3 + 4 + (-15)$
$\qquad\qquad\; -22 \qquad\qquad\qquad\qquad\qquad\qquad\qquad\qquad\qquad 7 + (-15)$
$\qquad\qquad\qquad\qquad\qquad\qquad\qquad\qquad\qquad\qquad\qquad\qquad -8$

Problem 4 $4 - (-3) - 12 - (-7) - 20$
$\qquad\qquad\; 4 + 3 + (-12) + 7 + (-20)$
$\qquad\qquad\; 7 + (-12) + 7 + (-20)$
$\qquad\qquad\; -5 + 7 + (-20)$
$\qquad\qquad\; 2 + (-20)$
$\qquad\qquad\; -18$

SECTION 1.3 *pages 13–16*

Problem 1   **A.** $-38 \cdot 51$ **B.** $-7(-8)(9)(-2)$ **Problem 2**   **A.** $(-135) \div (-9)$ **B.** $84 \div (-6)$
$\qquad\qquad\qquad\; -1938 \qquad\qquad\; 56(9)(-2) \qquad\qquad\qquad\qquad\qquad\qquad 15 \qquad\qquad\qquad\qquad -14$
$\qquad\qquad\qquad\qquad\qquad\qquad\; 504(-2)$
$\qquad\qquad\qquad\qquad\qquad\qquad\; -1008$

Problem 3

Strategy   To find the difference, subtract the temperature at which radon freezes from the temperature at which it boils.

Solution $-62 - (-71)$
$\qquad\qquad\;\; -62 + 71$
$\qquad\qquad\;\; 9$

The difference is 9°C.

Problem 4

Strategy   To find the average daily high temperature:
- Add the seven temperature readings.
- Divide by 7.

Solution $-5 + (-6) + 3 + 0 + (-4) + (-7) + (-2)$
$-11 + 3 + 0 + (-4) + (-7) + (-2)$
$-8 + 0 + (-4) + (-7) + (-2)$
$-8 + (-4) + (-7) + (-2)$
$-12 + (-7) + (-2)$
$-19 + (-2)$
-21

$-21 \div 7 = -3$

The average daily high temperature was $-3°C$.

SECTION 1.4   *pages 19–25*

Problem 1

$$
\begin{array}{r}
0.16 \\
25\overline{)4.00} \\
\underline{-2\ 5} \\
1\ 50 \\
\underline{-1\ 50} \\
0
\end{array}
$$

$\dfrac{4}{25} = 0.16$

Problem 2

$$
\begin{array}{r}
0.444 \\
9\overline{)4.000} \\
\underline{-3\ 6} \\
40 \\
\underline{-36} \\
40 \\
\underline{-36} \\
4
\end{array}
$$

$\dfrac{4}{9} = 0.\overline{4}$

Problem 3   Prime factorization of 9 and 12:
$9 = 3 \cdot 3$
$12 = 2 \cdot 2 \cdot 3$
LCM $= 2 \cdot 2 \cdot 3 \cdot 3 = 36$

$\dfrac{5}{9} - \dfrac{11}{12} = \dfrac{20}{36} - \dfrac{33}{36} = \dfrac{20}{36} + \dfrac{-33}{36} = \dfrac{20 - 33}{36} = \dfrac{-13}{36} = -\dfrac{13}{36}$

Problem 4 $-\dfrac{7}{8} - \dfrac{5}{6} + \dfrac{1}{2} = -\dfrac{21}{24} - \dfrac{20}{24} + \dfrac{12}{24} = \dfrac{-21}{24} + \dfrac{-20}{24} + \dfrac{12}{24}$

$= \dfrac{-21 - 20 + 12}{24} = \dfrac{-29}{24} = -\dfrac{29}{24}$

Problem 5

$$
\begin{array}{r}
3.097 \\
4.9 \\
+\ 3.09 \\
\hline
11.087
\end{array}
$$

Problem 6

$$
\begin{array}{r}
\overset{\overset{10}{8\ \cancel{0}\ 10}}{67.9\cancel{1}\cancel{0}} \\
-16.127 \\
\hline
51.783
\end{array}
$$

$16.127 - 67.91 = -51.783$

Problem 7 $-\dfrac{7}{12} \cdot \dfrac{9}{14} = -\dfrac{7 \cdot 9}{12 \cdot 14}$

$$= -\dfrac{\overset{1}{7} \cdot \overset{1}{\cancel{3}} \cdot 3}{2 \cdot 2 \cdot \underset{1}{\cancel{3}} \cdot 2 \cdot \underset{1}{\cancel{7}}} = -\dfrac{3}{8}$$

Problem 8 $-\dfrac{3}{8} \div \left(-\dfrac{5}{12}\right) = \dfrac{3}{8} \cdot \dfrac{12}{5} = \dfrac{3 \cdot 12}{8 \cdot 5}$

$$= \dfrac{3 \cdot \overset{1}{\cancel{2}} \cdot \overset{1}{\cancel{2}} \cdot 3}{\underset{1}{\cancel{2}} \cdot \underset{1}{\cancel{2}} \cdot 2 \cdot 5} = \dfrac{9}{10}$$

Problem 9
$$
\begin{array}{r}
5.44 \\
\times \quad 3.8 \\
\hline
4352 \\
1632 \\
\hline
20.672
\end{array}
$$

$(-5.44)(3.8) = -20.672$

Problem 10
$$
\begin{array}{r}
0.231 \\
1.7\,\overline{)0.3.940} \\
\underline{-3\,4} \\
54 \\
\underline{-51} \\
30 \\
\underline{-17} \\
13
\end{array}
$$

$-0.394 \div 1.7 \approx -0.23$

SECTION 1.5   *pages 28–31*

Problem 1 $(-5)^3 = (-5)(-5)(-5) = 25(-5) = -125$

$-5^3 = -(5 \cdot 5 \cdot 5) = -(25 \cdot 5) = -125$

Problem 2 $(-3)^3 = (-3)(-3)(-3) = 9(-3) = -27$

$(-3)^4 = (-3)(-3)(-3)(-3) = 9(-3)(-3) = -27(-3) = 81$

Problem 3 $(3^3)(-2)^3 = (3 \cdot 3 \cdot 3) \cdot (-2)(-2)(-2) = 27 \cdot (-8) = -216$

$\left(-\dfrac{2}{5}\right)^2 = \left(-\dfrac{2}{5}\right)\left(-\dfrac{2}{5}\right) = \dfrac{2 \cdot 2}{5 \cdot 5} = \dfrac{4}{25}$

Problem 4
$36 \div (8 - 5)^2 - (-3)^2 \cdot 2$
$36 \div (3)^2 - (-3)^2 \cdot 2$
$36 \div 9 - 9 \cdot 2$
$4 - 9 \cdot 2$
$4 - 18$
-14

Problem 5
$27 \div 3^2 + (-3)^2 \cdot 4$
$27 \div 9 + 9 \cdot 4$
$3 + 9 \cdot 4$
$3 + 36$
39

Problem 6
$4 - 3[4 - 2(6 - 3)] \div 2$
$4 - 3[4 - 2(3)] \div 2$
$4 - 3[4 - 6] \div 2$
$4 - 3(-2) \div 2$
$4 - (-6) \div 2$
$4 - (-3)$
7

SOLUTIONS to Chapter 2 Problems

SECTION 2.1   *pages 43–47*

Problem 1 -4

Problem 2
$2xy + y^2$
$2(-4)(2) + (2)^2$
$2(-4)(2) + 4$
$-8(2) + 4$
$-16 + 4$
-12

Problem 3
$$\frac{a^2 + b^2}{a + b}$$
$$\frac{(5)^2 + (-3)^2}{5 + (-3)}$$
$$\frac{25 + 9}{5 + (-3)}$$
$$\frac{34}{2}$$
17

Problem 4
$x^3 - 2(x + y) + z^2$
$(2)^3 - 2[2 + (-4)] + (-3)^2$
$(2)^3 - 2(-2) + (-3)^2$
$8 - 2(-2) + 9$
$8 + 4 + 9$
$12 + 9$
21

Problem 5 $7 + (-7) = 0$

Problem 6   The Associative Property of Addition

SECTION 2.2   *pages 50–54*

Problem 1
A. $9x + 6x$
$(9 + 6)x$
$15x$

B. $-4y - 7y$
$[-4 + (-7)]y$
$-11y$

Problem 2
A. $3a - 2b + 5a$
$3a + 5a - 2b$
$(3a + 5a) - 2b$
$8a - 2b$

B. $x^2 - 7 + 9x^2 - 14$
$x^2 + 9x^2 - 7 - 14$
$(x^2 + 9x^2) + (-7 - 14)$
$10x^2 - 21$

Problem 3
A. $-7(-2a)$
$[-7(-2)]a$
$14a$

B. $-\frac{5}{6}(-30y^2)$
$\left[-\frac{5}{6}(-30)\right]y^2$
$25y^2$

C. $(-5x)(-2)$
$(-2)(-5x)$
$[-2(-5)]x$
$10x$

Problem 4
A. $7(4 + 2y)$
$7(4) + 7(2y)$
$28 + 14y$

B. $-(5x - 12)$
$-1(5x - 12)$
$-1(5x) - (-1)(12)$
$-5x + 12$

C. $(3a - 1)5$
$(3a)(5) - (1)(5)$
$15a - 5$

D. $-3(6a^2 - 8a + 9)$
$-3(6a^2) + (-3)(-8a) + (-3)(9)$
$-18a^2 + 24a - 27$

Problem 5 $7(x - 2y) - 3(-x - 2y)$
$7x - 14y + 3x + 6y$
$10x - 8y$

Problem 6 $3y - 2[x - 4(2 - 3y)]$
$3y - 2[x - 8 + 12y]$
$3y - 2x + 16 - 24y$
$-2x - 21y + 16$

SECTION 2.3   *pages 57–60*

Problem 1 A. 18 <u>less than</u> the <u>cube</u> of x
$x^3 - 18$

B. y <u>decreased by</u> the <u>sum</u> of z and 9
$y - (z + 9)$

C. the <u>difference between</u> the <u>square</u> of q and the <u>sum</u> of r and t
$q^2 - (r + t)$

Problem 2 the unknown number: n
the square of the number: n^2
the product of five and the square of the number: $5n^2$

$5n^2 + n$

Problem 3 the unknown number: n
twice the number: $2n$
the sum of seven and twice the number: $7 + 2n$

$3(7 + 2n)$

Problem 4 the unknown number: n
the difference between twice the number and 17: $2n - 17$

$n - (2n - 17)$
$n - 2n + 17$
$-n + 17$

Problem 5 the unknown number: n
three-fourths of the number: $\frac{3}{4}n$
one-fifth of the number: $\frac{1}{5}n$

$\frac{3}{4}n + \frac{1}{5}n$
$\frac{15}{20}n + \frac{4}{20}n$
$\frac{19}{20}n$

SOLUTIONS to Chapter 3 Problems

SECTION 3.1 *pages 75–80*

Problem 1

$$5 - 4x = 8x + 2$$

$$\begin{array}{c|c} 5 - 4\left(\dfrac{1}{4}\right) & 8\left(\dfrac{1}{4}\right) + 2 \\ 5 - 1 & 2 + 2 \\ 4 = 4 \end{array}$$

Yes, $\dfrac{1}{4}$ is a solution.

Problem 2

$$10x - x^2 = 3x - 10$$

$$\begin{array}{c|c} 10(5) - (5)^2 & 3(5) - 10 \\ 50 - 25 & 15 - 10 \\ 25 \neq 5 \end{array}$$

No, 5 is not a solution.

Problem 3

$$x - \frac{1}{3} = -\frac{3}{4}$$

$$x - \frac{1}{3} + \frac{1}{3} = -\frac{3}{4} + \frac{1}{3}$$

$$x = -\frac{5}{12}$$

The solution is $-\dfrac{5}{12}$.

Check:

$$x - \frac{1}{3} = -\frac{3}{4}$$

$$\begin{array}{c|c} -\dfrac{5}{12} - \dfrac{1}{3} & -\dfrac{3}{4} \\ -\dfrac{3}{4} = -\dfrac{3}{4} \end{array}$$

Problem 4

$$-8 = 5 + x$$
$$-8 - 5 = 5 - 5 + x$$
$$-13 = x$$

The solution is -13.

Problem 5

$$-\frac{2x}{5} = 6$$

$$\left(-\frac{5}{2}\right)\left(-\frac{2}{5}x\right) = \left(-\frac{5}{2}\right)(6)$$

$$x = -15$$

The solution is -15.

Problem 6

$$6x = 10$$
$$\frac{6x}{6} = \frac{10}{6}$$
$$x = \frac{5}{3}$$

The solution is $\dfrac{5}{3}$.

Check:

$$6x = 10$$

$$\begin{array}{c|c} 6\left(\dfrac{5}{3}\right) & 10 \\ 10 = 10 \end{array}$$

Problem 7

$$4x - 8x = 16$$
$$-4x = 16$$
$$\frac{-4x}{-4} = \frac{16}{-4}$$
$$x = -4$$

The solution is -4.

SECTION 3.2   *pages 84–86*

Problem 1
$$5x + 7 = 10$$
$$5x + 7 - 7 = 10 - 7$$
$$5x = 3$$
$$\frac{5x}{5} = \frac{3}{5}$$
$$x = \frac{3}{5}$$

The solution is $\frac{3}{5}$.

Problem 2
$$11 = 11 + 3x$$
$$11 - 11 = 11 - 11 + 3x$$
$$0 = 3x$$
$$\frac{0}{3} = \frac{3x}{3}$$
$$0 = x$$

The solution is 0.

Problem 3

Strategy   To find the number of units made, replace each of the variables with its given value, and solve for N.

Solution
$$T = U \cdot N + F$$
$$8000 = 15N + 2000$$
$$8000 - 2000 = 15N + 2000 - 2000$$
$$6000 = 15N$$
$$\frac{6000}{15} = \frac{15N}{15}$$
$$400 = N$$

The number of units made was 400.

SECTION 3.3   *pages 89–94*

Problem 1
$$5x + 4 = 6 + 10x$$
$$5x - 10x + 4 = 6 + 10x - 10x$$
$$-5x + 4 = 6$$
$$-5x + 4 - 4 = 6 - 4$$
$$-5x = 2$$
$$\frac{-5x}{-5} = \frac{2}{-5}$$
$$x = -\frac{2}{5}$$

The solution is $-\frac{2}{5}$.

Problem 2
$$5x - 10 - 3x = 6 - 4x$$
$$2x - 10 = 6 - 4x$$
$$2x + 4x - 10 = 6 - 4x + 4x$$
$$6x - 10 = 6$$
$$6x - 10 + 10 = 6 + 10$$
$$6x = 16$$
$$\frac{6x}{6} = \frac{16}{6}$$
$$x = \frac{8}{3}$$

The solution is $\frac{8}{3}$.

Problem 3

$$5x - 4(3 - 2x) = 2(3x - 2) + 6$$
$$5x - 12 + 8x = 6x - 4 + 6$$
$$13x - 12 = 6x + 2$$
$$13x - 6x - 12 = 6x - 6x + 2$$
$$7x - 12 = 2$$
$$7x - 12 + 12 = 2 + 12$$
$$7x = 14$$
$$\frac{7x}{7} = \frac{14}{7}$$
$$x = 2$$

The solution is 2.

Problem 4

$$-2[3x - 5(2x - 3)] = 3x - 8$$
$$-2[3x - 10x + 15] = 3x - 8$$
$$-2[-7x + 15] = 3x - 8$$
$$14x - 30 = 3x - 8$$
$$14x - 3x - 30 = 3x - 3x - 8$$
$$11x - 30 = -8$$
$$11x - 30 + 30 = -8 + 30$$
$$11x = 22$$
$$\frac{11x}{11} = \frac{22}{11}$$
$$x = 2$$

The solution is 2.

Problem 5

Strategy To find the location of the fulcrum when the system balances, replace the variables F_1, F_2, and d in the lever system equation with the given values, and solve for x.

Solution
$$F_1 x = F_2(d - x)$$
$$80x = 560(8 - x)$$
$$80x = 4480 - 560x$$
$$80x + 560x = 4480 - 560x + 560x$$
$$640x = 4480$$
$$\frac{640x}{640} = \frac{4480}{640}$$
$$x = 7$$

The fulcrum is 7 ft from the 80-lb force.

Problem 6

Strategy To find the force when the system balances, replace the variables F_1, x, and d in the lever system equation with the given values, and solve for F_2.

Solution
$$F_1 x = F_2(d - x)$$
$$40(6) = F_2(14 - 6)$$
$$240 = F_2(8)$$
$$240 = 8F_2$$
$$\frac{240}{8} = \frac{8F_2}{8}$$
$$30 = F_2$$

A 30-lb force must be applied to the other end.

SECTION 3.4   *pages 97–100*

Problem 1   the unknown number: n

three more than a number	is equal to	the difference between seven and the number

$$n + 3 = 7 - n$$
$$n + n + 3 = 7 - n + n$$
$$2n + 3 = 7$$
$$2n + 3 - 3 = 7 - 3$$
$$2n = 4$$
$$\frac{2n}{2} = \frac{4}{2}$$
$$n = 2$$

The number is 2.

Problem 2   the unknown number: n

nine less than twice a number	is	five times the sum of the number and twelve

$$2n - 9 = 5(n + 12)$$
$$2n - 9 = 5n + 60$$
$$2n - 5n - 9 = 5n - 5n + 60$$
$$-3n - 9 = 60$$
$$-3n - 9 + 9 = 60 + 9$$
$$-3n = 69$$
$$\frac{-3n}{-3} = \frac{69}{-3}$$
$$n = -23$$

The number is -23.

Problem 3

Strategy   To find the number of carbon atoms, write and solve an equation using n to represent the number of carbon atoms.

Solution

eight	represents	twice the number of carbon atoms

$$8 = 2n$$
$$\frac{8}{2} = \frac{2n}{2}$$
$$4 = n$$

There are 4 carbon atoms in a butane molecule.

Problem 4

Strategy   To find the number of 10-speed bicycles made, write and solve an equation using n to represent the number of 10-speed bicycles and $160 - n$ to represent the number of 3-speed bicycles.

Solution

four times the number of 3-speed bicycles made	equals	30 less than the number of 10-speed bicycles made

$$4(160 - n) = n - 30$$
$$640 - 4n = n - 30$$
$$640 - 4n - n = n - n - 30$$
$$640 - 5n = -30$$
$$640 - 640 - 5n = -30 - 640$$
$$-5n = -670$$
$$\frac{-5n}{-5} = \frac{-670}{-5}$$
$$n = 134$$

There are 134 10-speed bicycles made each day.

SOLUTIONS to Chapter 4 Problems

SECTION 4.1 *pages 115–117*

Problem 1 $125\% = 125\left(\dfrac{1}{100}\right) = \dfrac{125}{100} = 1\dfrac{1}{4}$

Problem 2 $16\dfrac{2}{3}\% = 16\dfrac{2}{3}\left(\dfrac{1}{100}\right) = \dfrac{50}{3}\left(\dfrac{1}{100}\right) = \dfrac{1}{6}$

$125\% = 125(0.01) = 1.25$

Problem 3 $6.08\% = 6.08(0.01) = 0.0608$

Problem 4 $0.043 = 0.043(100\%) = 4.3\%$

Problem 5 $2.57 = 2.57(100\%) = 257\%$

Problem 6 $\dfrac{5}{9} = \dfrac{5}{9}(100\%) = \dfrac{500}{9}\% \approx 55.6\%$

Problem 7 $\dfrac{9}{16} = \dfrac{9}{16}(100\%) = \dfrac{900}{16}\% = 56\dfrac{1}{4}\%$

SECTION 4.2 *pages 119–121*

Problem 1
$$PB = A$$
$$P(60) = 27$$
$$\dfrac{P(60)}{60} = \dfrac{27}{60}$$
$$P = 0.45$$

27 is 45% of 60.

Problem 2
$$PB = A$$
$$P(50) = 12$$
$$\dfrac{P(50)}{50} = \dfrac{12}{50}$$
$$P = 0.24$$

12 is 24% of 50.

Problem 3

Strategy To find the percent of the questions answered correctly, solve the basic percent equation using $B = 80$ and $A = 72$. The percent is unknown.

Solution
$$PB = A$$
$$P(80) = 72$$
$$\dfrac{P(80)}{80} = \dfrac{72}{80}$$
$$P = 0.9$$

90% of the questions were answered correctly.

Problem 4

Strategy To find how many gallons are used efficiently, solve the basic percent equation using $B = 15$ and $P = 32\% = 0.32$. The amount is unknown.

Solution
$$PB = A$$
$$0.32(15) = A$$
$$4.8 = A$$

Out of 15 gal of gasoline, 4.8 gal are used efficiently.

SECTION 4.3 *pages 124–126*

Problem 1

Strategy   Given: $C = \$60$
$S = \$90$
Unknown markup rate: r
Use the equation $S = C + rC$.

Solution $S = C + rC$
$90 = 60 + 60r$
$30 = 60r$
$0.5 = r$

The markup rate is 50%.

Problem 2

Strategy   Given: $r = 40\% = 0.40$
$S = \$133$
Unknown cost: C
Use the equation $S = C + rC$.

Solution $S = C + rC$
$133 = C + 0.40C$
$133 = 1.40C$
$95 = C$

The cost is $95.

Problem 3

Strategy   Given: $R = \$29.80$
$S = \$22.35$
Unknown discount rate: r
Use the equation $S = R - rR$.

Solution $S = R - rR$
$22.35 = 29.80 - 29.80r$
$-7.45 = -29.80r$
$0.25 = r$

The discount rate is 25%.

Problem 4

Strategy   Given: $S = \$43.50$
$r = 25\% = 0.25$
Unknown regular price: R
Use the equation $S = R - rR$.

Solution $S = R - rR$
$43.50 = R - 0.25R$
$43.50 = 0.75R$
$58 = R$

The regular price is $58.

SECTION 4.4 *pages 129–131*

Problem 1

Strategy   ■ Additional amount: x

	Principal	·	Rate	=	Interest
Amount at 7%	2500	·	0.07	=	0.07(2500)
Amount at 10%	x	·	0.10	=	0.10x
Amount at 9%	2500 + x	·	0.09	=	0.09(2500 + x)

■ The sum of the interest earned by the two investments equals the interest earned on the total investment.

Solution $0.07(2500) + 0.10x = 0.09(2500 + x)$
$175 + 0.10x = 225 + 0.09x$
$175 + 0.01x = 225$
$0.01x = 50$
$x = 5000$

$5000 more must be invested at 10%.

SECTION 4.5   *pages 134–138*

Problem 1

Strategy   ■ Pounds of $.75 fertilizer: x

	Amount	Cost	Value
$.90 fertilizer	20	$.90	0.90(20)
$.75 fertilizer	x	$.75	0.75x
$.85 fertilizer	20 + x	$.85	0.85(20 + x)

■ The sum of the values before mixing equals the value after mixing.

Solution $0.90(20) + 0.75x = 0.85(20 + x)$
$18 + 0.75x = 17 + 0.85x$
$18 - 0.10x = 17$
$-0.10x = -1$
$x = 10$

10 lb of the $.75 fertilizer must be added.

Problem 2

Strategy   ■ Liters of water: x

	Amount	Percent	Quantity
Water	x	0	0x
10%	6	0.10	6(0.10)
8%	$x + 6$	0.08	0.08(x + 6)

■ The sum of the quantities before mixing is equal to the quantity after mixing.

Solution $0x + 6(0.10) = 0.08(x + 6)$
$$0.60 = 0.08x + 0.48$$
$$0.12 = 0.08x$$
$$1.5 = x$$

The pharmacist adds 1.5 L of water to the 10% solution to get an 8% solution.

SECTION 4.6   *pages 142–145*

Problem 1

Strategy   ■ Rate of the first train: r
Rate of the second train: $2r$

	Rate	Time	Distance
First train	r	3	$3r$
Second train	$2r$	3	$3(2r)$

The sum of the distances traveled by each train equals 306 mi.

Solution $3r + 3(2r) = 306$
$$3r + 6r = 306$$
$$9r = 306$$
$$r = 34$$

$2r = 2(34) = 68$

The first train is traveling at 34 mph.
The second train is traveling at 68 mph.

Problem 2

Strategy   ■ Time spent flying out: t
Time spent flying back: $7 - t$

	Rate	Time	Distance
Out	120	t	$120t$
Back	90	$7 - t$	$90(7 - t)$

■ The distance out equals the distance back.

Solution $120t = 90(7 - t)$
$120t = 630 - 90t$
$210t = 630$
$t = 3$ (The time out was 3 h.)

The distance $= 120t = 120(3) = 360$ mi.

The parcel of land was 360 mi away.

SECTION 4.7   *pages 148–150*

Problem 1

Strategy   ■ Width of the rectangle: w
Length of the rectangle: $w + 3$
■ Use the equation for the perimeter of a rectangle.

Solution $2l + 2w = P$
$2(w + 3) + 2w = 42$
$2w + 6 + 2w = 42$
$4w + 6 = 42$
$4w = 36$
$w = 9$

The width of the rectangle is 9 m.

Problem 2

Strategy   ■ Measure of the second angle: x
Measure of the first angle: $2x$
Measure of the third angle: $x - 8$
■ Use the equation $\angle A + \angle B + \angle C = 180°$.

Solution $x + 2x + (x - 8) = 180$
$4x - 8 = 180$
$4x = 188$
$x = 47$

$2x = 2(47) = 94$
$x - 8 = 47 - 8 = 39$

The measure of the first angle is 94°.
The measure of the second angle is 47°.
The measure of the third angle is 39°.

SECTION 4.8 *pages 153–158*

Problem 1

Strategy   ■ First consecutive integer: n
Second consecutive integer: $n + 1$
Third consecutive integer: $n + 2$
■ The sum of the three integers is -12.

Solution $n + (n + 1) + (n + 2) = -12$
$$3n + 3 = -12$$
$$3n = -15$$
$$n = -5$$

$$n + 1 = -5 + 1 = -4$$
$$n + 2 = -5 + 2 = -3$$

The three consecutive integers are -5, -4, and -3.

Problem 2

Strategy   ■ Number of dimes: x
Number of nickels: $5x$
Number of quarters: $x + 6$

Coin	Number	Value	Total value
Dime	x	10	$10x$
Nickel	$5x$	5	$5(5x)$
Quarter	$x + 6$	25	$25(x + 6)$

■ The sum of the total values of each denomination of coin equals the total value of all the coins (630 cents).

Solution $10x + 5(5x) + 25(x + 6) = 630$
$$10x + 25x + 25x + 150 = 630$$
$$60x + 150 = 630$$
$$60x = 480$$
$$x = 8$$

$$5x = 5(8) = 40$$
$$x + 6 = 8 + 6 = 14$$

The bank contains 8 dimes, 40 nickels, and 14 quarters.

Problem 3

Strategy ■ The number of years ago: x

	Present age	Past age
Half-dollar	35	$35 - x$
Dime	25	$25 - x$

■ At a past age, the half-dollar was twice as old as the dime.

Solution $35 - x = 2(25 - x)$
$35 - x = 50 - 2x$
$35 + x = 50$
$\quad\ x = 15$

Fifteen years ago, the half-dollar was twice as old as the dime.

SOLUTIONS to Chapter 5 Problems

SECTION 5.1 *pages 175–176*

Problem 1 $2x^2 + 4x - 3$
$\underline{5x^2 - 6x\qquad}$
$7x^2 - 2x - 3$

Problem 2 $(-4x^2 - 3xy + 2y^2) + (3x^2 - 4y^2)$
$(-4x^2 + 3x^2) - 3xy + (2y^2 - 4y^2)$
$-x^2 - 3xy - 2y^2$

Problem 3 $\begin{array}{l} 8y^2 - 4xy + \ x^2 \\ \underline{2y^2 - \ xy + 5x^2} \end{array}$ $\begin{array}{l} 8y^2 - 4xy + \ x^2 \\ \underline{-2y^2 + \ xy - 5x^2} \\ 6y^2 - 3xy - 4x^2 \end{array}$

Problem 4 $(-3a^2 - 4a + 2) - (5a^3 + 2a - 6)$
$(-3a^2 - 4a + 2) + (-5a^3 - 2a + 6)$
$-5a^3 - 3a^2 - 6a + 8$

SECTION 5.2 *pages 179–181*

Problem 1 $(3x^2)(6x^3) = (3 \cdot 6)(x^2 \cdot x^3) = 18x^5$

Problem 2 $(-3xy^2)(-4x^2y^3) = [(-3)(-4)](x \cdot x^2)(y^2 \cdot y^3) = 12x^3y^5$

Problem 3 $(3x)(2x^2y)^3 = (3x)(2^3x^6y^3) = (3x)(8x^6y^3) = (3 \cdot 8)(x \cdot x^6)y^3 = 24x^7y^3$

SECTION 5.3   *pages 182–187*

Problem 1   **A.** $(-2y + 3)(-4y) = 8y^2 - 12y$ **B.** $-a^2(3a^2 + 2a - 7) = -3a^4 - 2a^3 + 7a^2$

Problem 2

$$
\begin{array}{r}
2y^3 + 2y^2 - 3 \\
3y - 1 \\
\hline
-2y^3 - 2y^2 \quad\ + 3 \\
6y^4 + 6y^3 \qquad\ - 9y \\
\hline
6y^4 + 4y^3 - 2y^2 - 9y + 3
\end{array}
$$

Problem 3

$$
\begin{array}{r}
3x^3 - 2x^2 + x - 3 \\
2x + 5 \\
\hline
15x^3 - 10x^2 + 5x - 15 \\
6x^4 - 4x^3 + 2x^2 - 6x \\
\hline
6x^4 + 11x^3 - 8x^2 - x - 15
\end{array}
$$

Problem 4 $(4y - 5)(3y - 3) = 12y^2 - 12y - 15y + 15 = 12y^2 - 27y + 15$

Problem 5 $(3a + 2b)(3a - 5b) = 9a^2 - 15ab + 6ab - 10b^2 = 9a^2 - 9ab - 10b^2$

Problem 6 $(2a + 5c)(2a - 5c) = (2a)^2 - (5c)^2 = 4a^2 - 25c^2$

Problem 7 $(3x + 2y)^2 = (3x)^2 + 2(3x)(2y) + (2y)^2 = 9x^2 + 12xy + 4y^2$

Problem 8

Strategy   To find the area in terms of x, replace the variables L and W in the equation $A = LW$ with the given values and solve for A.

Solution $A = LW$
$A = (x + 8)(x - 5)$
$A = x^2 - 5x + 8x - 40$
$A = x^2 + 3x - 40$

The area is $x^2 + 3x - 40$.

Problem 9

Strategy   To find the area of the triangle in terms of x, replace the variables b and h in the equation $A = \frac{1}{2}bh$ with the given values and solve for A.

Solution $A = \frac{1}{2}bh$

$A = \frac{1}{2}(x + 3)(4x - 6)$

$A = \frac{1}{2}(4x^2 + 6x - 18)$

$A = 2x^2 + 3x - 9$

The area is $2x^2 + 3x - 9$.

SECTION 5.4 *pages 191–197*

Problem 1 $\dfrac{2^{-2}}{2^3} = 2^{-2-3} = 2^{-5} = \dfrac{1}{2^5} = \dfrac{1}{32}$

Problem 2 $\dfrac{b^8}{a^{-5}b^6} = a^5b^2$

Problem 3
A. $\dfrac{12x^{-8}y^4}{-16xy^{-3}} = -\dfrac{3x^{-9}y^7}{4} = -\dfrac{3y^7}{4x^9}$

B. $(-3ab)(2a^3b^{-2})^{-3} = (-3ab)(2^{-3}a^{-9}b^6)$
$$= -\dfrac{3a^{-8}b^7}{2^3}$$
$$= -\dfrac{3b^7}{8a^8}$$

Problem 4 $\dfrac{(6a^{-2}b^3)^{-1}}{(4a^3b^{-2})^{-2}} = \dfrac{6^{-1}a^2b^{-3}}{4^{-2}a^{-6}b^4}$
$$= \dfrac{4^2a^8b^{-7}}{6}$$
$$= \dfrac{16a^8}{6b^7}$$
$$= \dfrac{8a^8}{3b^7}$$

Problem 5 $\dfrac{4x^3y + 8x^2y^2 - 4xy^3}{2xy} = \dfrac{4x^3y}{2xy} + \dfrac{8x^2y^2}{2xy} - \dfrac{4xy^3}{2xy}$
$$= 2x^2 + 4xy - 2y^2$$

Problem 6 $\dfrac{24x^2y^2 - 18xy + 6y}{6xy} = \dfrac{24x^2y^2}{6xy} - \dfrac{18xy}{6xy} + \dfrac{6y}{6xy}$
$$= 4xy - 3 + \dfrac{1}{x}$$

Problem 7

$$\begin{array}{r}
x^2 + 2x + 2 \\
x - 2\overline{\smash{)}x^3 + 0x^2 - 2x - 4} \\
\underline{x^3 - 2x^2} \\
2x^2 - 2x \\
\underline{2x^2 - 4x} \\
2x - 4 \\
\underline{2x - 4} \\
0
\end{array}$$

$(x^3 - 2x - 4) \div (x - 2) = x^2 + 2x + 2$

SOLUTIONS to Chapter 6 Problems

SECTION 6.1 *pages 213–216*

Problem 1 $4x^6y = 2 \cdot 2 \cdot x^6 \cdot y$
$18x^2y^6 = 2 \cdot 3 \cdot 3 \cdot x^2 \cdot y^6$

$\text{GCF} = 2 \cdot x^2 \cdot y = 2x^2y$

Problem 2   A. $14a^2 = 2 \cdot 7 \cdot a^2$
$21a^4b = 3 \cdot 7 \cdot a^4 \cdot b$
The GCF is $7a^2$.

$14a^2 - 21a^4b = 7a^2(2) + 7a^2(-3a^2b) = 7a^2(2 - 3a^2b)$

B. $6x^4y^2 = 2 \cdot 3 \cdot x^4 \cdot y^2$
$9x^3y^2 = 3 \cdot 3 \cdot x^3 \cdot y^2$
$12x^2y^4 = 2 \cdot 2 \cdot 3 \cdot x^2 \cdot y^4$
The GCF is $3x^2y^2$.

$6x^4y^2 - 9x^3y^2 + 12x^2y^4 = 3x^2y^2(2x^2) + 3x^2y^2(-3x) + 3x^2y^2(4y^2)$
$= 3x^2y^2(2x^2 - 3x + 4y^2)$

Problem 3 $a(b - 7) + b(b - 7) = (b - 7)(a + b)$

Problem 4 $3y(5x - 2) - 4(2 - 5x) = 3y(5x - 2) + 4(5x - 2) = (5x - 2)(3y + 4)$

Problem 5 $y^5 - 5y^3 + 4y^2 - 20 = (y^5 - 5y^3) + (4y^2 - 20)$
$= y^3(y^2 - 5) + 4(y^2 - 5) = (y^2 - 5)(y^3 + 4)$

SECTION 6.2   *pages 218–222*

Problem 1

Factors of 15	Sum
−1, −15	−16
−3, −5	−8

$x^2 - 8x + 15 = (x - 3)(x - 5)$

Problem 2

Factors of −18	Sum
1, −18	−17
−1, 18	17
2, −9	−7
−2, 9	7
3, −6	−3
−3, 6	3

$x^2 + 3x - 18 = (x + 6)(x - 3)$

Problem 3   The GCF is $3b$.

$3a^2b - 18ab - 81b = 3b(a^2 - 6a - 27)$

Factors of −27	Sum
1, −27	−26
−1, 27	26
3, −9	−6
−3, 9	6

$3a^2b - 18ab - 81b = 3b(a + 3)(a - 9)$

Problem 4   The GCF is 4.

$4x^2 - 40xy + 84y^2 = 4(x^2 - 10xy + 21y^2)$

Factors of 21	Sum
−1, −21	−22
−3, −7	−10

$4x^2 - 40xy + 84y^2 = 4(x - 3y)(x - 7y)$

SECTION 6.3 *pages 225–232*

Problem 1

Factors of 6	Factors of 5
1, 6	−1, −5
2, 3	

Trial Factors	Middle Term
$(x - 1)(6x - 5)$	$-5x - 6x = -11x$

$6x^2 - 11x + 5 = (x - 1)(6x - 5)$

Problem 2

Factors of 8	Factors of −15
1, 8	1, −15
2, 4	−1, 15
	3, −5
	−3, 5

Trial Factors	Middle Term
$(x + 1)(8x - 15)$	$-15x + 8x = -7x$
$(x - 1)(8x + 15)$	$15x - 8x = 7x$
$(x + 3)(8x - 5)$	$-5x + 24x = 19x$
$(x - 3)(8x + 5)$	$5x - 24x = -19x$
$(2x + 1)(4x - 15)$	$-30x + 4x = -26x$
$(2x - 1)(4x + 15)$	$30x - 4x = 26x$
$(2x + 3)(4x - 5)$	$-10x + 12x = 2x$
$(2x - 3)(4x + 5)$	$10x - 12x = -2x$
$(8x + 1)(x - 15)$	$-120x + x = -119x$
$(8x - 1)(x + 15)$	$120x - x = 119x$
$(8x + 3)(x - 5)$	$-40x + 3x = -37x$
$(8x - 3)(x + 5)$	$40x - 3x = 37x$
$(4x + 1)(2x - 15)$	$-60x + 2x = -58x$
$(4x - 1)(2x + 15)$	$60x - 2x = 58x$
$(4x + 3)(2x - 5)$	$-20x + 6x = -14x$
$(4x - 3)(2x + 5)$	$20x - 6x = 14x$

$8x^2 + 14x - 15 = (4x - 3)(2x + 5)$

Problem 3

Factors of 24	Factors of 1
1, 24	−1, −1
2, 12	
3, 8	
4, 6	

Trial Factors	Middle Term
$(1 - y)(24 - y)$	$-y - 24y = -25y$
$(2 - y)(12 - y)$	$-2y - 12y = -14y$
$(3 - y)(8 - y)$	$-3y - 8y = -11y$
$(4 - y)(6 - y)$	$-4y - 6y = -10y$

$24 - 10y + y^2 = (4 - y)(6 - y)$

Problem 4

The GCF is $2a^2$.

$4a^2b^2 - 30a^2b + 14a^2 = 2a^2(2b^2 - 15b + 7)$

Factors of 2	Factors of 7
1, 2	−1, −7

Trial Factors	Middle Term
$(b - 1)(2b - 7)$	$-7b - 2b = -9b$
$(b - 7)(2b - 1)$	$-b - 14b = -15b$

$4a^2b^2 - 30a^2b + 14a^2 = 2a^2(b - 7)(2b - 1)$

Problem 5 $\quad a \cdot c = -14,\ -1(14) = -14,$
$\quad -1 + 14 = 13$

$2a^2 + 13a - 7 =$
$2a^2 - a + 14a - 7 =$
$(2a^2 - a) + (14a - 7) =$
$a(2a - 1) + 7(2a - 1) =$
$(2a - 1)(a + 7)$

Problem 6 $\quad a \cdot c = -12,\ 1(-12) = -12,\ 1 - 12 = -11$

$4a^2 - 11a - 3 =$
$4a^2 + a - 12a - 3 =$
$(4a^2 + a) - (12a + 3) =$
$a(4a + 1) - 3(4a + 1) =$
$(4a + 1)(a - 3)$

Problem 7   The GCF is $5x$.

$15x^3 + 40x^2 - 80x = 5x(3x^2 + 8x - 16)$

$-4(12) = -48, \ -4 + 12 = 8$

$3x^2 + 8x - 16 =$
$3x^2 - 4x + 12x - 16 =$
$(3x^2 - 4x) + (12x - 16) =$
$x(3x - 4) + 4(3x - 4) =$
$(3x - 4)(x + 4)$

$15x^3 + 40x^2 - 80x = 5x(3x^2 + 8x - 16)$
$\qquad\qquad\qquad\quad = 5x(3x - 4)(x + 4)$

SECTION 6.4   *pages 236–240*

Problem 1   **A.** $25a^2 - b^2 = (5a)^2 - b^2 = (5a + b)(5a - b)$
B. $6x^2 - 1$ is nonfactorable over the integers.
C. $n^8 - 36 = (n^4)^2 - 6^2 = (n^4 + 6)(n^4 - 6)$

Problem 2 $n^4 - 81 = (n^2 + 9)(n^2 - 9)$
$\qquad\qquad\quad = (n^2 + 9)(n + 3)(n - 3)$

Problem 3   **A.** $16y^2 = (4y)^2, \ 1 = 1^2$
$(4y + 1)^2 = (4y)^2 + 2(4y)(1) + (1)^2$
$\qquad\qquad\; = 16y^2 + 8y + 1$

$16y^2 + 8y + 1 = (4y + 1)^2$
B. $x^2 = (x)^2, \ 36 = 6^2$
$(x + 6)^2 = x^2 + 2(x)(6) + 6^2$
$\qquad\qquad = x^2 + 12x + 36$

The polynomial is not a perfect square.

$x^2 + 14x + 36$ is nonfactorable over the integers.

Problem 4   **A.** $12x^3 - 75x = 3x(4x^2 - 25)$
$\qquad\qquad\qquad\; = 3x(2x + 5)(2x - 5)$
B. $a^2b - 7a^2 - b + 7 = (a^2b - 7a^2) - (b - 7)$
$\qquad\qquad\qquad\qquad\; = a^2(b - 7) - (b - 7)$
$\qquad\qquad\qquad\qquad\; = (b - 7)(a^2 - 1)$
$\qquad\qquad\qquad\qquad\; = (b - 7)(a + 1)(a - 1)$
C. $4x^3 + 28x^2 - 120x = 4x(x^2 + 7x - 30)$
$\qquad\qquad\qquad\qquad\;\; = 4x(x + 10)(x - 3)$

SECTION 6.5 *pages 243–246*

Problem 1

$$2x^2 - 50 = 0$$
$$2(x^2 - 25) = 0$$
$$x^2 - 25 = 0$$
$$(x + 5)(x - 5) = 0$$

$x + 5 = 0$	$x - 5 = 0$
$x = -5$	$x = 5$

The solutions are -5 and 5.

Problem 2

$$(x + 2)(x - 7) = 52$$
$$x^2 - 5x - 14 = 52$$
$$x^2 - 5x - 66 = 0$$
$$(x - 11)(x + 6) = 0$$

$x - 11 = 0$	$x + 6 = 0$
$x = 11$	$x = -6$

The solutions are -6 and 11.

Problem 3

Strategy   First positive integer: n
Second positive integer: $n + 1$

The sum of the squares of the two integers is 85.

Solution
$$n^2 + (n + 1)^2 = 85$$
$$n^2 + n^2 + 2n + 1 = 85$$
$$2n^2 + 2n - 84 = 0$$
$$2(n^2 + n - 42) = 0$$
$$n^2 + n - 42 = 0$$
$$(n + 7)(n - 6) = 0$$

$n + 7 = 0$	$n - 6 = 0$	(-7 is not a positive integer.)
$n = -7$	$n = 6$	

$$n + 1 = 6 + 1 = 7$$

The two integers are 6 and 7.

Problem 4

Strategy   Width $= x$
Length $= 2x + 3$

The area of the rectangle is 90 in.^{2}.
Use the equation $A = lw$.

Solution $A = lw$
$$90 = (2x + 3)x$$
$$90 = 2x^2 + 3x$$
$$0 = 2x^2 + 3x - 90$$
$$0 = (2x + 15)(x - 6)$$

$2x + 15 = 0$	$x - 6 = 0$
$2x = -15$	$x = 6$
$x = -\dfrac{15}{2}$	(The width cannot be a negative number.)

$$2x + 3 = 2(6) + 3 = 12 + 3 = 15$$

The width is 6 in. The length is 15 in.

SOLUTIONS to Chapter 7 Problems

SECTION 7.1 *pages 261–265*

Problem 1 **A.** $\dfrac{6x^5y}{12x^2y^3} = \dfrac{\overset{1}{\cancel{2}} \cdot \overset{1}{\cancel{3}} \cdot x^5y}{\underset{1}{\cancel{2}} \cdot 2 \cdot \underset{1}{\cancel{3}} \cdot x^2y^3} = \dfrac{x^3}{2y^2}$

B. $\dfrac{x^2 + 2x - 24}{16 - x^2} = \dfrac{\overset{-1}{\cancel{(x-4)}}(x+6)}{\underset{1}{\cancel{(4-x)}}(4+x)} = -\dfrac{x+6}{x+4}$

Problem 2 **A.** $\dfrac{12x^2 + 3x}{10x - 15} \cdot \dfrac{8x - 12}{9x + 18} =$

$\dfrac{3x(4x + 1)}{5(2x - 3)} \cdot \dfrac{4(2x - 3)}{9(x + 2)} =$

$\dfrac{\overset{1}{\cancel{3}}x(4x + 1) \cdot 2 \cdot 2(2\cancel{x - 3})}{5(2\cancel{x - 3}) \cdot \underset{1}{\cancel{3}} \cdot 3(x + 2)} = \dfrac{4x(4x + 1)}{15(x + 2)}$

B. $\dfrac{x^2 + 2x - 15}{9 - x^2} \cdot \dfrac{x^2 - 3x - 18}{x^2 - 7x + 6} =$

$\dfrac{(x - 3)(x + 5)}{(3 - x)(3 + x)} \cdot \dfrac{(x + 3)(x - 6)}{(x - 1)(x - 6)} =$

$\dfrac{\overset{-1}{\cancel{(x-3)}}(x + 5) \cdot \overset{1}{\cancel{(x+3)}}\overset{1}{\cancel{(x-6)}}}{\underset{1}{\cancel{(3-x)}}\underset{1}{\cancel{(3+x)}} \cdot (x - 1)\underset{1}{\cancel{(x-6)}}} = -\dfrac{x + 5}{x - 1}$

Problem 3 **A.** $\dfrac{a^2}{4bc^2 - 2b^2c} \div \dfrac{a}{6bc - 3b^2} =$

$\dfrac{a^2}{4bc^2 - 2b^2c} \cdot \dfrac{6bc - 3b^2}{a} =$

$\dfrac{a^{\overset{1}{\cancel{2}}} \cdot 3b(2\cancel{c - b})}{2bc(2\cancel{c - b}) \cdot a} = \dfrac{3a}{2c}$

B. $\dfrac{3x^2 + 26x + 16}{3x^2 - 7x - 6} \div \dfrac{2x^2 + 9x - 5}{x^2 + 2x - 15} =$

$\dfrac{3x^2 + 26x + 16}{3x^2 - 7x - 6} \cdot \dfrac{x^2 + 2x - 15}{2x^2 + 9x - 5} =$

$\dfrac{\overset{1}{\cancel{(3x+2)}}(x + 8)}{\underset{1}{\cancel{(3x+2)}}\cancel{(x - 3)}} \cdot \dfrac{\overset{1}{\cancel{(x+5)}}\cancel{(x-3)}}{(2x - 1)\underset{1}{\cancel{(x+5)}}} = \dfrac{x + 8}{2x - 1}$

SECTION 7.2 *pages 268–270*

Problem 1 $8uv^2 = 2 \cdot 2 \cdot 2 \cdot u \cdot v \cdot v$
$12uw = 2 \cdot 2 \cdot 3 \cdot u \cdot w$

$\text{LCM} = 2 \cdot 2 \cdot 2 \cdot 3 \cdot u \cdot v \cdot v \cdot w$
$\quad\quad = 24uv^2w$

Problem 2 $m^2 - 6m + 9 = (m - 3)(m - 3)$
$m^2 - 2m - 3 = (m + 1)(m - 3)$

$\text{LCM} = (m - 3)(m - 3)(m + 1)$

Problem 3 The LCM is $36xy^2z$.

$\dfrac{x - 3}{4xy^2} = \dfrac{x - 3}{4xy^2} \cdot \dfrac{9z}{9z} = \dfrac{9xz - 27z}{36xy^2z}$

$\dfrac{2x + 1}{9y^2z} = \dfrac{2x + 1}{9y^2z} \cdot \dfrac{4x}{4x} = \dfrac{8x^2 + 4x}{36xy^2z}$

Problem 4 $\dfrac{2x}{25 - x^2} = \dfrac{2x}{-(x^2 - 25)} = -\dfrac{2x}{x^2 - 25}$

The LCM is $(x - 5)(x + 5)(x + 2)$.

$\dfrac{x + 4}{x^2 - 3x - 10} = \dfrac{x + 4}{(x + 2)(x - 5)} \cdot \dfrac{x + 5}{x + 5} = \dfrac{x^2 + 9x + 20}{(x + 2)(x - 5)(x + 5)}$

$\dfrac{2x}{25 - x^2} = -\dfrac{2x}{(x - 5)(x + 5)} \cdot \dfrac{x + 2}{x + 2} = -\dfrac{2x^2 + 4x}{(x + 2)(x - 5)(x + 5)}$

SECTION 7.3 *pages 273–277*

Problem 1   **A.** $\dfrac{3}{xy} + \dfrac{12}{xy} = \dfrac{3+12}{xy} = \dfrac{15}{xy}$ **B.** $\dfrac{2x^2}{x^2-x-12} - \dfrac{7x+4}{x^2-x-12} = \dfrac{2x^2-(7x+4)}{x^2-x-12} =$

$$\dfrac{2x^2-7x-4}{x^2-x-12} = \dfrac{(2x+1)\cancel{(x-4)}^{\,1}}{(x+3)\cancel{(x-4)}_{\,1}} = \dfrac{2x+1}{x+3}$$

Problem 2   **A.** The LCM of the denominators is $24y$.

$$\dfrac{z}{8y} - \dfrac{4z}{3y} + \dfrac{5z}{4y} = \dfrac{z}{8y}\cdot\dfrac{3}{3} - \dfrac{4z}{3y}\cdot\dfrac{8}{8} + \dfrac{5z}{4y}\cdot\dfrac{6}{6} = \dfrac{3z}{24y} - \dfrac{32z}{24y} + \dfrac{30z}{24y} =$$

$$\dfrac{3z-32z+30z}{24y} = \dfrac{z}{24y}$$

B. The LCM is $x-2$.

$$\dfrac{5x}{x-2} - \dfrac{3}{2-x} = \dfrac{5x}{x-2}\cdot\dfrac{1}{1} - \dfrac{3}{-(x-2)}\cdot\dfrac{-1}{-1} = \dfrac{5x}{x-2} - \dfrac{-3}{x-2} =$$

$$\dfrac{5x-(-3)}{x-2} = \dfrac{5x+3}{x-2}$$

Problem 3   **A.** The LCM is $(3x-1)(x+4)$.

$$\dfrac{4x}{3x-1} - \dfrac{9}{x+4} = \dfrac{4x}{3x-1}\cdot\dfrac{x+4}{x+4} - \dfrac{9}{x+4}\cdot\dfrac{3x-1}{3x-1} =$$

$$\dfrac{4x^2+16x}{(3x-1)(x+4)} - \dfrac{27x-9}{(3x-1)(x+4)} =$$

$$\dfrac{(4x^2+16x)-(27x-9)}{(3x-1)(x+4)} = \dfrac{4x^2+16x-27x+9}{(3x-1)(x+4)} =$$

$$\dfrac{4x^2-11x+9}{(3x-1)(x+4)}$$

B. The LCM is $(x+5)(x-5)$.

$$\dfrac{2x-1}{x^2-25} + \dfrac{2}{5-x} = \dfrac{2x-1}{(x+5)(x-5)} + \dfrac{2}{-(x-5)}\cdot\dfrac{-1(x+5)}{-1(x+5)} =$$

$$\dfrac{2x-1}{(x+5)(x-5)} + \dfrac{-2(x+5)}{(x+5)(x-5)} =$$

$$\dfrac{(2x-1)+(-2)(x+5)}{(x+5)(x-5)} = \dfrac{2x-1-2x-10}{(x+5)(x-5)} =$$

$$\dfrac{-11}{(x+5)(x-5)} = -\dfrac{11}{(x+5)(x-5)}$$

SECTION 7.4   *pages 281–283*

Problem 1   **A.**   The LCM of 3, x, 9, and x^2 is $9x^2$.

B.   The LCM of x and x^2 is x^2.

$$\frac{\dfrac{1}{3} - \dfrac{1}{x}}{\dfrac{1}{9} - \dfrac{1}{x^2}} = \frac{\dfrac{1}{3} - \dfrac{1}{x}}{\dfrac{1}{9} - \dfrac{1}{x^2}} \cdot \frac{9x^2}{9x^2} = \frac{\dfrac{1}{3} \cdot 9x^2 - \dfrac{1}{x} \cdot 9x^2}{\dfrac{1}{9} \cdot 9x^2 - \dfrac{1}{x^2} \cdot 9x^2} =$$

$$\frac{3x^2 - 9x}{x^2 - 9} = \frac{\overset{1}{\cancel{3x}}(\cancel{x - 3})}{(\cancel{x - 3})(x + 3)} = \frac{3x}{x + 3}$$

$$\frac{1 + \dfrac{4}{x} + \dfrac{3}{x^2}}{1 + \dfrac{10}{x} + \dfrac{21}{x^2}} = \frac{1 + \dfrac{4}{x} + \dfrac{3}{x^2}}{1 + \dfrac{10}{x} + \dfrac{21}{x^2}} \cdot \frac{x^2}{x^2} =$$

$$\frac{1 \cdot x^2 + \dfrac{4}{x} \cdot x^2 + \dfrac{3}{x^2} \cdot x^2}{1 \cdot x^2 + \dfrac{10}{x} \cdot x^2 + \dfrac{21}{x^2} \cdot x^2} =$$

$$\frac{x^2 + 4x + 3}{x^2 + 10x + 21} =$$

$$\frac{(x + 1)\overset{1}{\cancel{(x + 3)}}}{\underset{1}{\cancel{(x + 3)}}(x + 7)} = \frac{x + 1}{x + 7}$$

SECTION 7.5   *pages 285–290*

Problem 1 $x + \dfrac{1}{3} = \dfrac{4}{3x}$ The LCM of 3 and $3x$ is $3x$.

$$3x\left(x + \frac{1}{3}\right) = 3x\left(\frac{4}{3x}\right)$$

$$3x \cdot x + \frac{3x}{1} \cdot \frac{1}{3} = \frac{3x}{1} \cdot \frac{4}{3x}$$

$$3x^2 + x = 4$$

$$3x^2 + x - 4 = 0$$

$$(3x + 4)(x - 1) = 0$$

$$3x + 4 = 0 \qquad x - 1 = 0$$

$$3x = -4 \qquad\quad x = 1$$

$$x = -\frac{4}{3}$$

Both $-\dfrac{4}{3}$ and 1 check as solutions.

The solutions are $-\dfrac{4}{3}$ and 1.

Problem 2
$$\frac{5x}{x+2} = 3 - \frac{10}{x+2} \qquad \text{The LCM is } x+2.$$

$$\frac{(x+2)}{1} \cdot \frac{5x}{x+2} = \frac{(x+2)}{1}\left(3 - \frac{10}{x+2}\right)$$

$$5x = (x+2)3 - 10$$
$$5x = 3x + 6 - 10$$
$$5x = 3x - 4$$
$$2x = -4$$
$$x = -2$$

−2 does not check as a solution.
The equation has no solution.

Problem 3   **A.**
$$\frac{2}{x+3} = \frac{6}{5x+5}$$

$$(x+3)(5x+5)\frac{2}{x+3} = (x+3)(5x+5)\frac{6}{5x+5}$$
$$(5x+5)2 = (x+3)6$$
$$10x + 10 = 6x + 18$$
$$4x + 10 = 18$$
$$4x = 8$$
$$x = 2$$

The solution is 2.

   B.
$$\frac{5}{2x-3} = \frac{10}{x+3}$$

$$(x+3)(2x-3)\frac{5}{2x-3} = (x+3)(2x-3)\frac{10}{x+3}$$
$$(x+3)5 = (2x-3)10$$
$$5x + 15 = 20x - 30$$
$$-15x + 15 = -30$$
$$-15x = -45$$
$$x = 3$$

The solution is 3.

Problem 4

Strategy   To find the total area that 270 ceramic tiles will cover, write and solve a proportion using x to represent the number of square feet that 270 tiles will cover.

Solution
$$\frac{4}{9} = \frac{x}{270}$$

$$270\left(\frac{4}{9}\right) = 270\left(\frac{x}{270}\right)$$
$$120 = x$$

A 120-ft^2 area can be tiled using 270 ceramic tiles.

Problem 5

Strategy   Tō find the additional amount of medication required for a 180-lb adult, write and solve a proportion using x to represent the additional medication. Then $3 + x$ is the total amount required for a 180-lb adult.

Solution
$$\frac{120}{3} = \frac{180}{3 + x}$$
$$\frac{40}{1} = \frac{180}{3 + x}$$
$$(3 + x) \cdot 40 = (3 + x) \cdot \frac{180}{3 + x}$$
$$120 + 40x = 180$$
$$40x = 60$$
$$x = 1.5$$

1.5 additional ounces are required for a 180-lb adult.

SECTION 7.6   *pages 295–297*

Problem 1
$$5x - 2y = 10$$
$$5x - 5x - 2y = -5x + 10$$
$$-2y = -5x + 10$$
$$\frac{-2y}{-2} = \frac{-5x + 10}{-2}$$
$$y = \frac{5}{2}x - 5$$

Problem 2
$$s = \frac{A + L}{2}$$
$$2 \cdot s = 2\left(\frac{A + L}{2}\right)$$
$$2s = A + L$$
$$2s - A = A - A + L$$
$$2s - A = L$$

Problem 3
$$S = a + (n - 1)d$$
$$S = a + nd - d$$
$$S - a = a - a + nd - d$$
$$S - a = nd - d$$
$$S - a + d = nd - d + d$$
$$S - a + d = nd$$
$$\frac{S - a + d}{d} = \frac{nd}{d}$$
$$\frac{S - a + d}{d} = n$$

Problem 4
$$S = C + rC$$
$$S = (1 + r)C$$
$$\frac{S}{1 + r} = \frac{(1 + r)C}{1 + r}$$
$$\frac{S}{1 + r} = C$$

SECTION 7.7   *pages 300–304*

Problem 1

Strategy   ■ Time for one printer to complete the job: t

	Rate	Time	Part
1st printer	$\frac{1}{t}$	3	$\frac{3}{t}$
2nd printer	$\frac{1}{t}$	5	$\frac{5}{t}$

■ The sum of the parts of the task completed must equal 1.

Solution
$$\frac{3}{t} + \frac{5}{t} = 1$$
$$t\left(\frac{3}{t} + \frac{5}{t}\right) = t \cdot 1$$
$$3 + 5 = t$$
$$8 = t$$

Working alone, one printer takes 8 h to print the payroll.

Problem 2

Strategy
- Rate sailing across the lake: r
  Rate sailing back: $2r$

	Distance	Rate	Time
Across	6	r	$\frac{6}{r}$
Back	6	$2r$	$\frac{6}{2r}$

- The total time for the trip was 3 h.

Solution
$$\frac{6}{r} + \frac{6}{2r} = 3$$
$$2r\left(\frac{6}{r} + \frac{6}{2r}\right) = 2r(3)$$
$$2r \cdot \frac{6}{r} + 2r \cdot \frac{6}{2r} = 6r$$
$$12 + 6 = 6r$$
$$18 = 6r$$
$$3 = r$$

The rate across the lake was 3 km/h.

SOLUTIONS to Chapter 8 Problems

SECTION 8.1   *pages 321–326*

Problem 1

Problem 2 $A(4, 2)$, $B(-3, 4)$, $C(-3, 0)$, $D(0, 0)$

Problem 3 $y = -\frac{1}{2}x - 3$

$$
\begin{array}{c|c}
-4 & -\frac{1}{2}(2) - 3 \\
 & -1 - 3 \\
 & -4 \\
\end{array}
$$

$-4 = -4$

Yes, $(2, -4)$ is a solution of

$y = -\frac{1}{2}x - 3$.

Problem 4 $y = -\frac{1}{4}x + 1$

$$= -\frac{1}{4}(4) + 1$$

$$= -1 + 1$$

$$= 0$$

The ordered pair solution is $(4, 0)$.

Problem 5

Strategy   Graph the ordered pairs on a rectangular coordinate system where the reading on the horizontal axis represents the age of the car in years and the vertical axis represents the price paid for the car in hundreds of dollars.

Solution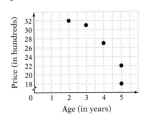

SECTION 8.2   *pages 329–334*

Problem 1

Problem 2

Problem 3   **A.** $5x - 2y = 10$

$-2y = -5x + 10$

$y = \frac{5}{2}x - 5$

B. $x - 3y = 9$

$-3y = -x + 9$

$y = \frac{1}{3}x - 3$

Problem 4   **A.**   **B.**

SECTION 8.3   *pages 336–344*

Problem 1   x-intercept: y-intercept:
$$4x - y = 4 \qquad 4x - y = 4$$
$$4x - 0 = 4 \qquad 4(0) - y = 4$$
$$4x = 4 \qquad -y = 4$$
$$x = 1 \qquad y = -4$$

(1, 0)   (0, −4)

Problem 2   x-intercept: y-intercept:
$$y = 3x - 6 \qquad y = 3x - 6$$
$$0 = 3x - 6 \qquad b = -6$$
$$-3x = -6$$
$$x = 2 \qquad (0, -6)$$

(2, 0)

Problem 3   **A.** $m = \dfrac{y_2 - y_1}{x_2 - x_1} = \dfrac{3 - 2}{1 - (-1)} = \dfrac{1}{2}$

The slope is $\dfrac{1}{2}$.

B. $m = \dfrac{y_2 - y_1}{x_2 - x_1} = \dfrac{-5 - 2}{4 - 1} = \dfrac{-7}{3}$

The slope is $-\dfrac{7}{3}$.

C. $m = \dfrac{y_2 - y_1}{x_2 - x_1} = \dfrac{7 - 3}{2 - 2} = \dfrac{4}{0}$

The slope is undefined.

D. $m = \dfrac{y_2 - y_1}{x_2 - x_1} = \dfrac{-3 - (-3)}{-5 - 1} = \dfrac{0}{-6} = 0$

The slope is 0.

Problem 4   y-intercept = (0, b) = (0, −1)
$$m = -\dfrac{1}{4}$$

Problem 5   Solve the equation for y.
$$x - 2y = 4$$
$$-2y = -x + 4$$
$$y = \dfrac{1}{2}x - 2$$

y-intercept = (0, b) = (0, −2)

$$m = \dfrac{1}{2}$$

SECTION 8.4   *pages 347–349*

Problem 1 $y = \frac{3}{2}x + b$

$-2 = \frac{3}{2}(4) + b$

$-2 = 6 + b$

$-8 = b$

$y = \frac{3}{2}x - 8$

Problem 2 $m = \frac{2}{5}$ $(x_1, y_1) = (5, 4)$

$y - y_1 = m(x - x_1)$

$y - 4 = \frac{2}{5}(x - 5)$

$y - 4 = \frac{2}{5}x - 2$

$y = \frac{2}{5}x + 2$

SOLUTIONS to Chapter 9 Problems

SECTION 9.1   *pages 365–368*

Problem 1

$2x - 5y = 8$	
$2(-1) - 5(-2)$	8
$-2 + 10$	8
$8 = 8$	

$-x + 3y = -5$	
$-(-1) + 3(-2)$	-5
$1 + (-6)$	-5
$-5 = -5$	

Yes, $(-1, -2)$ is a solution of the system of equations.

Problem 2   **A.**

$(-3, 2)$

The solution is $(-3, 2)$.

B.

No solution

The lines are parallel and therefore do not intersect.
The system of equations is inconsistent and has no solution.

SECTION 9.2   *pages 370–372*

Problem 1   (1) $7x - y = 4$
(2) $3x + 2y = 9$

$7x - y = 4$ ▶ Solve equation (1) for y.

$-y = -7x + 4$

$y = 7x - 4$

$$3x + 2y = 9$$ ▶ Substitute $7x - 4$ for y in equation (2).
$$3x + 2(7x - 4) = 9$$
$$3x + 14x - 8 = 9$$
$$17x - 8 = 9$$
$$17x = 17$$
$$x = 1$$

$$7x - y = 4$$ ▶ Substitute the value of x in equation (1).
$$7(1) - y = 4$$
$$7 - y = 4$$
$$-y = -3$$
$$y = 3$$

The solution is (1, 3).

Problem 2   (1) $3x - y = 4$
(2) $y = 3x + 2$

$$3x - y = 4$$
$$3x - (3x + 2) = 4$$
$$3x - 3x - 2 = 4$$
$$-2 = 4$$

The system of equations is inconsistent and has no solution.

Problem 3   (1) $y = -2x + 1$
(2) $6x + 3y = 3$

$$6x + 3y = 3$$
$$6x + 3(-2x + 1) = 3$$
$$6x - 6x + 3 = 3$$
$$3 = 3$$

The system of equations is dependent.
The solutions are the ordered pairs that satisfy the equation $y = -2x + 1$.

SECTION 9.3   *pages 374–378*

Problem 1   (1) $x - 2y = 1$
(2) $2x + 4y = 0$

$$2(x - 2y) = 2 \cdot 1$$ ▶ Eliminate y.
$$2x + 4y = 0$$

$$2x - 4y = 2$$
$$2x + 4y = 0$$
$$4x + 0y = 2$$ ▶ Add the equations.
$$4x = 2$$
$$x = \frac{2}{4} = \frac{1}{2}$$

$$2\left(\frac{1}{2}\right) + 4y = 0 \qquad \blacktriangleright \text{ Replace } x \text{ in equation (2).}$$

$$1 + 4y = 0$$
$$4y = -1$$
$$y = -\frac{1}{4}$$

The solution is $\left(\frac{1}{2}, -\frac{1}{4}\right)$.

Problem 2   (1) $2x - 3y = 4$
(2) $-4x + 6y = -8$

$$2(2x - 3y) = 2 \cdot 4 \qquad \blacktriangleright \text{ Eliminate } y.$$
$$-4x + 6y = -8$$

$$4x - 6y = 8$$
$$-4x + 6y = -8$$
$$0 + 0 = 0 \qquad \blacktriangleright \text{ Add the equations.}$$
$$0 = 0$$

The system of equations is dependent.
The solutions are the ordered pairs that satisfy the equation $2x - 3y = 4$.

SECTION 9.4   *pages 379–383*

Problem 1

Strategy   ■ Rate of the current: c
Rate of the canoeist in calm water: r

	Rate	Time	Distance
With current	$r + c$	3	$3(r + c)$
Against current	$r - c$	4	$4(r - c)$

■ The distance traveled with the current is 24 mi.
The distance traveled against the current is 24 mi.

Solution $3(r + c) = 24 \qquad \dfrac{3(r + c)}{3} = \dfrac{24}{3} \qquad r + c = 8$

$4(r - c) = 24 \qquad \dfrac{4(r - c)}{4} = \dfrac{24}{4} \qquad r - c = 6$

$$2r = 14$$
$$r = 7$$

$r + c = 8$
$7 + c = 8$
$c = 1$

The rate of the current is 1 mph.
The rate of the canoeist in calm water is 7 mph.

Problem 2

Strategy   ▪ The number of dimes in the first bank: d
The number of quarters in the first bank: q

First bank:

	Number	Value	Total value
Dimes	d	10	$10d$
Quarters	q	25	$25q$

Second bank:

	Number	Value	Total value
Dimes	$2d$	10	$20d$
Quarters	$\frac{1}{2}q$	25	$\frac{25}{2}q$

▪ The total value of the coins in the first bank is $3.90.
The total value of the coins in the second bank is $3.30.

Solution $10d + 25q = 390$ $10d + 25q = 390$ $10d + 25q = 390$

$20d + \dfrac{25}{2}q = 330$ $-2\left(20d + \dfrac{25}{2}q\right) = -2(330)$ $-40d - 25q = -660$

$-30d = -270$

$d = 9$

$10d + 25q = 390$
$10(9) + 25q = 390$
$90 + 25q = 390$
$25q = 300$
$q = 12$

There are 9 dimes and 12 quarters in the first bank.

SOLUTIONS to Chapter 10 Problems

SECTION 10.1   *pages 399–403*

Problem 1   **A.** $A = \{1, 3, 5, 7, 9, 11\}$ **Problem 2** $A \cup B = \{-2, -1, 0, 1, 2, 3, 4\}$

B. $A = \{2, 4, 6, 8, \ldots\}$

C. $A = \{-9, -7, -5, -3, -1\}$

Problem 3 $A \cap B = \{10, 16\}$ **Problem 4** $A \cap B = \varnothing$

Problem 5   **A.** $\{x|x < 59,\ x\ \text{is a positive even integer}\}$

B. $\{x|x > -3,\ x\ \text{is a real number}\}$

Problem 6   The solution set is the numbers greater than -2.

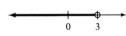

$-2 \quad 0$

Problem 7   **A.**   The solution set is the numbers greater than -1 and the numbers less than -3.

$-3 \quad -1\ 0$

B.   The solution set is the real numbers.

$-5 \qquad 0 \qquad 5$

C.   The solution set is the numbers less than or equal to 4 and greater than or equal to -4.

$-4 \qquad 0 \qquad 4$

SECTION 10.2   *pages 405–409*

Problem 1
$$x + 2 < -2$$
$$x + 2 - 2 < -2 - 2$$
$$x < -4$$

$-4 \qquad 0$

Problem 2
$$5x + 3 > 4x + 5$$
$$5x - 4x + 3 > 4x - 4x + 5$$
$$x + 3 > 5$$
$$x + 3 - 3 > 5 - 3$$
$$x > 2$$

Problem 3
$$3x < 9$$
$$\frac{3x}{3} < \frac{9}{3}$$
$$x < 3$$

$0 \qquad 3$

Problem 4
$$-\frac{3}{4}x \geq 18$$
$$-\frac{4}{3}\left(-\frac{3}{4}x\right) \leq -\frac{4}{3}(18)$$
$$x \leq -24$$

Problem 5

Strategy   To find the minimum selling price, write and solve an inequality using S to represent the selling price.

Solution
$$340 < 0.70S$$
$$\frac{340}{0.70} < \frac{0.70S}{0.70}$$
$$485.71429 < S$$

The minimum selling price is $485.72.

SECTION 10.3 *pages 414–416*

Problem 1
$$5 - 4x > 9 - 8x$$
$$5 - 4x + 8x > 9 - 8x + 8x$$
$$5 + 4x > 9$$
$$5 - 5 + 4x > 9 - 5$$
$$4x > 4$$
$$x > 1$$

Problem 2
$$8 - 4(3x + 5) \leq 6(x - 8)$$
$$8 - 12x - 20 \leq 6x - 48$$
$$-12 - 12x \leq 6x - 48$$
$$-12 - 12x - 6x \leq 6x - 6x - 48$$
$$-12 - 18x \leq -48$$
$$-12 + 12 - 18x \leq -48 + 12$$
$$-18x \leq -36$$
$$\frac{-18x}{-18} \geq \frac{-36}{-18}$$
$$x \geq 2$$

Problem 3

Strategy To find the maximum number of miles:
- Write an expression for the cost of each car, using x to represent the number of miles driven during the week.
- Write and solve an inequality.

Solution

Cost of a Company A car	is less than	Cost of a Company B car

$$9(7) + 0.10x < 12(7) + 0.08x$$
$$63 + 0.10x < 84 + 0.08x$$
$$63 + 0.10x - 0.08x < 84 + 0.08x - 0.08x$$
$$63 + 0.02x < 84$$
$$63 - 63 + 0.02x < 84 - 63$$
$$0.02x < 21$$
$$\frac{0.02x}{0.02} < \frac{21}{0.02}$$
$$x < 1050$$

The maximum number of miles is 1049.

SECTION 10.4 *pages 420–422*

Problem 1
$$x - 3y < 2$$
$$x - x - 3y < -x + 2$$
$$-3y < -x + 2$$
$$\frac{-3y}{-3} > \frac{-x + 2}{-3}$$
$$y > \frac{1}{3}x - \frac{2}{3}$$

Problem 2 $x < 3$

SOLUTIONS to Chapter 11 Problems

SECTION 11.1 *pages 434–439*

Problem 1 $-5\sqrt{32} = -5\sqrt{2^5} = -5\sqrt{2^4 \cdot 2} = -5\sqrt{2^4}\,\sqrt{2} = -5 \cdot 2^2\sqrt{2} = -20\sqrt{2}$

Problem 2 $\sqrt{216} = \sqrt{2^3 \cdot 3^3} = \sqrt{2^2 \cdot 3^2(2 \cdot 3)} = \sqrt{2^2 \cdot 3^2}\,\sqrt{2 \cdot 3} = 2 \cdot 3\sqrt{2 \cdot 3} = 6\sqrt{6}$

Problem 3 $\sqrt{y^{19}} = \sqrt{y^{18} \cdot y} = \sqrt{y^{18}}\,\sqrt{y} = y^9\sqrt{y}$

Problem 4 **A.** $\sqrt{45b^7} = \sqrt{3^2 \cdot 5 \cdot b^7} = \sqrt{3^2 b^6(5 \cdot b)} = \sqrt{3^2 b^6}\,\sqrt{5b} = 3b^3\sqrt{5b}$

           B. $3a\sqrt{28a^9 b^{18}} = 3a\sqrt{2^2 \cdot 7 \cdot a^9 \cdot b^{18}} = 3a\sqrt{2^2 a^8 b^{18}(7a)} =$

              $3a\sqrt{2^2 a^8 b^{18}}\,\sqrt{7a} = 3a \cdot 2a^4 b^9\sqrt{7a} = 6a^5 b^9\sqrt{7a}$

Problem 5 $\sqrt{25(a + 3)^2} = \sqrt{5^2(a + 3)^2} = 5(a + 3) = 5a + 15$

SECTION 11.2 *pages 442–444*

Problem 1 **A.** $9\sqrt{3} + 3\sqrt{3} - 18\sqrt{3} = -6\sqrt{3}$

           B. $2\sqrt{50} - 5\sqrt{32} = 2\sqrt{2 \cdot 5^2} - 5\sqrt{2^5} = 2\sqrt{5^2}\,\sqrt{2} - 5\sqrt{2^4}\,\sqrt{2} =$

              $2 \cdot 5\sqrt{2} - 5 \cdot 2^2\sqrt{2} = 10\sqrt{2} - 20\sqrt{2} = -10\sqrt{2}$

Problem 2 **A.** $y\sqrt{28y} + 7\sqrt{63y^3} = y\sqrt{2^2 \cdot 7y} + 7\sqrt{3^2 \cdot 7 \cdot y^3} =$

              $y\sqrt{2^2}\,\sqrt{7y} + 7\sqrt{3^2 \cdot y^2}\sqrt{7y} = y \cdot 2\sqrt{7y} + 7 \cdot 3y\sqrt{7y} =$

              $2y\sqrt{7y} + 21y\sqrt{7y} = 23y\sqrt{7y}$

           B. $2\sqrt{27a^5} - 4a\sqrt{12a^3} + a^2\sqrt{75a} =$

              $2\sqrt{3^3 \cdot a^5} - 4a\sqrt{2^2 \cdot 3 \cdot a^3} + a^2\sqrt{3 \cdot 5^2 \cdot a} =$

              $2\sqrt{3^2 \cdot a^4}\,\sqrt{3a} - 4a\sqrt{2^2 \cdot a^2}\,\sqrt{3a} + a^2\sqrt{5^2}\,\sqrt{3a} =$

              $2 \cdot 3a^2\sqrt{3a} - 4a \cdot 2a\sqrt{3a} + a^2 \cdot 5\sqrt{3a} =$

              $6a^2\sqrt{3a} - 8a^2\sqrt{3a} + 5a^2\sqrt{3a} = 3a^2\sqrt{3a}$

SECTION 11.3 *pages 446–450*

Problem 1 $\sqrt{5a}\,\sqrt{15a^3 b^4}\,\sqrt{3b^5} = \sqrt{225a^4 b^9} = \sqrt{3^2 5^2 a^4 b^9} = \sqrt{3^2 5^2 a^4 b^8}\,\sqrt{b} =$

             $3 \cdot 5a^2 b^4\sqrt{b} = 15a^2 b^4\sqrt{b}$

Problem 2 $\sqrt{5x}\,(\sqrt{5x} - \sqrt{25y}) = \sqrt{5^2 x^2} - \sqrt{5^3 xy} = \sqrt{5^2 x^2} - \sqrt{5^2}\,\sqrt{5xy} = 5x - 5\sqrt{5xy}$

Problem 3 $(3\sqrt{x} - \sqrt{y})(5\sqrt{x} - 2\sqrt{y}) = 15(\sqrt{x})^2 - 6\sqrt{xy} - 5\sqrt{xy} + 2(\sqrt{y})^2 =$

             $15x - 11\sqrt{xy} + 2y$

Problem 4 $(2\sqrt{x} + 7)(2\sqrt{x} - 7) = (2\sqrt{x})^2 - 7^2 = 4x - 49$

Problem 5 **A.** $\dfrac{\sqrt{15x^6y^7}}{\sqrt{3x^7y^9}} = \sqrt{\dfrac{15x^6y^7}{3x^7y^9}} = \sqrt{\dfrac{5}{xy^2}} = \dfrac{\sqrt{5}}{y\sqrt{x}} = \dfrac{\sqrt{5}}{y\sqrt{x}} \cdot \dfrac{\sqrt{x}}{\sqrt{x}} = \dfrac{\sqrt{5x}}{xy}$

B. $\dfrac{\sqrt{y}}{\sqrt{y}+3} = \dfrac{\sqrt{y}}{\sqrt{y}+3} \cdot \dfrac{\sqrt{y}-3}{\sqrt{y}-3} = \dfrac{y-3\sqrt{y}}{y-9}$

C. $\dfrac{5+\sqrt{y}}{1-2\sqrt{y}} = \dfrac{5+\sqrt{y}}{1-2\sqrt{y}} \cdot \dfrac{1+2\sqrt{y}}{1+2\sqrt{y}} = \dfrac{5+10\sqrt{y}+\sqrt{y}+2y}{1-4y} =$

$\dfrac{5+11\sqrt{y}+2y}{1-4y}$

SECTION 11.4 *pages 453–458*

Problem 1 $\sqrt{4x}+3 = 7$ Check: $\sqrt{4x}+3 = 7$
$\sqrt{4x} = 4$ $\sqrt{4\cdot 4}+3 = 7$
$(\sqrt{4x})^2 = 4^2$ $\sqrt{4^2}+3 = 7$
$4x = 16$ $4+3 = 7$
$x = 4$ $7 = 7$

The solution is 4.

Problem 2 $\sqrt{3x-2}-5 = 0$ Check: $\sqrt{3x-2}-5 = 0$
$\sqrt{3x-2} = 5$ $\sqrt{3\cdot 9-2}-5 = 0$
$(\sqrt{3x-2})^2 = 5^2$ $\sqrt{27-2}-5 = 0$
$3x-2 = 25$ $\sqrt{25}-5 = 0$
$3x = 27$ $\sqrt{5^2}-5 = 0$
$x = 9$ $5-5 = 0$
$0 = 0$

The solution is 9.

Problem 3 $\sqrt{x}+\sqrt{x+9} = 9$ Check: $\sqrt{x}+\sqrt{x+9} = 9$
$\sqrt{x} = 9-\sqrt{x+9}$ $\sqrt{16}+\sqrt{16+9} = 9$
$(\sqrt{x})^2 = (9-\sqrt{x+9})^2$ $4+\sqrt{25} = 9$
$x = 81-18\sqrt{x+9}+(x+9)$ $4+5 = 9$
$18\sqrt{x+9} = 90$ $9 = 9$
$\sqrt{x+9} = 5$
$(\sqrt{x+9})^2 = 5^2$
$x+9 = 25$
$x = 16$

The solution is 16.

Problem 4

Strategy To find the distance, use the Pythagorean Theorem. The hypotenuse is the length of the ladder. One leg is the distance from the bottom of the ladder to the base of the building. The distance along the building from the ground to the top of the ladder is the unknown leg.

Solution
$$a^2 = \sqrt{c^2 - b^2}$$
$$a^2 = \sqrt{(12)^2 - (5)^2}$$
$$a^2 = \sqrt{144 - 25}$$
$$a^2 = \sqrt{119}$$
$$a \approx 10.91$$

The distance is 10.91 ft.

Problem 5

Strategy   To find the length of the pendulum, replace T in the equation with the given value and solve for L.

Solution
$$T = 2\pi \sqrt{\frac{L}{32}}$$

$$1.5 = 2(3.14) \sqrt{\frac{L}{32}}$$

$$1.5 = 6.28 \sqrt{\frac{L}{32}}$$

$$\frac{1.5}{6.28} = \sqrt{\frac{L}{32}}$$

$$\left(\frac{1.5}{6.28}\right)^2 = \left(\sqrt{\frac{L}{32}}\right)^2$$

$$\frac{2.25}{39.4384} = \frac{L}{32}$$

$$(32)\left(\frac{2.25}{39.4384}\right) = (32)\left(\frac{L}{32}\right)$$

$$\frac{72}{39.4384} = L$$

$$1.83 \approx L$$

The length of the pendulum is 1.83 ft.

SOLUTIONS to Chapter 12 Problems

SECTION 12.1   *pages 473–476*

Problem 1
$$3t = t^2 - 10$$
$$0 = t^2 - 3t - 10$$
$$0 = (t + 2)(t - 5)$$

$$t + 2 = 0 \qquad t - 5 = 0$$
$$t = -2 \qquad t = 5$$

The solutions are -2 and 5.

Problem 2
$$(x + 5)(x - 1) = 3x + 1$$
$$x^2 + 4x - 5 = 3x + 1$$
$$x^2 + x - 6 = 0$$
$$(x + 3)(x - 2) = 0$$

$$x + 3 = 0 \qquad x - 2 = 0$$
$$x = -3 \qquad x = 2$$

The solutions are -3 and 2.

Problem 3 **A.** $4x^2 - 96 = 0$

$$4x^2 = 96$$
$$x^2 = 24$$
$$\sqrt{x^2} = \sqrt{24}$$
$$x = \pm\sqrt{24}$$
$$x = \pm 2\sqrt{6}$$

The solutions are $2\sqrt{6}$ and $-2\sqrt{6}$.

B. $x^2 + 81 = 0$

$$x^2 = -81$$
$$\sqrt{x^2} = \sqrt{-81} \quad \text{(Not a real number)}$$

The equation has no real number solution.

Problem 4

$$(x + 5)^2 = 20$$
$$\sqrt{(x + 5)^2} = \sqrt{20}$$
$$x + 5 = \pm\sqrt{20}$$
$$x + 5 = \pm 2\sqrt{5}$$

$$x + 5 = 2\sqrt{5} \qquad\qquad x + 5 = -2\sqrt{5}$$
$$x = -5 + 2\sqrt{5} \qquad\qquad x = -5 - 2\sqrt{5}$$

The solutions are $-5 + 2\sqrt{5}$ and $-5 - 2\sqrt{5}$.

SECTION 12.2 *pages 479–483*

Problem 1

$$3x^2 - 6x - 2 = 0$$
$$3x^2 - 6x = 2$$
$$\frac{1}{3}(3x^2 - 6x) = \frac{1}{3} \cdot 2$$
$$x^2 - 2x = \frac{2}{3}$$

Complete the square.

$$x^2 - 2x + 1 = \frac{2}{3} + 1$$
$$(x - 1)^2 = \frac{5}{3}$$

$$\sqrt{(x - 1)^2} = \sqrt{\frac{5}{3}}$$

$$x - 1 = \pm\sqrt{\frac{5}{3}}$$

$$x - 1 = \pm\frac{\sqrt{15}}{3}$$

$$x - 1 = \frac{\sqrt{15}}{3} \qquad\qquad x - 1 = -\frac{\sqrt{15}}{3}$$
$$x = 1 + \frac{\sqrt{15}}{3} \qquad\qquad x = 1 - \frac{\sqrt{15}}{3}$$
$$= \frac{3 + \sqrt{15}}{3} \qquad\qquad = \frac{3 - \sqrt{15}}{3}$$

The solutions are $\dfrac{3 + \sqrt{15}}{3}$ and $\dfrac{3 - \sqrt{15}}{3}$.

Problem 2

$$x^2 + 8x + 8 = 0$$
$$x^2 + 8x = -8$$
$$x^2 + 8x + 16 = -8 + 16$$
$$(x + 4)^2 = 8$$
$$\sqrt{(x + 4)^2} = \sqrt{8}$$
$$x + 4 = \pm\sqrt{8}$$
$$x + 4 = \pm 2\sqrt{2}$$

$x + 4 = 2\sqrt{2}$	$x + 4 = -2\sqrt{2}$
$x = -4 + 2\sqrt{2}$	$x = -4 - 2\sqrt{2}$
$\approx -4 + 2(1.414)$	$\approx -4 - 2(1.414)$
$\approx -4 + 2.828$	$\approx -4 - 2.828$
≈ -1.172	≈ -6.828

The solutions are approximately -1.172 and -6.828.

SECTION 12.3   *pages 485–488*

Problem 1

A. $3x^2 + 4x - 4 = 0$

$a = 3, b = 4, c = -4$

$$x = \frac{-(4) \pm \sqrt{(4)^2 - 4(3)(-4)}}{2 \cdot 3}$$

$$= \frac{-4 \pm \sqrt{16 + 48}}{6}$$

$$= \frac{-4 \pm \sqrt{64}}{6} = \frac{-4 \pm 8}{6}$$

$$x = \frac{-4 + 8}{6} \qquad x = \frac{-4 - 8}{6}$$

$$= \frac{4}{6} = \frac{2}{3} \qquad = \frac{-12}{6} = -2$$

The solutions are $\frac{2}{3}$ and -2.

B. $2x^2 - 3x + 4 = 0$

$a = 2, b = -3, c = 4$

$$x = \frac{-(-3) \pm \sqrt{(-3)^2 - 4(2)(4)}}{2 \cdot 2}$$

$$= \frac{3 \pm \sqrt{9 - 32}}{4}$$

$$= \frac{3 \pm \sqrt{-23}}{4}$$

$\sqrt{-23}$ is not a real number.

The equation has no real number solution.

Problem 2

$$x^2 + 2x = 1$$
$$x^2 + 2x - 1 = 0$$

$a = 1, b = 2, c = -1$

$$x = \frac{-(2) \pm \sqrt{(2)^2 - 4(1)(-1)}}{2 \cdot 1} = \frac{-2 \pm \sqrt{4 + 4}}{2}$$

$$= \frac{-2 \pm \sqrt{8}}{2} = \frac{-2 \pm 2\sqrt{2}}{2} = \frac{2(-1 \pm \sqrt{2})}{2} = -1 \pm \sqrt{2}$$

The solutions are $-1 + \sqrt{2}$ and $-1 - \sqrt{2}$.

SECTION 12.4   *pages 491–492*

Problem 1   **A.** $a = 1$. a is positive.
The parabola opens up.

B. $a = -1$. a is negative.
The parabola opens down.

SECTION 12.5   *pages 494–496*

Problem 1

Strategy   ■ This is a geometry problem.
■ Width of the rectangle: w
Length of the rectangle: $w + 3$
■ Use the equation $A = lw$.

Solution $A = lw$
$40 = (w + 3)w$
$40 = w^2 + 3w$
$0 = w^2 + 3w - 40$
$0 = (w + 8)(w - 5)$

$w + 8 = 0$ $w - 5 = 0$
$w = -8$ $w = 5$

The solution -8 is not possible.
The width is 5 m.

ANSWERS to Chapter 1 Exercises

SECTION 1.1   *pages 5–7*

1. $<$ **3.** $>$ **5.** $<$ **7.** $>$ **9.** $<$ **11.** $>$ **13.** $>$ **15.** $>$ **17.** $<$ **19.** $>$
21. $<$ **23.** $>$ **25.** $>$ **27.** $<$ **29.** $>$ **31.** $>$ **33.** $>$ **35.** $>$ **37.** $<$ **39.** $<$
41. $<$ **43.** $<$ **45.** $<$ **47.** -16 **49.** 3 **51.** -45 **53.** 88 **55.** 97 **57.** -450
59. 2 **61.** 6 **63.** 5 **65.** 1 **67.** -8 **69.** 0 **71.** 19 **73.** 22 **75.** -20
77. -18 **79.** 23 **81.** -27 **83.** -41 **85.** 25 **87.** 30 **89.** -34 **91.** -45 **93.** 36
95. 61 **97.** -52 **99.** -93 **101.** 119 **103.** 85 **105.** -48 **107.** $<$ **109.** $<$
111. $>$ **113.** $>$ **115.** $>$ **117.** $>$ **119.** $>$ **121.** $>$ **123.** $<$ **125.** $-21, -3, -1, 14$
127. $-|-7|, -4, 0, |-15|$ **129.** $-|-26|, -(5), -(-8), |-17|$ **131.** $-6; 6$ **133.** $-8; 2$
135. positive   **137.** true

SECTION 1.2   *pages 11–12*

1. -2 **3.** 20 **5.** -11 **7.** -9 **9.** -3 **11.** 1 **13.** -5 **15.** -30 **17.** 9 **19.** 1
21. -10 **23.** -28 **25.** -41 **27.** -392 **29.** 8 **31.** -7 **33.** 7 **35.** -2 **37.** -28
39. -13 **41.** 6 **43.** -9 **45.** 2 **47.** -138 **49.** -8 **51.** 5 **53.** 11 **55.** -4
57. 4 **59.** $-23, -27, -31$ **61.** $-9, -16, -24$ **63.** b **65.** a

SECTION 1.3   *pages 16–19*

1. 42 **3.** -24 **5.** 6 **7.** 18 **9.** -20 **11.** -16 **13.** 25 **15.** 0 **17.** -72
19. -102 **21.** 140 **23.** -228 **25.** -320 **27.** -156 **29.** -70 **31.** 162 **33.** 120
35. 36 **37.** 192 **39.** -108 **41.** -2100 **43.** 0 **45.** $-251,636$ **47.** -2 **49.** 8
51. 0 **53.** -9 **55.** -9 **57.** 9 **59.** -24 **61.** -12 **63.** 31 **65.** 17 **67.** 15
69. -13 **71.** -18 **73.** 19 **75.** 13 **77.** -19 **79.** 17 **81.** 26 **83.** 23 **85.** 25
87. -34 **89.** 11 **91.** -13 **93.** 13 **95.** 12 **97.** -14 **99.** $-11°C$ **101.** $13°C$
103. 7000 m   **105.** 6280 m   **107.** $-3°C$ **109.** 112 **111.** -364 **113.** $84,432$
115. 4 ($804, 834, 864, 894$)

SECTION 1.4   *pages 25–28*

1. $0.\overline{3}$ **3.** 0.25 **5.** 0.4 **7.** $0.1\overline{6}$ **9.** 0.125 **11.** $0.\overline{2}$ **13.** $0.\overline{45}$ **15.** $0.58\overline{3}$ **17.** $0.2\overline{6}$
19. 0.5625 **21.** $0.3\overline{8}$ **23.** 0.05 **25.** 0.24 **27.** $0.2\overline{3}$ **29.** 0.225 **31.** $0.36\overline{1}$ **33.** $0.68\overline{1}$
35. $0.458\overline{3}$ **37.** $0.\overline{15}$ **39.** $0.0\overline{81}$ **41.** $\dfrac{13}{12}$ **43.** $-\dfrac{5}{24}$ **45.** $-\dfrac{19}{24}$ **47.** $\dfrac{5}{26}$ **49.** $\dfrac{7}{24}$
51. 0 **53.** $-\dfrac{7}{16}$ **55.** $\dfrac{11}{24}$ **57.** 1 **59.** $\dfrac{11}{8}$ **61.** $\dfrac{169}{315}$ **63.** -38.8 **65.** -6.192
67. 13.355 **69.** 4.676 **71.** -10.03 **73.** -37.19 **75.** -17.5 **77.** 853.2594 **79.** $-\dfrac{3}{8}$
81. $\dfrac{1}{10}$ **83.** $\dfrac{15}{64}$ **85.** $\dfrac{3}{2}$ **87.** $-\dfrac{8}{9}$ **89.** $\dfrac{2}{3}$ **91.** 4.164 **93.** 4.347 **95.** -4.028
97. -2.22 **99.** 12.26448 **101.** 0.75 **103.** 2.32 **105.** 3.83 **107.** $56,201.08$

109. integer, negative integer, rational number, real number   **111.** rational number, real number
113. rational number, real number   **115.** 6.3   **117.** 7.17   **119.** 0   **121.** $\frac{11}{16}$ **123.** 53¢
125. 212

SECTION 1.5   *pages 31–33*

1. 36   **3.** −49   **5.** 9   **7.** 81   **9.** $\frac{1}{4}$ **11.** 0.09   **13.** 12   **15.** 0.216   **17.** −12
19. 16   **21.** −864   **23.** −1008   **25.** 3   **27.** −77,760   **29.** 9   **31.** 12   **33.** 1   **35.** 13
37. −36   **39.** 13   **41.** 4   **43.** 15   **45.** −1   **47.** 4   **49.** 0.51   **51.** 1.7   **53.** $\frac{17}{48}$
55. >   **57.** 30   **59.** −100   **61.** positive   **63.** 17 seconds   **65.** 6   **67.** 9

CALCULATORS AND COMPUTERS   *page 33*

1. 0.$\overline{1764705882352941}$ **3.** 0.$\overline{3913043478260869565217}$

THE KELVIN SCALE   *page 35*

1. 340 K   **3.** 235 K   **5.** 47°C   **7.** −84°C

CHAPTER REVIEW   *pages 37–39*

1. −4   (Objective 1.1.2)   **2.** 4   (Objective 1.2.2)   **3.** 17   (Objective 1.3.2)
4. 0.$\overline{7}$ (Objective 1.4.1)   **5.** −5.3578   (Objective 1.4.3)   **6.** 8   (Objective 1.5.2)
7. 4   (Objective 1.1.2)   **8.** −14   (Objective 1.2.2)   **9.** −9   (Objective 1.3.2)
10. 0.85   (Objective 1.4.1)   **11.** $-\frac{1}{2}$ (Objective 1.4.3)   **12.** 9   (Objective 1.5.2)
13. <   (Objective 1.1.1)   **14.** −16   (Objective 1.2.1)   **15.** 90   (Objective 1.3.1)
16. −6.881   (Objective 1.4.2)   **17.** 12   (Objective 1.5.1)   **18.** >   (Objective 1.1.1)
19. −3   (Objective 1.2.1)   **20.** −48   (Objective 1.3.1)   **21.** $\frac{1}{15}$ (Objective 1.4.2)
22. −108   (Objective 1.5.1)   **23.** 2   (Objective 1.1.2)   **24.** −14   (Objective 1.2.2)
25. −8   (Objective 1.3.2)   **26.** 0.35   (Objective 1.4.1)   **27.** $-\frac{7}{6}$ (Objective 1.4.3)
28. 12   (Objective 1.5.2)   **29.** 3   (Objective 1.1.2)   **30.** 18   (Objective 1.2.2)
31. 12   (Objective 1.3.2)   **32.** 0.$\overline{63}$ (Objective 1.4.1)   **33.** −11.5   (Objective 1.4.3)
34. 8   (Objective 1.5.2)   **35.** >   (Objective 1.1.1)   **36.** −8   (Objective 1.2.1)
37. 72   (Objective 1.3.1)   **38.** $-\frac{11}{24}$ (Objective 1.4.2)   **39.** −4   (Objective 1.5.1)
40. 3.561   (Objective 1.4.2)   **41.** −17   (Objective 1.1.2)   **42.** −27   (Objective 1.2.2)
43. $-\frac{1}{10}$ (Objective 1.4.3)   **44.** <   (Objective 1.1.1)   **45.** 0.72   (Objective 1.4.1)
46. −128   (Objective 1.5.1)   **47.** −18   (Objective 1.2.1)   **48.** 0   (Objective 1.3.1)
49. 9   (Objective 1.5.2)   **50.** $\frac{7}{8}$ (Objective 1.4.2)   **51.** 16   (Objective 1.3.2)

52. $<$ (Objective 1.1.1)   **53.** -3 (Objective 1.5.1)   **54.** -6 (Objective 1.2.1)
55. 300   (Objective 1.3.1)   **56.** 8°C   (Objective 1.3.3)   **57.** -2°C   (Objective 1.3.3)
58. 13°C   (Objective 1.3.3)   **59.** -6°C   (Objective 1.3.3)   **60.** 714°C   (Objective 1.3.3)

Note: The numbers in parentheses following each answer in the Chapter Review refer to the objective that corresponds with that problem. The first number of the reference indicates the chapter, and the second and third numbers indicate the objective. For example, the reference (Objective 1.2.1) stands for Chapter 1, Section 2, Objective 1. This notation will be used for all Chapter Reviews, Chapter Tests, and Cumulative Reviews throughout the text.

CHAPTER TEST   *pages 39–40*

1. -3 (Objective 1.2.2)   **2.** 0.85   (Objective 1.4.1)   **3.** $-\dfrac{1}{14}$ (Objective 1.4.3)

4. -15 (Objective 1.3.2)   **5.** $-\dfrac{8}{3}$ (Objective 1.5.1)   **6.** 2   (Objective 1.2.1)

7. 29   (Objective 1.1.2)   **8.** $>$ (Objective 1.1.1)   **9.** $-\dfrac{23}{18}$ (Objective 1.4.2)

10. 258   (Objective 1.3.1)   **11.** 3   (Objective 1.5.2)   **12.** $\dfrac{5}{6}$ (Objective 1.4.3)

13. 14   (Objective 1.2.2)   **14.** $0.4\overline{3}$ (Objective 1.4.1)   **15.** -2.43 (Objective 1.4.3)
16. 15   (Objective 1.3.2)   **17.** 640   (Objective 1.5.1)   **18.** -12 (Objective 1.2.1)
19. -34 (Objective 1.1.2)   **20.** $>$ (Objective 1.1.1)   **21.** -11.384 (Objective 1.4.2)
22. 160   (Objective 1.3.1)   **23.** -2 (Objective 1.5.2)   **24.** 4°C   (Objective 1.3.3)
25. -4°C   (Objective 1.3.3)

ANSWERS to Chapter 2 Exercises

SECTION 2.1   *pages 48–50*

1. $2x^2,\ 5x,\ \underline{-8}$ **3.** $-a^4,\ 6$ **5.** $7x^2y,\ 6xy^2$ **7.** $1,\ -9$ **9.** $1,\ -4,\ -1$ **11.** 10   **13.** 32
15. 21   **17.** 16   **19.** -9 **21.** $\overline{3}$ **23.** -7 **25.** 13   **27.** -15 **29.** 41   **31.** 1
33. 5   **35.** 1   **37.** 57   **39.** 5   **41.** 12   **43.** 6   **45.** 10   **47.** 8   **49.** -3 **51.** -2
53. -22 **55.** 4   **57.** 20   **59.** 24   **61.** 4.96   **63.** -5.68 **65.** 17   **67.** 4   **69.** -7
71. 1   **73.** 0   **75.** The Inverse Property of Multiplication   **77.** The Addition Property of Zero
79. The Distributive Property   **81.** The Multiplication Property of Zero
83. The Multiplication Property of One   **85.** 13   **87.** 7   **89.** 4

SECTION 2.2   *pages 54–57*

1. $14x$ **3.** $5a$ **5.** $-6y$ **7.** $-3b - 7$ **9.** $5a$ **11.** $-2ab$ **13.** $5xy$ **15.** 0   **17.** $-\dfrac{5}{6}x$

19. $-\dfrac{1}{24}x^2$ **21.** $11x$ **23.** $7a$ **25.** $-14x^2$ **27.** $-x + 3y$ **29.** $17x - 3y$ **31.** $-2a - 6b$

33. $-3x - 8y$ **35.** $-4x^2 - 2x$ **37.** $12x$ **39.** $-21a$ **41.** $6y$ **43.** $8x$ **45.** $-6a$ **47.** $12b$
49. $-15x^2$ **51.** x^2 **53.** a **55.** x **57.** n **59.** x **61.** y **63.** $3x$ **65.** $-2x$
67. $-8a^2$ **69.** $8y$ **71.** $4y$ **73.** $-2x$ **75.** $6a$ **77.** $-x - 7$ **79.** $10x - 35$ **81.** $-5a - 80$

83. $-15y + 35$ **85.** $20 - 14b$ **87.** $-35 + 50x$ **89.** $18x^2 + 12x$ **91.** $10x - 35$ **93.** $-14x + 49$
95. $-30x^2 - 15$ **97.** $-24y^2 + 96$ **99.** $5x^2 + 5y^2$ **101.** $-4x^2 + 20y^2$ **103.** $3x^2 + 6x - 18$
105. $-2y^2 + 4y - 8$ **107.** $-2a^2 - 4a + 6$ **109.** $10x^2 + 15x - 35$ **111.** $6x^2 + 3xy - 9y^2$
113. $-3a^2 - 5a + 4$ **115.** $-2x - 16$ **117.** $-12y - 9$ **119.** $7n - 7$ **121.** $-2x + 41$
123. $3y - 3$ **125.** $-a - 7b$ **127.** $-4x + 24$ **129.** $-2x - 16$ **131.** $-3x + 21$ **133.** $-7x + 24$

135. $-x + 50$ **137.** $20x - 41y$ **139.** $0.3C$ **141.** $-\frac{1}{2}x + \frac{5}{2}y$ **143.** $-a + b$

SECTION 2.3   *pages 61–65*

1. $19 - d$ **3.** $r - 12$ **5.** $28a$ **7.** $5(n - 7)$ **9.** $y - 3y$ **11.** $-6b$ **13.** $\dfrac{4}{p - 6}$ **15.** $\dfrac{x - 9}{2x}$

17. $-4s - 21$ **19.** $\dfrac{d + 8}{d}$ **21.** $\dfrac{3}{8}(t + 15)$ **23.** $w + \dfrac{7}{w}$ **25.** $d + (16d - 3)$ **27.** $\dfrac{n}{19}$

29. $n + 40$ **31.** $(n - 90)^2$ **33.** $\dfrac{4}{9}n + 20$ **35.** $n(n + 10)$ **37.** $7n + 14$ **39.** $\dfrac{12}{n + 2}$

41. $\dfrac{2}{n + 1}$ **43.** $60 - \dfrac{n}{50}$ **45.** $n^2 + 3n$ **47.** $(n + 3) + n^3$ **49.** $n^2 - \dfrac{1}{4}n$ **51.** $2n - \dfrac{7}{n}$

53. $n^3 - 12n$ **55.** $n + (n + 10); 2n + 10$ **57.** $n - (9 - n); 2n - 9$ **59.** $\dfrac{1}{5}n - \dfrac{3}{8}n; -\dfrac{7}{40}n$

61. $(n + 9) + 4; n + 13$ **63.** $2(3n + 40); 6n + 80$ **65.** $7(5n); 35n$ **67.** $17n + 2n; 19n$

69. $n + 12n; 13n$ **71.** $3(n^2 + 4); 3n^2 + 12$ **73.** $\dfrac{3}{4}(16n + 4); 12n + 3$

75. $16 - (n + 9); -n + 7$ **77.** $\dfrac{4n}{2} + 5; 2n + 5$ **79.** $6(n + 8); 6n + 48$ **81.** $7 - (n + 2); -n + 5$

83. $\dfrac{1}{3}(n + 6n); \dfrac{7n}{3}$ **85.** $(8 + n^3) + 2n^3; 3n^3 + 8$ **87.** $(n + 12) + (n - 6); 2n + 6$

89. $(n + 9) + (n - 20); 2n - 11$ **91.** $14 - (n - 3)10; -10n + 44$ **93.** even   **95.** even
97. even   **99.** even   **101.** odd   **103.** $640 + 24n$ **105.** $q + 10$

DIMENSIONAL ANALYSIS   *page 66*

1. 10.5 cm^2 **3.** 32 m **5.** 24 m^3 **7.** 90 kg

CHAPTER REVIEW   *pages 68–70*

1. y^2 (Objective 2.2.1)   **2.** $3x$ (Objective 2.2.2)   **3.** $-10a$ (Objective 2.2.2)
4. $-4x + 8$ (Objective 2.2.3)   **5.** $7x + 38$ (Objective 2.2.4)   **6.** 16 (Objective 2.1.1)
7. 9 (Objective 2.1.2)   **8.** $36y$ (Objective 2.2.2)   **9.** $6y - 18$ (Objective 2.2.3)
10. $-3x + 21y$ (Objective 2.2.4)   **11.** $-8x^2 + 12y^2$ (Objective 2.2.3)   **12.** $5x$ (Objective 2.2.1)
13. 22 (Objective 2.1.1)   **14.** $2x$ (Objective 2.2.2)   **15.** $15 - 35b$ (Objective 2.2.3)
16. $-7x + 33$ (Objective 2.2.4)   **17.** The Commutative Property of Multiplication (Objective 2.1.2)
18. $24 - 6x$ (Objective 2.2.3)   **19.** $5x^2$ (Objective 2.2.1)   **20.** $-7x + 14$ (Objective 2.2.4)
21. $-9y^2 + 9y + 21$ (Objective 2.2.3)   **22.** $2x + y$ (Objective 2.2.4)   **23.** 3 (Objective 2.1.1)
24. $36y$ (Objective 2.2.2)   **25.** $5x - 43$ (Objective 2.2.4)   **26.** $2x$ (Objective 2.2.2)
27. $-6x^2 + 21y^2$ (Objective 2.2.3)   **28.** 6 (Objective 2.1.1)   **29.** $-x + 6$ (Objective 2.2.4)
30. $-5a - 2b$ (Objective 2.2.1)   **31.** $-10x^2 + 15x - 30$ (Objective 2.2.3)
32. $-9x - 7y$ (Objective 2.2.1)   **33.** $6a$ (Objective 2.2.2)   **34.** $17x - 24$ (Objective 2.2.4)

35. $-2x - 5y$ (Objective 2.2.1)   **36.** $30b$ (Objective 2.2.2)   **37.** 21 (Objective 2.1.2)
38. $-2x^2 + 4x$ (Objective 2.2.1)   **39.** $-6x^2$ (Objective 2.2.2)   **40.** $15x - 27$ (Objective 2.2.4)
41. $-8a^2 + 3b^2$ (Objective 2.2.3)   **42.** The Multiplication Property of Zero   (Objective 2.1.2)

43. $b - 7b$ (Objective 2.3.1)   **44.** $n + 2n^2$ (Objective 2.3.2)   **45.** $\dfrac{6}{n} - 3$ (Objective 2.3.2)

46. $8 - \dfrac{n}{12}$ (Objective 2.3.2)   **47.** $\dfrac{10}{y - 2}$ (Objective 2.3.1)

48. $(3 + n)5 + 12;\ 5n + 27$ (Objective 2.3.3)   **49.** $8\left(\dfrac{2n}{16}\right);\ n$ (Objective 2.3.3)

50. $4(2 + 5n);\ 8 + 20n$ (Objective 2.3.3)

CHAPTER TEST   *pages 70–71*

1. $36y$ (Objective 2.2.2)   **2.** $4x - 3y$ (Objective 2.2.1)   **3.** $10n - 6$ (Objective 2.2.4)
4. 2 (Objective 2.1.1)   **5.** The Multiplication Property of One   (Objective 2.1.2)

6. $4x - 40$ (Objective 2.2.3)   **7.** $\dfrac{1}{12}x^2$ (Objective 2.2.1)   **8.** $4x$ (Objective 2.2.2)

9. $-24y^2 + 48$ (Objective 2.2.3)   **10.** 19 (Objective 2.1.2)   **11.** 6 (Objective 2.1.1)
12. $-3x + 13y$ (Objective 2.2.4)   **13.** b (Objective 2.2.1)   **14.** $78a$ (Objective 2.2.2)
15. $3x^2 - 15x + 12$ (Objective 2.2.3)   **16.** -32 (Objective 2.1.1)   **17.** $37x - 5y$ (Objective 2.2.4)

18. $\dfrac{n + 8}{17}$ (Objective 2.3.1)   **19.** $(a + b) - b^2$ (Objective 2.3.1)   **20.** $n + 5n^3$ (Objective 2.3.2)

21. $n^2 + 11n$ (Objective 2.3.2)   **22.** $2n - 6$ (Objective 2.3.2)
23. $20(n + 9);\ 20n + 180$ (Objective 2.3.3)   **24.** $(n + 2) + (n - 3);\ 2n - 1$ (Objective 2.3.3)

25. $n - \dfrac{1}{4}(2n);\ \dfrac{1}{2}n$ (Objective 2.3.3)

CUMULATIVE REVIEW   *page 72*

1. -7 (Objective 1.2.1)   **2.** 5 (Objective 1.2.2)   **3.** 24 (Objective 1.3.1)

4. -5 (Objective 1.3.2)   **5.** 1.25 (Objective 1.4.1)   **6.** $\dfrac{11}{48}$ (Objective 1.4.2)

7. $\dfrac{1}{6}$ (Objective 1.4.3)   **8.** $\dfrac{1}{4}$ (Objective 1.4.3)   **9.** $\dfrac{8}{3}$ (Objective 1.5.1)   **10.** -5 (Objective 1.5.2)

11. $\dfrac{53}{48}$ (Objective 1.5.2)   **12.** -8 (Objective 2.1.1)   **13.** $5x^2$ (Objective 2.2.1)

14. $-a - 12b$ (Objective 2.2.1)   **15.** $3a$ (Objective 2.2.2)   **16.** $20b$ (Objective 2.2.2)
17. $20 - 10x$ (Objective 2.2.3)   **18.** $6y - 21$ (Objective 2.2.3)   **19.** $-6x^2 + 8y^2$ (Objective 2.2.3)
20. $-8y^2 + 20y + 32$ (Objective 2.2.3)   **21.** $-10x + 15$ (Objective 2.2.4)
22. $5x - 17$ (Objective 2.2.4)   **23.** $13x - 16$ (Objective 2.2.4)   **24.** $6x + 29y$ (Objective 2.2.4)

25. $\dfrac{1}{2}b + b$ (Objective 2.3.1)   **26.** $\dfrac{8}{n - 3}$ (Objective 2.3.1)   **27.** $6 - 12n$ (Objective 2.3.2)

28. $n + (n + 2);\ 2n + 2$ (Objective 2.3.3)   **29.** $5 + (n - 7);\ n - 2$ (Objective 2.3.3)
30. $(5 + n)10 + 9;\ 10n + 59$ (Objective 2.3.3)

ANSWERS to Chapter 3 Exercises

SECTION 3.1   *pages 80–83*

1. Yes   **3.** No   **5.** No   **7.** Yes   **9.** Yes   **11.** Yes   **13.** No   **15.** No   **17.** Yes
19. No   **21.** Yes   **23.** No   **25.** No   **27.** Yes   **29.** Yes   **31.** No   **33.** Yes   **35.** 6
37. 16   **39.** 7   **41.** -2 **43.** 1   **45.** 0   **47.** 3   **49.** -10 **51.** -3 **53.** -14
55. 2   **57.** 11   **59.** -9 **61.** -1 **63.** -14 **65.** -5 **67.** -1 **69.** 1   **71.** $-\dfrac{1}{2}$
73. $-\dfrac{3}{4}$ **75.** $\dfrac{1}{12}$ **77.** $-\dfrac{7}{12}$ **79.** $\dfrac{5}{14}$ **81.** $\dfrac{14}{15}$ **83.** $\dfrac{8}{9}$ **85.** 0.6529   **87.** -0.283
89. 9.257   **91.** 3   **93.** -4 **95.** -2 **97.** 9   **99.** 5   **101.** -4 **103.** 0   **105.** 4
107. -8 **109.** -7 **111.** -9 **113.** -7 **115.** 12   **117.** -18 **119.** -45 **121.** 30
123. -24 **125.** 15   **127.** -20 **129.** -12 **131.** 15   **133.** 25   **135.** $\dfrac{8}{3}$ **137.** $\dfrac{1}{3}$
139. $\dfrac{15}{7}$ **141.** 4   **143.** 3   **145.** 4.48   **147.** 2.06   **149.** -2.1 **151.** 11.742   **153.** 1.09
155. 36   **157.** 80   **159.** 48   **161.** -15 **163.** 22   **165.** -7 **167.** -7 **169.** -10
171. -20

SECTION 3.2   *pages 86–89*

1. 3   **3.** 6   **5.** -1 **7.** 1   **9.** -9 **11.** 3   **13.** 3   **15.** 3   **17.** 2   **19.** 2   **21.** 4
23. -6 **25.** $\dfrac{6}{7}$ **27.** $\dfrac{2}{3}$ **29.** $\dfrac{13}{9}$ **31.** -1 **33.** $\dfrac{3}{4}$ **35.** $\dfrac{4}{9}$ **37.** $\dfrac{1}{3}$ **39.** $-\dfrac{1}{2}$
41. $-\dfrac{3}{4}$ **43.** $-\dfrac{1}{7}$ **45.** $\dfrac{1}{3}$ **47.** $-\dfrac{1}{6}$ **49.** 1   **51.** 1   **53.** 0   **55.** $\dfrac{13}{10}$ **57.** $\dfrac{2}{5}$
59. $-\dfrac{4}{3}$ **61.** $-\dfrac{3}{2}$ **63.** 18   **65.** 8   **67.** -16 **69.** 25   **71.** 21   **73.** 15   **75.** -16
77. -21 **79.** $\dfrac{15}{2}$ **81.** $-\dfrac{18}{5}$ **83.** 2   **85.** 3   **87.** 1   **89.** -2 **91.** -0.66 **93.** 0
95. 19   **97.** -1 **99.** -11 **101.** 8 ft/s   **103.** 2 years   **105.** 31.8 in.   **107.** 1952
109. 51,000 people   **111.** 6   **113.** 62

SECTION 3.3   *pages 94–97*

1. 2   **3.** 3   **5.** 3   **7.** -1 **9.** -1 **11.** 2   **13.** -2 **15.** -3 **17.** -8 **19.** 0
21. -1 **23.** -3 **25.** -1 **27.** 4   **29.** -2 **31.** 3   **33.** -6 **35.** -2 **37.** $\dfrac{2}{3}$
39. $\dfrac{5}{6}$ **41.** $-\dfrac{2}{3}$ **43.** 2   **45.** 4   **47.** -17 **49.** 41   **51.** 8   **53.** 1   **55.** 4   **57.** -1
59. -1 **61.** 2   **63.** -4 **65.** $-\dfrac{7}{10}$ **67.** $\dfrac{1}{4}$ **69.** 5   **71.** $\dfrac{20}{3}$ **73.** 2   **75.** -1.5
77. 26   **79.** -2 **81.** No   **83.** 10 ft   **85.** 1770 lb   **87.** 190 telephones   **89.** 3000 bats
91. No solution   **93.** No solution   **95.** 4   **97.** -7 **99.** -21 **101.** No solution
103. No solution

SECTION 3.4   *pages 100–104*

1. $n - 15 = 7$; $n = 22$ **3.** $7n = -21$; $n = -3$ **5.** $3n - 4 = 5$; $n = 3$ **7.** $4(2n + 3) = 12$; $n = 0$
9. $12 = 6(n - 3)$; $n = 5$ **11.** $4n + 7 = 3$; $n = -1$ **13.** $22 = 6n - 2$; $n = 4$ **15.** $3n = 2n + 4$; $n = 4$
17. $4n + 7 = 2n + 3$; $n = -2$ **19.** $5n - 8 = 8n + 4$; $n = -4$ **21.** $2(n - 25) = 3n$; $n = -50$
23. $3n = 2(20 - n)$; 8 and 12   **25.** $3n - 4 = 2(24 - n) - 12$; 8 and 16   **27.** $3n + 2(18 - n) = 44$; 8 and 10
29. \$142   **31.** 32 megahertz   **33.** 4 field goals   **35.** 150 research assistants   **37.** 15 lb
39. \$117.75   **41.** 5 h   **43.** \$2000 and \$3000   **45.** length, 80 ft; width, 50 ft   **47.** 200 pixels
49. 8 ft   **51.** 3 ft   **53.** $\frac{1}{3}$ **55.** 15 min   **57.** 60 coins

DENSITY   *page 106*

1. 15 g/cm^3 **3.** 1.3 g/cm^3 **5.** 680 g   **7.** 300 g

CHAPTER REVIEW   *pages 107–109*

1. No (Objective 3.1.1)   **2.** 20 (Objective 3.1.2)   **3.** -7 (Objective 3.1.3)
4. 7 (Objective 3.2.1)   **5.** 4 (Objective 3.3.1)   **6.** No (Objective 3.1.1)
7. $-\frac{1}{5}$ (Objective 3.3.2)   **8.** 15 (Objective 3.3.1)   **9.** -3 (Objective 3.2.1)
10. $\frac{1}{3}$ (Objective 3.3.1)   **11.** Yes (Objective 3.1.1)   **12.** -2 (Objective 3.3.2)
13. -1 (Objective 3.3.2)   **14.** -9 (Objective (3.1.3)   **15.** 5 (Objective 3.2.1)
16. 2.5 (Objective 3.1.2)   **17.** -49 (Objective 3.1.3)   **18.** $\frac{1}{2}$ (Objective 3.2.1)
19. -1 (Objective 3.3.1)   **20.** 21 (Objective 3.1.2)   **21.** 0 (Objective 3.3.2)
22. 10 (Objective 3.3.2)   **23.** Yes (Objective 3.1.1)   **24.** 20 (Objective 3.1.3)
25. $\frac{5}{6}$ (Objective 3.1.2)   **26.** 6 (Objective 3.2.1)   **27.** 4 (Objective 3.3.1)
28. 7 (Objective 3.1.2)   **29.** 4 (Objective 3.1.3)   **30.** 3 (Objective 3.3.2)
31. $-\frac{6}{7}$ (Objective 3.2.1)   **32.** -5 (Objective 3.3.1)   **33.** 110° (Objective 3.2.2)
34. 24 lb (Objective 3.3.3)   **35.** 3 s (Objective 3.2.2)   **36.** $5n - 4 = 16$; $n = 4$ (Objective 3.4.1)
37. 14 in. (Objective 3.4.2)   **38.** 8 and 13 (Objective 3.4.1)
39. $6(n + 3) = 2n - 10$; $n = -7$ (Objective 3.4.1)   **40.** 3 ft from the 25-lb force (Objective 3.3.3)
41. 24 ft (Objective 3.2.2)   **42.** 6 and 30 (Objective 3.4.1)   **43.** \$79.25 (Objective 3.2.2)
44. $\frac{n}{9} = n - 24$; $n = 27$ (Objective 3.4.1)   **45.** 993 ft (Objective 3.4.2)   **46.** 80 ft (Objective 3.2.2)
47. $15 = \frac{2}{3}n + 3$; $n = 18$ (Objective 3.4.1)   **48.** 60 in. (Objective 3.4.2)   **49.** 7 ft (Objective 3.4.2)
50. 8 h (Objective 3.4.2)

CHAPTER TEST *pages 110–111*

1 −12 (Objective 3.1.3)   **2.** −$\frac{1}{2}$ (Objective 3.3.1)   **3.** −3 (Objective 3.2.1)

4. No (Objective 3.1.1)   **5.** $\frac{1}{8}$ (Objective 3.1.2)   **6.** −$\frac{1}{3}$ (Objective 3.3.2)

7. Yes (Objective 3.1.1)   **8.** 5 (Objective 3.2.1)   **9.** −5 (Objective 3.1.2)

10. $\frac{1}{2}$ (Objective 3.3.1)   **11.** −$\frac{40}{3}$ (Objective 3.1.3)   **12.** −5 (Objective 3.3.1)

13. 2 (Objective 3.3.2)   **14.** −$\frac{22}{7}$ (Objective 3.3.2)   **15.** −3 (Objective 3.3.2)

16. $\frac{12}{11}$ (Objective 3.3.2)   **17.** 200 calculators (Objective 3.2.2)

18. $6n + 13 = 3n − 5$; $n = −6$ (Objective 3.4.1)   **19.** 60°C (Objective 3.3.3)
20. $5n + 6 = 3(n + 12)$; $n = 15$ (Objective 3.4.1)   **21.** $3n − 15 = 27$; $n = 14$ (Objective 3.4.1)
22. 8 and 10 (Objective 3.4.1)   **23.** 7 h (Objective 3.4.2)   **24.** 200 cameras (Objective 3.2.2)
25. 6 ft and 12 ft (Objective 3.4.2)

CUMULATIVE REVIEW *pages 111–112*

1. 6 (Objective 1.2.2)   **2.** −48 (Objective 1.3.1)   **3.** −$\frac{19}{48}$ (Objective 1.4.2)

4. −2 (Objective 1.4.3)   **5.** 54 (Objective 1.5.1)   **6.** 24 (Objective 1.5.2)
7. 6 (Objective 2.1.1)   **8.** −17x (Objective 2.2.1)   **9.** −5a − 2b (Objective 2.2.1)
10. 2x (Objective 2.2.2)   **11.** 36y (Objective 2.2.2)   **12.** $2x^2 + 6x − 4$ (Objective 2.2.3)
13. −4x + 14 (Objective 2.2.4)   **14.** 6x − 34 (Objective 2.2.4)   **15.** Yes (Objective 3.1.1)
16. No (Objective 3.1.1)   **17.** −5 (Objective 3.1.2)   **18.** −25 (Objective 3.1.3)
19. −3 (Objective 3.2.1)   **20.** 3 (Objective 3.2.1)   **21.** 13 (Objective 3.3.2)

22. −2 (Objective 3.1.3)   **23.** −3 (Objective 3.3.1)   **24.** −$\frac{9}{2}$ (Objective 3.3.2)

25. 250 cameras (Objective 3.2.2)   **26.** 60°C (Objective 3.3.3)   **27.** $12 − 5n = −18$; $n = 6$
(Objective 3.4.1)   **28.** $8n + 12 = 4n$; $n = −3$ (Objective 3.4.1)   **29.** 600 ft² (Objective 3.4.2)
30. 5 ft and 11 ft (Objective 3.4.2)

ANSWERS to Chapter 4 Exercises

SECTION 4.1 *pages 117–118*

1. $\frac{3}{4}$; 0.75   **3.** $\frac{1}{2}$; 0.5   **5.** $\frac{16}{25}$; 0.64   **7.** $1\frac{3}{4}$; 1.75   **9.** $\frac{19}{100}$; 0.19   **11.** $\frac{1}{20}$; 0.05   **13.** $4\frac{1}{2}$; 4.5

15. $\frac{2}{25}$; 0.08   **17.** $\frac{1}{9}$ **19.** $\frac{1}{8}$ **21.** $\frac{5}{16}$ **23.** $\frac{1}{400}$ **25.** $\frac{23}{400}$ **27.** $\frac{1}{16}$ **29.** 0.073

31. 0.158   **33.** 0.003   **35.** 0.0915   **37.** 0.1823   **39.** 0.0015   **41.** 15%   **43.** 5%
45. 17.5%   **47.** 115%   **49.** 62%   **51.** 316.5%   **53.** 0.8%   **55.** 6.5%   **57.** 54%

59. 33.3%   **61.** 45.5%   **63.** 87.5%   **65.** 166.7%   **67.** 128.6%   **69.** 34%   **71.** $37\frac{1}{2}$%

73. $35\frac{5}{7}\%$ **75.** $18\frac{3}{4}\%$ **77.** 125%   **79.** $155\frac{5}{9}\%$ **81.** $\frac{14}{100}$; 14%   **83.** $\frac{55}{100}$; 55%

85. $\frac{80}{100}$; 80%   **87.** 1.06x

SECTION 4.2   *pages 121–123*

1. 24%   **3.** 7.2   **5.** 400   **7.** 9   **9.** 25%   **11.** 5   **13.** 200%   **15.** 400   **17.** 7.7
19. 200   **21.** 400   **23.** 20   **25.** 80%   **27.** $18,000   **29.** 4536 L   **31.** 1000%

33. 9000 seats   **35.** $83\frac{1}{3}\%$ **37.** 60%   **39.** 14.8%   **41.** 62.5%   **43.** $22,250   **45.** $45

47. No, lower.

SECTION 4.3   *pages 127–129*

1. $35   **3.** 87.5%   **5.** $196   **7.** 40%   **9.** $69.75   **11.** 23.3%   **13.** $41.25   **15.** 25%
17. $24.50   **19.** 40%   **21.** $218.50   **23.** 26%   **25.** $18   **27.** 40%   **29.** $230   **31.** $55

SECTION 4.4   *pages 131–134*

1. $2500   **3.** $9000 at 7%; $6000 at 6.5%   **5.** $1500   **7.** $200,000 at 10%; $100,000 at 8.5%
9. $3000   **11.** $40,500 at 8%; $13,500 at 12%   **13.** $650,000   **15.** $500,000   **17.** $45,000
19. $3040   **21.** $3831.87

SECTION 4.5   *pages 138–142*

1. 2 lb of the diet supplement; 3 lb of the vitamin supplement   **3.** $6.98   **5.** 56 oz of the alloy that
costs $4.30 per ounce; 144 oz of the alloy that costs $1.80 per ounce   **7.** $2.90   **9.** 10 kg   **11.** 30 lb
of the meat that costs $2.20 per pound; 20 lb of the meat that costs $4.20 per pound   **13.** 8 kg   **15.** 37
lb of almonds; 63 lb of walnuts   **17.** $.70   **19.** 9.6 lb   **21.** 20 ml of the 13% solution; 30 ml of the
18% solution   **23.** 50%   **25.** 30 lb   **27.** 0.74%   **29.** 100 ml of the 7% solution; 200 ml of the 4%
solution   **31.** 25 oz   **33.** 27%   **35.** 10 oz   **37.** 3 lb of the 20% jasmine; 2 lb of the 15% jasmine
39. 1.2 oz   **41.** $3.65   **43.** 10 oz   **45.** 75 g

SECTION 4.6   *pages 145–147*

1. first plane, 105 mph; second plane, 130 mph   **3.** 3 h   **5.** 0.5 h   **7.** 300 mi   **9.** $\frac{1}{3}$ h

11. 570 mi   **13.** 2 h at 115 mph and 3 h at 125 mph   **15.** 33 mi   **17.** passenger train, 60 mph;
freight train, 45 mph   **19.** 180 mi   **21.** 14 mph   **23.** 1 h   **25.** 10:15 A.M.   **27.** 2:15 P.M.
29. It is impossible to average 60 mph.

SECTION 4.7   *pages 151–153*

1. length, 50 ft; width, 25 ft   **3.** length, 130 ft; width, 52 ft   **5.** 13 ft, 12 ft, and 14 ft   **7.** 11 ft,
6 ft, and 13 ft   **9.** 60 ft, 60 ft, and 15 ft   **11.** length, 13.5 m; width, 12.8 m   **13.** 36°, 36°, and 108°
15. 32°, 58°, and 90°   **17.** 38°, 38°, and 104°   **19.** 60°, 45°, and 75°   **21.** 105°, 40°, and 35°
23. 57°, 19°, and 104°   **25.** length, 9 cm; width, 3 cm   **27.** length, 9 cm; width, 4 cm   **29.** 11x

SECTION 4.8   *pages 158–162*

1. 17, 18, and 19   **3.** 26, 28, and 30   **5.** 17, 19, and 21   **7.** 8 and 10   **9.** 7 and 9   **11.** −9, −8, and −7   **13.** 10, 12, and 14   **15.** 4, 6, and 8   **17.** 12 dimes and 15 quarters   **19.** eight 20¢ stamps and thirty-two 25¢ stamps   **21.** 5 stamps   **23.** 28 quarters   **25.** 20 one-dollar bills and 6 five-dollar bills   **27.** 11 pennies   **29.** 27 stamps   **31.** thirteen 3¢ stamps, eighteen 7¢ stamps, and nine 12¢ stamps   **33.** eighteen 6¢ stamps, six 8¢ stamps, and twenty-four 15¢ stamps   **35.** 4 years   **37.** wool tapestry, 84 years; linen tapestry, 52 years   **39.** 14 years   **41.** 6 years   **43.** mosaic, 141 years; engraving, 67 years   **45.** 44 years   **47.** china plate, 10 years; glass plate, 6 years   **49.** older ship, 8 years; newer ship, 4 years   **51.** 5¢ stamp, 7 years; 8¢ stamp, 17 years   **53.** −12, −10, −8, and −6   **55.** 8, 11, and 14 years   **57.** oil painting, 39 years; watercolor, 21 years   **59.** $6.70   **61.** any three consecutive odd integers

ACCELERATION   *page 165*

1. 2 m/s^2 **3.** −5 m/s^2 **5.** 23 m/s   **7.** 55 m/s

CHAPTER REVIEW   *pages 167–169*

1. $\frac{11}{20}$ (Objective 4.1.1)   **2.** $288\frac{8}{9}\%$ (Objective 4.1.2)   **3.** 25   (Objective 4.2.1)

4. $\frac{159}{200}$ (Objective 4.1.1)   **5.** $16\frac{2}{3}\%$ (Objective 4.2.1)   **6.** 16   (Objective 4.2.1)

7. 2.4   (Objective 4.1.1)   **8.** 159%   (Objective 4.1.2)   **9.** 4%   (Objective 4.2.1)
10. 3.42   (Objective 4.1.1)   **11.** 62.5%   (Objective 4.1.2)   **12.** 15   (Objective 4.2.1)
13. 0.07   (Objective 4.1.1)   **14.** 277.8%   (Objective 4.1.2)   **15.** 25   (Objective 4.2.1)
16. 0.062   (Objective 4.1.1)   **17.** 23.1%   (Objective 4.1.2)   **18.** 125   (Objective 4.2.1)

19. 67.2%   (Objective 4.1.2)   **20.** $69\frac{13}{23}\%$ (Objective 4.1.2)   **21.** 67.5%   (Objective 4.2.1)

22. 0.2%   (Objective 4.1.2)   **23.** $54\frac{2}{7}\%$ (Objective 4.1.2)   **24.** 405   (Objective 4.2.1)

25. $33\frac{1}{3}\%$ (Objective 4.3.2)   **26.** $8000 at 6%; $7000 at 7%   (Objective 4.4.1)

27. $1.84   (Objective 4.5.1)   **28.** 45 mph   (Objective 4.6.1)   **29.** 16°, 82°, and 82°   (Objective 4.7.2)
30. 400%   (Objective 4.2.2)   **31.** 70%   (Objective 4.3.1)   **32.** 14%   (Objective 4.5.2)
33. length, 80 ft; width, 20 ft   (Objective 4.7.1)   **34.** 30 lb   (Objective 4.2.2)
35. $5600   (Objective 4.4.1)   **36.** 9 and 10   (Objective 4.8.1)   **37.** 1 L   (Objective 4.5.2)
38. 10 years   (Objective 4.8.3)   **39.** 75°, 60°, and 45°   (Objective 4.7.2)
40. 30 one-dollar bills   (Objective 4.8.2)   **41.** $36   (Objective 4.2.2)   **42.** $32   (Objective 4.3.2)
43. cranberry juice, 7 qt; apple juice, 3 qt   (Objective 4.5.1)   **44.** $595   (Objective 4.3.1)
45. $671.25   (Objective 4.3.1)   **46.** 4 min   (Objective 4.6.1)   **47.** 8 in., 12 in., and 15 in.   (Objective 4.7.1)   **48.** −17, −15, and −13   (Objective 4.8.1)   **49.** grandparent, 48 years; child, 6 years   (Objective 4.8.3)   **50.** 5 dimes and 7 quarters   (Objective 4.8.2)

CHAPTER TEST   *pages 170–171*

1. $\frac{3}{5}$, 0.60   (Objective 4.1.1)   **2.** 125%   (Objective 4.2.1)   **3.** 37.5%   (Objective 4.1.2)

4. $\frac{5}{8}$ (Objective 4.1.1)   **5.** $87\frac{1}{2}\%$ (Objective 4.1.2)   **6.** 6.4   (Objective 4.2.1)

7. $\frac{4}{5}$, 0.80 (Objective 4.1.1)   **8.** 8% (Objective 4.1.2)   **9.** 40 (Objective 4.2.1)

10. $\frac{1}{6}$ (Objective 4.1.1)   **11.** 20 (Objective 4.2.1)   **12.** 7.5% (Objective 4.1.2)

13. $200 (Objective 4.3.1)   **14.** 20% (Objective 4.3.2)   **15.** 20 gal (Objective 4.5.2)
16. length, 14 m; width, 5 m   (Objective 4.7.1)   **17.** 5, 7, 9 (Objective 4.8.1)
18. 10 years (Objective 4.8.3)   **19.** $3000 (Objective 4.2.2)   **20.** $5000 at 10%; $2000 at
15% (Objective 4.4.1)   **21.** 8 lb at $7/lb; 4 lb at $4/lb (Objective 4.5.1)   **22.** first plane, 225 mph;
second plane, 125 mph (Objective 4.6.1)   **23.** 48°, 33°, 99° (Objective 4.7.2)   **24.** 15 nickels; 35
quarters (Objective 4.8.2)   **25.** $1400 at 6.75%; $1000 at 9.45% (Objective 4.4.1)

CUMULATIVE REVIEW   *pages 171–172*

1. 6 (Objective 1.2.2)   **2.** $-\frac{1}{6}$ (Objective 1.5.1)   **3.** $-\frac{11}{6}$ (Objective 1.5.2)

4. -18 (Objective 1.1.2)   **5.** -24 (Objective 2.1.1)   **6.** $9x + 4y$ (Objective 2.2.1)
7. $-12 + 8x + 20x^3$ (Objective 2.2.3)   **8.** $4x + 4$ (Objective 2.2.4)   **9.** $6x^2$ (Objective 2.2.1)
10. No (Objective 3.1.1)   **11.** -3 (Objective 3.1.2)   **12.** -15 (Objective 3.1.3)

13. 3 (Objective 3.2.1)   **14.** -10 (Objective 3.3.2)   **15.** $\frac{2}{5}$ (Objective 4.1.1)

16. $\frac{5}{6}$ (Objective 4.1.1)   **17.** 2.5% (Objective 4.1.2)   **18.** 12% (Objective 4.1.2)

19. 3 (Objective 4.2.1)   **20.** 45 (Objective 4.2.1)   **21.** 4, 11 (Objective 3.4.1)
22. 5 h (Objective 3.4.2)   **23.** 20% (Objective 4.2.2)   **24.** $2000 (Objective 4.4.1)
25. 75% (Objective 4.3.1)   **26.** 60 g (Objective 4.5.1)   **27.** 30 oz (Objective 4.5.2)
28. 47° (Objective 4.7.2)   **29.** 14 (Objective 4.8.1)   **30.** 7 dimes (Objective 4.8.2)

ANSWERS to Chapter 5 Exercises

SECTION 5.1   *pages 177–178*

1. $-2x^2 + 3x$ **3.** $y^2 - 8$ **5.** $5x^2 + 7x + 20$ **7.** $x^3 + 2x^2 - 6x - 6$ **9.** $2a^3 - 3a^2 - 11a + 2$
11. $5x^2 + 8x$ **13.** $7x^2 + xy - 4y^2$ **15.** $3a^2 - 3a + 17$ **17.** $5x^3 + 10x^2 - x - 4$
19. $3r^3 + 2r^2 - 11r + 7$ **21.** $-2x^3 + 3x^2 + 10x + 11$ **23.** $4x$ **25.** $3y^2 - 4y - 2$ **27.** $-7x - 7$
29. $4x^3 + 3x^2 + 3x + 1$ **31.** $y^3 - y^2 + 6y - 6$ **33.** $-y^2 - 13xy$ **35.** $2x^2 - 3x - 1$
37. $-2x^3 + x^2 + 2$ **39.** $3a^3 - 2$ **41.** $4y^3 - 2y^2 + 2y - 4$ **43.** binomial   **45.** monomial
47. $x^2 + x - 1$ **49.** No   **51.** Yes   **53.** Yes   **55.** $2x^3 + x^2 - 5x + 6$ **57.** $x^3 - x^2 + 2x + 6$

SECTION 5.2   *pages 181–182*

1. $2x^2$ **3.** $12x^2$ **5.** $6a^7$ **7.** x^3y^5 **9.** $-10x^9y$ **11.** x^7y^8 **13.** $-6x^3y^5$ **15.** x^4y^5z
17. $a^3b^5c^4$ **19.** $-a^5b^8$ **21.** $-6a^5b$ **23.** $40y^{10}z^6$ **25.** $-20a^2b^3$ **27.** $x^3y^5z^3$ **29.** $-12a^{10}b^7$
31. 64   **33.** 4   **35.** -64 **37.** x^9 **39.** x^{14} **41.** x^4 **43.** $4x^2$ **45.** $-8x^6$ **47.** x^4y^6
49. $9x^4y^2$ **51.** $27a^8$ **53.** $-8x^7$ **55.** x^8y^4 **57.** a^4b^6 **59.** $16x^{10}y^3$ **61.** $-18x^3y^4$
63. $-8a^7b^5$ **65.** $-54a^9b^3$ **67.** $-72a^5b^5$ **69.** $32x^3$ **71.** $-3a^3b^3$ **73.** $13x^5y^5$ **75.** $27a^5b^3$
77. $-5x^7y^4$ **79.** a^{2n} **81.** a^{2n} **83.** $12ab$

SECTION 5.3 *pages 187–190*

1. $x^2 - 2x$　　**3.** $-x^2 - 7x$　　**5.** $3a^3 - 6a^2$　　**7.** $-5x^4 + 5x^3$　　**9.** $-3x^5 + 7x^3$　　**11.** $12x^3 - 6x^2$
13. $6x^2 - 12x$　　**15.** $3x^2 + 4x$　　**17.** $-x^3y + xy^3$　　**19.** $2x^4 - 3x^2 + 2x$　　**21.** $2a^3 + 3a^2 + 2a$
23. $3x^6 - 3x^4 - 2x^2$　　**25.** $-6y^4 - 12y^3 + 14y^2$　　**27.** $-2a^3 - 6a^2 + 8a$　　**29.** $6y^4 - 3y^3 + 6y^2$
31. $x^3y - 3x^2y^2 + xy^3$　　**33.** $x^3 + 4x^2 + 5x + 2$　　**35.** $a^3 - 6a^2 + 13a - 12$　　**37.** $-2b^3 + 7b^2 + 19b - 20$
39. $-6x^3 + 31x^2 - 41x + 10$　　**41.** $x^3 - 3x^2 + 5x - 15$　　**43.** $x^4 - 4x^3 - 3x^2 + 14x - 8$
45. $15y^3 - 16y^2 - 70y + 16$　　**47.** $5a^4 - 20a^3 - 15a^2 + 62a - 8$　　**49.** $y^4 + 4y^3 + y^2 - 5y + 2$
51. $x^2 + 4x + 3$　　**53.** $a^2 + a - 12$　　**55.** $y^2 - 5y - 24$　　**57.** $y^2 - 10y + 21$　　**59.** $2x^2 + 15x + 7$
61. $3x^2 + 11x - 4$　　**63.** $4x^2 - 31x + 21$　　**65.** $3y^2 - 2y - 16$　　**67.** $9x^2 + 54x + 77$　　**69.** $21a^2 - 83a + 80$
71. $15b^2 + 47b - 78$　　**73.** $2a^2 + 7ab + 3b^2$　　**75.** $6a^2 + ab - 2b^2$　　**77.** $2x^2 - 3xy - 2y^2$
79. $10x^2 + 29xy + 21y^2$　　**81.** $6a^2 - 25ab + 14b^2$　　**83.** $2a^2 - 11ab - 63b^2$　　**85.** $100a^2 - 100ab + 21b^2$
87. $15x^2 + 56xy + 48y^2$　　**89.** $14x^2 - 97xy - 60y^2$　　**91.** $56x^2 - 61xy + 15y^2$　　**93.** $y^2 - 25$
95. $4x^2 - 9$　　**97.** $x^2 + 2x + 1$　　**99.** $9a^2 - 30a + 25$　　**101.** $9x^2 - 49$　　**103.** $4a^2 + 4ab + b^2$
105. $x^2 - 4xy + 4y^2$　　**107.** $16 - 9y^2$　　**109.** $25x^2 + 20xy + 4y^2$　　**111.** $10x^2 - 35x$　　**113.** $8x^2 + 34x - 9$
115. $x^2 + 14x + 49$　　**117.** $4x^2 + 10x$　　**119.** $\pi x^2 + 8\pi x + 16\pi$　　**121.** $4ab$
123. $9a^4 - 24a^3 + 28a^2 - 16a + 4$　　**125.** $24x^3 - 3x^2$　　**127.** $x^{2n} + x^n$　　**129.** $x^{2n} + 2x^n + 1$
131. $2x^2 + 11x - 6$　　**133.** 1024

SECTION 5.4 *pages 197–201*

1. $\dfrac{1}{5^2} = \dfrac{1}{25}$　　**3.** $8^2 = 64$　　**5.** $\dfrac{1}{3^3} = \dfrac{1}{27}$　　**7.** $2^0 = 1$　　**9.** $\dfrac{1}{x^2}$　　**11.** a^6　　**13.** $\dfrac{x^2}{y^3}$　　**15.** $\dfrac{1}{xy^2}$

17. x　　**19.** $\dfrac{1}{a^8}$　　**21.** $\dfrac{y}{x^3}$　　**23.** $\dfrac{a^4}{b^3}$　　**25.** $\dfrac{b}{a^2}$　　**27.** $\dfrac{1}{y^{12}}$　　**29.** 1　　**31.** $\dfrac{y^4}{x^4}$

33. $\dfrac{y^2}{x^4}$　　**35.** $-\dfrac{8x^3}{y^6}$　　**37.** $\dfrac{16x^4}{y^6}$　　**39.** $\dfrac{2}{x^4}$　　**41.** $-\dfrac{5}{a^8}$　　**43.** $\dfrac{1}{a^5b^6}$　　**45.** $\dfrac{1}{4x^3}$　　**47.** $\dfrac{16}{3a^5}$

49. $\dfrac{2y}{x^3}$　　**51.** $\dfrac{a^2}{2b^9}$　　**53.** $-2y^3$　　**55.** $\dfrac{x}{3}$　　**57.** $-\dfrac{9}{4y^2}$　　**59.** $\dfrac{1}{x^6y}$　　**61.** $-\dfrac{a^4}{3}$　　**63.** $-\dfrac{1}{6x^3}$

65. $\dfrac{2a^{18}}{b^3}$　　**67.** $\dfrac{a^4}{y^4}$　　**69.** $\dfrac{4a^2}{9b^3}$　　**71.** $-\dfrac{3x^5}{4y^3}$　　**73.** $\dfrac{x^{10}y^2}{8}$　　**75.** $\dfrac{s^8t^4}{4r^{12}}$　　**77.** $x + 1$　　**79.** $2a - 5$

81. $3a + 2$　　**83.** $4b^2 - 3$　　**85.** $x - 2$　　**87.** $-x + 2$　　**89.** $x^2 + 3x - 5$　　**91.** $x^4 - 3x^2 - 1$

93. $xy + 2$　　**95.** $-3y^3 + 5$　　**97.** $3x - 2 + \dfrac{1}{x}$　　**99.** $-3x + 7 - \dfrac{6}{x}$　　**101.** $4a - 5 + 6b$

103. $9x + 6 - 3y$　　**105.** $x + 5$　　**107.** $b - 7$　　**109.** $y - 5$　　**111.** $2y - 7$　　**113.** $2y + 6 + \dfrac{25}{y - 3}$

115. $x - 2 + \dfrac{8}{x + 2}$　　**117.** $3y - 5 + \dfrac{20}{2y + 4}$　　**119.** $6x - 12 + \dfrac{19}{x + 2}$　　**121.** $b - 5 - \dfrac{24}{b - 3}$

123. $3x + 17 + \dfrac{64}{x - 4}$　　**125.** $5y + 3 + \dfrac{1}{2y + 3}$　　**127.** $4a + 1$　　**129.** $2a + 9 + \dfrac{33}{3a - 1}$　　**131.** $x^2 - 5x + 2$

133. $2a^2 + a + 1 + \dfrac{6}{2a + 3}$　　**135.** $2b^2 - 3b + 4 - \dfrac{17}{2b + 3}$　　**137.** $\dfrac{3}{64}$　　**139.** 0.008　　**141.** 0.0016

143. $2ab^2c$　　**145.** $4x + 4$　　**147.** $\dfrac{x^6y}{18}$　　**149.** $\dfrac{1}{15x^2z^4}$　　**151.** $\dfrac{5}{4}$　　**153.** a　　**155.** 0　　**157.** $4y^2$

159. $x^3 - 4x^2 + 11x - 2$

CALCULATORS AND COMPUTERS *page 203*

1. 27　　**3.** 73　　**5.** 25

SCIENTIFIC NOTATION   *page 204*

1. $9.7 \cdot 10^7$ **3.** $2.3 \cdot 10^{-14}$ **5.** $3.3 \cdot 10^{-3}$ **7.** $3 \cdot 10^{16}$

CHAPTER REVIEW   *pages 205–207*

1. $21y^2 + 4y - 1$ (Objective 5.1.1)   **2.** $-20x^3y^5$ (Objective 5.2.1)
3. $-8x^3 - 14x^2 + 18x$ (Objective 5.3.1)   **4.** $10a^2 + 31a - 63$ (Objective 5.3.3)
5. $6x^2 - 7x + 10$ (Objective 5.4.2)   **6.** $2x^2 + 3x - 8$ (Objective 5.1.2)   **7.** -729 (Objective 5.2.2)
8. $x^3 - 6x^2 + 7x - 2$ (Objective 5.3.2)   **9.** $a^2 - 49$ (Objective 5.3.4)   **10.** 1296 (Objective 5.4.1)
11. $x - 6$ (Objective 5.4.3)   **12.** $2x^3 + 9x^2 - 3x - 12$ (Objective 5.1.1)   **13.** $18a^8b^6$ (Objective 5.2.1)
14. $3x^4y - 2x^3y + 12x^2y$ (Objective 5.3.1)   **15.** $8b^2 - 2b - 15$ (Objective 5.3.3)
16. $-4y + 8$ (Objective 5.4.2)   **17.** $13y^3 - 12y^2 - 5y - 1$ (Objective 5.1.2)   **18.** 64 (Objective 5.2.2)
19. $6y^3 + 17y^2 - 2y - 21$ (Objective 5.3.2)   **20.** $4b^2 - 81$ (Objective 5.3.4)
21. $\dfrac{b^6c^2}{a^4}$ (Objective 5.4.1)   **22.** $2y - 9$ (Objective 5.4.3)   **23.** 14 (Objective 5.1.1)
24. $x^4y^8z^4$ (Objective 5.2.1)   **25.** $-12y^5 + 4y^4 - 18y^3$ (Objective 5.3.1)
26. $18x^2 - 48x + 24$ (Objective 5.3.3)   **27.** $a^3 - 12a^2 + 3a$ (Objective 5.4.2)
28. $-7a^2 - a + 4$ (Objective 5.1.2)   **29.** $9x^4y^6$ (Objective 5.2.2)
30. $12a^3 - 8a^2 - 9a + 6$ (Objective 5.3.3)   **31.** $25y^2 - 70y + 49$ (Objective 5.3.4)
32. $\dfrac{x^4y^6}{9}$ (Objective 5.4.1)   **33.** $x + 5 + \dfrac{4}{x + 12}$ (Objective 5.4.3)
34. $10a^2 + 12a - 22$ (Objective 5.1.1)   **35.** $a^6b^{11}c^9$ (Objective 5.2.1)
36. $8a^3b^3 - 4a^2b^4 + 6ab^5$ (Objective 5.3.1)   **37.** $6x^2 - 7xy - 20y^2$ (Objective 5.3.3)
38. $4b^4 + 12b^2 - 1$ (Objective 5.4.2)   **39.** $-4b^2 - 13b + 28$ (Objective 5.1.2)
40. $100a^{15}b^{13}$ (Objective 5.2.2)   **41.** $12b^5 - 4b^4 - 6b^3 - 8b^2 + 5$ (Objective 5.3.2)
42. $36 - 25x^2$ (Objective 5.3.4)   **43.** $\dfrac{2y^2}{3x^4}$ (Objective 5.4.1)   **44.** $a^2 - 2a + 6$ (Objective 5.4.3)
45. $4b^3 - 5b^2 - 9b + 7$ (Objective 5.1.1)   **46.** $-54a^{13}b^5c^7$ (Objective 5.2.1)
47. $-18x^4 - 27x^3 + 63x^2$ (Objective 5.3.1)   **48.** $30y^2 - 109y + 30$ (Objective 5.3.3)
49. $3x^5 + 2x + \dfrac{2}{5x}$ (Objective 5.4.2)   **50.** $-2y^2 + y - 5$ (Objective 5.1.2)
51. $144x^{14}y^{18}z^{16}$ (Objective 5.2.2)   **52.** $-12x^4 - 17x^3 - 2x^2 - 33x - 27$ (Objective 5.3.2)
53. $64a^2 + 16a + 1$ (Objective 5.3.4)   **54.** $\dfrac{1}{b^8}$ (Objective 5.4.1)
55. $b^2 + 5b + 2 + \dfrac{7}{b - 7}$ (Objective 5.4.3)   **56.** $12x^2 - 21x$ (Objective 5.3.5)
57. $25x^2 + 40x + 16$ (Objective 5.3.5)   **58.** $9x^2 - 4$ (Objective 5.3.5)
59. $\pi x^2 - 12\pi x + 36\pi$ (Objective 5.3.5)   **60.** $15x^2 - 28x - 32$ (Objective 5.3.5)

CHAPTER TEST   *pages 208–209*

1. $3x^3 + 6x^2 - 8x + 3$ (Objective 5.1.1)   **2.** $-4x^4 + 8x^3 - 3x^2 - 14x + 21$ (Objective 5.3.2)
3. $4x^3 - 6x^2$ (Objective 5.3.1)   **4.** $-8a^6b^3$ (Objective 5.2.2)   **5.** $-4x^6$ (Objective 5.4.1)
6. $\dfrac{6b}{a}$ (Objective 5.4.1)   **7.** $-5a^3 + 3a^2 - 4a + 3$ (Objective 5.1.2)
8. $a^2 + 3ab - 10b^2$ (Objective 5.3.3)   **9.** $4x^4 - 2x^2 + 5$ (Objective 5.4.2)

10. $2x + 3 + \dfrac{2}{2x - 3}$ (Objective 5.4.3) **11.** $-6x^3y^6$ (Objective 5.2.1)

12. $6y^4 - 9y^3 + 18y^2$ (Objective 5.3.1) **13.** $\dfrac{9y^6}{x}$ (Objective 5.4.1)

14. $4x^2 - 20x + 25$ (Objective 5.3.4) **15.** $10x^2 - 43xy + 28y^2$ (Objective 5.3.3)

16. $x^3 - 7x^2 + 17x - 15$ (Objective 5.3.2) **17.** $\dfrac{a^4}{b^6}$ (Objective 5.4.1)

18. $x + 7$ (Objective 5.4.3) **19.** $16y^2 - 9$ (Objective 5.3.4)
20. $3y^3 + 2y^2 - 10y$ (Objective 5.1.2) **21.** $9a^6b^4$ (Objective 5.2.2)
22. $10a^3 - 39a^2 + 20a - 21$ (Objective 5.3.2) **23.** $9b^2 + 12b + 4$ (Objective 5.3.4)

24. $\dfrac{1}{2b^2}$ (Objective 5.4.1) **25.** $4x + 8 + \dfrac{21}{2x - 3}$ (Objective 5.4.3) **26.** a^3b^7 (Objective 5.2.1)

27. $a^2 + ab - 12b^2$ (Objective 5.3.3) **28.** $4x - 1 + \dfrac{3}{x^2}$ (Objective 5.4.2)

29. $4x^2 + 12x + 9$ (Objective 5.3.5) **30.** $\pi x^2 - 10\pi x + 25\pi$ (Objective 5.3.5)

CUMULATIVE REVIEW *pages 209–210*

1. $\dfrac{1}{144}$ (Objective 1.4.2) **2.** $\dfrac{25}{9}$ (Objective 1.5.1) **3.** $\dfrac{4}{5}$ (Objective 1.5.2)

4. 87 (Objective 1.1.2) **5.** 0.775 (Objective 1.4.1) **6.** $-\dfrac{27}{4}$ (Objective 2.1.1)

7. $-x - 4xy$ (Objective 2.2.1) **8.** $-12x$ (Objective 2.2.2) **9.** $-22x + 20$ (Objective 2.2.4)
10. 8 (Objective 2.1.2) **11.** -18 (Objective 3.1.3) **12.** 16 (Objective 3.3.1)
13. 12 (Objective 3.3.2) **14.** 80 (Objective 3.3.1) **15.** 24% (Objective 4.2.1)
16. $5b^3 - 7b^2 + 8b - 10$ (Objective 5.1.2) **17.** $15x^3 - 26x^2 + 11x - 4$ (Objective 5.3.2)
18. $20b^2 - 47b + 24$ (Objective 5.3.3) **19.** $25b^2 + 30b + 9$ (Objective 5.3.4)

20. $-\dfrac{9a^5b^8}{4}$ (Objective 5.4.1) **21.** $5y - 4 + \dfrac{1}{y}$ (Objective 5.4.2) **22.** $a - 7$ (Objective 5.4.3)

23. $\dfrac{9y^2}{x^6}$ (Objective 5.4.1) **24.** $5(n - 12); 5n - 60$ (Objective 2.3.3) **25.** $8n - 2n = 18; n = 3$

(Objective 3.4.1) **26.** length, 15 m; width, 6 m (Objective 4.7.1) **27.** \$43.20 (Objective 4.3.1)
28. 28% (Objective 4.5.2) **29.** 25 mi (Objective 4.6.1) **30.** $9x^2 + 12x + 4$ (Objective 5.3.5)

ANSWERS to Chapter 6 Exercises

SECTION 6.1 *pages 216–218*

1. x^3 **3.** xy^4 **5.** xy^4z^2 **7.** ab^2c^3 **9.** $3x^2$ **11.** $2a$ **13.** $7a^3$ **15.** 1 **17.** $3a^2b^2$
19. ab **21.** $2x$ **23.** $3x$ **25.** $5(a + 1)$ **27.** $8(2 - a^2)$ **29.** $4(2x + 3)$ **31.** $6(5a - 1)$
33. $x(7x - 3)$ **35.** $a^2(3 + 5a^3)$ **37.** $y(14y + 11)$ **39.** $2x(x^3 - 2)$ **41.** $2x^2(5x^2 - 6)$
43. $4a^5(2a^3 - 1)$ **45.** $xy(xy - 1)$ **47.** $3xy(xy^3 - 2)$ **49.** $xy(x - y^2)$ **51.** $2a^5b + 3xy^3$
53. $6b^2(a^2b - 2)$ **55.** $2abc(3a + 2b)$ **57.** $9(2x^2y^2 - a^2b^2)$ **59.** $6x^3y^3(1 - 2x^3y^3)$ **61.** $x(x^2 - 3x - 1)$
63. $2(x^2 + 4x - 6)$ **65.** $b(b^2 - 5b - 7)$ **67.** $4(2y^2 - 3y + 8)$ **69.** $5y(y^2 - 4y + 2)$
71. $3y^2(y^2 - 3y - 2)$ **73.** $3y(y^2 - 3y + 8)$ **75.** $a^2(6a^3 - 3a - 2)$ **77.** $ab(2a - 5ab + 7b)$
79. $2b(2b^4 + 3b^2 - 6)$ **81.** $x^2(8y^2 - 4y + 1)$ **83.** $x^3y^3(4x^2y^2 - 8xy + 1)$ **85.** $(a + b)(x + 2)$
87. $(b + 2)(x - y)$ **89.** $(x - 2)(a - 5)$ **91.** $(y - 3)(b - 3)$ **93.** $(x - y)(a + 2)$ **95.** $(x + 4)(x^2 + 3)$
97. $(y + 2)(2y^2 + 3)$ **99.** $(a + 3)(b - 2)$ **101.** $(a - 2)(x^2 - 3)$ **103.** $(a - b)(3x - 2y)$

105. $(x - 3)(x + 4a)$ **107.** $(x - 5)(y - 2)$ **109.** $(7x + 2y)(3x - 7)$ **111.** $(2r + a)(a - 1)$
113. $(4x + 3y)(x - 4)$ **115.** $(2y - 3)(5xy + 3)$ **117. a.** $(x + 3)(2x + 5)$ **b.** $(2x + 5)(x + 3)$
119. a. $(a - b)(2a - 3b)$ **b.** $(2a - 3b)(a - b)$ **121.** 28   **123.** *P* doubles

---

SECTION 6.2   *pages 222–225*

1. $(x + 1)(x + 2)$ **3.** $(x + 1)(x - 2)$ **5.** $(a + 4)(a - 3)$ **7.** $(a - 1)(a - 2)$ **9.** $(a + 2)(a - 1)$
11. $(b - 3)(b - 3)$ **13.** $(b + 8)(b - 1)$ **15.** $(y + 11)(y - 5)$ **17.** $(y - 2)(y - 3)$ **19.** $(z - 5)(z - 9)$
21. $(z + 8)(z - 20)$ **23.** $(p + 3)(p + 9)$ **25.** $(x + 10)(x + 10)$ **27.** $(b + 4)(b + 5)$
29. $(x + 3)(x - 14)$ **31.** $(b + 4)(b - 5)$ **33.** $(y + 3)(y - 17)$ **35.** $(p + 3)(p - 7)$
37. Nonfactorable over the integers   **39.** $(x - 5)(x - 15)$ **41.** $(x - 7)(x - 8)$ **43.** $(x + 8)(x - 7)$
45. $(a + 3)(a - 24)$ **47.** $(a - 3)(a - 12)$ **49.** $(z + 8)(z - 17)$ **51.** $(c + 9)(c - 10)$
53. $(z + 4)(z + 11)$ **55.** $(c + 2)(c + 17)$ **57.** $(x + 8)(x - 12)$ **59.** $(x - 8)(x - 14)$
61. $(b + 15)(b - 7)$ **63.** $(a + 3)(a - 12)$ **65.** $(b - 6)(b - 17)$ **67.** $(a + 3)(a + 24)$
69. $(x + 12)(x + 13)$ **71.** $(x + 6)(x - 16)$ **73.** $2(x + 1)(x + 2)$ **75.** $3(a + 3)(a - 2)$
77. $a(b + 5)(b - 3)$ **79.** $x(y - 2)(y - 3)$ **81.** $z(z - 3)(z - 4)$ **83.** $3y(y - 2)(y - 3)$
85. $3(x + 4)(x - 3)$ **87.** $5(z + 4)(z - 7)$ **89.** $2a(a + 8)(a - 4)$ **91.** $(x - 2y)(x - 3y)$
93. $(a - 4b)(a - 5b)$ **95.** $(x + 4y)(x - 7y)$ **97.** Nonfactorable over the integers
99. $z^2(z - 5)(z - 7)$ **101.** $b^2(b - 10)(b - 12)$ **103.** $2y^2(y + 3)(y - 16)$ **105.** $x^2(x + 8)(x - 1)$
107. $4y(x + 7)(x - 2)$ **109.** $8(y - 1)(y - 3)$ **111.** $c(c + 3)(c + 10)$ **113.** $3x(x - 3)(x - 9)$
115. $(x - 3y)(x - 5y)$ **117.** $(a - 6b)(a - 7b)$ **119.** $(y + z)(y + 7z)$ **121.** $3y(x + 21)(x - 1)$
123. $3x(x + 4)(x - 3)$ **125.** $4z(z + 11)(z - 3)$ **127.** $4x(x + 3)(x - 1)$ **129.** $5(p + 12)(p - 7)$
131. $p^2(p + 12)(p - 3)$ **133.** $(a + 3b)(a - 11b)$ **135.** $15a(b + 4)(b - 1)$ **137.** $(c + 4)(c + 5)$
139. $a^2(b - 5)(b - 9)$ **141.** 36, 12, −12, −36   **143.** 22, 10, −10, −22

SECTION 6.3   *pages 232–235*

1. $(x + 1)(2x + 1)$ **3.** $(y + 3)(2y + 1)$ **5.** $(a - 1)(2a - 1)$ **7.** $(b - 5)(2b - 1)$ **9.** $(x + 1)(2x - 1)$
11. $(x - 3)(2x + 1)$ **13.** Nonfactorable over the integers   **15.** $(2t - 1)(3t - 4)$ **17.** $(x + 4)(8x + 1)$
19. $(b - 4)(3b - 4)$ **21.** $(z - 14)(2z + 1)$ **23.** $(p + 8)(3p - 2)$ **25.** $(2x - 3)(3x - 4)$
27. $(b + 7)(5b - 2)$ **29.** $(2a - 3)(3a + 8)$ **31.** $(3t + 1)(6t - 5)$ **33.** $(b + 12)(6b - 1)$
35. $(3x + 2)(3x + 2)$ **37.** $(3a + 7)(5a - 3)$ **39.** $(2y - 5)(4y - 3)$ **41.** $(2z + 3)(4z - 5)$
43. $(x + 5)(3x - 1)$ **45.** $(3x + 4)(4x + 3)$ **47.** $(2z + 7)(6z - 5)$ **49.** $b(a - 4)(3a - 4)$
51. Nonfactorable over the integers   **53.** $(x + y)(3x - 2y)$ **55.** $(4 + z)(7 - z)$ **57.** $(1 - x)(8 + x)$
59. $3(x + 5)(3x - 4)$ **61.** $4(2x - 3)(3x - 2)$ **63.** $a^2(5a + 2)(7a - 1)$ **65.** $5(b - 7)(3b - 2)$
67. $(3a - 2b)(5a + 7b)$ **69.** $z(3 - z)(11 + z)$ **71.** $2x(x + 1)(5x + 1)$ **73.** $2y(y - 4)(5y - 2)$
75. $yz(z + 2)(4z - 3)$ **77.** $b^2(4b + 5)(5b + 4)$ **79.** $xy(3x + 2)(3x + 2)$ **81.** $ab(a - 5b)(2a - b)$
83. $y^2(3x - 2y)(4x - 3y)$ **85.** $(t + 2)(2t - 5)$ **87.** $(p - 5)(3p - 1)$ **89.** $(3y - 1)(4y - 1)$
91. Nonfactorable over the integers   **93.** $(3y + 1)(4y + 5)$ **95.** $(a + 7)(7a - 2)$ **97.** $(z + 2)(4z + 3)$
99. $(2p + 5)(11p - 2)$ **101.** $(y + 1)(8y + 9)$ **103.** $(2b - 3)(3b - 2)$ **105.** $(3b + 5)(11b - 7)$
107. $(3y - 4)(6y - 5)$ **109.** Nonfactorable over the integers   **111.** $(2z - 5)(5z - 2)$
113. $(6z + 5)(6z + 7)$ **115.** $(2y - 3)(7y - 4)$ **117.** $(x + 6)(6x - 1)$ **119.** $3(x + 3)(4x - 1)$
121. $10(y + 1)(3y - 2)$ **123.** $x(x + 1)(2x - 5)$ **125.** $(a - 3b)(2a - 3b)$ **127.** $(y + z)(2y + 5z)$
129. $(1 + x)(2 - x)$ **131.** $6(2y + 1)(2y - 3)$ **133.** $z(2z - 5)(3z - 4)$ **135.** $y(x - 3)(8x - 3)$
137. $(2x + 3y)(2x + 5y)$ **139.** $4(9y + 1)(10y - 1)$ **141.** $(1 + x)(18 - x)$ **143.** $8(t + 4)(2t - 3)$
145. $p(2p + 1)(3p + 1)$ **147.** $3(2z - 5)(5z - 2)$ **149.** $3a(2a + 3)(7a - 3)$ **151.** $y(3x - 5y)(3x - 5y)$
153. $xy(3x - 4y)(3x - 4y)$ **155.** $2y(y - 3)(4y - 1)$ **157.** $ab(a + 4)(a - 6)$ **159.** $5t(3 + 2t)(4 - t)$
161. $(y + 3)(2y + 1)$ **163.** $(2x - 1)(5x + 7)$ **165.** 7, 5, −7, −5
167. 7, −7, 5, −5   **169.** 11, −11, 7, −7

SECTION 6.4 *pages 240–243*

1. $(x + 2)(x - 2)$ **3.** $(a + 9)(a - 9)$ **5.** $(2x + 1)(2x - 1)$ **7.** $(y + 1)^2$ **9.** $(a - 1)^2$
11. Nonfactorable over the integers **13.** $(x^3 + 3)(x^3 - 3)$ **15.** $(5x + 1)(5x - 1)$ **17.** $(1 + 7x)(1 - 7x)$
19. $(x + y)^2$ **21.** $(2a + 1)^2$ **23.** $(8a - 1)^2$ **25.** Nonfactorable over the integers
27. $(x^2 + y)(x^2 - y)$ **29.** $(3x + 4y)(3x - 4y)$ **31.** $(4b + 1)^2$ **33.** $(2b + 7)^2$ **35.** $(5a + 3b)^2$
37. $(xy + 2)(xy - 2)$ **39.** $(4 + xy)(4 - xy)$ **41.** $(2y - 9z)^2$ **43.** $(3ab - 1)^2$ **45.** $2(2y + 1)(2y - 1)$
47. $3a(a + 1)^2$ **49.** $(m^2 + 16)(m + 4)(m - 4)$ **51.** $(x + 1)(9x + 4)$ **53.** $4y^2(2y + 3)^2$
55. $(y^4 + 9)(y^2 + 3)(y^2 - 3)$ **57.** $(5 - 2p)^2$ **59.** $2(x + 3)(x - 3)$ **61.** $x^2(x + 7)(x - 5)$
63. $5(b + 3)(b + 12)$ **65.** Nonfactorable over the integers **67.** $2y(x + 11)(x - 3)$ **69.** $x(x^2 - 6x - 5)$
71. $3(y^2 - 12)$ **73.** $(2a + 1)(10a + 1)$ **75.** $y^2(x + 1)(x - 8)$ **77.** $5(a + b)(2a - 3b)$
79. $2(5 + x)(5 - x)$ **81.** $ab(3a - b)(4a + b)$ **83.** $2(x - 1)(a + b)$ **85.** $3a(2a - 1)^2$ **87.** $3(81 + a^2)$
89. $2a(2a - 5)(3a - 4)$ **91.** $(x - 2)(x + 1)(x - 1)$ **93.** $a(2a + 5)^2$ **95.** $3b(3a - 1)^2$
97. $6(4 + x)(2 - x)$ **99.** $(x + 2)(x - 2)(a + b)$ **101.** $x^2(x + y)(x - y)$ **103.** $2a(3a + 2)^2$
105. $b(2 - 3a)(1 + 2a)$ **107.** $(x - 5)(2 + x)(2 - x)$ **109.** $8x(3y + 1)^2$ **111.** $y^2(5 + x)(3 - x)$
113. $y(y + 3)(y - 3)$ **115.** $2x^2y^2(x + 1)(x - 1)$ **117.** $x^5(x^2 + 1)(x + 1)(x - 1)$ **119.** $2xy(3x - 2)(4x + 5)$
121. $x^2y^2(2x - 5)^2$ **123.** $(x - 2)^2(x + 2)$ **125.** $4xy^2(1 - x)(2 - 3x)$ **127.** $2ab(2a - 3b)(9a - 2b)$
129. $x^2y^2(1 - 3x)(5 + 4x)$ **131.** $2(y^2 - 2)(a + b)$ **133.** $x^2y^2(3x - 5y)(5x + 4y)$ **135.** $10, -10$
137. $16, -16$ **139.** 4 **141.** 25 **143.** $(a - 4)(a^2 + 4a + 16)$ **145.** $(2z + 3)(4z^2 - 6z + 9)$

SECTION 6.5 *pages 246–250*

1. $-3, -2$ **3.** $7, 3$ **5.** $0, 5$ **7.** $0, 9$ **9.** $0, -\dfrac{3}{2}$ **11.** $0, \dfrac{2}{3}$ **13.** $-2, 5$ **15.** $9, -9$

17. $\dfrac{7}{4}, -\dfrac{7}{4}$ **19.** $3, 5$ **21.** $8, -9$ **23.** $-2, 5$ **25.** $-\dfrac{1}{2}, 1$ **27.** $-\dfrac{2}{3}, -4$ **29.** $0, 7$

31. $-1, -4$ **33.** $2, 3$ **35.** $\dfrac{1}{2}, -4$ **37.** $\dfrac{1}{3}, 4$ **39.** $3, 9$ **41.** $9, -2$ **43.** $-1, -2$

45. $5, -9$ **47.** $4, -7$ **49.** $-2, -3$ **51.** $-8, -5$ **53.** $1, 3$ **55.** $-12, 5$ **57.** $3, 4$

59. $-\dfrac{1}{2}, -4$ **61.** 7 **63.** $4, 5$ **65.** $5, 6$ **67.** $4, 8$ **69.** $15, 16$ **71.** base, 18 ft; height, 6 ft

73. length, 40 in.; width, 10 in. **75.** 6 in. **77.** 138.2 in.^{2} **79.** 4 in. by 7 in. **81.** 10 s

83. 12 **85.** 8 teams **87.** $-\dfrac{3}{2}, -5$ **89.** $9, -3$ **91.** $0, 9$ **93.** $-6, 3$ **95.** 48 or 3

97. -25 **99.** length, 20 in.; width, 10 in.

PRIME AND COMPOSITE NUMBERS *pages 251–252*

1. 101, 103, 107, 109, 113, 127, 131, 137, 139, 149, 151, 157, 163, 167, 173, 179, 181, 191, 193, 197, 199
3. a. 2 and 3 **b.** No. Every other pair of consecutive natural numbers consists of an even number, and no even number except 2 is prime. **5.** $3 = 2^2 - 1$

CHAPTER REVIEW *pages 253–255*

1. $7y^3(2y^6 - 7y^3 + 1)$ (Objective 6.1.1) **2.** $(a - 4)(3a + b)$ (Objective 6.1.2)
3. $(c + 2)(c + 6)$ (Objective 6.2.1) **4.** $a(a - 2)(a - 3)$ (Objective 6.2.2)
5. $(2x - 7)(3x - 4)$ (Objective 6.3.1/6.3.2) **6.** $(y + 6)(3y - 2)$ (Objective 6.3.1/6.3.2)
7. $(3a + 2)(6a - 5)$ (Objective 6.3.1/6.3.2) **8.** $(ab + 1)(ab - 1)$ (Objective 6.4.1)

9. $4(y - 2)^2$ (Objective 6.4.2)   **10.** $0, -\dfrac{1}{5}$ (Objective 6.5.1)   **11.** $3ab(4a + b)$ (Objective 6.1.1)

12. $(b - 3)(b - 10)$ (Objective 6.2.1)   **13.** $(2x + 5)(5x + 2y)$ (Objective 6.1.2)

14. $3(a + 2)(a - 7)$ (Objective 6.2.2)   **15.** $n^2(n + 1)(n - 3)$ (Objective 6.2.2)

16. Nonfactorable over the integers   (Objective 6.3.1/6.3.2)

17. $(2x - 1)(3x - 2)$ (Objective 6.3.1/6.3.2)   **18.** Nonfactorable over the integers   (Objective 6.4.1)

19. $2, \dfrac{3}{2}$ (Objective 6.5.1)   **20.** $7(x + 1)(x - 1)$ (Objective 6.4.2)

21. $x^3(3x^2 - 9x - 4)$ (Objective 6.1.1)   **22.** $(x - 3)(4x + 5)$ (Objective 6.1.2)

23. $(a + 7)(a - 2)$ (Objective 6.2.1)   **24.** $(y + 9)(y - 4)$ (Objective 6.2.1)

25. $5(x - 12)(x + 2)$ (Objective 6.2.2)   **26.** $-3, 7$ (Objective 6.5.1)

27. $(a + 2)(7a + 3)$ (Objective 6.3.1/6.3.2)   **28.** $(x + 20)(4x + 3)$ (Objective 6.3.1/6.3.2)

29. $(3y^2 + 5z)(3y^2 - 5z)$ (Objective 6.4.1)   **30.** $5(x + 2)(x - 3)$ (Objective 6.2.2)

31. $-\dfrac{3}{2}, \dfrac{2}{3}$ (Objective 6.5.1)   **32.** $2b(2b - 7)(3b - 4)$ (Objective 6.3.1/6.3.2)

33. $5x(x^2 + 2x + 7)$ (Objective 6.1.1)   **34.** $(x - 2)(x - 21)$ (Objective 6.2.1)

35. $(3a + 2)(a - 7)$ (Objective 6.1.2)   **36.** $(2x - 5)(4x - 9)$ (Objective 6.3.1/6.3.2)

37. $10x(a - 4)(a - 9)$ (Objective 6.2.2)   **38.** $(a - 12)(2a + 5)$ (Objective 6.3.1/6.3.2)

39. $(3a - 5b)(7x + 2y)$ (Objective 6.1.2)   **40.** $(a^3 + 10)(a^3 - 10)$ (Objective 6.4.1)

41. $(4a + 1)^2$ (Objective 6.4.1)   **42.** $\dfrac{1}{4}, -7$ (Objective 6.5.1)

43. $10(a + 4)(2a - 7)$ (Objective 6.3.1/6.3.2)   **44.** $6(x - 3)$ (Objective 6.1.1)

45. $x^2y(3x^2 + 2x + 6)$ (Objective 6.1.1)   **46.** $(d - 5)(d + 8)$ (Objective 6.2.1)

47. $2(2x - y)(6x - 5)$ (Objective 6.1.2)   **48.** $4x(x + 1)(x - 6)$ (Objective 6.2.2)

49. $-2, 10$ (Objective 6.5.1)   **50.** $(x - 5)(3x - 2)$ (Objective 6.3.1/6.3.2)

51. $(2x - 11)(8x - 3)$ (Objective 6.3.1/6.3.2)   **52.** $(3x - 5)^2$ (Objective 6.4.1)

53. $(2y + 3)(6y - 1)$ (Objective 6.3.1/6.3.2)   **54.** $3(x + 6)^2$ (Objective 6.4.2)

55. length, 100 yd; width, 50 yd   (Objective 6.5.2)   **56.** length, 100 yd; width, 60 yd   (Objective 6.5.2)

57. 4, 5   (Objective 6.5.2)   **58.** 20 ft   (Objective 6.5.2)   **59.** 15 ft   (Objective 6.5.2)

60. 20 ft   (Objective 6.5.2)

CHAPTER TEST   *page 256*

1. $3y^2(2x^2 + 3x + 4)$ (Objective 6.1.1)   **2.** $2x(3x^2 - 4x + 5)$ (Objective 6.1.1)

3. $(p + 2)(p + 3)$ (Objective 6.2.1)   **4.** $(x - 2)(a - b)$ (Objective 6.1.2)

5. $\dfrac{3}{2}, -7$ (Objective 6.5.1)   **6.** $(a - 3)(a - 16)$ (Objective 6.2.1)

7. $x(x + 5)(x - 3)$ (Objective 6.2.2)   **8.** $4(x + 4)(2x - 3)$ (Objective 6.3.1/6.3.2)

9. $(b + 6)(a - 3)$ (Objective 6.1.2)   **10.** $\dfrac{1}{2}, -\dfrac{1}{2}$ (Objective 6.5.1)

11. $(2x + 1)(3x + 8)$ (Objective 6.3.1/6.3.2)   **12.** $(x + 3)(x - 12)$ (Objective 6.2.1)

13. $2(b + 4)(b - 4)$ (Objective 6.4.2)   **14.** $(2a - 3b)^2$ (Objective 6.4.1)

15. $(p + 1)(x - 1)$ (Objective 6.1.2)   **16.** $5(x^2 - 9x - 3)$ (Objective 6.1.1)

17. Nonfactorable over the integers   (Objective 6.3.1/6.3.2)   **18.** $(2x + 7y)(2x - 7y)$ (Objective 6.4.1)

19. 3, 5   (Objective 6.5.1)   **20.** $(p + 6)^2$ (Objective 6.4.1)   **21.** $2(3x - 4y)^2$ (Objective 6.4.2)

22. $2y^2(y + 1)(y - 8)$ (Objective 6.2.2)   **23.** length, 15 cm; width, 6 cm   (Objective 6.5.2)

24. 12 in.   (Objective 6.5.2)   **25.** 3, 7   (Objective 6.5.2)

CUMULATIVE REVIEW   *pages 257–258*

1. −8   (Objective 1.2.2)   **2.** −8.1   (Objective 1.4.3)   **3.** 4   (Objective 1.5.2)
4. −31   (Objective 2.1.1)   **5.** The Associative Property of Addition   (Objective 2.1.2)
6. $18x^2$ (Objective 2.2.2)   **7.** $-6x + 24$ (Objective 2.2.4)   **8.** $\dfrac{2}{3}$ (Objective 3.1.3)

9. 5   (Objective 3.3.2)   **10.** $\dfrac{7}{4}$ (Objective 3.3.1)   **11.** 3   (Objective 3.3.2)

12. 35   (Objective 4.2.1)   **13.** $3y^3 - 3y^2 - 8y - 5$ (Objective 5.1.1)   **14.** $-27a^{12}b^6$ (Objective 5.2.2)

15. $x^3 - 3x^2 - 6x + 8$ (Objective 5.3.2)   **16.** $4x + 8 + \dfrac{21}{2x - 3}$ (Objective 5.4.3)

17. $\dfrac{y^6}{x^{12}}$ (Objective 5.4.1)   **18.** $(a - b)(3 - x)$ (Objective 6.1.2)

19. $(x + 5y)(x - 2y)$ (Objective 6.2.2)   **20.** $2a^2(3a + 2)(a + 3)$ (Objective 6.3.1)/6.3.2)
21. $(5a + 6b)(5a - 6b)$ (Objective 6.4.1)   **22.** $3(2x - 3y)^2$ (Objective 6.4.2)

23. $\dfrac{4}{3}, -5$ (Objective 6.5.1)   **24.** −3°C   (Objective 1.3.3)   **25.** 20 years   (Objective 4.8.3)

26. length, 15 cm; width, 6 cm   (Objective 4.7.1)   **27.** 4 ft and 6 ft   (Objective 3.4.2)
28. $6500   (Objective 4.4.1)   **29.** 40%   (Objective 4.3.2)   **30.** 10, 12, and 14   (Objective 4.8.1)

ANSWERS to Chapter 7 Exercises

SECTION 7.1   *pages 265–268*

1. $\dfrac{3}{4x}$ **3.** $\dfrac{1}{x + 3}$ **5.** −1   **7.** $\dfrac{2}{3y}$ **9.** $-\dfrac{3}{4x}$ **11.** $\dfrac{a}{b}$ **13.** $-\dfrac{2}{x}$ **15.** $\dfrac{y - 2}{y - 3}$ **17.** $\dfrac{x + 5}{x + 4}$

19. $\dfrac{x + 4}{x - 3}$ **21.** $-\dfrac{x + 2}{x + 5}$ **23.** $\dfrac{2(x + 2)}{x + 3}$ **25.** $\dfrac{2x - 1}{2x + 3}$ **27.** $\dfrac{2}{3xy}$ **29.** $\dfrac{8xy^2ab}{3}$ **31.** $\dfrac{2}{9}$ **33.** $\dfrac{y^2}{x}$

35. $\dfrac{y(x + 4)}{x(x + 1)}$ **37.** $\dfrac{x^3(x - 7)}{y^2(x - 4)}$ **39.** $-\dfrac{y}{x}$ **41.** $\dfrac{x + 3}{x + 1}$ **43.** $\dfrac{x - 5}{x + 3}$ **45.** $-\dfrac{x + 3}{x + 5}$

47. $\dfrac{12x^4}{(x + 1)(2x + 1)}$ **49.** $-\dfrac{x + 3}{x - 12}$ **51.** $\dfrac{x + 2}{x + 4}$ **53.** $\dfrac{2x - 5}{2x - 1}$ **55.** $\dfrac{3x - 4}{2x + 3}$ **57.** $\dfrac{2xy^2ab^2}{9}$

59. $\dfrac{5}{12}$ **61.** $3x$ **63.** $\dfrac{y(x + 3)}{x(x + 1)}$ **65.** $\dfrac{x + 7}{x - 7}$ **67.** $-\dfrac{4ac}{y}$ **69.** $\dfrac{x - 5}{x - 6}$ **71.** 1

73. $-\dfrac{x + 6}{x + 5}$ **75.** $\dfrac{2x + 3}{x - 6}$ **77.** $\dfrac{4x + 3}{2x - 1}$ **79.** $\dfrac{(2x + 5)(4x - 1)}{(2x - 1)(4x + 5)}$ **81.** −6, 1   **83.** −1, 1

85. 3, −2   **87.** −3   **89.** 0, 4   **91.** $-\dfrac{2}{3}, 4$ **93.** $\dfrac{xy}{2}$ **95.** $\dfrac{72x}{y}$ **97.** $-\dfrac{b}{a - 3}$ **99.** 1

101. $\dfrac{(x + 6)(x + 6)}{(x + 7)(x + 3)}$

SECTION 7.2   *pages 270–273*

1. $24x^3y^2$ **3.** $30x^4y^2$ **5.** $8x^2(x + 2)$ **7.** $6x^2y(x + 4)$ **9.** $40x^3(x - 1)^2$ **11.** $4(x - 3)^2$
13. $(2x - 1)(2x + 1)(x + 4)$ **15.** $(x - 7)^2(x + 2)$ **17.** $(x + 4)(x - 3)$ **19.** $(x + 5)(x - 2)(x + 7)$
21. $(x - 7)(x - 3)(x - 5)$ **23.** $(x + 2)(x + 5)(x - 5)$ **25.** $(2x - 1)(x - 3)(x + 1)$

27. $(2x - 5)(x - 2)(x + 3)$ **29.** $(x + 3)(x - 5)$ **31.** $(x + 6)(x - 3)$ **33.** $\dfrac{4x}{x^2}, \dfrac{3}{x^2}$ **35.** $\dfrac{4x}{12y^2}, \dfrac{3yz}{12y^2}$

37. $\dfrac{xy}{x^2(x - 3)}, \dfrac{6x - 18}{x^2(x - 3)}$ **39.** $\dfrac{9x}{x(x - 1)^2}, \dfrac{6x - 6}{x(x - 1)^2}$ **41.** $\dfrac{3x}{x(x - 3)}, \dfrac{5}{x(x - 3)}$ **43.** $\dfrac{3}{(x - 5)^2}, -\dfrac{2x - 10}{(x - 5)^2}$

45. $\dfrac{3x}{x^2(x + 2)}, \dfrac{4x + 8}{x^2(x + 2)}$ **47.** $\dfrac{x^2 - 6x + 8}{(x + 3)(x - 4)}, \dfrac{x^2 + 3x}{(x + 3)(x - 4)}$ **49.** $\dfrac{3}{(x + 2)(x - 1)}, \dfrac{x^2 - x}{(x + 2)(x - 1)}$

51. $\dfrac{5}{(2x - 5)(x - 2)}, \dfrac{x^2 - 3x + 2}{(2x - 5)(x - 2)}$ **53.** $\dfrac{x^2 - 3x}{(x + 3)(x - 3)(x - 2)}, \dfrac{2x^2 - 4x}{(x + 3)(x - 3)(x - 2)}$

55. $-\dfrac{x^2 - 3x}{(x - 3)^2(x + 3)}, \dfrac{x^2 + 2x - 3}{(x - 3)^2(x + 3)}$ **57.** $\dfrac{3x^2 + 12x}{(x - 5)(x + 4)}, \dfrac{x^2 - 5x}{(x - 5)(x + 4)}, -\dfrac{3}{(x - 5)(x + 4)}$

59. $\dfrac{300}{10^4}, \dfrac{5}{10^4}$ **61.** $\dfrac{b^2}{b}, \dfrac{5}{b}$ **63.** $\dfrac{y - 1}{y - 1}, \dfrac{y}{y - 1}$ **65.** $\dfrac{x^2 + 1}{(x - 1)^3}, \dfrac{x^2 - 1}{(x - 1)^3}, \dfrac{x^2 - 2x + 1}{(x - 1)^3}$

67. $\dfrac{2b}{8(a + b)(a - b)}, \dfrac{a^2 + ab}{8(a + b)(a - b)}$ **69.** $\dfrac{x - 2}{(x + y)(x + 2)(x - 2)}, \dfrac{x + 2}{(x + y)(x + 2)(x - 2)}$

71. The LCM of two expressions is equal to their product when they have no common factors.

SECTION 7.3   *pages 277–281*

1. $\dfrac{11}{y^2}$ **3.** $-\dfrac{7}{x + 4}$ **5.** $\dfrac{8x}{2x + 3}$ **7.** $\dfrac{5x + 7}{x - 3}$ **9.** $\dfrac{2x - 5}{x + 9}$ **11.** $\dfrac{-3x - 4}{2x + 7}$ **13.** $\dfrac{1}{x + 5}$

15. $\dfrac{1}{x - 6}$ **17.** $\dfrac{3}{2y - 1}$ **19.** $\dfrac{1}{x - 5}$ **21.** $\dfrac{7b + 5a}{ab}$ **23.** $\dfrac{11}{12a}$ **25.** $\dfrac{11}{12y}$ **27.** $\dfrac{120 + 7y}{20y^2}$

29. $\dfrac{21b + 4a}{12ab}$ **31.** $\dfrac{14x + 3}{12x}$ **33.** $\dfrac{8x - 3}{6x}$ **35.** $\dfrac{7}{36}$ **37.** $\dfrac{-3x^2 - 11x - 2}{3x^2}$ **39.** $\dfrac{6x^2 + 5x + 10}{6x^2}$

41. $\dfrac{4x^2 + 9x + 9}{24x^2}$ **43.** $\dfrac{3x - 1 - 2xy - 3y}{xy^2}$ **45.** $\dfrac{20x^2 + 28x - 12xy + 9y}{24x^2y^2}$ **47.** $\dfrac{9x^2 - 3x - 2xy - 10y}{18xy^2}$

49. $\dfrac{7x - 23}{(x - 3)(x - 4)}$ **51.** $\dfrac{-y - 33}{(y + 6)(y - 3)}$ **53.** $\dfrac{3x^2 + 20x - 8}{(x - 4)(x + 6)}$ **55.** $\dfrac{3(4x^2 + 5x - 5)}{(x + 5)(2x + 3)}$ **57.** $\dfrac{-4x + 5}{x - 6}$

59. $\dfrac{2(y + 2)}{(y + 4)(y - 4)}$ **61.** $\dfrac{3x + 4}{(x + 4)(x - 2)}$ **63.** $\dfrac{11}{(a - 3)(a - 4)}$ **65.** $\dfrac{-x - 1}{(x + 2)(x - 5)}$ **67.** $\dfrac{x + 2y}{2xy}$

69. $\dfrac{a + 3b}{3ab}$ **71.** $-\dfrac{x + 2}{4x}$ **73.** $\dfrac{x}{(2x - 3)(x + 1)}$ **75.** $-\dfrac{2}{2x + 1}$ **77.** $\dfrac{2(x^2 - x - 3)}{(x + 2)(x + 3)(x - 3)}$

79. $\dfrac{-2y^2 + 3y - 5}{(y - 1)(y + 1)(y - 3)}$ **81.** $\dfrac{3x^2 - x - 12}{3x(x - 2)}$ **83.** $-\dfrac{2x - 3}{6x}$ **85.** $\dfrac{x + 3}{x + 5}$ **87.** $\dfrac{x - 2}{x + 4}$

89. $\dfrac{x^2 - 9x + 12}{12x(x - 4)}$ **91.** $\dfrac{4(2x - 1)}{(x + 5)(x - 2)}$ **93.** $\dfrac{7 - 2n^2}{2n}$ **95.** $\dfrac{7a - 2}{a - 1}$ **97.** $\dfrac{3x - y}{x - y}$

99. $\dfrac{2y^2 + 9y + 1}{y + 5}$ **101.** $\dfrac{8y}{(y + 2)^2}$ **103.** $\dfrac{4}{x - 7}$ **105.** $\dfrac{x^2 - 9x + 30}{(x + 5)(x + 1)}$

SECTION 7.4   *pages 283–285*

1. $\dfrac{x}{x - 3}$ **3.** $\dfrac{2}{3}$ **5.** $\dfrac{y + 3}{y - 4}$ **7.** $\dfrac{2(2x + 13)}{5x + 36}$ **9.** $\dfrac{3}{4}$ **11.** $\dfrac{x - 2}{x + 2}$ **13.** $\dfrac{x + 2}{x + 3}$ **15.** $\dfrac{x - 6}{x + 5}$

17. $-\dfrac{x - 2}{x + 1}$ **19.** $x - 1$ **21.** $\dfrac{1}{2x - 1}$ **23.** $\dfrac{x - 3}{x + 5}$ **25.** $\dfrac{x - 7}{x - 8}$ **27.** $\dfrac{2y - 1}{2y + 1}$ **29.** $\dfrac{x - 2}{2x - 5}$

31. $\dfrac{x - 2}{x + 1}$ **33.** $\dfrac{x - 1}{x + 4}$ **35.** $\dfrac{-x - 1}{4x - 3}$ **37.** $\dfrac{x + 1}{2(5x - 2)}$ **39.** $\dfrac{b + 11}{4b - 21}$ **41.** $\dfrac{5}{3}$

43. $-\dfrac{1}{x - 1}$ **45.** $\dfrac{ab}{b + a}$ **47.** $\dfrac{y + 4}{2(y - 2)}$

SECTION 7.5   *pages 291–295*

1. 3   **3.** 1   **5.** 9   **7.** 1   **9.** $\frac{1}{4}$ **11.** 1   **13.** -3 **15.** $\frac{1}{2}$ **17.** 8   **19.** 5   **21.** -1

23. 5   **25.** No solution   **27.** 2 and 4   **29.** $-\frac{3}{2}$ and 4   **31.** 3   **33.** -1 **35.** 4   **37.** 15

39. 9   **41.** 36   **43.** 10   **45.** 113   **47.** -2 **49.** 15   **51.** 4   **53.** 20,000 people
55. 140 air vents   **57.** 6 c   **59.** 800 fish   **61.** 14 vials   **63.** Yes   **65.** 2 additional gallons

67. 10 additional acres   **69.** 40 ft^2 **71.** 750 mi   **73.** 160 ml   **75.** 0   **77.** $-\frac{2}{5}$

79. $-\frac{1}{4}$ and $-\frac{1}{2}$ **81.** -3 **83.** $\frac{4}{3}$ or $\frac{3}{4}$ **85.** -3 and -4 **87.** $\frac{7}{10}$ **89.** 210 foul shots

91. $75

SECTION 7.6   *pages 297–300*

1. $y = -3x + 10$ **3.** $y = 4x - 3$ **5.** $y = -\frac{3}{2}x + 3$ **7.** $y = \frac{2}{5}x - 2$ **9.** $y = -\frac{2}{7}x + 2$

11. $y = -\frac{1}{3}x + 2$ **13.** $y = \frac{1}{3}x - 3$ **15.** $y = \frac{2}{9}x - 2$ **17.** $y = 2x + 5$ **19.** $x = -6y + 10$

21. $x = \frac{1}{2}y + 3$ **23.** $x = -\frac{3}{4}y + 3$ **25.** $x = 4y + 3$ **27.** $x = -\frac{5}{3}y - 5$ **29.** $x = \frac{8}{5}y - 2$

31. $b = P - a - c$ **33.** $R = \frac{E}{I}$ **35.** $h = \frac{A}{b}$ **37.** $C = \frac{5F - 160}{9}$ **39.** $F = \frac{9C + 160}{5}$

41. $C = R - P$ **43.** $R = Pn + C$ **45.** $m = \frac{T}{f - g}$ **47.** $S = \frac{a}{1 - r}$ **49. a.** $S = \frac{F + BV}{B}$

b. $180   **c.** $75   **51. a.** $R_1 = \frac{RR_2}{R_2 - R}$ **b.** 20 ohms   **c.** 10 ohms

SECTION 7.7   *pages 305–309*

1. 6 h   **3.** 3 h   **5.** 30 h   **7.** 6 min   **9.** 15 h   **11.** 10 min   **13.** 6 h   **15.** 3 h
17. 40 h   **19.** 28 h   **21.** 5 mph   **23.** jogger, 8 mph; cyclist, 20 mph   **25.** 360 mph
27. 150 mph   **29.** 48 mph   **31.** 3 mph   **33.** 2 mph   **35.** 6 mph   **37.** 55 mph

39. 5 mph   **41.** $1\frac{1}{19}$ h   **43.** 2 h   **45.** 60 mph

INTENSITY OF ILLUMINATION   *page 311*

1. 4 lm   **3.** 320 cd   **5.** 2 m   **7.** 64 cd

CHAPTER REVIEW   *pages 313–314*

1. $\frac{by^3}{6ax^2}$ (Objective 7.1.2)   **2.** $\frac{22x - 1}{(3x - 4)(2x + 3)}$ (Objective 7.3.2)   **3.** $y = -\frac{4}{9}x + 2$ (Objective 7.6.1)

4. $\frac{2x^4}{3y^7}$ (Objective 7.1.1)   **5.** $\frac{1}{x^2}$ (Objective 7.1.3)   **6.** $\frac{x - 2}{3x - 10}$ (Objective 7.4.1)

7. $72a^3b^5$ (Objective 7.2.1)   **8.** $\frac{4x}{3x + 7}$ (Objective 7.3.1)   **9.** $\frac{3x}{16x^2}, \frac{10}{16x^2}$ (Objective 7.2.2)

10. $\dfrac{x-9}{x-3}$ (Objective 7.1.1) **11.** $\dfrac{x-4}{x+3}$ (Objective 7.1.3) **12.** $\dfrac{2y-3}{5y-7}$ (Objective 7.3.2)

13. $\dfrac{1}{x}$ (Objective 7.1.2) **14.** $\dfrac{1}{x+3}$ (Objective 7.3.1) **15.** $15x^4(x-7)^2$ (Objective 7.2.1)

16. 10 (Objective 7.5.1) **17.** No solution (Objective 7.5.1) **18.** $-\dfrac{4a}{5b}$ (Objective 7.1.1)

19. $\dfrac{1}{x+3}$ (Objective 7.3.1) **20.** $\dfrac{2x-6}{(x+3)(x-3)}, \dfrac{7x+21}{(x+3)(x-3)}$ (Objective 7.2.2)

21. 5 (Objective 7.5.1) **22.** $\dfrac{x+9}{4x}$ (Objective 7.4.1) **23.** $\dfrac{3x-1}{x-5}$ (Objective 7.3.2)

24. $\dfrac{x-3}{x+3}$ (Objective 7.1.3) **25.** $-\dfrac{x+6}{x+3}$ (Objective 7.1.1) **26.** 2 (Objective 7.5.1)

27. $x-2$ (Objective 7.4.1) **28.** $\dfrac{x+3}{x-4}$ (Objective 7.1.2) **29.** 15 (Objective 7.5.1)

30. 12 (Objective 7.5.1) **31.** $\dfrac{x}{x-7}$ (Objective 7.4.1) **32.** $x=2y+15$ (Objective 7.6.1)

33. $\dfrac{6b+9a}{ab}$ (Objective 7.3.2) **34.** $(5x-3)(2x-1)(4x-1)$ (Objective 7.2.1)

35. $c=\dfrac{100m}{i}$ (Objective 7.6.1) **36.** 40 (Objective 7.5.1) **37.** 3 (Objective 7.5.1)

38. $\dfrac{7x+22}{60x}$ (Objective 7.3.2) **39.** $\dfrac{8a+3}{4a-3}$ (Objective 7.1.2)

40. $\dfrac{3x^2-x}{(6x-1)(2x+3)(3x-1)}, \dfrac{24x^3-4x^2}{(6x-1)(2x+3)(3x-1)}$ (Objective 7.2.2) **41.** $\dfrac{2}{ab}$ (Objective 7.3.1)

42. 62 (Objective 7.5.1) **43.** 6 (Objective 7.5.1) **44.** $\dfrac{b^3y}{10ax}$ (Objective 7.1.3)

45. 6 h (Objective 7.7.1) **46.** 8 in. (Objective 7.5.2) **47.** 20 mph (Objective 7.7.2)
48. 16 additional ounces (Objective 7.5.2) **49.** 6 h (Objective 7.7.1) **50.** 45 mph (Objective 7.7.2)

CHAPTER TEST *pages 315–316*

1. $\dfrac{x+5}{x+4}$ (Objective 7.1.3) **2.** $\dfrac{2}{x+5}$ (Objective 7.3.1) **3.** $3(2x-1)(x+1)$ (Objective 7.2.1)

4. -1 (Objective 7.5.1) **5.** $\dfrac{x+1}{x^3(x-2)}$ (Objective 7.1.2) **6.** $\dfrac{x-3}{x-2}$ (Objective 7.4.1)

7. $\dfrac{3x+6}{x(x-2)(x+2)}, \dfrac{x^2}{x(x-2)(x+2)}$ (Objective 7.2.2) **8.** $y=\dfrac{3}{8}x-2$ (Objective 7.6.1)

9. 2 (Objective 7.5.1) **10.** $\dfrac{5}{(2x-1)(3x+1)}$ (Objective 7.3.2) **11.** 1 (Objective 7.1.3)

12. $\dfrac{3}{x+8}$ (Objective 7.3.1) **13.** $6x(x+2)^2$ (Objective 7.2.1) **14.** $-\dfrac{x-2}{x+5}$ (Objective 7.1.1)

15. 4 (Objective 7.5.1) **16.** $t=\dfrac{f-v}{a}$ (Objective 7.6.1) **17.** $\dfrac{2x^3}{3y^5}$ (Objective 7.1.1)

18. $\dfrac{3}{(2x-1)(x+1)}$ (Objective 7.3.2) **19.** 3 (Objective 7.5.1) **20.** $\dfrac{x^3(x+3)}{y(x+2)}$ (Objective 7.1.2)

21. $-\dfrac{3xy+3y}{x(x+1)(x-1)}, \dfrac{x^2}{x(x+1)(x-1)}$ (Objective 7.2.2) **22.** $\dfrac{x+3}{x+5}$ (Objective 7.4.1)

23. 2 lb (Objective 7.5.2) **24.** 20 mph (Objective 7.7.2) **25.** 6 min (Objective 7.7.1)

CUMULATIVE REVIEW   *pages 317–318*

1. −17   (Objective 1.1.2)   2.   −6   (Objective 1.5.1)   3. $\dfrac{31}{30}$ (Objective 1.5.2)

4. 21   (Objective 2.1.1)   5. $5x − 2y$ (Objective 2.2.1)   6. $−8x + 26$ (Objective 2.2.4)

7. −20   (Objective 3.2.1)   8.   −12   (Objective 3.3.2)   9.   10   (Objective 4.2.1)

10. $−6x^4y^5$ (Objective 5.2.1)   11. $a^{20}b^{15}$ (Objective 5.2.2)   12. $\dfrac{a^3}{b^2}$ (Objective 5.4.1)

13. $a^2 + ab − 12b^2$ (Objective 5.3.3)   14. $3b^3 − b + 2$ (Objective 5.4.2)

15. $x^2 + 2x + 4$ (Objective 5.4.3)   16. $(4x + 1)(3x − 1)$ (Objective 6.3.1/6.3.2)

17. $(y − 6)(y − 1)$ (Objective 6.2.1)   18. $a(2a − 3)(a + 5)$ (Objective 6.3.1/6.3.2)

19. $4(b + 5)(b − 5)$ (Objective 6.4.2)   20. $−3$ and $\dfrac{5}{2}$ (Objective 6.5.1)

21. $−\dfrac{x + 7}{x + 4}$ (Objective 7.1.1)   22.   1   (Objective 7.1.3)   23. $\dfrac{8}{(3x − 1)(x + 1)}$ (Objective 7.3.2)

24. 1   (Objective 7.5.1)   25. $a = \dfrac{f − v}{t}$ (Objective 7.6.1)   26. $5x − 18 = −3; x = 3$ (Objective 3.4.1)

27. $5000   (Objective 4.4.1)   28.   70%   (Objective 4.5.2)

29. base, 10 in.; height, 6 in.   (Objective 6.5.2)   30.   8 min   (Objective 7.7.1)

ANSWERS to Chapter 8 Exercises

SECTION 8.1   *pages 326–329*

1.   3.   5.   7. $A(2, 3), B(4, 0), C(−4, 1), D(−2, −2)$

9. $A(−2, 5), B(3, 4), C(0, 0), D(−3, −2)$ 11. a.   2; −4   b.   1; −3   13.   Yes   15.   No   17.   No

19.   Yes   21.   No   23.   (3, 7)   25.   (6, 3)   27.   (0, 1)   29.   (−5, 0)   31.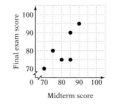

33. 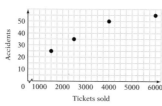   35. 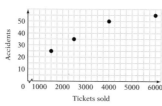   37.   Quadrant IV

39.   Quadrant III   41.   4 units   43.   2 units   45.   5 units   47. $m = −3, b = −2$

49. $m = −1, b = 0$ 51. $y = \dfrac{5}{2}x − 5$

SECTION 8.2   *pages 334–336*

1.

3.

5.

7.

9.

11.

13.

15.

17.

19.

21.

23.

25.

27.

29.

31.

33.

35.

37. a. $y = 2x + 5$ **b.** $(-4, -3)$

39. 2 units

SECTION 8.3   *pages 344–346*

1. $(3, 0), (0, -3)$ **3.** $(3, 0), (0, -6)$ **5.** $(10, 0), (0, -2)$ **7.** $(-4, 0), (0, 12)$ **9.** $(0, 0), (0, 0)$

11. $(-6, 0), (0, 3)$ **13.**   **15.**   **17.**

19. -2 **21.** $\dfrac{1}{3}$ **23.** $-\dfrac{5}{2}$ **25.** $-\dfrac{1}{2}$ **27.** -1 **29.** undefined   **31.** 0   **33.** $-\dfrac{1}{3}$ **35.** 0

37. -5 **39.** undefined   **41.** $-\dfrac{2}{3}$

43.

45.

47.

49.

51.

53.

55.

57.

59.

61. a. $y = -\frac{3}{4}x + 3$ **b.** (4, 0) and (0, 3)   **63.** It increases the slope.   **65.** It increases the point at which the graph crosses the y-axis (the y-intercept).

SECTION 8.4   *pages 349–352*

1. $y = 2x + 2$ **3.** $y = -3x - 1$ **5.** $y = \frac{1}{3}x$ **7.** $y = \frac{3}{4}x - 5$ **9.** $y = -\frac{3}{5}x$ **11.** $y = \frac{1}{4}x + \frac{5}{2}$

13. $y = -\frac{2}{3}x - 7$ **15.** $y = 2x - 3$ **17.** $y = -2x - 3$ **19.** $y = \frac{2}{3}x$ **21.** $y = \frac{1}{2}x + 2$

23. $y = -\frac{3}{4}x - 2$ **25.** $y = \frac{3}{4}x + \frac{5}{2}$ **27.** $y = -\frac{4}{3}x - 9$ **29.** $y = -x + 6$ **31.** No   **33.** 7

35. −1   **37.** $F = \frac{9}{5}C + 32$

GRAPHS OF MOTION   *page 354*

1. a. 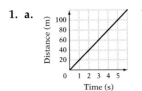   **b.** 20   **c.** $y = 20x$ **d.**   **e.** 60 m

3. a. Yes   **b.** The object has not moved any distance.

CHAPTER REVIEW   *pages 356–358*

1. (3, 0) (Objective 8.1.2)   **2.** $y = 3x - 1$ (Objective 8.4.1/8.4.2)

3. 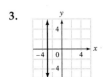   (Objective 8.2.2)   **4.**   (Objective 8.1.1)

5. (−2, 0), (0, 1) (Objective 8.3.1)   **6.** 0 (Objective 8.3.2)   **7.**   (Objective 8.2.1)

8. 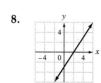   (Objective 8.2.2)   **9.** $y = -\frac{2}{3}x + \frac{4}{3}$ (Objective 8.4.1/8.4.2)

10. (2, 0), (0, −3)   (Objective 8.3.1)   **11.**   (Objective 8.3.3)

12.   (Objective 8.2.2)   **13.** $y = \frac{2}{3}x + 3$ (Objective 8.4.1/8.4.2)

14. 2   (Objective 8.3.2)   **15.**   (Objective 8.2.1)

16.   (Objective 8.3.3)   **17.**   (Objective 8.1.1)

18.   (Objective 8.2.1)   **19.** (6, 0), (0, −4)   (Objective 8.3.1)

20. $y = 2x + 2$ (Objective 8.4.1/8.4.2)   **21.**   (Objective 8.2.2)

22.   (Objective 8.2.1)   **23.** $-\frac{7}{5}$ (Objective 8.3.2)

24. $y = \frac{1}{2}x + 2$ (Objective 8.4.1/8.4.2)   **25.**   (Objective 8.3.3)

26.   (Objective 8.2.2)   **27.** (−2, −5)   (Objective 8.1.2)

28. $y = -3x + 2$ (Objective 8.4.1/8.4.2)   **29.** Undefined   (Objective 8.3.2)

30. $y = \frac{1}{2}x - 2$ (Objective 8.4.1/8.4.2)   **31.** Yes   (Objective 8.1.2)   **32.** $(2, -1)$ (Objective 8.1.2)

33. $(0, 0); (0, 0)$ (Objective 8.3.1)   **34.** $y = 3$ (Objective 8.4.1/8.4.2)

35.   (Objective 8.2.1)   **36.**   (Objective 8.3.3)

37. $y = 3x - 4$ (Objective 8.4.1/8.4.2)   **38.** 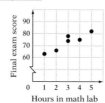   (Objective 8.2.2)

39. 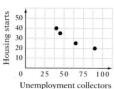   (Objective 8.1.3)   **40.**   (Objective 8.1.3)

CHAPTER TEST   *pages 359–360*

1. $y = -\frac{1}{3}x$ (Objective 8.4.1/8.4.2)   **2.** $\frac{7}{11}$ (Objective 8.3.2)   **3.** $(8, 0); (0, -12)$ (Objective 8.3.1)

4. $(9, -13)$ (Objective 8.1.2)   **5.**   (Objective 8.2.2)   **6.** (Objective 8.2.1)

7. $y = 4x - 7$ (Objective 8.4.1/8.4.2)   **8.** No   (Objective 8.1.2)

9.   (Objective 8.3.3)   **10.** (Objective 8.2.2)

11.   (Objective 8.1.1)   **12.** (Objective 8.3.3)

13. 0   (Objective 8.3.2)   **14.** $y = -\frac{2}{5}x + 7$ (Objective 8.4.1/8.4.2)

15. (Objective 8.2.1)   **16.** (Objective 8.2.1)

17. 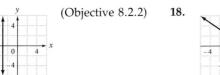   (Objective 8.2.2)   **18.** 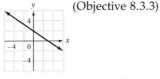   (Objective 8.3.3)

19. $y = \frac{3}{5}x - 3$ (Objective 8.4.1/8.4.2)   **20.**   (Objective 8.1.3)

Damage (in thousands) / Distance (in miles)

CUMULATIVE REVIEW   *pages 360–362*

1. -12 (Objective 1.5.2)   **2.** $-\frac{5}{8}$ (Objective 2.1.1)   **3.** $-17x + 28$ (Objective 2.2.4)

4. $\frac{3}{2}$ (Objective 3.2.1)   **5.** 1 (Objective 3.3.2)   **6.** $\frac{1}{15}$ (Objective 4.1.1)

7. $\frac{1}{125}$ (Objective 5.4.1)   **8.** $-32x^8y^7$ (Objective 5.2.2)   **9.** $-3x^9$ (Objective 5.4.1)

10. $x + 3$ (Objective 5.4.3)   **11.** $5xy(x - 4y)$ (Objective 6.1.1)   **12.** $5(x + 2)(x + 1)$ (Objective 6.2.2)

13. $(a + 2)(x + y)$ (Objective 6.1.2)   **14.** 4 and -2 (Objective 6.5.1)   **15.** $\frac{x^3(x + 3)}{y(x + 2)}$ (Objective 7.1.2)

16. $\frac{3}{x + 8}$ (Objective 7.3.1)   **17.** 2 (Objective 7.5.1)   **18.** $y = \frac{4}{5}x - 3$ (Objective 7.6.1)

19. $(-2, -7)$ (Objective 8.1.2)   **20.** 0 (Objective 8.3.2)   **21.** $(4, 0); (0, 10)$ (Objective 8.3.1)

22. $y = -x + 5$ (Objective 8.4.1/8.4.2)   **23.**   (Objective 8.2.1)

24.   (Objective 8.2.2)   **25.** 7 and 17 (Objective 3.4.1)

26. 22 ft   (Objective 4.7.1)   **27.** $\$62.30$ (Objective 4.3.2)

28. gold coin, 135 years; silver coin, 75 years   (Objective 4.8.3)

29. $\$110,000$ (Objective 7.5.2)   **30.** $3\frac{3}{4}$ h   (Objective 7.7.1)

ANSWERS to Chapter 9 Exercises

SECTION 9.1   *pages 368–369*

1. Yes   **3.** Yes   **5.** No   **7.** No   **9.** No   **11.** Yes   **13.** Yes

15.

The solution is (4, 1).

17.

The solution is (4, 1).

19.

The solution is (4, 3).

21.

The solution is (3, −2).

23.

The solution is (2, −2).

25.

The system of equations is inconsistent and has no solution.

27.

The system of equations is dependent. The solutions are the ordered pairs that satisfy the equation $y = 2x - 2$.

29.

The solution is (1, −4).

31.

The solution is (0, 0).

33.

The system of equations is inconsistent and has no solution.

35.
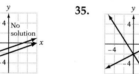
The solution is (0, −2).

37.

The solution is (1, −1).

39. $y = 2$
$y = x + 4$

41. $y = x$
$y = -x + 2$

SECTION 9.2   *pages 372–374*

1. (2, 1)   **3.** (4, 1)   **5.** (−1, 1)   **7.** (3, 1)   **9.** (1, 1)   **11.** (−1, 1)
13. The system of equations is inconsistent and has no solution.

15. The system of equations is inconsistent and has no solution.   **17.** $\left(-\dfrac{3}{4}, -\dfrac{3}{4}\right)$ **19.** $\left(\dfrac{9}{5}, \dfrac{6}{5}\right)$

21. (−7, −23)   **23.** (1, 1)   **25.** (2, 0)   **27.** (2, 1)   **29.** $\left(\dfrac{9}{19}, -\dfrac{13}{19}\right)$ **31.** (5, 7)   **33.** (1, 7)

35. $\left(\dfrac{17}{5}, -\dfrac{7}{5}\right)$ **37.** $\left(-\dfrac{6}{11}, \dfrac{31}{11}\right)$ **39.** (2, 3)   **41.** (0, 0)   **43.** The system of equations is dependent.
The solutions are the ordered pairs that satisfy the equation $3x + y = 4$.   **45.** $\left(\dfrac{20}{17}, -\dfrac{15}{17}\right)$

47. (5, 2)   **49.** (−17, −8)   **51.** $\left(-\dfrac{5}{7}, \dfrac{13}{7}\right)$ **53.** (3, −2)   **55.** $5x - 8y = 7$

57. $129x + 340y = 850$ **59.** $492x - 340y = 702$ **61.** (3, 2)   **63.** (−1, −1)   **65.** (1, 5)
67. (−5, 2)   **69.** (2, 100)   **71.** 1

SECTION 9.3   *pages 378–379*

1. (5, −1)   **3.** (1, 3)   **5.** (1, 1)   **7.** (3, −2)   **9.** The system of equations is dependent. The solutions are the ordered pairs that satisfy the equation $2x - y = 1$.   **11.** (3, 1)   **13.** The system of equations is dependent. The solutions are the ordered pairs that satisfy the equation $2x - 3y = 1$.
15. $\left(-\dfrac{13}{17}, -\dfrac{24}{17}\right)$ **17.** (2, 0)   **19.** (0, 0)   **21.** (5, −2)   **23.** $\left(\dfrac{32}{19}, -\dfrac{9}{19}\right)$ **25.** (3, 4)
27. (1, −1)   **29.** The system of equations is dependent. The solutions are the ordered pairs that satisfy the equation $5x + 15y = 20$.   **31.** (3, 1)   **33.** (−1, 2)   **35.** (1, 1)   **37.** $\left(\dfrac{1}{2}, -\dfrac{1}{2}\right)$ **39.** $\left(\dfrac{2}{3}, \dfrac{1}{9}\right)$
41. $\left(\dfrac{7}{25}, -\dfrac{1}{25}\right)$ **43.** (5, 2)   **45.** (3, −2)   **47.** (4, −1)   **49.** (3, 1)   **51.** $A = 3, B = -3$
53. $A = 5, B = 5$ **55.** $A = 1$

SECTION 9.4   *pages 384–387*

1. rate of the plane in calm air, 400 mph; rate of the wind, 50 mph   **3.** rate of the boat in calm water, 7 mph; rate of the current, 3 mph   **5.** rate of the plane in calm air, 125 mph; rate of the wind, 25 mph   **7.** rate of the plane in calm air, 100 mph; rate of the wind, 20 mph   **9.** rate of the plane in calm air, 110 mph; rate of the wind, 30 mph   **11.** rate of the plane in calm air, 180 km/h; rate of the wind, 20 km/h   **13.** $245   **15.** dividend per share of the oil company, $.25; dividend per share of the movie company, $.45   **17.** number of two-point baskets, 21; number of three-point baskets, 15   **19.** 8 nickels; 10 quarters   **21.** 8 dimes; 10 quarters   **23.** adult, 34 years; child, 10 years   **25.** 55° and 125°   **27.** gold coin, 20 years; silver coin, 25 years   **29.** $6000 at 9%; $4000 at 8%   **31.** It is impossible to earn $600 in interest.

BREAK-EVEN ANALYSIS   *page 389*

1. a. $R = 125N; T = 25N + 20,000$ **b.**   **c.** 200 watches

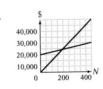

3. a. 500 calculators   **b.** 400 calculators

CHAPTER REVIEW   *pages 390–393*

1. (−1, 1) (Objective 9.2.1)   **2.** (2, −3) (Objective 9.1.1)   **3.** (−3, 1) (Objective 9.3.1)
4. (1, 6) (Objective 9.2.1)   **5.** (1, 1) (Objective 9.1.1)   **6.** (1, −5) (Objective 9.3.1)
7. (3, 2) (Objective 9.2.1)   **8.** The system of equations is dependent. The solutions are the ordered pairs that satisfy the equation $8x - y = 25$. (Objective 9.3.1)   **9.** (3, −1) (Objective 9.3.1)
10. (1, −3) (Objective 9.1.1)   **11.** (−2, −7) (Objective 9.3.1)   **12.** (4, 0) (Objective 9.2.1)
13. Yes (Objective 9.1.1)   **14.** $\left(-\dfrac{5}{6}, \dfrac{1}{2}\right)$ (Objective 9.3.1)   **15.** (−1, −2) (Objective 9.3.1)
16. (3, −3) (Objective 9.1.1)   **17.** The system of equations is inconsistent and has no solution. (Objective 9.3.1)   **18.** $\left(-\dfrac{1}{3}, \dfrac{1}{6}\right)$ (Objective 9.2.1)   **19.** The system of equations is dependent. The solutions are the ordered pairs that satisfy the equation $y = 2x - 4$. (Objective 9.1.1)

20. The system of equations is dependent. The solutions are the ordered pairs that satisfy the equation $3x + y = -2$. (Objective 9.3.1) **21.** $(-2, -13)$ (Objective 9.3.1) **22.** The system of equations is dependent. The solutions are the ordered pairs that satisfy the equation $4x + 3y = 12$. (Objective 9.2.1) **23.** $(4, 2)$ (Objective 9.1.1) **24.** The system of equations is inconsistent and has no solution. (Objective 9.3.1) **25.** $\left(\frac{1}{2}, -1\right)$ (Objective 9.2.1) **26.** No (Objective 9.1.1)

27. $\left(\frac{2}{3}, -\frac{1}{6}\right)$ (Objective 9.3.1) **28.** The system of equations is inconsistent and has no solution. (Objective 9.2.1) **29.** $\left(\frac{1}{3}, 6\right)$ (Objective 9.2.1) **30.** The system of equations is inconsistent and has no solution. (Objective 9.1.1) **31.** $(0, -1)$ (Objective 9.3.1) **32.** $(-1, -3)$ (Objective 9.2.1) **33.** $(0, -2)$ (Objective 9.2.1) **34.** $(2, 0)$ (Objective 9.3.1) **35.** $(-4, 2)$ (Objective 9.3.1) **36.** $(1, 6)$ (Objective 9.2.1) **37.** rate of the plane in calm air, 180 mph; rate of the wind, 20 mph (Objective 9.4.1) **38.** 60 adult's tickets and 140 children's tickets (Objective 9.4.2) **39.** rate of the canoeist in calm water, 8 mph; rate of the current, 2 mph (Objective 9.4.1) **40.** 130 mailings (Objective 9.4.2) **41.** rate of the boat in calm water, 14 km/h; rate of the current, 2 km/h (Objective 9.4.1) **42.** four $15 compact discs and six $10 compact discs (Objective 9.4.2) **43.** rate of the plane in calm air, 125 km/h; rate of the wind, 15 km/h (Objective 9.4.1) **44.** rate of the boat in calm water, 3 mph; rate of the current, 1 mph (Objective 9.4.1) **45.** 350 bushels of lentils and 200 bushels of corn (Objective 9.4.2) **46.** rate of the plane in calm air, 105 mph; rate of the wind, 15 mph (Objective 9.4.1) **47.** 4 dimes (Objective 9.4.2) **48.** rate of the plane in calm air, 325 mph; rate of the wind, 25 mph (Objective 9.4.1) **49.** 1300 shares at $6 per share; 200 shares at $25 per share (Objective 9.4.2) **50.** rate of the sculling team in calm water, 9 mph; rate of the current, 3 mph (Objective 9.4.1)

CHAPTER TEST *pages 393–394*

1. $(3, 1)$ (Objective 9.2.1) **2.** $(2, 1)$ (Objective 9.3.1) **3.** Yes (Objective 9.1.1)

4. $(1, -1)$ (Objective 9.2.1) **5.** $\left(\frac{1}{2}, -1\right)$ (Objective 9.3.1) **6.** $(-2, 6)$ (Objective 9.1.1)

7. The system of equations is inconsistent and has no solution. (Objective 9.2.1)

8. $(2, -1)$ (Objective 9.2.1) **9.** $(2, -1)$ (Objective 9.3.1) **10.** $\left(\frac{22}{7}, -\frac{5}{7}\right)$ (Objective 9.2.1)

11. $(1, -2)$ (Objective 9.3.1) **12.** No (Objective 9.1.1) **13.** $(2, 1)$ (Objective 9.2.1) **14.** $(2, -2)$ (Objective 9.3.1) **15.** $(2, 0)$ (Objective 9.1.1) **16.** $(4, 1)$ (Objective 9.2.1) **17.** $(3, -2)$ (Objective 9.3.1) **18.** The system of equations is dependent. The solutions are the ordered pairs that satisfy the equation $3x + 6y = 2$. (Objective 9.1.1) **19.** $(1, 1)$ (Objective 9.2.1) **20.** $(4, -3)$ (Objective 9.3.1) **21.** $(-6, 1)$ (Objective 9.2.1) **22.** $(1, -4)$ (Objective 9.3.1) **23.** rate of plane in calm air, 100 mph; rate of wind, 20 mph (Objective 9.4.1) **24.** rate of boat in calm water, 14 mph; rate of current, 2 mph (Objective 9.4.1) **25.** 40 dimes; 30 nickels (Objective 9.4.2)

CUMULATIVE REVIEW *pages 394–396*

1. $\frac{3}{2}$ (Objective 2.1.1) **2.** $-\frac{3}{2}$ (Objective 3.1.3) **3.** $-\frac{7}{2}$ (Objective 3.3.2)

4. $-\frac{2}{9}$ (Objective 3.3.2) **5.** $-6a^3 + 13a^2 - 9a + 2$ (Objective 5.3.2) **6.** $-2x^5y^2$ (Objective 5.4.1)

7. $2b - 1 + \dfrac{1}{2b - 3}$ (Objective 5.4.3)   **8.** $-\dfrac{4y^9}{x^3}$ (Objective 5.4.1)   **9.** $4y^2(xy + 4)(xy - 4)$
(Objective 6.4.2)   **10.** 4 and -1 (Objective 6.5.1)   **11.** $x - 2$ (Objective 7.1.3)
12. $\dfrac{x^2 + 2}{(x + 2)(x - 1)}$ (Objective 7.3.2)   **13.** $\dfrac{x - 3}{x + 1}$ (Objective 7.4.1)   **14.** $-\dfrac{1}{5}$ (Objective 7.5.1)
15. $\dot{r} = \dfrac{A - P}{Pt}$ (Objective 7.6.1)   **16.** (4, 0), (0, -2) (Objective 8.3.1)

17.   (Objective 8.2.2)   **18.**   (Objective 8.2.2)

19. $-\dfrac{7}{5}$ (Objective 8.3.2)   **20.** $y = -\dfrac{3}{2}x$ (Objective 8.4.1/8.4.2)   **21.** Yes (Objective 9.1.1)
22. $(-2, 1)$ (Objective 9.2.1)   **23.** $(0, 2)$ (Objective 9.1.1)   **24.** $(2, 1)$ (Objective 9.3.1)
25. $3750 at 9.6\%; $5000 at 7.2\% (Objective 4.4.1)   **26.** freight train, 48 mph; passenger train, 56 mph
(Objective 4.6.1)   **27.** 8 in. (Objective 6.5.2)   **28.** 30 mph (Objective 7.7.2)
29. 10 mph (Objective 9.4.1)   **30.** 40 dimes (Objective 9.4.2)

ANSWERS to Chapter 10 Exercises

SECTION 10.1   *pages 403–404*

1. $A = \{16, 17, 18, 19, 20, 21\}$ **3.** $A = \{9, 11, 13, 15, 17\}$ **5.** $A = \{b, c\}$
7. $A = \{0, 1, 4, 9, 16, 25, 36, 49\}$ **9.** $A \cup B = \{3, 4, 5, 6\}$ **11.** $A \cup B = \{-10, -9, -8, 8, 9, 10\}$
13. $A \cup B = \{a, b, c, d, e, f\}$ **15.** $A \cup B = \{1, 3, 7, 9, 11, 13\}$ **17.** $A \cap B = \{4, 5\}$ **19.** $A \cap B = \varnothing$
21. $A \cap B = \{c, d, e\}$ **23.** $A \cap B = \{7, 11\}$ **25.** $\{x | x > -5, x \text{ is a negative integer}\}$
27. $\{x | x > 30, x \text{ is an integer}\}$ **29.** $\{x | x > 5, x \text{ is an even integer}\}$ **31.** $\{x | x > 8, x \text{ is a real number}\}$
33. $\{x | x > -5, x \text{ is a real number}\}$ **35.**   **37.**

39.   **41.**   **43.**

45. $A \cap B = \{11, 12, 13, 14\}$ **47.** $A \cup B = \{x | x > 8, x \text{ is a positive integer}\}$

49.   **51.**

53.   **55.**

SECTION 10.2   *pages 410–414*

1. $x < 2$ **3.** $x > 3$ **5.** $n \geq 3$

7. $x \leq -4$ **9.** $x \geq -1$ **11.** $y \geq -9$ **13.** $x < 12$

15. $x \geq 5$ **17.** $x < -11$ **19.** $x \leq 10$ **21.** $x \geq -6$ **23.** $x > 2$ **25.** $d < -\dfrac{1}{6}$ **27.** $x \geq -\dfrac{31}{24}$

29. $x < \dfrac{5}{8}$ **31.** $x < \dfrac{5}{4}$ **33.** $x > \dfrac{5}{24}$ **35.** $x < -3.8$ **37.** $x \le -1.2$ **39.** $x < 5.6$

41. $x > -1.48$ **43.** $x < 4$ 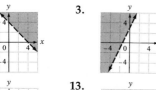   **45.** $y \ge 3$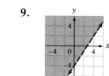

47. $x \le 1$ **49.** $x < -1$ 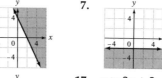   **51.** $b < -4$

53. $y \le -4$ **55.** $x > \dfrac{2}{7}$ **57.** $x \ge \dfrac{20}{3}$ **59.** $x \le -\dfrac{2}{7}$ **61.** $x > \dfrac{3}{7}$ **63.** $x \le -\dfrac{5}{2}$ **65.** $x < -18$

67. $x > -16$ **69.** $y \ge 6$ **71.** $x \ge -6$ **73.** $b \le 33$ **75.** $n < \dfrac{3}{4}$ **77.** $x \le -\dfrac{6}{7}$ **79.** $x > \dfrac{10}{7}$

81. $y \le \dfrac{5}{6}$ **83.** $x \le \dfrac{27}{28}$ **85.** $y > -\dfrac{3}{2}$ **87.** $n \ge -\dfrac{4}{5}$ **89.** $y \le -\dfrac{2}{7}$ **91.** $n \le -\dfrac{32}{15}$ **93.** $x \ge \dfrac{12}{35}$

95. $y < -\dfrac{7}{29}$ **97.** $x > -0.5$ **99.** $y \ge -0.8$ **101.** $x \le 4.2$ **103.** $d < -2.1$ **105.** $m > -8$

107. $x < -2.1$ **109.** $x \ge -0.4$ **111.** $n > -0.7$ **113.** $b > -2.9$ **115.** $x < 14$
117. 2.87 in. $< d <$ 2.95 in.   **119.** \$3150 or more   **121.** 55 or more units   **123.** 440 or more pounds
125. 78 or better   **127.** \$5715 or more   **129.** {1, 2, 3, 4, 5, 6}   **131.** {1, 2}   **133.** {1, 2, 3, 4}
135. $\{c \mid c < 0, c \text{ is a real number}\}$ **137.** $\{c \mid c < 0, c \text{ is a real number}\}$

SECTION 10.3   *pages 417–420*

1. $x < 4$ **3.** $x < -4$ **5.** $x \ge 1$ **7.** $x \le 5$ **9.** $x < 0$ **11.** $x < 20$ **13.** $x > 500$ **15.** $x > 2$

17. $x \le -5$ **19.** $y \le \dfrac{5}{2}$ **21.** $x < \dfrac{25}{11}$ **23.** $x > 11$ **25.** $n \le \dfrac{11}{18}$ **27.** $x \ge 6$ **29.** $x \le \dfrac{2}{5}$

31. $t < 1$ **33.** $n > \dfrac{7}{10}$ **35.** 11   **37.** 201 or more hits   **39.** 8 or more terminals

41. 75 or less pounds   **43.** less than 60,000 gal   **45.** between 8 and 12 aircraft
47. 1, 3, and 5 or 3, 5, and 7   **49.** $3 < 5$ and $-3 > -5$ **51.** $-8 < -4$ and $8 > 4$
53. $19 < 36$ and $-19 > -36$ **55.** $>$ **57.** {1, 2}   **59.** {1, 2, 3, 4, 5}   **61.** {1, 2}
63. {−4, −3, −2}   **65.** {−1, 0, 1}   **67.** 467 calendars

SECTION 10.4   *pages 422–423*

1. 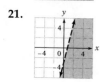   **3.**   **5.**   **7.**   **9.**

11. 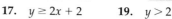   **13.**   **15.**   **17.** $y \ge 2x + 2$ **19.** $y > 2$

21.   **23.**   **25.**

MEASUREMENTS AS APPROXIMATIONS *page 425*

1. $23.385 \text{ m} \le P < 23.415 \text{ m}$ **3.** $49.9875 \text{ cm}^2 \le A < 51.4175 \text{ cm}^2$

CHAPTER REVIEW *pages 426–429*

1. $\{x \mid x > -8, x \text{ is an odd integer}\}$ (Objective 10.1.2)   **2.** $A = \{1, 3, 5, 7\}$ (Objective 10.1.1)
3. $x < -18$ (Objective 10.2.1)   **4.** $A \cup B = \{2, 4, 6, 8, 10\}$ (Objective 10.1.1)

5.   (Objective 10.1.3)   **6.** $x > 2$ (Objective 10.2.1)

7.   (Objective 10.1.3)   **8.** $x < -4$ (Objective 10.2.1)

9. $x \ge 4$ (Objective 10.3.1)   **10.** $x < \dfrac{9}{5}$ (Objective 10.2.2)

11. $\{x \mid x < 5, x \text{ is an even integer}\}$ (Objective 10.1.2)   **12.** $x \ge 4$ (Objective 10.3.1)

13.   (Objective 10.4.1)   **14.**   (Objective 10.4.1)

15. $A \cap B = \varnothing$ (Objective 10.1.1)   **16.** $\{x \mid x < 14, x \text{ is a real number}\}$ (Objective 10.1.2)

17. $x \le -2$ (Objective 10.2.2)   **18.**   (Objective 10.1.3)

19. $x > -18$ (Objective 10.3.1)   **20.** $x \le -\dfrac{3}{8}$ (Objective 10.2.2)

21.   (Objective 10.1.3)   **22.** $x > -4$

23. $x \ge 12$ (Objective 10.2.1)   **24.** $A = \{-4, -3, -2, -1\}$ (Objective 10.1.1)
25. $x < 12$ (Objective 10.3.1)   **26.** $C \cap D = \{1, 5, 9\}$ (Objective 10.1.1)

27.   (Objective 10.1.3)   **28.** $x < -\dfrac{8}{9}$ (Objective 10.2.2)

29.   (Objective 10.1.3)   **30.** $x \le 1$ (Objective 10.2.1)

31. $x > 18$ (Objective 10.2.1)   **32.** $x > 5$ (Objective 10.3.1)
33. $\{x \mid x < 20, x \text{ is a positive integer}\}$ (Objective 10.1.2)   **34.** $x \ge -3$ (Objective 10.2.2)

35.   (Objective 10.4.1)   **36.**   (Objective 10.4.1)

37. $x < \dfrac{1}{2}$ (Objective 10.3.1)   **38.**   (Objective 10.2.2)

39. $x \ge -4$ (Objective 10.3.1)   **40.** $E \cup F = \{1, 2, 3, 4, 5\}$ (Objective 10.1.1)
41. $A = \{6, 8, 10, 12\}$ (Objective 10.1.1)   **42.** $\{x \mid x > -5, x \text{ is a negative integer}\}$ (Objective 10.1.2)

43. ──────┼┼──┼────→   (Objective 10.1.3)   **44.**   (Objective 10.4.1)
       −1 0   3

45. −6   (Objective 10.3.2)   **46.**   5 or more residents   (Objective 10.3.2)
47. 72   (Objective 10.2.3)   **48.** $27.24 or more   (Objective 10.2.3)

49. $-\dfrac{64}{3}$ (Objective 10.2.3)   **50.**   24 ft   (Objective 10.3.2)

CHAPTER TEST   *pages 429–430*

1. $\{x|x > -23,\ x \text{ is a real number}\}$ (Objective 10.1.2)   **2.** $x > \dfrac{1}{2}$ (Objective 10.2.1)

3. ←───⊕──┼────→   (Objective 10.1.3)   **4.** $x \ge 3$ ←────┼───●──→   (Objective 10.2.2)
        −2 0   0   3

5. $x \le -\dfrac{22}{7}$ (Objective 10.3.1)   **6.** $x \le -\dfrac{9}{2}$ (Objective 10.3.1)   **7.** $A = \{4,\ 6,\ 8\}$ (Objective 10.1.1)

8. $x \ge -16$ (Objective 10.2.2)   **9.**   (Objective 10.4.1)

10.   (Objective 10.4.1)   **11.** $A \cap B = \{12\}$ (Objective 10.1.1)

12. $x \le -3$ (Objective 10.3.1)   **13.** $\{x|x < 50,\ x \text{ is a positive integer}\}$ (Objective 10.1.2)
14. $A \cup B = \{-2,\ -1,\ 0,\ 1,\ 2\}$ (Objective 10.1.1)   **15.** $A = \{1,\ 3,\ 5,\ 7,\ 9\}$ (Objective 10.1.1)
16. $x > 2$ (Objective 10.3.1)   **17.** ←───⊕──┼────→   (Objective 10.2.2)
                                                 −2 0
18. ←───⊕─────⊕→   (Objective 10.1.3)   **19.** ←───⊕──┼●──→   (Objective 10.1.3)
          0   5   −2   0 1
20. ←────●──┼────→   (Objective 10.2.1)   **21.**   (Objective 10.4.1)
           −1 0

22. −9   (Objective 10.2.3)   **23.**   11 ft   (Objective 10.3.2)
24. −26   (Objective 10.2.3)   **25.**   359 mi   (Objective 10.3.2)

CUMULATIVE REVIEW   *pages 430–432*

1. 10   (Objective 1.5.2)   **2.** −8   (Objective 2.1.1)   **3.** $40a - 28$ (Objective 2.2.4)

4. $\dfrac{1}{8}$ (Objective 3.2.1)   **5.** 4   (Objective 3.3.2)   **6.** $-12a^7b^4$ (Objective 5.2.2)

7. $6y^4 + 12y^3 - 15y^2$ (Objective 5.3.1)   **8.** $-b^4$ (Objective 5.4.1)

9. $4x - 2 - \dfrac{4}{4x - 1}$ (Objective 5.4.3)   **10.** $(3x + 8y)^2$ (Objective 6.4.1)

11. $(4x - 1)(x - 5)$ (Objective 6.3.1/6.3.2)   **12.** $3a^2(3x + 1)(3x - 1)$ (Objective 6.4.2)

13. $\dfrac{1}{x + 2}$ (Objective 7.1.3)   **14.** $\dfrac{18a}{(2a - 3)(a + 3)}$ (Objective 7.3.2)   **15.** $-\dfrac{5}{9}$ (Objective 7.5.1)

16. $C = S + Rt$ (Objective 7.6.1)   **17.** 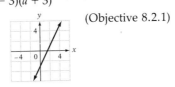 (Objective 8.2.1)

18.   (Objective 8.2.2)   **19.** $-\dfrac{9}{5}$ (Objective 8.3.2)

20. $y = -\dfrac{3}{2}x - \dfrac{3}{2}$ (Objective 8.4.1/8.4.2)   **21.** $(4, 1)$ (Objective 9.2.1)

22. $(1, -4)$ (Objective 9.3.1)   **23.** (Objective 10.1.3)

24. (Objective 10.2.2)   **25.** $x < \dfrac{1}{5}$ (Objective 10.3.1)

26. (Objective 10.4.1)   **27.** $79.20 (Objective 4.3.2)

28. seven 13¢ stamps (Objective 4.8.2)   **29.** 5000 fish (Objective 7.5.2)

30. 24 ft (Objective 10.3.2)

ANSWERS to Chapter 11 Exercises

SECTION 11.1 *pages 440–442*

1. 4   **3.** 7   **5.** $4\sqrt{2}$ **7.** $2\sqrt{2}$ **9.** $18\sqrt{2}$ **11.** $10\sqrt{10}$ **13.** $\sqrt{15}$ **15.** $\sqrt{29}$

17. $-54\sqrt{2}$ **19.** $3\sqrt{5}$ **21.** 0   **23.** $48\sqrt{2}$ **25.** $\sqrt{105}$ **27.** 30   **29.** $30\sqrt{5}$ **31.** $5\sqrt{10}$

33. $4\sqrt{6}$ **35.** 18   **37.** 15.492   **39.** 16.971   **41.** 16   **43.** 16.583   **45.** 15.652

47. 18.762   **49.** x^3 **51.** $y^7\sqrt{y}$ **53.** a^{10} **55.** x^2y^2 **57.** $2x^2$ **59.** $2x\sqrt{6}$ **61.** $xy^3\sqrt{xy}$

63. $ab^5\sqrt{ab}$ **65.** $2x^2\sqrt{15x}$ **67.** $7a^2b^4$ **69.** $3x^2y^3\sqrt{2xy}$ **71.** $2x^5y^3\sqrt{10xy}$ **73.** $4a^4b^5\sqrt{5a}$

75. $8ab\sqrt{b}$ **77.** x^3y **79.** $8a^2b^3\sqrt{5b}$ **81.** $6x^2y^3\sqrt{3y}$ **83.** $4x^3y\sqrt{2y}$ **85.** $5a + 20$

87. $2x^2 + 8x + 8$ **89.** $x + 2$ **91.** $y + 1$ **93.** $0.4a^2$ **95.** $\sqrt{x}$ **97.** $-6x^2y^2$

99. $xy\sqrt{y} + x$ **101.** $\{3, 4, 5, 6, 7, 8, 9\}$ **103.** 8.7 cm   **105.** $x \geq 0$ **107.** $x \geq -5$

109. $x \leq \dfrac{5}{2}$ **111.** All real numbers

SECTION 11.2 *pages 444–445*

1. $3\sqrt{2}$ **3.** $-\sqrt{7}$ **5.** $-11\sqrt{11}$ **7.** $10\sqrt{x}$ **9.** $-2\sqrt{y}$ **11.** $-11\sqrt{3b}$ **13.** $2x\sqrt{2}$
15. $-3a\sqrt{3a}$ **17.** $-5\sqrt{xy}$ **19.** $8\sqrt{5}$ **21.** $8\sqrt{2}$ **23.** $15\sqrt{2}-10\sqrt{3}$ **25.** $\sqrt{x}$
27. $-12x\sqrt{3}$ **29.** $2xy\sqrt{x}-3xy\sqrt{y}$ **31.** $-9x\sqrt{3x}$ **33.** $-13y^2\sqrt{2y}$ **35.** $4a^2b^2\sqrt{ab}$ **37.** $7\sqrt{2}$
39. $6\sqrt{x}$ **41.** $-3\sqrt{y}$ **43.** $-45\sqrt{2}$ **45.** $13\sqrt{3}-12\sqrt{5}$ **47.** $32\sqrt{3}-3\sqrt{11}$ **49.** $6\sqrt{x}$
51. $-34\sqrt{3x}$ **53.** $10a\sqrt{3b}+10a\sqrt{5b}$ **55.** $-2xy\sqrt{3}$ **57.** $-7b\sqrt{ab}+4a^2\sqrt{b}$
59. $3ab\sqrt{2a}-ab+4ab\sqrt{3b}$ **61.** $8\sqrt{x}+2$ **63.** $2x\sqrt{2y}$ **65.** $2ab\sqrt{6ab}$ **67.** $(6\sqrt{3}+2\sqrt{15})$ cm
69. 22.4 cm **71.** No

SECTION 11.3 *pages 451–453*

1. 5 **3.** 6 **5.** x **7.** x^3y^2 **9.** $3ab^6\sqrt{2a}$ **11.** $12a^4b\sqrt{b}$ **13.** $2-\sqrt{6}$ **15.** $x-\sqrt{xy}$
17. $5\sqrt{2}-\sqrt{5x}$ **19.** $4-2\sqrt{10}$ **21.** $x-6\sqrt{x}+9$ **23.** $3a-3\sqrt{ab}$ **25.** $10abc$
27. $15x-22y\sqrt{x}+8y^2$ **29.** $x-y$ **31.** $10x+13\sqrt{xy}+4y$ **33.** 4 **35.** 7 **37.** 3 **39.** $x\sqrt{5}$
41. $\dfrac{a^2}{7}$ **43.** $\dfrac{\sqrt{3}}{3}$ **45.** $\sqrt{3}$ **47.** $\dfrac{\sqrt{3x}}{x}$ **49.** $\dfrac{\sqrt{2x}}{x}$ **51.** $\dfrac{3\sqrt{x}}{x}$ **53.** $\dfrac{2\sqrt{y}}{xy}$ **55.** $\dfrac{4\sqrt{b}}{7b}$
57. $\dfrac{5\sqrt{6}}{27}$ **59.** $\dfrac{\sqrt{y}}{3xy}$ **61.** $\dfrac{2\sqrt{3x}}{3y}$ **63.** $-\dfrac{\sqrt{2}+3}{7}$ **65.** $\dfrac{15-3\sqrt{5}}{20}$ **67.** $\dfrac{x\sqrt{y}+y\sqrt{x}}{x-y}$
69. $\dfrac{\sqrt{3xy}}{3y}$ **71.** $-2-\sqrt{6}$ **73.** $-\dfrac{\sqrt{10}+5}{3}$ **75.** $\dfrac{x-3\sqrt{x}}{x-9}$ **77.** $-\dfrac{1}{2}$ **79.** 7 **81.** -5
83. $\dfrac{7}{4}$ **85.** $-\dfrac{42-26\sqrt{3}}{11}$ **87.** $\dfrac{14-9\sqrt{2}}{17}$ **89.** $\dfrac{a-5\sqrt{a}+4}{2a-2}$ **91.** $\dfrac{y+5\sqrt{y}+6}{y-9}$
93. 0.6 **95.** $\dfrac{5}{8}$ **97.** $-\dfrac{4}{9}$ **99.** $\dfrac{4}{3}$ **101.** $\dfrac{3}{2}$ **103.** $-\dfrac{5}{2}$
105.

$$\sqrt{x+2}+\sqrt{x-1}=3$$

$$\begin{array}{c|c}\sqrt{2+2}+\sqrt{2-1} & 3 \\ \sqrt{4}+\sqrt{1} & \\ 2+1 & \\ 3=3 & \end{array}$$

107.

$$\begin{array}{c|c}x^2-2x-5=0 & \\ \hline (1+\sqrt{6})^2-2(1+\sqrt{6})-5 & 0 \\ 1+2\sqrt{6}+6-2-2\sqrt{6}-5 & \\ & 0=0\end{array}$$

$$\begin{array}{c|c}x^2-2x-5=0 & \\ \hline (1-\sqrt{6})^2-2(1-\sqrt{6})-5 & 0 \\ 1-2\sqrt{6}+6-2+2\sqrt{6}-5 & \\ & 0=0\end{array}$$

SECTION 11.4 *pages 458–461*

1. 25 **3.** 144 **5.** 5 **7.** 16 **9.** 8 **11.** No solution **13.** 6 **15.** 24 **17.** -1
19. $-\dfrac{2}{5}$ **21.** $\dfrac{4}{3}$ **23.** 15 **25.** 5 **27.** 1 **29.** 1 **31.** 1 **33.** 2 **35.** No solution
37. 10.30 cm **39.** 9.75 ft **41.** less than halfway **43.** 8.71 in. **45.** 1 **47.** 11
49. 8.16 ft **51.** 64 ft **53.** 7.30 ft **55.** 6 **57.** 4 **59.** 3 **61.** 30 units

THE SQUARE ROOT ALGORITHM *page 463*

1. 7.211 **3.** 9.327

CHAPTER REVIEW *pages 464–467*

1. $-11\sqrt{3}$ (Objective 11.2.1) **2.** $-x\sqrt{3}-x\sqrt{5}$ (Objective 11.3.2) **3.** $x+8$ (Objective 11.1.2)
4. No solution (Objective 11.4.1) **5.** 1 (Objective 11.3.2) **6.** $7\sqrt{7}$ (Objective 11.2.1)
7. $12a+28\sqrt{ab}+15b$ (Objective 11.3.1) **8.** $7x^2+42x+63$ (Objective 11.1.2)

9. 12 (Objective 11.1.1)   **10.** 16 (Objective 11.4.1)   **11.** $4x\sqrt{5}$ (Objective 11.2.1)
12. $5ab - 7$ (Objective 11.3.1)   **13.** 3 (Objective 11.4.1)   **14.** $\sqrt{35}$ (Objective 11.1.1)

15. $7x^2y\sqrt{15xy}$ (Objective 11.2.1)   **16.** $\dfrac{x}{3y^2}$ (Objective 11.3.2)

17. $9x - 6\sqrt{xy} + y$ (Objective 11.3.1)   **18.** $a + 4$ (Objective 11.1.2)

19. $20\sqrt{3}$ (Objective 11.1.1)   **20.** $26\sqrt{3x}$ (Objective 11.2.1)   **21.** $\dfrac{8\sqrt{x} + 24}{x - 9}$ (Objective 11.3.2)

22. $3a\sqrt{2} + 2a\sqrt{3}$ (Objective 11.3.1)   **23.** $9a^2\sqrt{2ab}$ (Objective 11.1.2)
24. $-6\sqrt{30}$ (Objective 11.1.1)   **25.** $-4a^2b^4\sqrt{5ab}$ (Objective 11.2.1)

26. $7x^2y^4$ (Objective 11.3.2)   **27.** $\dfrac{3}{5}$ (Objective 11.4.1)   **28.** c^9 (Objective 11.1.2)

29. $15\sqrt{2}$ (Objective 11.1.1)   **30.** $18a\sqrt{5b} + 5a\sqrt{b}$ (Objective 11.2.1)   **31.** $\dfrac{16\sqrt{a}}{a}$ (Objective 11.3.2)

32. 8 (Objective 11.4.1)   **33.** $a^5b^3c^2$ (Objective 11.3.1)   **34.** $21\sqrt{70}$ (Objective 11.1.1)
35. $2y^4\sqrt{6}$ (Objective 11.1.2)   **36.** 5 (Objective 11.3.2)   **37.** 5 (Objective 11.4.1)
38. $8y + 10\sqrt{5y} - 15$ (Objective 11.3.1)   **39.** 99.499 (Objective 11.1.1)   **40.** 7 (Objective 11.3.1)
41. $25x^4\sqrt{6x}$ (Objective 11.1.2)   **42.** $-6x^3y^2\sqrt{2y}$ (Objective 11.2.1)   **43.** $3a$ (Objective 11.3.2)

44. $20\sqrt{10}$ (Objective 11.1.1)   **45.** 20 (Objective 11.4.1)   **46.** $\dfrac{7a}{3}$ (Objective 11.3.2)

47. 10 (Objective 11.3.1)   **48.** $6x^2y^2\sqrt{y}$ (Objective 11.1.2)   **49.** $36x^8y^5\sqrt{3xy}$ (Objective 11.1.2)
50. 3 (Objective 11.4.1)   **51.** 20 (Objective 11.1.1)   **52.** $-2\sqrt{2x}$ (Objective 11.2.1)
53. 6 (Objective 11.4.1)   **54.** 3 (Objective 11.3.1)   **55.** 51 (Objective 11.4.2)
56. 144 lb (Objective 11.4.2)   **57.** 14.25 cm (Objective 11.4.2)   **58.** 25 ft (Objective 11.4.2)
59. 100 ft (Objective 11.4.2)   **60.** 26.25 ft (Objective 11.4.2)

CHAPTER TEST   *pages 467–468*

1. $11x^4y$ (Objective 11.1.2)   **2.** $-5\sqrt{2}$ (Objective 11.2.1)   **3.** $6x^2\sqrt{xy}$ (Objective 11.3.1)
4. $3\sqrt{5}$ (Objective 11.1.1)   **5.** $6x^3y\sqrt{2x}$ (Objective 11.1.2)   **6.** $6\sqrt{2y} + 3\sqrt{2x}$ (Objective 11.2.1)
7. $y + 8\sqrt{y} + 15$ (Objective 11.3.1)   **8.** $x - 4$ (Objective 11.3.2)   **9.** 9 (Objective 11.3.2)
10. 11 (Objective 11.4.1)   **11.** 22.361 (Objective 11.1.1)   **12.** $4a^2b^5\sqrt{2ab}$ (Objective 11.1.2)
13. $a - \sqrt{ab}$ (Objective 11.3.1)   **14.** $4x^2y^2\sqrt{5y}$ (Objective 11.3.1)   **15.** $7ab\sqrt{a}$ (Objective 11.3.2)
16. 25 (Objective 11.4.1)   **17.** $8x^6y^2\sqrt{3xy}$ (Objective 11.1.2)   **18.** $4ab\sqrt{2ab} - 5ab\sqrt{ab}$
(Objective 11.2.1)   **19.** $a - 4$ (Objective 11.3.1)   **20.** $-6 - 3\sqrt{5}$ (Objective 11.3.2)
21. 5 (Objective 11.4.1)   **22.** $3\sqrt{2} - x\sqrt{3}$ (Objective 11.3.1)   **23.** $-6\sqrt{a}$ (Objective 11.2.1)
24. $6\sqrt{3}$ (Objective 11.1.1)   **25.** 7.937 (Objective 11.1.1)   **26.** $6ab\sqrt{a}$ (Objective 11.3.2)
27. No solution (Objective 11.4.1)   **28.** 40 (Objective 11.4.2)   **29.** 1.82 ft (Objective 11.4.2)
30. 9.80 ft (Objective 11.4.2)

CUMULATIVE REVIEW   *pages 468–470*

1. $-\dfrac{1}{12}$ (Objective 1.5.2)   **2.** $2x + 18$ (Objective 2.2.4)   **3.** $\dfrac{1}{13}$ (Objective 3.3.2)

4. $\dfrac{6x^5}{y^3}$ (Objective 5.4.1)   **5.** $-2b^2 + 1 - \dfrac{1}{3b^2}$ (Objective 5.4.2)   **6.** $3x^2y^2(4x - 3y)$ (Objective 6.1.1)

7. $(3b + 5)(3b - 4)$ (Objective 6.3.1/6.3.2) **8.** $2a(a - 3)(a - 5)$ (Objective 6.2.2)

9. $\dfrac{1}{4(x + 1)}$ (Objective 7.1.2) **10.** $\dfrac{x - 5}{x - 3}$ (Objective 7.4.1) **11.** $\dfrac{x + 3}{x - 3}$ (Objective 7.3.2)

12. $\dfrac{5}{3}$ (Objective 7.5.1) **13.** $-\dfrac{4}{3}$ (Objective 8.3.2) **14.** $y = \dfrac{1}{2}x - 2$ (Objective 8.4.1/8.4.2)

15. $(1, 1)$ (Objective 9.2.1) **16.** $(3, -2)$ (Objective 9.3.1) **17.** $(2, -1)$ (Objective 9.1.1)

18. $A \cup B = \{-8, -4, -2, 0, 2, 4, 8\}$ (Objective 10.1.1) **19.** $x \leq -\dfrac{9}{2}$ (Objective 10.3.1)

20. (Objective 10.4.1) **21.** $\{x \mid x > 30, x \text{ is a real number}\}$ (Objective 10.1.2)

22. $38\sqrt{3a} - 35\sqrt{a}$ (Objective 11.2.1) **23.** 8 (Objective 11.3.2) **24.** 14 (Objective 11.4.1)
25. \$24.50 (Objective 4.3.1) **26.** 56 oz (Objective 4.5.2) **27.** 8 and 13 (Objective 6.5.2)
28. 48 h (Objective 7.7.1) **29.** -33 (Objective 10.2.3) **30.** 24 (Objective 11.4.2)

ANSWERS to Chapter 12 Exercises

SECTION 12.1 *pages 477–478*

1. -5 and 3 **3.** 1 and 3 **5.** -1 and -2 **7.** 3 **9.** 0 and $\dfrac{3}{2}$ **11.** -2 and 6

13. $-\dfrac{1}{2}$ and 2 **15.** $\dfrac{4}{3}$ and -3 **17.** $-\dfrac{3}{2}$ and $\dfrac{4}{3}$ **19.** -3 and $\dfrac{4}{5}$ **21.** $\dfrac{1}{3}$ **23.** -4 and 4

25. $-\dfrac{2}{3}$ and $\dfrac{2}{3}$ **27.** -3 and 6 **29.** 0 and 12 **31.** -6 and 6 **33.** -1 and 1 **35.** $-\dfrac{7}{2}$ and $\dfrac{7}{2}$

37. $-\dfrac{2}{3}$ and $\dfrac{2}{3}$ **39.** $-\dfrac{3}{4}$ and $\dfrac{3}{4}$ **41.** No real number solution **43.** $-2\sqrt{6}$ and $2\sqrt{6}$

45. -5 and 7 **47.** -3 and -7 **49.** 4 and -6 **51.** -1 and -9 **53.** 16 and 2

55. $-\dfrac{3}{2}$ and $-\dfrac{9}{2}$ **57.** $-\dfrac{1}{3}$ and $\dfrac{7}{3}$ **59.** $-\dfrac{2}{7}$ and $-\dfrac{12}{7}$ **61.** $4 + 2\sqrt{5}$ and $4 - 2\sqrt{5}$

63. No real number solution **65.** $\dfrac{1}{2} + \sqrt{6}$ and $\dfrac{1}{2} - \sqrt{6}$ **67.** $\dfrac{8}{3}$ and $-\dfrac{4}{3}$ **69.** $a = 2, b = 0, c = -5$

71. $x^2 - 4x - 5 = 0$ **73.** $4x^2 - 4x - 5 = 0$ **75.** $2x^2 - x - 20 = 0$ **77.** $\dfrac{2}{3}$ and 2 **79.** $\dfrac{3}{2}$ and -6

81. $\sqrt{\dfrac{b}{a}}$ and $-\sqrt{\dfrac{b}{a}}$

SECTION 12.2 *pages 483–485*

1. -3 and 1 **3.** -2 and 8 **5.** 2 **7.** No real number solution **9.** -1 and -4 **11.** -8 and 1
13. $-2 + \sqrt{3}$ and $-2 - \sqrt{3}$ **15.** $-3 + \sqrt{14}$ and $-3 - \sqrt{14}$ **17.** $1 + \sqrt{2}$ and $1 - \sqrt{2}$

19. $\dfrac{-3 + \sqrt{13}}{2}$ and $\dfrac{-3 - \sqrt{13}}{2}$ **21.** 1 and 2 **23.** $4 + \sqrt{17}$ and $4 - \sqrt{17}$ **25.** $4 + \sqrt{14}$ and $4 - \sqrt{14}$

27. $-1 + \sqrt{5}$ and $-1 - \sqrt{5}$ **29.** $\dfrac{-3 + \sqrt{13}}{2}$ and $\dfrac{-3 - \sqrt{13}}{2}$ **31.** $\dfrac{1 + \sqrt{5}}{2}$ and $\dfrac{1 - \sqrt{5}}{2}$

33. $\dfrac{5 + \sqrt{13}}{2}$ and $\dfrac{5 - \sqrt{13}}{2}$ **35.** $\dfrac{-1 + \sqrt{13}}{2}$ and $\dfrac{-1 - \sqrt{13}}{2}$ **37.** $-5 + 4\sqrt{2}$ and $-5 - 4\sqrt{2}$

39. $\dfrac{3 + \sqrt{29}}{2}$ and $\dfrac{3 - \sqrt{29}}{2}$ **41.** $\dfrac{1 + \sqrt{17}}{2}$ and $\dfrac{1 - \sqrt{17}}{2}$ **43.** No real number solution   **45.** 1 and $\dfrac{1}{2}$

47. -3 and $\dfrac{1}{2}$ **49.** 2 and $\dfrac{3}{2}$ **51.** 1 and $-\dfrac{1}{2}$ **53.** -2 and $\dfrac{1}{3}$ **55.** -2 and $-\dfrac{2}{3}$ **57.** $\dfrac{1}{2}$ and $-\dfrac{3}{2}$

59. $\dfrac{1}{3}$ and $-\dfrac{3}{2}$ **61.** $-\dfrac{1}{2}$ and $\dfrac{4}{3}$ **63.** $\dfrac{1 + \sqrt{2}}{2}$ and $\dfrac{1 - \sqrt{2}}{2}$ **65.** $\dfrac{2 + \sqrt{5}}{2}$ and $\dfrac{2 - \sqrt{5}}{2}$ **67.** 2 and 4

69. $1 + \sqrt{6}$ and $1 - \sqrt{6}$ **71.** 0.373 and -5.373 **73.** 1.712 and -3.212 **75.** 0.652 and -1.152

77. $1 + \sqrt{7}$ and $1 - \sqrt{7}$ **79.** $\dfrac{3 + \sqrt{21}}{3}$ and $\dfrac{3 - \sqrt{21}}{3}$ **81.** 2   **83.** 0 and -1 **85.** 4 and 2

87. $6 + \sqrt{58}$ and $6 - \sqrt{58}$

SECTION 12.3   *pages 488–490*

1. -1 and 5   **3.** -3 and 5   **5.** -7 and 1   **7.** -3 and 2   **9.** -1 and 3   **11.** -5 and 1

13. $-\dfrac{1}{2}$ and 1   **15.** No real number solution   **17.** 0 and 1   **19.** $-\dfrac{3}{2}$ and $\dfrac{3}{2}$ **21.** $-\dfrac{5}{2}$ and $\dfrac{3}{2}$

23. $-\dfrac{2}{3}$ and 3   **25.** $\dfrac{4}{5}$ and -3 **27.** $-\dfrac{1}{2}$ and $\dfrac{2}{3}$ **29.** No real number solution

31. $1 + \sqrt{6}$ and $1 - \sqrt{6}$ **33.** $-3 + \sqrt{10}$ and $-3 - \sqrt{10}$ **35.** $2 + \sqrt{13}$ and $2 - \sqrt{13}$

37. $\dfrac{1 + \sqrt{2}}{2}$ and $\dfrac{1 - \sqrt{2}}{2}$ **39.** $-3 + 2\sqrt{2}$ and $-3 - 2\sqrt{2}$ **41.** $\dfrac{1}{4}$ and -1 **43.** 1 and $\dfrac{2}{3}$

45. $\dfrac{3 + 2\sqrt{6}}{5}$ and $\dfrac{3 - 2\sqrt{6}}{5}$ **47.** $\dfrac{7 + \sqrt{17}}{8}$ and $\dfrac{7 - \sqrt{17}}{8}$ **49.** -1 **51.** $\dfrac{1 + \sqrt{21}}{10}$ and $\dfrac{1 - \sqrt{21}}{10}$

53. $\dfrac{3 + \sqrt{129}}{12}$ and $\dfrac{3 - \sqrt{129}}{12}$ **55.** $\dfrac{-3 + \sqrt{65}}{4}$ and $\dfrac{-3 - \sqrt{65}}{4}$ **57.** $\dfrac{7 + \sqrt{89}}{2}$ and $\dfrac{7 - \sqrt{89}}{2}$

59. $\dfrac{3 + 2\sqrt{6}}{2}$ and $\dfrac{3 - 2\sqrt{6}}{2}$ **61.** $\dfrac{-1 + \sqrt{2}}{3}$ and $\dfrac{-1 - \sqrt{2}}{3}$ **63.** $-\dfrac{1}{2}$ **65.** No real number solution

67. $\dfrac{-4 + \sqrt{5}}{2}$ and $\dfrac{-4 - \sqrt{5}}{2}$ **69.** $\dfrac{1 + 2\sqrt{3}}{2}$ and $\dfrac{1 - 2\sqrt{3}}{2}$ **71.** $\dfrac{-5 + \sqrt{2}}{3}$ and $\dfrac{-5 - \sqrt{2}}{3}$

73. 5.690 and -3.690 **75.** 7.690 and -1.690 **77.** 4.590 and -1.090 **79.** 0.118 and -2.118

81. 1.105 and -0.905 **83.** $\dfrac{-2 + \sqrt{10}}{3}$ and $\dfrac{-2 - \sqrt{10}}{3}$ **85.** 11   **87.** 11 and -1 **89.** Yes

91. No   **93.** No

SECTION 12.4   *pages 493–494*

1. 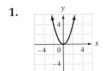   **3.**   **5.**   **7.**

9.   **11.**   **13.**   **15.**   **17.**

19. Down   **21.** Down   **23.** $y = 3x^2 + 6x + 5$ **25.** 2 and -2 **27.** 0 and 1   **29.** 1 and 3
31. 3   **33.** No

SECTION 12.5   *pages 496–498*

1. length, 6 ft; width, 4 ft   **3.** length, 100 ft; width, 50 ft   **5.** 7 and 9   **7.** 5 and 7   **9.** 0 or 2
11. silver coin, 10 years; gold coin, 5 years   **13.** first coin, 8 years; second coin, 6 years   **15.** first
computer, 35 min; second computer, 14 min   **17.** first engine, 12 h; second engine, 6 h   **19.** 100 mph
21. 25 mph   **23.** $-6, -5, -4,$ and -3, or 3, 4, 5, and 6   **25.** 7 in.

PROFIT   *page 501*

1. a. 20 or 60 guitars   **b.** $2000

CHAPTER REVIEW   *pages 502–504*

1. 4 and -4 (Objective 12.1.2)   **2.** $\dfrac{1 + \sqrt{13}}{2}$ and $\dfrac{1 - \sqrt{13}}{2}$ (Objective 12.2.1/12.3.1)

3. $\dfrac{3 + \sqrt{29}}{2}$ and $\dfrac{3 - \sqrt{29}}{2}$ (Objective 12.2.1/12.3.1)   **4.** $\dfrac{5}{7}$ and $-\dfrac{5}{7}$ (Objective 12.1.2)

5.   (Objective 12.4.1)   **6.**   (Objective 12.4.1)

7. $5 + \sqrt{23}$ and $5 - \sqrt{23}$ (Objective 12.2.1/12.3.1)   **8.** $-\dfrac{1}{2}$ and $-\dfrac{1}{3}$ (Objective 12.1.1)

9. No real number solution   (Objective 12.1.2)   **10.** $\dfrac{-10 + 2\sqrt{10}}{5}$ and $\dfrac{-10 - 2\sqrt{10}}{5}$

(Objective 12.2.1/12.3.1)   **11.** $2 + \sqrt{3}$ and $2 - \sqrt{3}$ (Objective 12.2.1/12.3.1)   **12.** 6 and -5

(Objective 12.1.1)   **13.** $\dfrac{4}{3}$ and $-\dfrac{7}{2}$ (Objective 12.1.1)   **14.** $2\sqrt{10}$ and $-2\sqrt{10}$ (Objective 12.1.2)

15. $\dfrac{2 + \sqrt{7}}{3}$ and $\dfrac{2 - \sqrt{7}}{3}$ (Objective 12.2.1/12.3.1)   **16.** $1 + \sqrt{11}$ and $1 - \sqrt{11}$ (Objective 12.2.1/12.3.1)

17. 3 and 9   (Objective 12.1.1)   **18.** 16 and -2 (Objective 12.1.2)

19.   (Objective 12.4.1)   **20.**   (Objective 12.4.1)

21. 1 and -9 (Objective 12.1.2)   **22.** $\dfrac{-4 + \sqrt{23}}{2}$ and $\dfrac{-4 - \sqrt{23}}{2}$ (Objective 12.2.1/12.3.1)

23. $-\dfrac{1}{6}$ and $-\dfrac{5}{4}$ (Objective 12.1.1)   **24.** No real number solution   (Objective 12.2.1/12.3.1)

25. $4 + \sqrt{21}$ and $4 - \sqrt{21}$ (Objective 12.2.1/12.3.1)   **26.** $\dfrac{4}{7}$ (Objective 12.1.1)

27. 2 and −1   (Objective 12.1.2)   **28.** $\frac{3}{8}$ and $\frac{3}{2}$ (Objective 12.1.1)   **29.** $\frac{-7 + \sqrt{61}}{2}$ and $\frac{-7 - \sqrt{61}}{2}$

(Objective 12.2.1/12.3.1)   **30.** 2 and $\frac{5}{12}$ (Objective 12.1.1)   **31.** $3 + \sqrt{5}$ and $3 - \sqrt{5}$ (Objective 12.1.2)

32. $-4 + \sqrt{19}$ and $-4 - \sqrt{19}$ (Objective 12.2.1/12.3.1)

33.   (Objective 12.4.1)   **34.**   (Objective 12.4.1)

35. −7 and −10   (Objective 12.1.1)   **36.** $2 + 2\sqrt{6}$ and $2 - 2\sqrt{6}$ (Objective 12.1.2)
37. No real number solution   (Objective 12.2.1/12.3.1)

38. $-3 + \sqrt{11}$ and $-3 - \sqrt{11}$ (Objective 12.2.1/12.3.1)   **39.** 3 and $-\frac{1}{9}$ (Objective 12.1.1)

40. $\frac{-5 + \sqrt{33}}{4}$ and $\frac{-5 - \sqrt{33}}{4}$ (Objective 12.2.1/12.3.1)   **41.**   (Objective 12.4.1)

42. $\frac{9 + \sqrt{69}}{2}$ and $\frac{9 - \sqrt{69}}{2}$ (Objective 12.2.1/12.3.1)   **43.** 1 and $\frac{5}{2}$ (Objective 12.1.1)

44.   (Objective 12.4.1)   **45.** 75 mph   (Objective 12.5.1)

46. height, 10 m; base, 4 m   (Objective 12.5.1)   **47.** 3 and 5   (Objective 12.5.1)
48. 5 mph   (Objective 12.5.1)   **49.** first car, 14 years; second car, 12 years   (Objective 12.5.1)
50. smaller drain, 12 h; larger drain, 4h   (Objective 12.5.1)

CHAPTER TEST   *pages 504–505*

1. $-4 + 2\sqrt{5}$ and $-4 - 2\sqrt{5}$ (Objective 12.1.2)   **2.** $\frac{-4 + \sqrt{22}}{2}$ and $\frac{-4 - \sqrt{22}}{2}$ (Objective 12.2.1/12.3.1)

3. −4 and $\frac{5}{3}$ (Objective 12.1.1)   **4.** $\frac{3 + \sqrt{33}}{2}$ and $\frac{3 - \sqrt{33}}{2}$ (Objective 12.2.1/12.3.1)

5. $-2 + 2\sqrt{5}$ and $-2 - 2\sqrt{5}$ (Objective 12.2.1/12.3.1)   **6.**    (Objective 12.4.1)

7. $-2 + \sqrt{2}$ and $-2 - \sqrt{2}$ (Objective 12.2.1/12.3.1)
8. $\frac{-3 + \sqrt{41}}{2}$ and $\frac{-3 - \sqrt{41}}{2}$ (Objective 12.2.1/12.3.1)   **9.** $-\frac{1}{2}$ and 3   (Objective 12.1.1)

10. $\frac{3 + \sqrt{7}}{2}$ and $\frac{3 - \sqrt{7}}{2}$ (Objective 12.2.1/12.3.1)   **11.** $\frac{1 + \sqrt{13}}{6}$ and $\frac{1 - \sqrt{13}}{6}$ (Objective 12.2.1/12.3.1)

12. $5 + 3\sqrt{2}$ and $5 - 3\sqrt{2}$ (Objective 12.1.2)   **13.** $3 + \sqrt{14}$ and $3 - \sqrt{14}$ (Objective 12.2.1/12.3.1)

14. $\dfrac{5 + \sqrt{29}}{2}$ and $\dfrac{5 - \sqrt{29}}{2}$ (Objective 12.2.1/12.3.1)   **15.** $\dfrac{5 + \sqrt{33}}{2}$ and $\dfrac{5 - \sqrt{33}}{2}$ (Objective 12.2.1/12.3.1)

16. $\dfrac{5}{2}$ and $\dfrac{1}{3}$ (Objective 12.1.1)   **17.** $\dfrac{-3 + \sqrt{37}}{2}$ and $\dfrac{-3 - \sqrt{37}}{2}$ (Objective 12.2.1/12.3.1)

18. $\dfrac{2 + \sqrt{14}}{2}$ and $\dfrac{2 - \sqrt{14}}{2}$ (Objective 12.2.1/12.3.1)   **19.** $-\dfrac{1}{2}$ and 2   (Objective 12.1.1)

20.   (Objective 12.4.1)   **21.** $\dfrac{1 + \sqrt{10}}{3}$ and $\dfrac{1 - \sqrt{10}}{3}$ (Objective 12.2.1/12.3.1)

22. length, 8 ft; width, 5 ft   (Objective 12.5.1)   **23.** 11 mph   (Objective 12.5.1)
24. 5 or −5   (Objective 12.5.1)   **25.** 4 mph   (Objective 12.5.1)

CUMULATIVE REVIEW   *pages 505–506*

1. $-28x + 27$ (Objective 2.2.4)   **2.** $\dfrac{3}{2}$ (Objective 3.1.3)   **3.** 3   (Objective 3.3.2)

4. $-\dfrac{4a}{3b^2}$ (Objective 5.4.1)   **5.** $x + 2 - \dfrac{4}{x - 2}$ (Objective 5.4.3)   **6.** $(x - 4)(4y - 3)$ (Objective 6.1.2)

7. $x(3x - 4)(x + 2)$ (Objective 6.3.1/6.3.2)   **8.** $\dfrac{9x^2(x - 2)(x - 4)}{(2x - 3)^2(x + 2)}$ (Objective 7.1.3)

9. $\dfrac{x + 2}{2(x + 1)}$ (Objective 7.3.2)   **10.** $\dfrac{x - 4}{2x + 5}$ (Objective 7.4.1)   **11.** -3 and 6   (Objective 7.5.1)

12. $(3, 0)$ and $(0, -4)$ (Objective 8.3.1)   **13.** $y = -\dfrac{4}{3}x - 2$ (Objective 8.4.1/8.4.2)

14. $(2, 1)$ (Objective 9.2.1)   **15.** $(2, -2)$ (Objective 9.3.1)   **16.** $x > \dfrac{1}{9}$ (Objective 10.3.1)

17. $a - 2$ (Objective 11.3.1)   **18.** $2x - 3$ (Objective 11.3.2)   **19.** 5   (Objective 11.4.1)

20. $\dfrac{1}{2}$ and 3   (Objective 12.1.1)   **21.** $2 + 2\sqrt{3}$ and $2 - 2\sqrt{3}$ (Objective 12.1.2)

22. $\dfrac{2 + \sqrt{19}}{3}$ and $\dfrac{2 - \sqrt{19}}{3}$ (Objective 12.2.1/12.3.1)   **23.**   (Objective 10.4.1)

24.   (Objective 12.4.1)   **25.** $2.25   (Objective 4.5.1)

26. 250 additional shares   (Objective 7.5.2)   **27.** rate of the plane in calm air, 200 mph; rate of the wind, 40 mph   (Objective 9.4.1)   **28.** 77 or better   (Objective 10.2.3)   **29.** 31.62 m   (Objective 11.4.2)   **30.** −7 or 7   (Objective 12.5.1)

FINAL EXAM   *pages 507–510*

1. -3 (Objective 1.1.2)   **2.** -6 (Objective 1.2.2)   **3.** -256 (Objective 1.5.1)

4. -11 (Objective 1.5.2)   **5.** $-\dfrac{15}{2}$ (Objective 2.1.1)   **6.** $9x + 6y$ (Objective 2.2.1)

7. $6z$ (Objective 2.2.2)   **8.** $16x - 52$ (Objective 2.2.4)   **9.** -50 (Objective 3.1.3)

10. -3 (Objective 3.3.2)   **11.** 12.5% (Objective 4.1.2)   **12.** 15.2 (Objective 4.2.1)

13. $-3x^2 - 3x + 8$ (Objective 5.1.2)   **14.** $81x^4y^{12}$ (Objective 5.2.2)

15. $6x^3 + 7x^2 - 7x - 6$ (Objective 5.3.2)   **16.** $-\dfrac{x^8y}{2}$ (Objective 5.4.1)

17. $3x^2 - 4x - \dfrac{5}{x}$ (Objective 5.4.2)   **18.** $5x - 12 + \dfrac{23}{x + 2}$ (Objective 5.4.3)   **19.** $\dfrac{16y^7}{x^8}$ (Objective 5.4.1)

20. $(2x - 3)(x + 1)$ (Objective 6.3.1/6.3.2)   **21.** $(x - 6)(x + 1)$ (Objective 6.2.1)

22. $(3x + 2)(2x - 3)$ (Objective 6.3.1/6.3.2)   **23.** $4x(2x - 1)(x - 3)$ (Objective 6.3.1/6.3.2)

24. $(5x + 4)(5x - 4)$ (Objective 6.4.1)   **25.** $2(4 - x)(a + 3)$ (Objective 6.1.2)

26. $3y(5 + 2x)(5 - 2x)$ (Objective 6.4.2)   **27.** $\dfrac{1}{2}$ and 3 (Objective 6.5.1)

28. $\dfrac{2(x + 1)}{x - 1}$ (Objective 7.1.2)   **29.** $\dfrac{-3x^2 + x - 25}{(2x - 5)(x + 3)}$ (Objective 7.3.2)

30. $x + 1$ (Objective 7.4.1)   **31.** 2 (Objective 7.5.1)   **32.** $a = b$ (Objective 7.6.1)

33. $\dfrac{2}{3}$ (Objective 8.3.2)   **34.** $y = -\dfrac{2}{3}x - 2$ (Objective 8.4.1/8.4.2)   **35.** $(6, 17)$ (Objective 9.2.1)

36. $(2, -1)$ (Objective 9.3.1)   **37.** $x \le -3$ (Objective 10.3.1)   **38.** $y \ge \dfrac{5}{2}$ (Objective 10.3.1)

39. $7x^3$ (Objective 11.1.2)   **40.** $38\sqrt{3a}$ (Objective 11.2.1)   **41.** $\sqrt{15} + 2\sqrt{3}$ (Objective 11.3.2)

42. 5 (Objective 11.4.1)   **43.** -1 and $\dfrac{4}{3}$ (Objective 12.1.1)

44. $\dfrac{1 + \sqrt{5}}{4}$ and $\dfrac{1 - \sqrt{5}}{4}$ (Objective 12.2.1/12.3.1)   **45.**   (Objective 8.3.3)

46.   (Objective 12.4.1)   **47.** $2x + 3(x - 2)$; $5x - 6$ (Objective 2.3.3)

48. $\$3000$ (Objective 4.2.2)   **49.** 65% (Objective 4.3.1)   **50.** $\$6000$ (Objective 4.4.1)

51. $\$3$ (Objective 4.5.1)   **52.** 36% (Objective 4.5.2)   **53.** 215 km   (Objective 4.6.1)

54. $60°$, $50°$, and $70°$ (Objective 4.7.2)   **55.** 60 dimes   (Objective 4.8.2)

56. length, 10 m; width, 5 m   (Objective 6.5.2)   **57.** 16 oz   (Objective 7.5.2)

58. 0.6 h   (Objective 7.7.1)   **59.** rate of the boat in calm water, 15 mph; rate of the current, 5 mph   (Objective 9.4.1)   **60.** 25 mph   (Objective 12.5.1)

INDEX

Table of Properties

Properties of Real Numbers

The Associative Property of Addition

If a, b, and c are real numbers, then
$(a + b) + c = a + (b + c)$.

The Associative Property of Multiplication

If a, b, and c are real numbers, then
$(a \cdot b) \cdot c = a \cdot (b \cdot c)$.

The Commutative Property of Addition

If a and b are real numbers, then
$a + b = b + a$.

The Commutative Property of Multiplication

If a and b are real numbers, then
$a \cdot b = b \cdot a$.

The Addition Property of Zero

If a is a real number, then
$a + 0 = 0 + a = a$.

The Multiplication Property of One

If a is a real number, then
$a \cdot 1 = 1 \cdot a = a$.

The Multiplication Property of Zero

If a is a real number, then
$a \cdot 0 = 0 \cdot a = 0$.

The Inverse Property of Multiplication

If a is a real number and $a \neq 0$, then
$a \cdot \dfrac{1}{a} = \dfrac{1}{a} \cdot a = 1$.

The Inverse Property of Addition

If a is a real number, then
$a + (-a) = (-a) + a = 0$.

Distributive Property

If a, b, and c are real numbers, then $a(b + c) = ab + ac$ or $(b + c)a = ba + ca$.

Properties of Exponents

If m and n are integers, then
$x^m \cdot x^n = x^{m+n}$.

If m and n are integers, then
$(x^m)^n = x^{mn}$.

If $x \neq 0$, then
$x^0 = 1$.

If m and n are integers and $x \neq 0$, then
$\dfrac{x^m}{x^n} = x^{m-n}$.

If m, n, and p are integers, then
$(x^m \cdot y^n)^p = x^{mp}y^{np}$.

If n is a positive integer and $x \neq 0$, then
$x^{-n} = \dfrac{1}{x^n}$ and $\dfrac{1}{x^{-n}} = x^n$.

If m, n, and p are integers and $y \neq 0$, then
$\left(\dfrac{x^m}{y^n}\right)^p = \dfrac{x^{mp}}{y^{np}}$.